The Ecology of Running Waters

H. B. N. HYNES
Professor of Biology,
University of Waterloo, Ontario, Canada

LIVERPOOL UNIVERSITY PRESS 1972

Published by
LIVERPOOL UNIVERSITY PRESS
123 Grove Street, Liverpool L7 7AF

Printed in Great Britain by
Richard Clay (The Chaucer Press) Ltd
Bungay, Suffolk

ISBN 0 85323 100 1

First published 1970
Second impression 1972

The Ecology of Running Waters

To Professor David G. Frey

who gave me this assignment in 1961

Limnol. Oceanogr. **6**, *98*

Preface

This book has been written in an attempt to provide a comprehensive and critical review of the literature on the biology of rivers and streams, as these have been given only very brief treatment in recent texts on limnology, e.g. Hutchinson (1957), and the more recent second volume which appeared when the manuscript of this book was complete, and Ruttner (1963).

A number of early limnological works dealt fairly extensively with what was known at the time about running water (Steinmann, 1907, 1909; Thienemann, 1925; Carpenter, 1928*b*), and over the years several books of a more or less popular nature, or concerned with rather narrow aspects of the subject, have been published (Needham, 1938; Macan and Worthington, 1951; Coker, 1954; Hynes, 1960; Reid, 1961; Illies, 1961*c*; Bardach, 1964). There has, however, been no recent attempt to make a thorough review of present-day knowledge.

Some years ago Dr. David Frey of Indiana University urged me to try to fill the gap, and this book is my effort to do so. I am sure that it contains errors and inevitably it is a zoologist's book; it will also be clear that my main interests are in the invertebrates. For the errors I apologize and I hope that readers will point them out to me; for the zoological bias I feel that no apology is needed.

Although much less ecological work has been done on rivers and streams than on lakes, there are nevertheless many hundreds of papers which deal directly or indirectly with them, and it is hardly possible for one individual to read more than a proportion of this great volume of literature. Cummins (1966) has recently published a list of over 500 references, and I have read about five times that number. Even so I am sure that I have failed to obtain a really world-wide picture of the subject because of my lack of knowledge of Slavonic and Asian languages. I have leaned heavily on Shadin's (1956) review of Russian work for an account of earlier studies in his country which were published only in

Russian, and I think that it is likely that I have overlooked some important Japanese studies, primarily because of my inability to glance through the journals and pick out titles which might be worth having translated. It is indeed pleasing to note that in recent years some Russian and Japanese journals have taken to giving tables of contents in one of the world languages that uses a Latin script. While I am among those who regard other languages as fun, I feel that there is a limit to the time that scientists can be expected to divert towards learning to read them, and as more and more countries take an interest in limnology it is to be hoped that we shall not approach closer and closer to Babel. I therefore also commend the recent practice of Czechoslovakian and Polish scientists of publishing, in the East German journal *Limnologica* in a western language, annual lists of their countries' limnological literature. Earlier Czechoslovakian lists (1960–3) were published in *Sb. vys. Šk. chem.-technol. Praze, Technol. vody*, volumes 4–7. Would it be too much to hope that our Japanese and Russian colleagues might do something similar and so save the rest of the world from battling with their peculiar scripts? No scientist anywhere can practise his profession effectively in this century without being able to read the Latin alphabet, and if one can read a title one can often understand it, even if it is in an unfamiliar language. We should then all be able to find out what had been done elsewhere.

The bibliography of this book contains over 1,500 entries of papers which have been cited in the text. The decision to cite a reference has not been particularly selective as the aim has been to supply a large and comprehensive bibliography, but there has been no selection of recent papers as opposed to older ones. Indeed, if anything, I have tended to stress the classical papers and to point to the very great contribution of some of the earlier workers in Europe and North America. Even so an analysis of the entries shows that over half the papers cited have been published since the end of 1955. If one adds to the total the approximately sixty papers published in 1966 and 1967, which have come to my notice since the manuscript was completed, it is possible to make the statement that half the literature has been published in the last decade. If one also adds the fact that during that decade there has been an enormous number of publications on pollution and water quality, a great many of which are concerned with running water, it is evident that there is a rapid increase in interest in rivers and streams. It is hoped, therefore, that this book may serve as a summary of what is known about unpolluted waters at the present time, and that it will indicate areas where more work is urgently needed.

Many new drawings have been made to illustrate organisms which are referred to frequently and to indicate the various anatomical adaptations which are discussed in the text. Most of these are original and were drawn from specimens collected by me or begged or borrowed from my limnological friends. A few have been redrawn from published figures, some have been reused from my previous book (Hynes, 1960) and some have been borrowed from other sources. The sources of drawings which are not original are noted in the legends. In many instances the continent of origin of the specimen from which the drawing was made has been stated; this does not, however, imply that the organism is confined to that continent.

I have a great many acknowledgements to make; this book would have been impossible to write without the help of a large number of people.

I was greatly aided in my work by the excellent libraries of Indiana University, the University of Liverpool, and the Freshwater Biological Association, and I have received much bibliographic help from the libraries of the National Research Council of Canada and the University of Waterloo. I am also indebted to several libraries in London, especially those of the Royal Entomological Society, the Natural History Museum, and the Zoological Society, where I was able to consult many old and obscure works. Great numbers of my fellow limnologists have kept me supplied with reprints of their papers for which I am very grateful, and I am particularly grateful to Professor V. I. Shadin of Leningrad for a copy of his large paper on life in rivers of the U.S.S.R. This was obviously a key work for my purpose, and after 1960 it seemed to be quite unobtainable on the market and impossible to borrow in the United Kingdom.

To my mother, Mrs. A. M. L. Hynes, I am enormously grateful for many hours spent on translations of Russian papers on subjects which must have been of little interest to her. Not only did she translate verbatim all 144 pages of Shadin's review, but she translated all or part of many of the Russian papers cited in the bibliography, as well as some others which I have not quoted. Without her help I should have been very largely excluded from considering the great volume of Russian work.

To my wife, Mary, I am very grateful as always for checking my mathematics, my physics, and my arithmetic, and above all for seeing to it that the final draft said what I meant it to say in acceptable syntactical form. She started this most helpful practice early in our career, on my Ph.D. thesis, and I have rarely since committed anything to print without her final critical perusal.

This brings me to a problem of language faced by all Britons who live in North America and by all authors who try to write English to be read on more than one continent. Most of this book was written in Canada where, like the Americans, they call petrol gasoline and flats apartments. Most people know this, but what does one do about mussels and clams, stickles and riffles, shingle and rubble, bullheads and sculpins, minnows and cyprinids, brooks and creeks, crayfish and crawfish, or even fish and fishes? I have avoided most terms which are unknown in America, such as stickle and shingle, and those which, like bullhead, creek, and crawfish, mean different things on the two sides of the Atlantic. But I have adopted the American usage of referring to small cyprinids as minnows, and I have used clam and mussel and autumn and fall indiscriminately. To my British readers I should perhaps explain that in this book clam does not mean *Mya arenaria* and to my American ones that a spate is a period of high discharge. I also feel that there is good biblical precedent for the use of the word fishes quite apart from the fact that it sometimes avoids ambiguity.

I am grateful to several people for the loan or gift of specimens. These include Dr. R. O. Brinkhurst of the University of Toronto, Dr. H. C. Clifford of the University of Alberta, Dr. G. F. Edmunds Jr. of the University of Utah, Dr. M. G. George and Dr. J. A. Spence of the University of Waterloo, Dr. H. D. Kennedy of the U.S. Fish and Wildlife Service, California, Dr. W. Schmitz of the Landesstelle für Gewässerkunde und Wasserwirtschaftliche Planung, Baden-Württemberg, Mr. T. R. Williams of the University of Liverpool, and Dr. W. D. Williams of Monash University. I am also indebted to the last three gentlemen, and to Dr. J. F. Hanson of the University of Massachusetts, Dr. R. C. Harrel of Lamar State College of Technology, Dr. N. K. Kaushik, Mr. H. T. Hui and Mrs. Mary Coleman of the University of Waterloo, and Dr. Khoo Soo Ghee of the University of Malaya, for unpublished information which they have allowed me to use.

Several authors and publishers have graciously allowed me to reuse published figures. Acknowledgement of this is made in the legends but I express gratitude here. In particular I am indebted to Mr. H. C. Gilson of the Freshwater Biological Association, Ambleside, for the loan of the original of Mrs. Jean Mackereth's excellent drawing of *Eriocheir* (Fig. VIII, 22), and to Dr. K. Müller of Messaure, Sweden, for the loan of the original graphs used in Figs. XIII, 1 and 2. I am also grateful to Dr. W. L. Minckley of Arizona State University for permission to use a graph from his doctoral thesis (Fig. II, 1).

Mrs. Helena Hahn, Miss Heather Luke, Miss Stella Piotrowski, Mrs. Pat Toogood, and Mrs. Marianne Buzza, all of the University of Waterloo, have done most of the typing, and Mr. K. Hahn has made available to me the services of the University's Photographic Department for the reproduction of a number of figures. I am most grateful to all these people for their indispensable help, and to the National Research Council of Canada whose research grant to me provided the funds necessary for the photocopying.

Finally, but by no means least, I wish to thank Mr. J. G. O'Kane, Secretary of Liverpool University Press, and his staff, for all their excellent work in the production of this book.

H. B. N. HYNES

Waterloo, Ontario

Contents

Figures

Tables

CHAPTER I

Water flow and stream channels

Although rivers and streams are major features of most types of landscape their total area is only about one-thousandth of that of the land surface, as compared with the 1·8 per cent of that area which is occupied by lakes. Nevertheless they annually carry about 37,000 cu. km. of water to the sea, which is equivalent to about 25 cm. of rainfall spread over the entire land surface (Schmitz, 1961). This is a considerable proportion of the water present at any one time in all freshwater lakes, for which the figures 200,000 cu. km. and 123,000 cu. km. (30,000 cu. miles) are variously quoted by Schmitz (1961) and Leopold *et al.* (1964). An enormous volume of water is therefore involved, and Schmitz has calculated that, taking the mean height of the land surface as 825 m. above sea level, the amount of energy expended by running water in a year is 10^{14} kW. hr., or about 100 times that produced by man from all sources in 1950.

Large though these figures are, only 2·8 per cent of the world's total water (326,000 cu. miles = 1,337,000 cu. km.) occurs on the land, and of this most (2·24 per cent) is locked up in polar ice-caps and glaciers. Lakes account for only 0·009 per cent of the total, and at any given time only 0·0001 per cent of the water is in river channels. This is only one-tenth of the amount in the atmosphere, and very much less than occurs in the groundwater (0·61 per cent) (Leopold *et al.*, 1964). Despite their size and grandeur, and their importance in erosion, even the great rivers of the world are thus relatively insignificant features of our planet.

In human terms, however, their significance is enormous. Shadin (1956) gives the figures shown overleaf.

These few large rivers therefore drain a substantial proportion of the 147,600,000 sq. km. (57,510,000 sq. miles) of the land surface, and the list is by no means complete. It does not include the Danube, the Hwang Ho, the Mackenzie, the Parana, the Murray-Darling, the Indus, the Ganges, the Congo, the Niger, or the Zambesi. There are over twenty

River	Length in km.	Drainage area (1,000 sq. km.)
Mississippi and Missouri	7,000	3,275
Nile	6,400	2,803
Amazon	6,300	7,000
Ob and Irtysh	5,206	2,946
Yangtze	5,200	1,800
Lena	4,599	2,582
Amur and Argune	4,478	1,937
Yennisey	4,011	2,549
Volga	3,700	1,402

rivers longer than 2,700 km., and each is a great erosive agent. It is calculated that the Mississippi alone carries 136 million tons of dissolved matter and 340 million tons of suspended matter into the sea each year (Reid, 1961), and the much smaller, but very silty, Amu-Daria carries 570 million tons of alluvium into the Aral Sea (Shadin, 1956).

The water that flows in streams and rivers comes, of course, from precipitation, but only a fraction of the rain or snow which falls actually reaches the watercourses. Some is immediately evaporated back into the atmosphere from rocks, soil, or vegetation; some enters the soil, where it is taken up by plants and transpired; some is lost into the deep groundwater; and only the remainder enters stream channels as run-off on or through the soil, or as shallow seepage. These points are illustrated diagrammatically in Fig. 1, 1.

Rainfall varies with season and from year to year, and because most of the losses outlined above are, to some extent, independent of rainfall the quantity which enters the streams is extremely variable. It is, as it were, the remnant of a variable from which a more or less constant amount is deducted, and it is thus even more irregular than the precipitation itself. For instance, Leopold (1962) calculates that of the average annual rainfall of 30 in. (76 cm.) which falls on the United States, only about 9 in. (23 cm.) gets into the rivers; so that in a year in which precipitation varies only 1 in., about 3 per cent from the average, stream flow is likely to alter by about 10 per cent.

In fact much of the water which does ultimately reach the streams goes first into the shallow underground reservoir of the water-table and is released more or less slowly into the channels, and this smoothing out of variations in flow is much influenced by vegetation (Rutter, 1958). Forests, because they increase the area from which evaporation can occur, and because trees transpire enormous amounts of water, reduce the total volume which finds its way into the channels. They also, however, cut down immediate run-off and tend to maintain moist soil, so that they usually result in more steady stream flows. Since forest is the natural vegetation of at least the narrow riparian strip of the valley of a permanent stream, even in dry or otherwise severe climates, its clearance inevitably alters the pattern of stream discharge. Many, probably most, of the rivers and streams of the world have had their flow regime altered

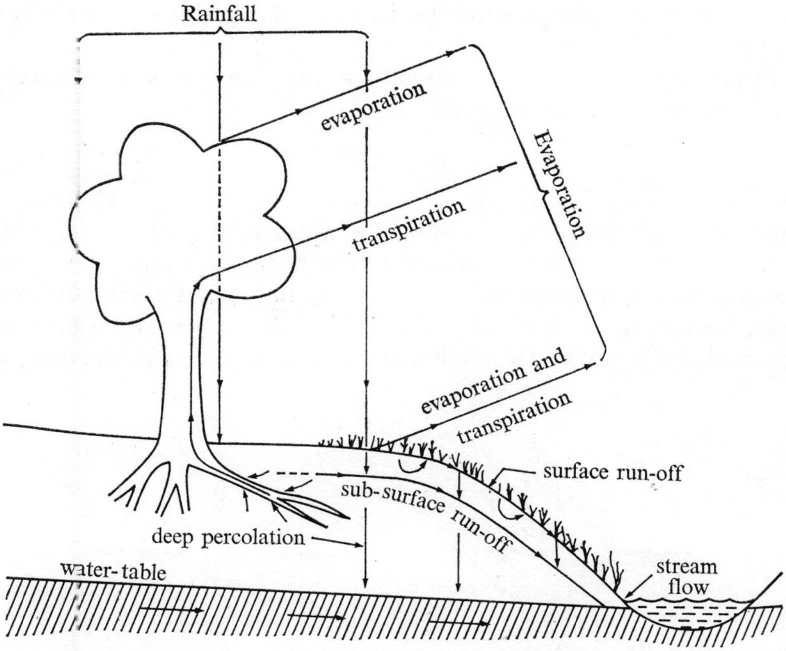

FIG. 1, 1. Diagram of the various routes back into the atmosphere or into streams taken by the products of precipitation. Partly after Rutter (1958).

by man in this way, and much of this has occurred during only the past few centuries.

The way in which water flows and the fact that it actually forms the patterns of rivers and streams and their beds are fundamental to most of the properties of biotic habitats in running water. These are aspects which have been largely ignored in biological studies despite their great interest to engineers and geologists. Thus most of this chapter is devoted to a general discussion of these topics and it leans heavily upon the excellent book of Leopold *et al.* (1964) to which the reader is referred for a much fuller treatment. Indeed I would go further in my recommendation and say that it should be required reading for all biologists who have any aspiration to work on running water.

THE FLOW OF WATER IN OPEN CHANNELS

Once water enters the channel of a stream it flows downhill, and the amount passing a given point, the discharge (D) is the variable quantity. This can be measured in a number of ways, the crudest of which is simply to multiply the cross-sectional area by the speed of flow as measured by a floating object. Quite frequently an orange has been used as a float, as it is conspicuous and it travels along almost entirely submerged.

Davis (1938) in his instructions for carrying out stream surveys suggests the use of the formula:

$$D = \frac{wdal}{t}$$

where D = discharge, w = width, d = mean depth, and l = distance over which the float travels in time t. The term a is a coefficient which varies from 0·8, if the stream bed is rough, to 0·9 if it is smooth (mud, sand, hard-pan, or bedrock). Somewhat more accurate results are obtained if the transverse section is divided into several sections, say

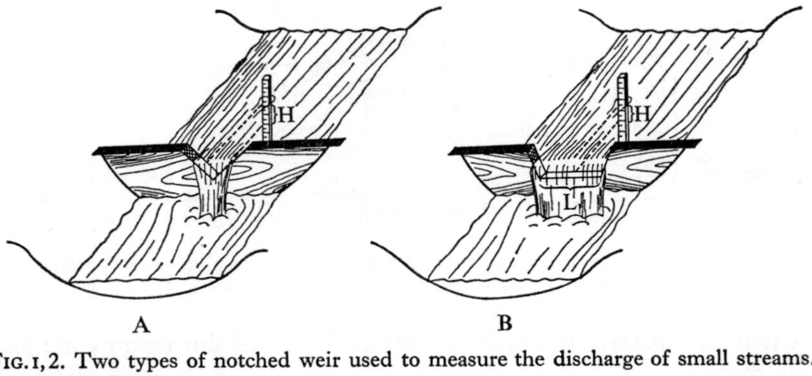

FIG. 1, 2. Two types of notched weir used to measure the discharge of small streams. For further details see text.

each one-third of the width, and the formula is applied to each section in turn (Robins and Crawford, 1954).

The incorporation of the term a is in recognition of the fact that friction with the stream bed is an important factor; and, as it has to be estimated, it can lead to considerable error. Thus, where it is important to make accurate measurements of discharge, it is necessary to measure the current at a large number of points, or to instal broad-crested weirs and measure the head of water flowing over them (Davis, 1938).

For small streams a suitable pattern is a 90° V-shaped notch (Fig. 1, 2A), and the head (H) above the tip of the notch is measured at a distance about 3 H upstream. Then:

$$D = 2{\cdot}52\, H^{2{\cdot}47} \text{ cu. ft./sec. when } H \text{ is in feet,}$$

or

$$D = 0{\cdot}925\, H^{2{\cdot}47} \text{ l./min. when } H \text{ is in centimetres.}$$

For larger streams a wider notch is used of which the sides diverge outwards at a slope of 0·25 to 1 (Fig. 1, 2B). Then when H is the head as before and L the length of the bottom of the notch:

$$D = 3{\cdot}367\, LH^{1{\cdot}5} \text{ cu. ft./sec.,}$$

or

$$D = 1{\cdot}12\, LH^{1{\cdot}5} \text{ l./min.,}$$

when H and L are in the appropriate units, feet or centimetres respectively. In large rivers the discharge is usually calculated from measurements taken at a considerable number of points with a current meter.

Discharge itself is, however, of little direct interest in most biological studies; much more usually biologists are interested in the rate of flow of the water where animals and plants actually live. Many types of large, and accurate, current meter are in use for measuring rates of flow, but very few organisms, apart from plankton which just travels with the water, live in places where large meters can be used. Most are found very close to the substratum, in shelter among plants, or downstream of obstructions.

Various biologists have, therefore, tried to devise methods of measuring the current close to the organisms. These have ranged from the release of small drops of coloured water from pipettes close to animals and plants and the timing of their movement over short distances, to tiny propeller-meters such as that of Edington and Molyneux (1960) in which the spinning head is only 1 cm. across. Other devices which have been used are small Pitot tubes—a rather neat one is described by Grenier (1949)—measurement of the cooling of a wire carrying an electric current (Ruttner, 1963), comparison of temperatures registered by two electrically warmed thermistors when both are immersed in the water but only one is exposed to the current (Kalman, 1966), and timing the rate of solution of standard pellets of salt (McConnell and Sigler, 1959). None is entirely satisfactory, although the thermistor device will detect very slow movement, but all do give some idea of the rate of flow near to the places where organisms actually live.

A particularly ingenious and very simple meter was designed by Gessner (1950, 1955*b*) and resembles a larger device which has been used in Russia. It deserves to be more widely known, because, although crude, it gives results which are probably almost as reliable as those of more complex, and costly, micro-meters. The apparatus consists of a truncated cone with an opening 5–10 mm. in diameter at the apex. To the base of the cone is attached a thin rubber bag, which is protected by an open-ended cylinder (Fig. 1, 3). With the bag in the collapsed condition the opening of the cone is closed with a finger, and the apparatus is

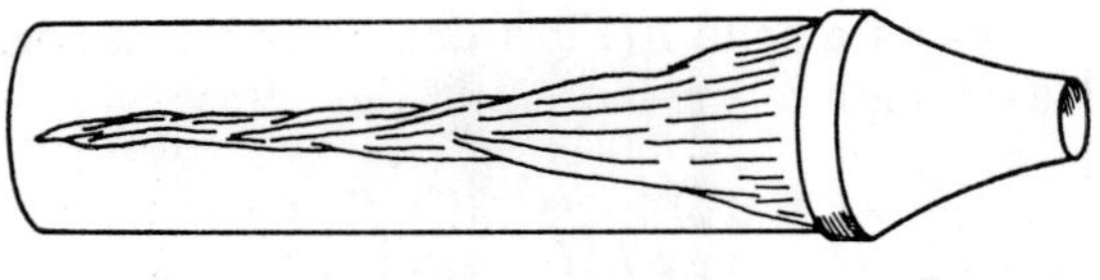

FIG. 1, 3. A simple rubber-bag current meter. Redrawn from Gessner (1955*b*).

placed, facing into the current, at the point where the measurement is to be made. The finger is then removed for a few, timed, seconds; water flows into the bag and the finger is replaced. The amount of water collected is then measured, and, from the area of the hole and the time of exposure, it is possible to work out the rate of flow. Gessner claims that if

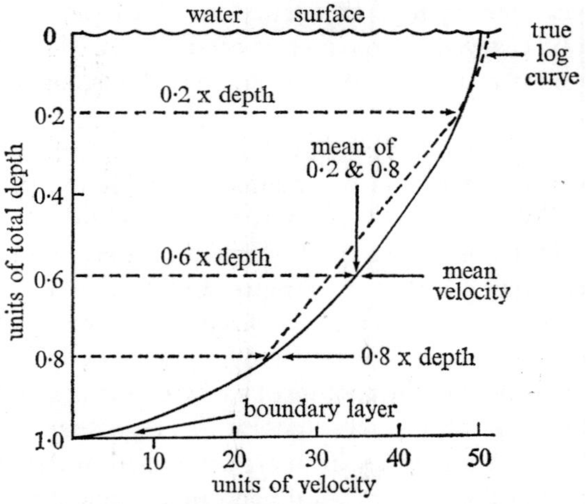

FIG. 1, 4. The rate of flow of water at different depths in an open channel, and depths at which the mean flow can be measured.

the measurement is repeated ten or twenty times at the same point consistent average results can be obtained.

Stream flow is a complex process, and in natural water-bodies it is always turbulent. The critical speed for the changeover from laminar to turbulent flow in a smooth pipe 1 cm. in diameter is 18 cm./sec., but in a pipe 100 cm. in diameter it is only 1·8 mm./sec. In an open channel the critical speed is only half these amounts, so we can conclude that, for practical purposes, real laminar flow does not occur in rivers and streams (Ambühl, 1959, 1962).

The velocity of flow at any point in a channel is nearly inversely proportional to the logarithm of the depth (Fig. 1, 4), and the steepness of the gradient of velocity towards the bottom depends upon the roughness of the bed. This is the reason for the term *a* in the first equation above, where it has been placed in empirical recognition of this fact. The banks, and the surface tension of the water, also exert a drag, and the drag of the surface distorts the curve from a truly logarithmic form to such an extent that the zone of maximum flow is sometimes, particularly in small streams, below the surface. Whether this occurs, and the depth of the point of maximum flow, depend, of course, on the relative resistances of the bed and the surface. In small streams these can change rapidly from place to place, and on bends the zone of fastest flow tends to swing towards the concave bank. Fig. 1, 5 illustrates how these points lead to differing rates of flow in various parts of the cross-sections of streams.

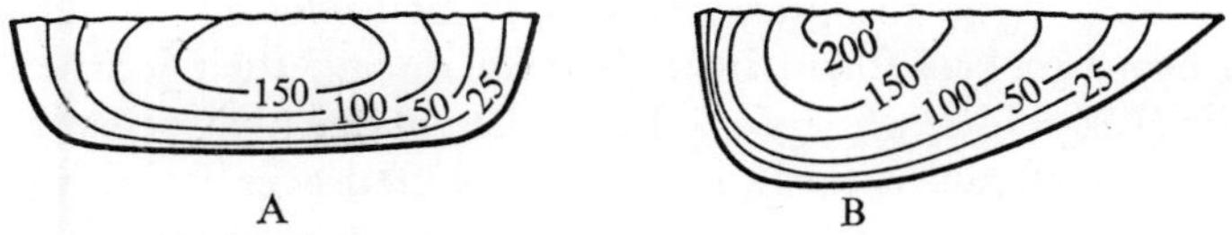

FIG. 1, 5. Idealized diagrams of the patterns of flow in cross-sections of open channels: A on a straight reach, B on a bend. The units on the lines of equal flow-rate could be cm./sec.

As a consequence of the almost logarithmic relationship between the depth and the rate of flow the mean speed of flow occurs at about 0·6 of the depth; also the mean of the speeds of flow at 0·2 of the total depth and 0·8 of that depth is the mean speed of flow of the water (Fig. 1, 4) (Grover and Harrington, 1943; King *et al.*, 1948). Another consequence is that the rate of flow decreases rapidly at the bottom, and there is a boundary layer right on the bed in which it declines very rapidly to zero. This layer is thicker because the flow is turbulent than it would be if it were laminar (Schmitz, 1961).

Hubault (1927) showed long ago, by measurements made during his field studies of stream animals, that the current becomes progressively less as one approaches the bottom, and this aspect of flow has been very thoroughly studied by Ambühl (1959, 1961, 1962) with particular reference to benthic animals. He demonstrated, by means of photography of suspended particles of acetyl cellulose in a flashing light, the existence of a boundary layer 1–3 mm. thick on the tops of stones, in which the current is very small, and that the layer becomes thinner the faster the mean rate of flow. This is, of course, merely a practical demonstration of the points made above, that the stream bed is a source of friction to the water and that the logarithmic depth/speed ratio holds true right to the bottom. One of Ambühl's very graphic pictures is reproduced here as Fig. 1, 6. His study also demonstrated very clearly, as can be seen from the figure, the existence of a zone of dead water on the downstream side of obstructions on the river bed. We shall see that these points are of considerable importance to many running-water organisms.

The fact that the stream banks, and also to some extent the water surface, retard the flow also results in a tendency for the surface water to be drawn towards the centre of the stream. This is discovered, even if it is not recognized, by every child who floats sticks in a ditch, and it is a superficial manifestation of the existence of rotational patterns of movement which are inherent in the arrangement of a number of layers of water flowing together at different speeds. These complex circulation-cells carrying water towards the centre actually cause a slightly convex surface on the river, and they have been exploited on the Mississippi where it used to be the practice to moor houseboats to water-soaked logs in order to keep them in midstream during their journey down the river (Leopold *et al.*, 1964). They also ensure that organic debris getting into a stream is fairly evenly distributed over its width before it sinks.

The mean rate of flow is, of course, related to the discharge, and to the width, depth, and roughness of the stream bed, but even at times of flood it rarely exceeds 300 cm./sec. (Einsele, 1960). Above about 200 cm./sec. most rivers begin to enlarge their beds by erosion, unless they are hemmed in by rock or man-made structures, and at speeds in excess of 300 cm./sec. air resistance begins to play a significant role. Even free-falling water does not normally exceed 600 cm./sec. (Nielsen, 1950*b*), although some part of the water may, because of restrictions or rapids, exceed this speed. Leopold *et al.* record that 22 ft./sec. (660 cm./sec.) was measured in a rock gorge of the Potomac River near Washington during a flood, and that speeds up to 30 ft./sec. (810 cm./sec.) are known.

It can be shown theoretically that the mean velocity of the water in a channel is proportional to the square root of the product of the hydraulic radius and the slope:

$$v = c\sqrt{Rs}$$

where v = mean velocity, c = a constant, R = the hydraulic radius (cross-sectional area divided by the length of the wetted perimeter), and

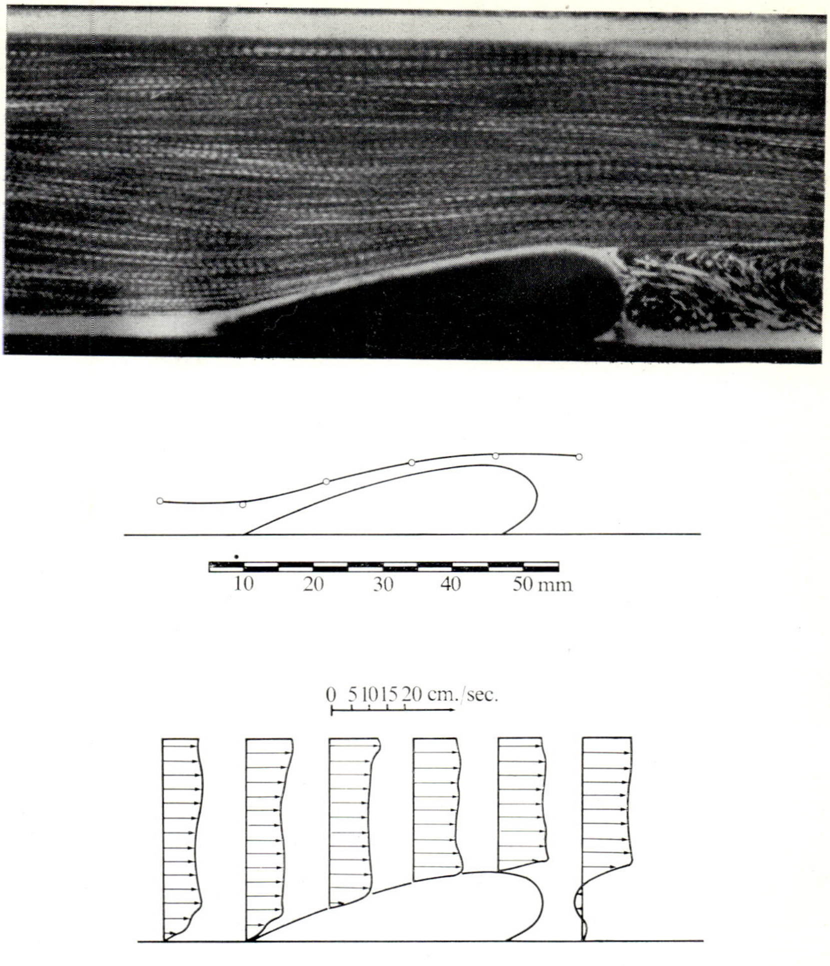

FIG. I, 6. Photograph, taken in flashing light through the glass side of a channel, of particles of acetyl cellulose carried by water flowing over an artificial stone. The middle diagram shows the scale, the upper limit of the boundary layer and the dead water downstream of the stone. The lower diagram shows the speed of the water at various points and depths, as measured by the distances moved by the particles during the timed series of flashes. From Jaag and Ambühl (1964) with permission.

s = the slope (effectively that of the water surface). In wide shallow channels this is almost equivalent to:

$$v = c\sqrt{ds}$$

where d = the mean depth.

The constant c, however, varies with the roughness (resistance to flow) of the stream bed, and this, under conditions of turbulent flow, is related to the depth, and hence to the hydraulic radius. So width, depth, mean current speed, slope, and resistance to flow can all adjust themselves relative to one another in a complex manner, which is fully discussed by Leopold *et al.* (1964). Here we are concerned only with the facts that overcoming the resistance to flow represents loss of energy from the system, and that a shear stress is exerted by the water on the banks and the bed.

In most situations in natural channels there is no over-all acceleration of flow over any considerable distance, although, of course, local accelerations and decelerations occur. The potential energy of the water is therefore lost, and is converted to heat; a fact that is well demonstrated when meltwater at 0 °C. carves channels in ice over which it flows. The heat is generated by friction with the stream bed and it causes an upward migration of turbulent eddies, and it is this upward migration of turbulence which supports and lifts suspended sediment.

In a natural stream, the resistance to flow, and hence the energy dissipation, varies with obstructions such as narrows, bends, rapids, tree trunks, etc., and the complex interactions of the channel usually result in a steepening of the slope of the water surface in the more resistant reaches. Thus the slope is steeper over a riffle than it is over a pool, and it is steeper on a curve than it is on a straight stretch. In some situations, as at the base of a waterfall or at the foot of a run, when the slope is suddenly changed by geological features or man-made structures, energy is released by the intense turbulent mixing sometimes known as the hydraulic jump. The turbulence, as a sort of rebound, produces a standing wave which rolls horizontally towards the upstream direction incorporating some of the downstream surface water, and usually also great quantities of bubbles (Fig. 1,7) (Braden, 1958). Such features in streams are very effective mixing devices and are important in the biology of fishes. The wave is stationary when the upward surge produces a wave of height h such that:

$$h = \frac{v}{g}$$

where v is the speed of flow of the water and g is the gravitational acceleration; but because the turbulence in a swiftly flowing stream varies widely—20 per cent or more—the wave is unstable and it constantly rises, falls, and wavers. It may take the form of a simple undulation or of a breaker with its breaking face upstream, or it may be completely submerged below a moving sheet of water. But, whatever its condition, it is the point from which fishes leap at falls, as they can there make use of the upward thrust of the turbulence (Fig. 1, 7A) (Stuart, 1962). Similar standing waves of lesser dimensions are caused by isolated obstructions on the bed, and they are associated with intense turbulent mixing, which is often manifest on the water surface as little whirlpools. They also draw water through the substratum if it is porous, as there is a definite downward component of flow at the bottom under the

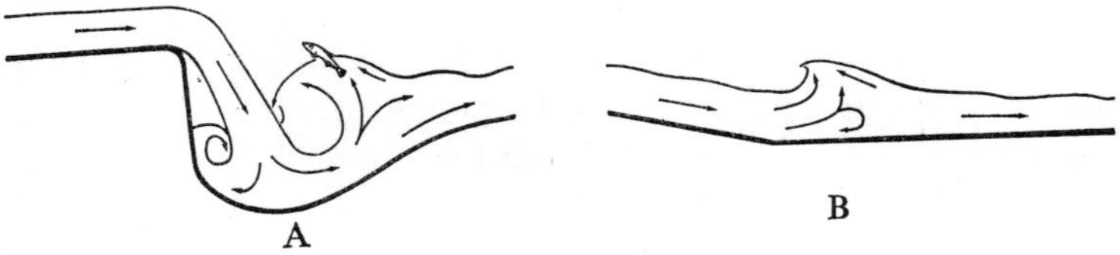

FIG. 1, 7. Diagrams of the patterns of flow and the standing waves: A, below a waterfall, B, at the foot of a slope.

crests and an upward one under associated troughs (Vaux, 1962). This leads to exchange of water in the gravel and is important to many animals.

Very swift turbulent flow and waterfalls incorporate numbers of air bubbles producing 'white water'. The amount of entrained air varies more or less as the speed of flow, and it can be as much as 60 per cent (Stuart, 1962). This is, of course, important for gaseous exchange, but it also greatly reduces the density of the water where it occurs. This again is important to fishes as it reduces their swimming thrust, and it is another reason why they leap from standing waves away from the masses of bubbles.

The shear stress on the bed, which can be thought of as a force tending to roll particles on the substratum, is correlated with the square of the velocity of flow, but the relationship is complex. For a thoughtful discussion the reader is once again referred to Leopold *et al.* For our purposes it is sufficient to be aware that increasing velocity picks up, or rolls along the bed, mineral particles of increasing size, and that these are then carried downstream, but at a rate which is less than that of the mean velocity of the water. Their presence in the water, however, cuts down the turbulence, and hence reduces the resistance to flow offered by the stream bed. Thus silty, or sandy, water picks up more silt or sand only if it flows more rapidly, although, of course, individual particles are constantly exchanged between the water and the bed.

A further complication is introduced by the packing coefficient of the particles. Compact clay is less readily carried up into the water than is

Type of bed	Critical mean current velocity cm./sec. Clear water	Muddy water
Fine-grained clay	30	50
Sandy clay	30	50
Hard clay	60	100
Fine sand	20	30
Coarse sand	30–50	45–70
Fine gravel	60	80
Medium gravel	60–80	80–100
Coarse gravel	100–140	140–190
Angular stones	170	180

TABLE 1, 1. The mean current velocity of clear and muddy water required to initiate movement along a stream bed of various types of bottom deposit. Modified from Schmitz (1961).

sand, even though, ultimately, its particles are much smaller. Tables which show the mean rates of flow which will just begin to move particles of given sizes are, therefore, for a variety of reasons, only generalizations,

Speed of flow cm./sec.	Diameter of mineral particles moved, cm.
10	0·2
25	1·3
50	5·0
75	11·0
100	20·0
150	45·0
200	80·0
300	180·0

TABLE 1, 2. The speeds of flow required to move mineral particles of different sizes. From Nielsen (1950*b*).

but they can serve as rough guides to the biologist. Two such, from Schmitz (1961) and Nielsen (1950*b*), are given in Tables 1, 1 and 1, 2, but with the comment that they do not agree very closely, and the warning that they are both only approximations. It should also be noted that the minimum current speed which will erode a material, for example compacted clay, is often different from that at which the resulting suspended particles will begin to settle. In general, however, it can be said that fine material (silt and mud) will settle out at mean current speeds up to about 20 cm./sec., and that where velocity does not normally greatly overstep the range 20–40 cm./sec. the bed is likely to be composed of sand (Einsele, 1960). Larger particles, if they happen to be there, are, however, likely to persist for some time, and it will be clear from our consideration of the complexities of flow on rough stony bottoms that large stones provide shelter for smaller ones and even, sometimes, for considerable quantities of sand. Sand and gravel can therefore persist among larger stones.

THE MORPHOLOGY OF CHANNEL SYSTEMS

A stream or river system normally consists of a pattern of tributaries joining one another and coalescing to form the main stream, and study of any map shows that there is a definite type of morphological pattern to which most drainage systems conform. Leopold *et al.*, discuss the way in which such patterns have developed, and they show that, starting with a more or less even slope drained by a series of parallel rills, the results of tributary capture when individual rills overflow and run into one another are such that the final type of drainage pattern, with various classes of tributary, is indeed the most probable. Even 'drainage patterns' constructed on squared paper by randomizing the direction (north, south, east, or west) towards which each square will 'drain' share many characteristics with natural systems. If one adds to such a model a bias in one direction, such as would be supplied by a general slope of the landscape, the analogy becomes very close indeed. Of course it must be recognized that not all stream systems will conform closely to a random system, because they have imposed upon them all sorts of factors in the geological history of the area which may have a marked influence. It is, however, possible to generalize, and generalization may prove of value to biologists to whom the practice is no stranger.

The idea of 'drainage analysis' was proposed by Robert A. Horton, and it is the study of the relationship of stream length to stream order. The last term is a measure of the position of a stream in the hierarchy of tributaries. First-order streams have no tributaries, but when two meet they become a second-order stream. Two second-order streams meet to form one of third-order and so on, with the proviso that acquisition of extra tributaries of an order lower than that of the receiving stream does not increase the order of that stream. A third-order stream may receive several first- or second-order tributaries without becoming a fourth-order stream.

In Horton's original scheme the highest-order stream was considered to extend, usually by the straightest route, up to the top of one of the small headwaters, originally considered to be a first-order stream. Later workers have not followed this practice, but it makes little difference to the analysis (Leopold *et al.*, 1964). Probably for biological work the later modification is the more meaningful, and Fig. 1, 8 is the map of a stream treated in this way. It also illustrates one of the weaknesses of applying

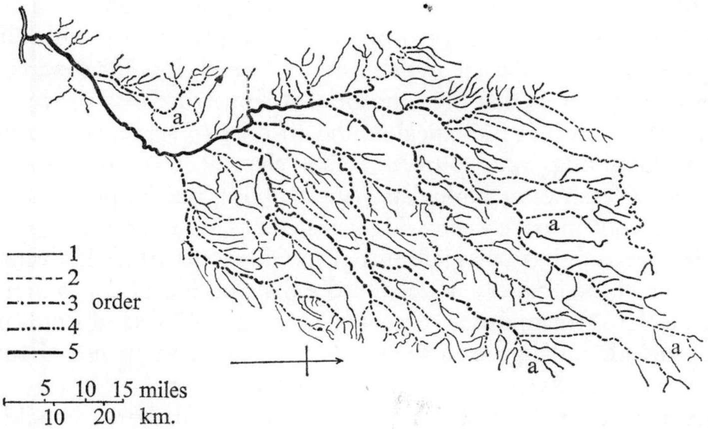

FIG. 1, 8. A sketch-map of the East Fork of White River, Indiana, showing stream orders. The significance of points marked 'a' is discussed in the text.

such a scheme to the natural situation, namely that, as at points marked 'a' in the figure, streams with similar drainage areas are placed in different orders because of the presence of very small tributaries. In biological work it seems probable that some allowance should be made for such situations, but apart from this problem the scheme does permit the objective subdivision of a stream network.

If the numbers, lengths, and drainage areas of the streams in each order are determined it is found that there are usually some fairly clear relationships between them. Stream order plotted on semi-logarithmic paper against the number of streams in each order, the mean length of stream in each order, and the average drainage area of streams of each order, gives in each instance a straight line. It is, in fact, usually found that there are between three and four times as many streams of order $n-1$ as there are of order n, and that each is, on average, rather less than half as long and drains rather more than one-fifth of the area. On the basis of analyses of this kind, which can, of course, be extrapolated, Leopold *et al.* calculate that there are about 3¼ million miles of stream channel in the United States.

Of course, the definition of a first-order channel has to be fairly precise, since it could refer to tiny temporary drainage-rills which are not shown on any map. Leopold *et al.* suggest that for many purposes first-order streams can be considered to be the smallest ones marked on a 1 : 24,000-scale map. Their approach is, however, primarily geomorphological; for the biologist the most useful criterion would seem to be perennial streams or those that persist for at least long enough to develop biota. The scale of the map to be used would then vary with climate, being about 1 : 250,000 for moist areas but much less for deserts.

It is quite possible that the concept of stream order may prove to be of considerable value in biological studies if only to serve as an objective way of classifying watercourses. It has already been presented to an audience of freshwater biologists (Abell, 1961), but its possible usefulness seems not to have been widely realized. One doctoral thesis based on this approach has been submitted to Oklahoma State University by R. C. Harrel, one paper has related the distribution of fishes in a Kentucky creek to stream order (Kuehne, 1962), and Illies and Botosaneanu (1963) include a diagram in their classification of the zones of running water which somewhat resembles the Horton system.

Carrying the analysis of drainage pattern still further, Leopold *et al.* have shown that for both small streams and large rivers there is a relationship between the length of channel (L) and the area drained (A) such that, on average:

$$L = 1{\cdot}4\,A^{0{\cdot}6}.$$

This implies that not only, on average, is there a definite length of channel for each area to be drained, but that, as the ratio $A : L$ increases more slowly than does L, the area drained tends to become longer and relatively narrower the larger it is. The larger river is relatively as well as absolutely longer, a point which may be important for migratory fish. It also implies that about 1 sq. km. is sufficient area to maintain about 1·4 km. of headwater, but whether this is 1·4 km. of effective stream habitat from the biologist's point of view depends, of course, on the amount of time that water actually flows along the channel. In areas which are not, however, subject to long, dry, periods this relationship probably holds for the 'biological' stream also.

The stream or river channel is itself a product of the water that flows along it, and is more or less adjusted to the pattern of discharge. We have seen that the shear stress on the material over which the water flows is proportional to the square of the velocity, which itself depends upon the roughness of the substratum, the slope, and the hydraulic radius (effectively the depth in many channels). For the channel to be stable this stress must equal the stability of the material into which it is cut. It can be shown that in a completely non-cohesive material, like loose sand, the cross-section of the channel would resemble a parabola; and such a channel would maintain itself at constant flow without carrying sediment, although it might migrate laterally by simultaneous erosion and deposition. Channels of this type are, of course, rare in nature, but they may occur below springs in desert areas, and the larger the flow the wider they become.

Most stream banks are, however, more cohesive than sand if only because of riparian vegetation. Natural streams thus tend to be relatively deeper than channels in sand, and to be rectangular in section although skew on curves (Fig. 1, 5). They also tend to have alternating deep and shallow areas—pools and riffles—especially where there are coarse constituents in the substratum. Riffles tend to be spaced at more or less regular distances of five to seven stream-widths apart and to be most characteristic of gravel-bed streams, although there seems to be a tendency for them to form even in rocky mountain streams. They do not normally form in sandy streams, since their presence seems to be connected with some degree of heterogeneity of particle size. In most streams the larger particles (boulders, stones, and gravel) tend to congregate on the bars that form the riffles, although this may not always be immediately obvious; and the bars themselves slope first towards one bank and then towards the other. This produces some sinuosity of flow even in a straight channel.

The reasons for the more or less regular spacing of riffles are unknown, but Leopold *et al.*, who made extensive studies on them, found that they do not move, although the stones that compose them migrate individually downstream and may reach the next riffle during a single small spate, being, however, replaced by others. They have also established that riffles are superficial features, with the largest stones in the upper layers. This cannot be explained merely in terms of the removal of smaller stones from the surface, as nearly all the larger stones are at or near the surface. The physics of this phenomenon and the reasons why stones tend to accumulate in groups to form riffles do not concern us here; but that riffles occur, and that almost all the large stones lie more or less on their surfaces, are of great importance to the stream biota.

At any given point in a channel, width, mean depth, and mean velocity vary as a power of the discharge in such a way that the sum of the exponents equals one (Leopold *et al.*, 1964).

$$w = pD^x \quad d = qD^y \quad v = rD^z$$

where w = width, d = mean depth, v = mean velocity, and D = discharge. $wdv = D$ by definition so:

$$pD^x \times qD^y \times rD^z = D$$

therefore:

$$p \times q \times r = 1 \text{ and } x + y + z = 1.$$

It follows that, as all the exponents are one or less—and they are all usually less than 0·5 in natural streams—width, depth, and velocity increase much less rapidly than does discharge. Which of them increases most depends on the erodability of the banks and thus on whether the channel is wide or ditch-like.

It can also be shown, empirically, that the suspended load of solids, which is a measure of the erosive power, varies as some power (usually between 2 and 3) of the discharge, and that the larger the ratio z/y the higher is this power. Thus the greater the ratio of velocity to depth the more tendency there is for a stream to erode its bed and so to increase the depth of the channel.

As we saw earlier (p. 9) mean velocity, mean slope, and roughness of the stream bed are related mathematically. With variations in discharge the slope alters little, but it is found that the resistance to flow (or effective roughness) decreases with increasing discharge (Leopold *et al.*, 1964). One reason for this, in a riffle and pool stream, is that, as the water deepens, the riffles no longer impose a sinuous flow on the main stream, and another is, of course, that a greater proportion of the water is moving far above the drag of the bottom.

This, however, applies primarily to streams with hard beds. In sandy rivers and streams the situation is complicated by the damping of turbulence by the suspended sand. As the flow increases some sand is brought into suspension, and this reduces the turbulence and hence the carrying capacity of the water. Sand is therefore redeposited, and a steady state is attained only if the deposits form ripples or dunes and so increase the resistance to flow. With changes in discharge therefore, sandy river beds are in constant motion, and ripples and dunes are normal features of their structure. In large rivers, such as the Mississippi, the shifting dunes may be very large and a hazard to shipping, as was well known to Mark Twain; and Sioli (1965) has reported serried ranks of dunes in the Amazon in depths down to 45 m., some of which are 200 m. long and up to 8 m. high.

Sandy beds, and others of readily erodable material, tend to scour during floods, but as the discharge declines they are refilled by material from further upstream. The depths to which such scour and refill can penetrate are astonishing; Leopold *et al.* give figures for some rivers in the United States which run to tens of metres, and a graphic example of one of their studies is given in Fig. 1, 9. Clearly phenomena of this magnitude can have enormous biological effects, and scour and refill can occur

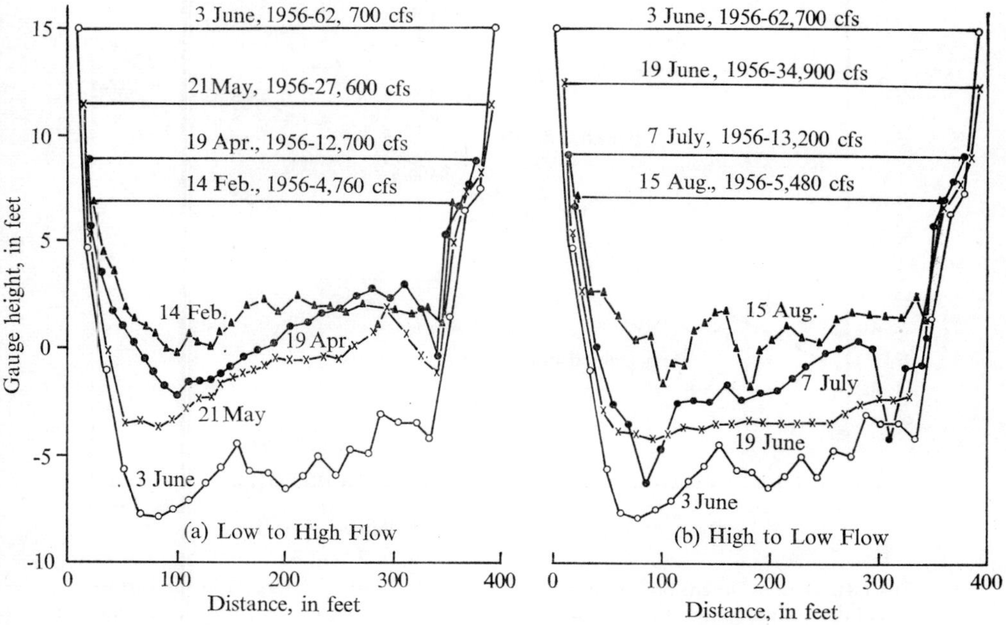

FIG. 1, 9. Scour and subsequent refill during the passage of a flood at Lees Ferry, Colorado River, Arizona, in 1956. From *Fluvial Processes in Geomorphology* by Luna B. Leopold, M. Gordon Wolman, and John P. Miller, W. H. Freeman and Company. Copyright © 1964.

even on gravel beds in some large rivers. Even though it does not apparently involve the whole channel simultaneously, and the material may not be moved very far, much of the bed is churned over, and light materials, including plants and animals, have little chance of remaining.

It would seem that the form of a channel is determined and maintained by discharges that would just about fill it. In many rivers and streams this discharge is reached about every $1\frac{1}{2}$ years on an average (Fig. 1, 10), and in their studies on Seneca Creek, Maryland, Leopold *et al.* found that median-sized gravel on the riffles began to move when the depth was about 0·75 bankfull. This occurs about once a year. So, about once a year, the flow is sufficient to cause erosion to fit the channel to the load of water which it carries.

Because of climatic irregularities much greater discharges than bankfull do occasionally occur, but, although they can make great alterations, the water escapes from the channel and hence flows relatively slowly

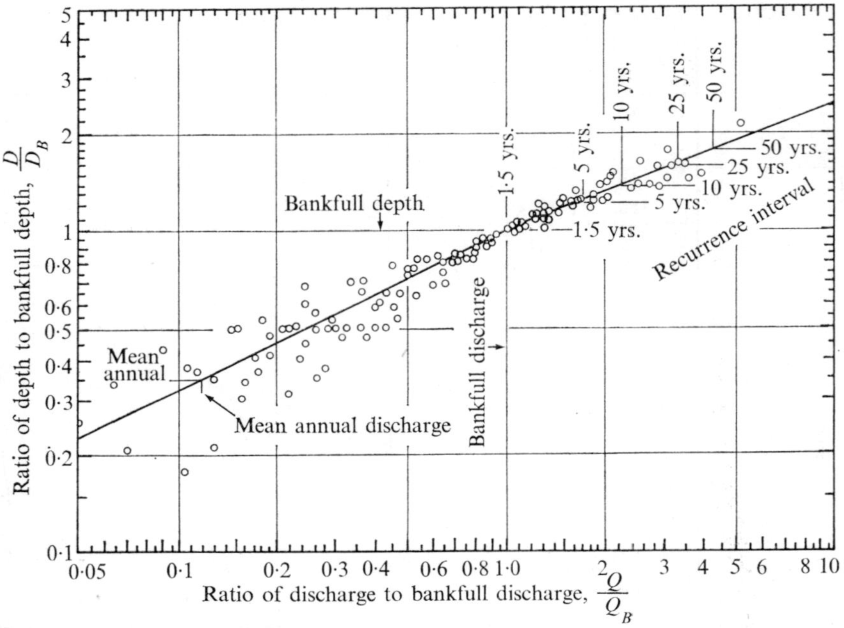

FIG. I, 10. A non-dimensional rating curve for thirteen gauging stations in the eastern half of the United States. From *Fluvial Processes in Geomorphology* by Luna B. Leopold, M. Gordon Wolman, and John P. Miller, W. H. Freeman and Company. Copyright © 1964.

(width increases very rapidly) (Einsele, 1960), and the occurrence interval of such floods is quite long. Fig. I, 10 is a graph made by Leopold *et al.* from data from a number of gauging stations in the eastern United States, an area of fairly high flood potential, from which one can read that the likelihood of occurrence of 'flood' of:

1·2	bankfull	depth	is	once in	about	5 years.
1·4	,,	,,	,,	,,		10 years.
1·6	,,	,,	,,	,,		25 years.
1·8	,,	,,	,,	,,		50 years.

As one proceeds downstream, the discharge in most rivers increases because of the accession of tributaries; and the same general formulae apply in a downstream direction as at a single point, except that discharge (D) then represents some definite value, such as bankfull discharge or mean annual discharge.

Using data on the mean annual discharge from a large number of rivers in the United States, Leopold *et al.* found that the width tends to increase as the square root of the mean discharge in most rivers:

$$w = pD^{0\cdot5}.$$

Depth increases much less rapidly, slope decreases markedly, but the mean velocity usually shows a slight rise. This last point is surprising for those who, like me, have always thought, for no very sound reason, that slope is the main controlling component of current speed, and that the nearer the river is to base level the slower is the current. The increase in current speed with the size of channel at a constant slope is well illustrated by some calculated figures given by Einsele (1960) and shown in Table 1, 3. This sort of discrepancy goes some way towards explaining

Slope m./km.	200 m. wide 4 m. deep	20 m. wide 0·5 m. deep	2 m. wide 0·25 m. deep
0·5	150	50	25
1·0	(250)	60	35
2·0		80	50
5·0		130	70
10·0		180	100

TABLE 1, 3. The calculated mean velocities in cm./sec. of water in three channels of differing dimensions at various slopes. From Einsele (1960). The figure 250 is bracketed because it is above the normal threshold of enlargement of channels (200 cm./sec.).

biotic differences between small and large streams in the same general area and type of landscape.

The exponential decrease of slope downstream indicates, in effect, that river profiles are usually concave, and this applies even to those rivers which, like the Nile, receive no tributaries in their lower reaches. The reasons for this are discussed by Leopold *et al.*, and need not concern us here. Suffice it to state that they are at least in part, but by no means simply, concerned with the particle size of the bed material and its influence on current resistance. Coarse material is associated with steep slopes and finer material with flatter ones, and this fact influences the resistance to flow in a complex way. Of course, few rivers have ideal concave profiles because, among the other reasons, they may acquire new loads of coarse materials from tributaries or themselves cross areas of hard rock. Such features tend to influence the profile and be associated with humps; a 'nick-point' where a river is cutting back into hard rock is such a feature on a grand scale.

A river channel is thus an almost infinitely adjustable complex of interrelations between discharge, width, depth, rate of flow, bed resistance, and sediment transport. Any change in one tends to be countered by adjustment in the others, and the whole system tends towards conservative dynamic equilibrium. Leopold *et al.* state that there is 'little doubt of the validity of the two postulated tendencies—uniform distribution of energy dissipation and minimum total work in the system'.

CHANNEL PATTERN

So far we have considered only simple channels with no complications other than riffles and sand dunes. Even in relatively straight reaches the

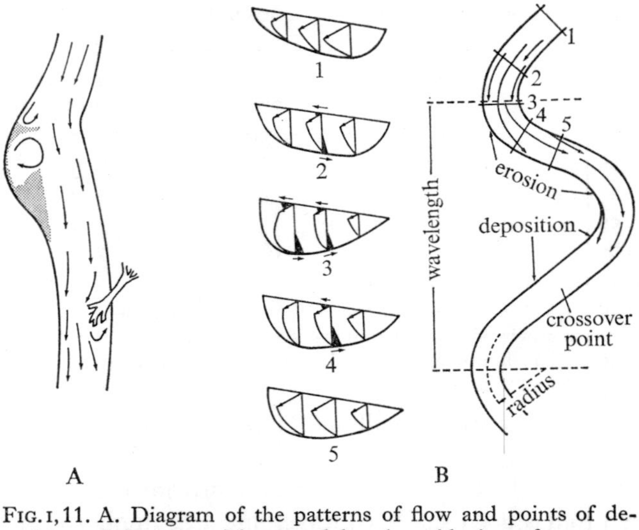

FIG. I, 11. A. Diagram of the patterns of flow and points of deposition of fine materials caused by the widening of a stream channel and a minor obstruction such as a fallen tree. B. Diagram of a meander and the patterns of flow at the surface and in transverse sections at various points (1–5). Note the transverse components of flow in sections 2–4 which are shown in black, and the points of erosion and deposition. Mainly after Leopold *et al.* (1964).

deep portion, or main stream, tends to swing from side to side of the channel, as we have seen happens on riffles. This tendency, which is little understood, can lead to more complex patterns, and changes in width can also produce special features.

When a stream swings round a curve the current stream hugs the concave bank, the channel is deeper there (Fig. 1, 5B), and the momentum of the water causes its surface to become banked like a curved road. But the water in contact with the bank is slowed by friction and so flows downwards and inwards carrying erosion products with it (Cunningham, 1937). This material is then, at least in part, deposited further downstream on the same bank, when the main stream once more swings across the channel. Quite often when, for some local reason, a channel widens, a complex of eddies and bars is formed, and debris collects in the eddies (Fig. 1, 11A) (Braden, 1951). Such features are unimportant in terms of river morphology; they are probably transient, and may be quite small, as when a fallen tree or a bridge piling diverts the current from the bank, but they are biologically significant. The organic debris which collects in such places offers a special habitat for certain creatures which occur rarely elsewhere, and all these minor irregularities afford shelter for many animals at times of high water. More fundamental differences of pattern are shown by braided and meandering channels.

Braided (or reticulate, to use a word more familiar to biologists) rivers and streams are composed of a series of anastomosing channels separated by islands, and they are usually formed in loose readily shifted material such as glacier debris or moraine gravel. The channels are, like those in loose sand, wide and shallow, and any small obstruction, such as a group of slightly larger stones, tends to divide the stream and to pile up more material on its downstream side. The diverted streams on each side are then somewhat constrained, so they erode their beds, and the obstruction emerges as an island which continues to grow at the downstream end. Often such structures are very numerous and the channel becomes very complex. Islands may last for a long or a short time before a rise in discharge sweeps them away, or they may become fixed by vegetation and persist for long periods. Because there are several channels instead of one large one the resistance to flow is increased by braiding, and the slope is correspondingly greater. Leopold *et al.*, however, stress that such streams can be in an equilibrium state even if the actual pattern is not, and that braiding is a response by running water to passage through an area of very readily eroded bank-material and a large debris load. Because of their constantly shifting beds, reticulate channels are often remarkably devoid of living organisms.

Meandering channels have a regularity of curving pattern which is shared by streams of all sizes to such an extent that one cannot distinguish the map of a small stream from that of a great river without reference to the scale. Of course, a decision as to when a series of bends qualify for the term meander is an arbitrary one, but it can confidently be applied to reaches exhibiting considerable symmetry.

Leopold *et al.* show from study of large amounts of data that the wavelength of meanders is related to the width of the river, being 7–10 times the width, and that the path measured along the channel itself is 11–16 times the width. It is interesting to note that this is about twice the 5–7 widths that occur between riffles, because in a typical meander there are two relatively shallow points per wavelength where the main stream swings across the river between the curves (Fig. 1, 11B, 1 and 5). This indicates some, as yet unexplained, feature common to the formation of both structures. There is also considerable constancy in the ratio of the radius of curvature of the individual bends to the width. This value is usually between 2 and 3, and it accounts very largely for the similarity in appearance of meanders between streams of widely different sizes. The amplitude varies widely, and it seems to be related to the erodability of the banks, although this relationship may be complicated by earlier geological history. For instance, meanders on the Rhine and the Moselle have quite large amplitudes but their present channels are cut into hard rock.

Within each meander the flow pattern is a little more complex than the simple pattern of the S-curve (Fig. 1, 11B). The main point of erosion is a little downstream of the mid-point of the curve on the concave bank, and the main point of deposition is similarly downstream of the mid-point of the curve on the convex bank. The whole system tends therefore to move slowly down the valley. The swing of the water round the bends results, as we have already seen, in some transverse components of flow (Fig. 1, 11B), which are important in moving erosion products. Similarly the point at which the channel is symmetrical is a little below the point of crossover. Recent studies show that the meander curve is the curve of minimal total work in bending the channel and that it is thus the most probable form that will appear in an erodable plain (Leopold and Langbein, 1966). It is also the form of channel that results in the most uniform loss of energy, and for these reasons meanders are significant features in many types of landscape. From the point of view of the biologist they provide a variety of habitat which is not present in a simpler channel. A point of interest is that, because of the alternating resistances to flow caused by the bends and the shallows, a meandering stream is steeper than a simple channel carrying the same discharge.

We are not, in this book, concerned with many features of the flood plain, which is composed of a series of deposits laid down by a stream or river as it swings back and forth across a wide valley, but it is important to note that many flood plains maintain areas of still water which at times of high discharge become linked with the river. These include ox-bow lakes and similar stretches of old channel, which acquire muddy bottoms by deposition of silt at times of flood, and pools behind levees. Levees are banks of eroded material along the margins of rivers, which are formed when flood-waters overtop the banks and, then being suddenly slowed, deposit their suspended load. In some places, such as the Mississippi delta, they can be quite high, and swamps and pools form behind them (Reid, 1961). In the Amazon, levees deposited round the perimeters of islands lead to the formation of 'lakes within the river', with margins standing up out of the water like atolls (Gessner, 1961). Such water-bodies, when linked to the channel at times of high water, supply plankton to the river, are used as spawning areas by some fishes and are nursery grounds for many others. Here again, as with riparian forests, many of these features have been eliminated by modern man in his quest for good agricultural land.

THE MATERIALS OF THE STREAM BED

The mineral particles on the beds of rivers and streams range from boulders to clay, and to date most biologists have been content to be merely descriptive about the nature of the substrata they have worked upon. Cummins (1962), in a thoughtful discussion of the problems of biological work in running water, advocates that we should be more numerical, and he has put forward a series of suggestions as to how this might be done.

For physical analysis he advocates using a modified Wentworth scale of particle size and the recording of data on the linear phi-scale ('phi' = the negative logarithm to the base 2 of the smallest particle in each size-group). The essentials of his ideas are shown in Table 1, 4. He advocates initial elutriation of the silt and clay from a sample of the stream bed, and their subsequent separation by the settling out of the silt. The clay is then centrifuged from an aliquot. The rest of the sample is dried and sieved in a mechanical shaker, and all the results are recorded as weights. Undoubtedly biologists should be doing this sort of recording more often, to give precision to their descriptions of substrata. He also advocates that amounts of organic matter in deposits should be determined by dry combustion, after treatment with acid to remove carbonates, and recorded as carbon dioxide produced. For samples in coarse materials, exceeding 4 mm. in diameter, Leopold *et al.* advocate the much simpler technique of picking up and measuring 100 stones located on a grid. They claim that this gives as consistent results as taking samples with a grab.

We have seen that the greater the velocity of the current the larger the particle which it can move, and that, because of the shelter they provide,

Name of particle	Range of size in mm.	phi-scale	Mesh size and methods of measurement: mm.	Approx. in. equivalent	U.S. sieve no.	Tyler sieve no.
Boulder	>256	−8	Direct measurement			
Cobble	64–256	−6–7	Individual wire square			
Pebble	32–64	−5	Individual wire square			
	16–32	−4	16	0·625 (15·9 mm.)		
Gravel	8–16	−3	8	0·312 (7·93 mm.)		
	4–8	−2	4	0·157	5	5
	2–4	−1	2	0·0787	10	9
Very coarse sand	1–2	0	1	0·0394	18	16
Coarse sand	0·5–1	1	0·5	0·0197	35	32
Medium sand	0·25–0·5	2	0·25	0·0098	60	60
Fine sand	0·125–0·25	3	0·125	0·0049	120	115
Very fine sand	0·0625–0·125	4	0·0625	0·0024	230	250
Silt	0·0039–0·0625	5, 6, 7, 8	Separated by settling			
Clay	<0·0039	9	Separated by centrifuge			

TABLE 1,4. Suggested terminology, categories, and methods for particle-size analysis. Modified from Cummins (1962).

large particles can protect smaller ones from being entrained. But we have also seen that, despite the reduction in slope, the current at lower altitudes on any given river does not usually decrease. Nevertheless, it is true of many, probably most, river systems, that the mean particle size decreases in a downstream direction, and that there is thus a general correlation between the particle size and the slope. Part of the explanation for this probably lies in the fact that the shear stress on the bottom, and hence the competence to move particles, decreases with increasing discharge, so that the winnowing effect declines downstream. But it is also undoubtedly true that the large stones derived from steep slopes, which are steep because they are composed of hard rocks, are rapidly broken up in running water. Were this not so they would persist in the beds of large rivers, whereas these, unless large stones are being derived locally or are brought in by tributaries, are usually composed of sand or small gravel. At this point it is as well to stress that this is so, even in lowland reaches which are generally thought of as muddy. Large amounts of silt occur only in backwaters and shallows or as a temporary thin sheet over sand during periods of low flow; silt is certainly not a major component of the substratum in the main channels of the great majority of even base-level rivers. The only very fine-grained material which does sometimes make up a large proportion of the substratum of rivers, as in the Irtysh in Siberia (Shadin, 1956), is hard clay. This is derived locally, and, as we have seen (Table I, 1), it is less erodable than sand.

Obviously the distance downstream over which large particles persist depends also upon their composition. Granite breaks up more readily than basalt, and limestone more readily than quartzite; and another factor is the degree of weathering to which they are exposed both before and after entering the stream. This depends on climate, and here it is worth stressing that, once it has entered a stream, a stone is much more liable to be alternately wetted and dried, and is more likely to be wet when exposed to frost, than is one on land. Leopold *et al.* also make the very pertinent point that a sediment particle deposited in a flood plain is likely to remain there for a very long period before it once again enters the stream. During all this time it is exposed to weathering and so is likely to disintegrate more rapidly when it is again exposed to abrasion. In any given river the further downstream, the wider is the flood plain, and thus the longer the average storage period and, presumably, the more friable is the stone when it finally re-enters the water. Many factors therefore combine to make it probable that the further down a river the smaller is the general size of the particles forming the bed. This would tend to lead to increasing uniformity of the substratum were it not for curves, meanders, and similar irregularities which, by causing local differences in current, tend to aggregate the various sizes of particles which are present. Because of this in many rivers, for example the Volga, there is actually an increase in the variety of types of substratum as one moves downstream (Lastochkin, 1943).

FLOW REGIMES

One thing that is certain about stream discharge is that it is irregular and that any regularity of pattern that it does show is largely a statistical phenomenon. We have seen that bankfull flow is equalled or exceeded about once every 1·5 years, and this is a fairly characteristic feature in most streams. Schmitz (1961) gives a figure which shows that the upper Weser reached at least bankfull flow twenty-four times in the period 1900–46. However, only one of these floods occurred in summertime, and in many climates there is more chance that high water will occur at some seasons than at others. As we have seen (p. 18) it is possible on available data to predict the likelihood of floods of given depths in terms of bankfull depth. Irregular therefore though they are, many rivers do show an annual mean pattern of flow which is related to the climate. Very high floods are infrequent even though they may have great effects, and the periods during which rivers work on their banks and beds (about 0·75–1 bankfull) are very short. Even the mean annual flow is equalled or exceeded only about 25 per cent of the time (Leopold *et al.*, 1964).

It is possible to classify the expected average discharge regime on the basis of the source of most of the water (Shadin, 1956; Kresser, 1961). Rivers that are fed largely by snow-melt tend to be high in the spring, and glacier-fed streams flow at a high level in the summer; near to the glacier they may even show a diurnal pulse of discharge. Cold winters result in low discharges, and fairly strictly seasonal rainfall, as in many parts of Africa, is also reflected in variations of discharge. Larger rivers, since they represent a summation of many minor local effects, tend towards greater regularity than small ones. The meltwater flood irregularity becomes less and less obvious the further one moves down the Volga (Behning, 1928), the Danube shows a regular spring rise because of meltwater from the Alps (Kresser, 1961), and the lower Amazon has a fairly regular period of high water between May and August (Gessner, 1961). Great rivers require a large amount of water to alter their pattern of discharge; the summer thunderstorm which will swamp a small valley covers so small a fraction of the drainage area of a large river that it produces little effect. Smaller streams are, therefore, far less stable than large ones, but even they can be classified into groups by graphing their *average* monthly discharges as a fraction of their mean annual flow (Browzin, 1962). Graphs of this type fall into definite patterns which reflect the local climate, although they are modified from the pattern of precipitation by such things as snow-melt, run-off from frozen ground and very dry soil, and evaporation and transpiration particularly during the summer.

The mean annual discharge, because of its relatively rare occurrence, appears to be quite high water to the casual observer. Baxter (1961) found that it was not attained during 250–60 days in the year, even in rivers in Scotland, Wales, and western England, an area which is famous for its consistently rainy weather. On about eighty-five of those days the discharge ranged from one-half to one-quarter of the mean annual discharge. This means that, even in a climate where the rainfall is unusually evenly distributed, stream organisms live for a great proportion of the time at relatively low current speeds. In larger rivers these long periods of low discharge do not greatly decrease the area of wetted bed because the water is relatively deep. Baxter found that most of the beds of Scottish salmon rivers were still covered by water at even one-eighth of the mean annual discharge. But small stony streams began to contract at one-half of mean discharge, and at one-eighth, only one-third to one-half of the bed remained wet. It will be appreciated that differences of this kind between small and large streams can have important biological consequences.

One last point, which is really a strictly biological one, is that great quantities of rooted vegetation may develop in some shallow slow streams during the summer. These not only alter the substratum by accumulating silt, but they impede the flow and so to some extent offset the effects of low summer-discharge. Berg (1943) found that the Susaa in Denmark maintained a high summer-level in this way, and in many parts of Britain it is common practice to cut the 'weeds' in such streams to lessen the threat of flooding by summer thunderstorms.

CHAPTER II

Physical characteristics of flowing water

The peculiar physical properties of water which are of such importance in freshwater biological studies are discussed by Hutchinson (1957) and Ruttner (1963), to whom the reader is referred for details. Discussion here will be confined to those properties which are of particular importance in running water.

TEMPERATURE

The temperatures of rivers and streams vary much more rapidly than those of lakes, but quite often this variation is over a much smaller range than that of at least the shallower parts of still water.

Several observers have kept a stretch of stream under observation for a period of time and have found that, superimposed upon the seasonal changes, there are diurnal cycles. These may amount to 6 °C. in small streams in summertime (Macan, 1958*c*; Madsen, 1962; Edington, 1966), with lower values in larger rivers. Maxima usually occur in the afternoon and minima in the latter half of the night. In small streams the deeper the water the less is the daily variation, which is caused primarily by radiation into and out of the water (Eckel, 1953), and when a stream is spring-fed the daily variation declines towards the source. It will be appreciated, therefore, that shallow streams a few metres wide, especially if they are not shaded from the sun, are particularly subject to short-term variation.

Temperature also varies along the lengths of the valleys. Large rivers, and streams at some considerable distance from their sources, are usually at more or less the mean monthly air-temperature at the point of measurement. This may vary from a few degrees at high latitudes and altitudes to quite high temperatures in summertime or in warm lands. For instance the temperature of the Amazon at Manaus in Brazil is usually 29–29·5 °C. (Gessner, 1961), and the River Zeraushan in

Uzbekistan can reach 31 °C. (Shadin, 1956). In wintertime, however, ice and snow form an insulating layer, and, even in extreme climates such as that of Alaska, the water temperature does not fall below 0 °C. (Sheridan, 1961). The range of mean monthly temperature in rivers, as in lakes, is therefore often much less than that of the land surface.

Superimposed upon the annual liaison with the mean air-temperature there are often special features. For instance, in springtime snow melt-water may keep the temperature below that of the air for quite some time (Sheridan, 1961); and Macan (1958*c*) observed that sunshine after heavy rain resulted in high temperatures, presumably because water from the warm soil was continuing to flow into the stream.

Several European workers have shown that the summertime temperature increases downstream in such a way that the rise is more or less proportional to the logarithm of the distance from the source (Schmitz and Volkert, 1959; Schmitz, 1961; Eckel, 1953). However, Dorris *et al.* (1963) observed that in the Mississippi increased discharge results in lowered temperatures, presumably because colder upstream water is being brought down faster than it can be warmed up. Spates can therefore move the whole pattern in a downstream direction.

In summertime the headwaters are relatively cool, either because they contain spring water from deep underground, and hence at the annual mean soil temperature, or because they come from high ground and are only slowly warmed by the air, the sun, and conduction from the ground. This warming continues downstream at a decreasing rate until the mean ambient temperature is attained, and the water then flows on influenced mainly by the weather and the daily cycle.

In winter the reverse may apply in spring-fed streams which start relatively warm and become cooler. Minckley (1963) found that 5 km. from its source in a large spring, the temperature of Doe Run, Kentucky, varied from 6·1 °C. in winter to 20 °C. in summer, while the temperature at the source remained between 13 and 13·5 °C. at all seasons. This emphasizes the biologically important fact that a stream which is cool in summer because it is spring-fed is warmer in winter than streams fed by run-off. The 'summer-cold' stream of the German limnologists is also a 'winter-warm' one, and in a biological context the latter property may be the more important.

Even in the tropics stream temperatures increase downstream until they reach equilibrium with the air temperature. The Marowijne River in Surinam rises at 22 °C. and reaches 31 °C. at its mouth (Geijskes, 1942). In Central Africa streams on the high mountains may arise from ice-water at 0 °C. and they warm up as they descend (Hynes and Williams, 1962).

But the simple picture of temperature change over distance can be altered by local conditions. Sometimes a river continues to receive groundwater along its length and its temperature therefore remains fairly constant for a long distance; this occurs for instance in the River Luhe in Germany and the Suwannee River of the song-book (Schmitz and Volkert, 1959; Beck, 1965). Streams flowing underground or through man-made culverts may be cooled or warmed in the process according to the season, and wind or shade may cause considerable changes (Kamler, 1965, 1966). Macan (1958*c*) reports a fall from 21·6 to 14 °C. in a small stream flowing through woodland on an east-facing slope during a period of easterly wind, and Edington (1966) showed that shaded reaches in streams in northern England are warmer in winter, cooler in summer, and reach their maximum temperature later in the year, than exposed reaches. Lakes on the course of a river also have profound effects which vary with the season (Eckel, 1953; Edington, 1966). It is therefore possible only to generalize about stream temperatures, although broad patterns are apparent.

In contrast to lakes, rivers normally show little stratification because of their turbulent flow. Shadin (1956), however, states that in water more than 15 m. deep there may be slight differences between the surface and the bed; and, in both large rivers and small streams, small differences between sides and centre, or between one side and the other, are sometimes observed. These may be caused by sunshine on the shallows, inflowing groundwater or by the water from a tributary hugging the bank on which it entered. Neel (1951) found small differences between parts of pools in Boone Creek, Kentucky, when warm water flowed over cool or cool under warm. Such differences are, however, small, and are probably of little biological significance as they do not endure for long periods.

One effect of temperature on water is to alter its viscosity, and this causes silt to sink twice as fast at 23 °C. as it does at 0 °C. Shadin (1956) gives the following figures for the coefficient of friction of water at various temperatures.

0·2°C.	0·01858
10·8	0·01317
20·0	0·01102
30·0	0·00800

Thus warmer water carries less silt than colder, and it flows a little faster (0·5 per cent faster for each 1 °C. rise between 4 and 20 °C.). It also produces a thinner boundary layer on the bottom (Ruttner, 1963). As yet the possible biological significance of these phenomena has not been investigated.

Any discussion of temperature inevitably involves the question of methods of measurement. It is often not possible to instal elaborate recording instruments at field sites, and from what has been said previously about diurnal and sporadic variation it is clear that isolated readings are of limited value. Macan (1958*c*), after his careful study of continuous records from a small English stream, concluded, however, that average weekly temperatures obtained from the means of daily, or even weekly, readings of a maximum and minimum thermometer do not differ greatly from the true value. It is fairly easy to instal such a thermometer in a stream, a satisfactory way to do so being to place it inside a tile drain, which protects it from both abrasion and the public. Schmitz and Volkert (1959) describe a method whereby the rate of the sugar-inversion reaction, which like all chemical reactions is temperature-controlled, can be used to estimate mean temperature. Tubes of sugar solution are placed in the water and left for a time, after which the change in optical rotation of the solution is measured with a polarimeter. Either of these simple methods could probably be used with advantage more often than has been the practice.

ICE

During cold weather ice may form on the water surface, although it does so less readily on running than on still water. This is, of course, because the turbulent flow continuously mixes the water, so the whole mass has to reach 0 °C. before any considerable amount of ice can form. It will be recalled that in lakes and ponds the peculiar physical properties of water ensure that most of the underlying water usually remains at 4 °C. or a little below (Hutchinson, 1957).

Once ice has been formed snow can accumulate on it, and together they form an excellent insulator against further heat loss (Sheridan, 1961; Gard, 1963). In many climates ice-cover is of short duration, but in the northern temperate regions of continents it may last for many months, and in mountain areas streams may freeze solid. Kamler (1965) reports that this occurs in the Tatra Mountains in southern Poland, and I have observed it even in small streams on the Pennines in England, a country in which severe cold is a rarity. Very probably, however, even where the water is frozen down to the bed, free water remains between and under the stones and in the deeper places. This is an aspect of stream study which might well be worth further investigation, as it seems to be fairly well established that most stream animals cannot survive being frozen solid.

Complete ice-cover, of course, prevents gaseous exchange between the atmosphere and the water, but de-oxygenation is rare in rivers for the following reasons. Very often small areas over riffles remain open and entrain air—river ice is notoriously treacherous to the unwary skater because of thin patches and holes—and long periods of cold weather are periods of low and reducing discharge, so that after the ice has formed the water level often falls. In small streams the ice may then be left as a bridge with air below it, and in large rivers fluctuations in level usually break the ice after a while, forming aeration holes. Indeed, to my knowledge, the only place where severe de-oxygenation has been reported as occurring naturally, that is to say not because of pollution by man, is the River Ob in Siberia (Mosevich, 1947). Even the River Varzuga in the Kolsky Peninsula of northern Russia, which is frozen over for half the year, is full of fish (Shadin, 1956), as are the rivers of northern Canada. One can therefore infer that long ice-cover is not a negative biological factor. The situation in the Ob is peculiar in that it, and its tributary the Irtysh, drain the vast swampy area of the West Siberian plain, and all through the winter de-oxygenated water from the swamps drains into the river, and the oxygen falls to about 5 per cent of saturation.

Surface ice formation is, therefore, of little biological significance except in special situations. It is, indeed, even beneficial to riverine organisms as it prevents further cooling, although when it breaks up in the spring it may scour the bottom in shallow water as it bumps downstream. Often, however, the ice 'goes out' on rising water produced by the thaw, and so it is raised off the substratum as it breaks up. A brief early thaw which is just sufficient to break up the ice may therefore do much more damage than a delayed and final one.

Underwater ice is of far greater biological significance. It forms during clear cold nights, and only in water which is not frozen over. Reports of the air temperatures needed to cause its formation vary from —15·6 to —23 °C. (Benson, 1955; Needham and Jones, 1959), and it is caused by radiant loss of heat from the dark-coloured substratum under water at 0 °C.

There are two forms of underwater ice (Coker, 1954; Brown *et al.*, 1953; Maciolek and Needham, 1951). *Frazil*, or slush, ice results from the formation of crystals in supercooled water or from the persistence of entrained snow-flakes (Pomeisl, 1953*a*). It occurs in both shallow and deep water, and the slush can reach considerable thicknesses; in places it can scour the substratum like sand carried by flood-water. *Anchor* ice forms in shallow water, usually at depths of less than 50 cm., and it is normally only a few centimetres thick, although Brown *et al.* (1953) report having observed up to 60 cm. in the West Gallatin River,

Montana. It begins to form on riffles, first on the upstream faces of the larger stones, and from there it may spread to cover much of the bottom and extend into the pools. At first it is a soft white deposit, but as it grows it becomes denser and more transparent.

Anchor ice may impede the flow of the water in the shallows and raise the level of the pools at night, but during the day the radiation usually detaches it and it is washed downstream. This not only scours the substratum, especially as it may carry stones with it, but it results in a diurnal pulse of discharge. Maciolek and Needham (1951), observed that, at 2,200 m. in Convict Creek in the Sierra Nevada, the daytime flow could be twice that occurring at night because of the formation and release of anchor ice.

In deeper water anchor ice is a rarity, although it has been reported on occasion. Shadin (1956) notes that it can be a nuisance in Russia as it forms readily on good heat-conductors, and masses forming on chains often raise anchors and set boats adrift on cold nights before the freeze-up in early winter.

TURBIDITY AND LIGHT

Although the word limpid is sometimes poetically associated with streams, limpidity is not often a property of running water. Rivers and streams are indeed normally considerably more turbid than still water, and many are always markedly cloudy. Such are most of the rivers of Central Africa; the White Nile derives its name from its persistent turbidity, and the same effect produces the white waters of the Amazon and the Apuro (Gessner, 1961, 1965). Shadin (1956) repeatedly comments upon the muddy waters of the Amu-Daria in Siberia during his review of Russian work on rivers, and the same impression would be made by the Fraser in Canada and the Missouri in its pre-impoundment era. Glacier-fed rivers and streams are consistently cloudy and appear as pale-coloured streaks on distant mountain landscapes; and in heavily cultivated areas, such as East Africa, the American Middle West, and southern Ontario, small streams are slightly turbid even at times of very low discharge.

At times of low water, however, most streams and rivers are normally fairly clear, although never as clear as lakes, and they become turbid during floods when great amounts of suspended matter may be carried. In the middle Mississippi for example, Dorris *et al.* (1963), who made a long series of measurements, found a good relationship between the discharge and the turbidity, and this is a fairly general phenomenon.

Lakes or dams on the course of a river allow great amounts of suspended solids to settle out, and they thus clarify the water (Symons *et al.*, 1964). This has been observed recently on the Missouri (Neel *et al.*, 1963), but it applies primarily at times of high water or in very turbid rivers. At times of low flow the production of plankton in the lake may actually lower the transparency of the water which passes through it, as has been the effect of the Tsimlyansk Reservoir on the Russian River Don (Shelomov and Spichak, 1960).

At most times, therefore, the percentage of light which penetrates to a given depth in river and stream water is lower than that reaching the same depth in still water. Schmitz (1961) gives figures showing that in the River Ybbs near Lunz, Austria, the amount reaching a depth of 1 m. varies from 80 to 100 per cent of the incident light, depending upon the

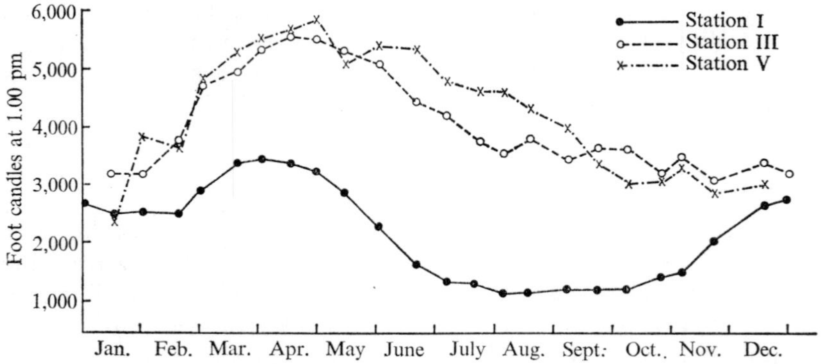

FIG. II, 1. The average level of light at about 1·00 p.m. at the water surface at three stations on Doe Run, Kentucky, at different seasons of the year. Reproduced, with permission from W. L. Minckley, 1962. 'Studies on the ecology of a spring stream: Doe Run, Meade County, Kentucky'. Thesis presented to the University of Louisville.

wavelength, at times of low water, and that it is reduced to almost nothing at times of flood.

As occurs in still water, the amount reflected from the surface increases with increasing angle of incidence, and hence depends upon the latitude and the time of day. At large angles, however, this applies less to a ruffled surface than to a smooth one, and hence less to turbulently flowing than to still water (Dimhirn, 1953). Thus the turbulence which ruffles the water allows more light to penetrate early and late in the day, and to a limited extent this offsets the effects of turbidity.

A further complication is introduced by riparian vegetation. As has been previously stressed, the natural vegetation of river and stream banks is, in most climates, trees. The shallow water near the banks is, or was before man's intervention, normally shaded for some of the time, and small streams, in their natural condition, are completely roofed over by branches. The penetration of much light down to the bottom in running water has therefore been, until fairly recently, a rather rare phenomenon, at least in summertime. We shall see that this has apparently been important in the evolution of the flora of running water.

Minckley (1963), who made a series of light measurements at the surface of Doe Run, Kentucky, a stream whose banks are still in a fairly natural state, found that the illumination reached a maximum in April as the sun rose higher in the sky, but it then fell steadily, as the leaves opened on the trees, to levels comparable with those of midwinter (Fig. II, 1). Indeed, in the narrow wooded gorge at the head of the stream (station I in the figure) the intensity of light at the stream surface was at its minimum during the summer, and it rose during the late autumn after leaf-fall. This must be a normal state of affairs in small unaltered streams where the canopy of branches meets over the water. We can conclude therefore that in temperate climates, in unaltered streams and in most places in the shallows of large rivers, the periods of maximum illumination are spring and fall, and that in the tropics the stream bed was always a fairly dark place before man cleared the bush. It should not be forgotten that these are the conditions under which the biota of running water evolved.

CHAPTER III

Chemical characteristics of flowing water

As with the physical characteristics discussed in the last chapter reference should be made to such fundamental texts on limnology as those of Hutchinson (1957) and Ruttner (1963) for a general discussion of the chemistry of fresh water. Here consideration will be confined to those aspects which are of particular interest in biological studies of rivers and streams.

The chemical content of running water varies, of course, enormously from region to region, and is a reflection of the local geography and climate. It is not, however, possible to work out the ratios of the various materials in solution from a knowledge of the mineral structure of the drainage basin; this is because of the differential solubilities of minerals and the fact that the biota of the water exert selective effects on many dissolved substances.

It should also be noted that rainwater is not pure water. It contains appreciable amounts of sodium, potassium, calcium, magnesium, chloride, and sulphate, as well as the dissolved gases, carbon dioxide, oxygen, and nitrogen, from the atmosphere. Gorham (1955, 1961) has shown that the ions originate from sea spray carried high into the atmosphere, from air pollution, which contributes sulphate in particular, and from dust, and that their proportions vary with the origin of the air masses contributing the rain. Sodium, chloride, and much of the magnesium are apparently derived from sea spray and are carried far inland on to the continents; calcium, some magnesium, and potassium are derived primarily from dust. Dust also contributes organic matter and probably most of the nitrate and phosphate. It was, for instance, observed during an 18-month study at Hamilton, Ontario, that rain contributed 6·5 kg./ha. per year of saline nitrogen (as N). Moreover, 61 per cent of this fell on only 25 per cent of the days on which it rained, which would seem to indicate that much of it came from dust in the atmosphere (Matheson, 1951).

Rain, because of its high content of carbon dioxide, and because sulphate is often present as free acid, normally has a fairly low pH. Usually this is neutralized as the water percolates through the soil and picks up calcium and magnesium ions, but in boggy areas, and in some types of tropical rain forest, the water enters streams without much contact with mineral soils. Streams in such areas have very low mineral contents, not many times that of rainwater, and they tend to be even more acid than rainwater because of base exchange with the organic soil and the loss of such cations as were originally present. They are also often brown because of dissolved organic matter, and examples of this are the coffee-coloured streams of the Scottish Highlands, many parts of Ireland and the Canadian northlands, and the Rio Negro of Brazil (Gessner, 1960).

More usually, however, much water enters streams as subsurface run-off in mineral soil, and only when there is heavy precipitation or a considerable amount of snow-melt does a large proportion enter as direct surface run-off. As a result there is a fairly clear inverse relationship between the discharge and the concentration of dissolved salts in the water. Also, up to a point, the *total* amount of dissolved material carried off the landscape, as opposed to the *concentration*, increases with the annual precipitation. In the United States the total amount increases up to an annual run-off equivalent of about 25 cm. of rainfall. When run-off attains or exceeds this level about 55 tons of dissolved matter are carried off per square kilometre per year (Leopold *et al.*, 1964), though this, of course, varies with the solubility of the rocks involved.

In any given stream the dissolved load carried generally varies as some power of the discharge:

$$T = KD^f$$

where T is load of total dissolved solids, D is the discharge, K is a constant, and f is less than one (Leopold *et al.*, 1964). Which is to say, as we have already noted, that the concentration declines with increasing discharge; but this is not true everywhere. Höll (1955) studied a number of small spring-fed woodland streams in north Germany, and he found that their chemical conditions were remarkably constant over long periods of time, despite variations in discharge caused by rainfall. This is probably a reflection of the fact that the streams in question are all spring-fed and that the samples were collected not far from the sources. Only a small proportion of the water even at times of high discharge would therefore be direct run-off. It is, however, of interest that so small was the variation that there was as much during some single days as over the whole year. Possibly small spring-fed streams which have not been much altered by man normally do present exceptionally stable chemical conditions, and do not follow the general rule.

One further point, which is related to the proportion of discharge which is direct run-off, is that in areas of damp climate, where the landscape is well covered with vegetation, the proportion of the total erosional load which is carried in the dissolved state, as opposed to particles in suspension, is higher than it is in drier areas (Leopold *et al.*, 1964). This is, of course, merely a consequence of the well-known fact that vegetation decreases run-off and hence the particulate load. It thus allows more time for the water to percolate into, and through, the soil and so to acquire more dissolved matter. It has, however, some importance in relation to modern man's tendency to clear the land for agriculture and to leave it bare for long periods. We are not only making streams more turbid, we may be softening their water to some extent.

DISTRIBUTION OF DISSOLVED MATERIAL

In most rivers and streams the turbulent mixing ensures a uniform distribution of dissolved substances, although, as we have seen, small, temporary, temperature discontinuities may occur in pools and deep places. In some large rivers, however, lateral differences caused by the entry of tributaries often persist over long distances as the inflowing water tends to follow the bank on which it entered. This has been observed in several rivers in the U.S.S.R. (Shadin, 1956), and Dolgoff (1929) found that along at least half a stretch of 2,120 km. the two sides of the Volga showed chemical differences attributable to this cause. He concluded that the persistence of this kind of difference depends upon many factors, including the sinuosity of the river, the roughness of the bed, the relative rates of flow of the rivers and the angle of entry of the tributary into the main stream. These are, of course, all factors which affect turbulent mixing, and the distance over which differences can be detected depends to a great extent upon how different the dissolved contents of the two rivers are before their confluence. The water of the Amazon, which is rich in electrolytes and has a high electrical conductivity except at times of high water caused by heavy rain, is distinguishable from the water of the Rio Negro, which is always poor in electrolytes, for about 100 km. below the confluence of these two great rivers, and the same applies to the Apuro and Caroni Rivers for 40 km. below their junction to form the Orinoco (Gessner, 1960, 1965).

It is interesting to note that the same lack of immediate mixing can be detected by the study of micro-organisms. In her detailed investigation of the variously polluted River Oka, a tributary of the Volga, Rodina

(1964) found that the number of micro-organisms suspended in water (which she calls bacterioplankton, although there were yeasts and some actinomycetes in addition to bacteria) was usually highest along the banks of the river, where, of course, the pollution entered. She was also able to detect the lack of mixing of relatively heavily polluted waters, such as those of the Moskva River which drains the Soviet capital, for some distance below their confluence with the Oka because of increased amounts of bacterioplankton along the side of the river on which they had entered.

We mentioned earlier that in large deep rivers there are sometimes small vertical variations in temperature. These demonstrate incomplete vertical mixing, and they also can be correlated with small differences in chemical content. Rodina found that at most places in the Oka the bacterioplankton count increased with depth, and that this was negatively correlated with dissolved-oxygen content. Possibly the lowered oxygen was the result of bacterial respiration; or the slightly denser polluted water—denser because of its organic-matter content and the presence of the micro-organisms—tends to flow downwards from the banks where it enters, and to occupy the lower layers. At two stations, however, the situation was reversed, and this may have been caused by slight differences in density resulting from differences in temperature or dissolved substances. We quote these examples to stress that it should not be assumed that conditions are uniform from surface to bottom in large deep rivers, even though the differences are never as important as they are in lakes and are undoubtedly far less stable.

Another factor which may cause local chemical differences is the activity of plants. It is well established that beds of aquatic plants in still water impede water movement and vertical circulation (e.g. Buscemi, 1958). This can lead to supersaturation with oxygen in the daytime, and to de-oxygenation, caused by respiration, at night. This phenomenon has also been observed in rivers, and even benthic algae can cause vertical stratification of dissolved gases in sluggishly moving water. Berg (1943) found differences of up to 1·33 cc./l. of oxygen between surface and bottom caused by plants in the Danish River Susaa during the summer.

DISSOLVED GASES

Apart from minor or local differences such as those noted above, which are probably usually of little biological importance, the gases dissolved in running water are usually more or less in equilibrium with the atmosphere. The three most important are oxygen, nitrogen, and carbon dioxide. Of these, nitrogen is not much affected by the biota nor does it enter into complicated relationships with other dissolved substances. As it is normally present at about 100 per cent saturation, and as there is little reason to suppose that it ordinarily departs far from this value in running water, it will not be considered further here. Oxygen and carbon dioxide are usually inversely related to one another because of the photosynthetic and respiratory activities of the biota, so it is not really justifiable to consider them separately. However, in the interests of clarity this will be done, although it should not be forgotten that, under normal circumstances, when the oxygen content of running water is high the carbon dioxide content is low, and vice versa.

In small turbulent streams the oxygen content is normally near, or above, saturation. Höll (1955), during his lengthy study of the chemistry of small forest streams recorded values that ranged from 100 to 129 per cent saturation. We have seen above, however, that in sluggish water, such as parts of the River Susaa, greater variation may occur and lower values may be recorded.

In fact, even in torrential streams the oxygen content varies seasonally and from source to mouth. Hubault (1927) studied a number of small streams in south-eastern France, and was able to show that in the autumn there was more oxygen at high altitudes than at low ones, and that this situation was reversed in the winter. This presumably reflects an oxygen demand created by fallen leaves in the autumn and greater photosynthesis at lower altitudes during the cold months. In springtime the photosynthesis increased, and caused supersaturation; this occurred first at lower altitudes and then steadily further and further upstream, and then, during the summer, the percentage saturation declined. This would appear to be a fairly general pattern, and seasonal swings of this kind have been reported elsewhere, e.g. in Boone Creek, Kentucky (Neel, 1951). It should be noted, however, that the variations are not large in a normal stream, and that they seem to be almost entirely produced by the biota.

In larger rivers similar, but rather greater, variations occur, and very often the over-riding factor is discharge. In the Mississippi and the Amazon for example, high water is accompanied by lowered oxygen-concentrations, and these are brought about by the wash-in of organic matter and the decrease of photosynthesis caused by turbidity (Dorris *et al.*, 1963; Gessner, 1961).

Winter ice-cover on rivers may also reduce the oxygen content, presumably by reducing photosynthesis and preventing the entry of oxygen from the air. For instance, levels as low as 2·35 mg./l. (at 0 °C.) have been recorded in the River Oka in March, although they rise somewhat before the ice breaks up, presumably because of increasing photosynthesis caused by the increasing light in the spring (Shadin, 1956). It should, though, be noted that the Oka is, as already stated, somewhat polluted, so the oxygen demand of its water is unnaturally high. The wintertime lowering of the oxygen content is therefore probably to a large extent a man-made phenomenon.

Under some conditions, however, very low oxygen-concentrations may occur seasonally for quite natural reasons. We have already mentioned the special instance of the River Ob in Siberia where swamp water may reduce the oxygen concentration under the ice to 5 per cent saturation or less (p. 32). This is analogous to man-made pollution, and similar effects can be produced by autumn leaf-fall. In the North American South-East and Middle West regions, and doubtless in other areas with similar hot and dry summers, small streams become reduced to a series of disconnected pools with little more than seepage between them. During the autumn great quantities of leaves are shed by riparian trees into the pools. The water acquires an inky black colour because of organic matter leached from the leaves, the oxygen falls to very low levels, and these conditions persist until rain falls (Schneller, 1955; Slack, 1964). Any organism which survives in such streams must be either very resistant to lowered oxygen, or have a life history which ensures that no specimens are active in the water during the late summer and autumn.

In many streams there is also a diurnal variation in oxygen content. The solubility of the gas varies inversely with the temperature, and this would lead to low values (in terms of concentration but not, of course, in percentage saturation) in the daytime and higher values at night; but the daily variation is more usually in the other direction—high in the daytime, usually highest in the afternoon, and lowest at night, shortly before dawn. It is, therefore, clearly related to photosynthesis and to the respiration of organisms. In most places where extensive diurnal variation has been observed there were dense growths of aquatic plants, e.g. in several rivers in England where the phenomenon was first described (Butcher *et al.*, 1930, 1937); and it is often particularly associated with pollution which encourages algal growth in the presence of a fairly high oxygen-demand (Schmassmann, 1951). But even in the absence of pollution, as in the English River Yare, variations of up to 10 mg./l. during 24 hours have been observed (Owens and Edwards, 1964). So, where large growths of plants are present, some downward trend from afternoon to dawn is probably normal, and in densely overgrown rivers the fluctuation may be important. This applies especially during the summer when both the light levels and the temperatures are high, but some variation has been observed even during the winter in the warm and sunny climate of Louisiana (Bamforth, 1962). Only where vegetation is sparse or absent, as in the stream studied by Laurie (1942) in Wales, is there a downward trend in the daytime caused entirely by the diurnal changes of temperature.

Very often underground water is almost or totally devoid of oxygen and contains large amounts of carbon dioxide. This is because of its exposure to organic matter and bacterial respiration in the soil. Spring water is therefore often totally de-oxygenated, and downstream of the spring very active exchange of gases with the atmosphere occurs (Entz, 1961). Odum and Caldwell (1955) found no oxygen at all in the water emerging at Beecher springs, Florida, and similar observations have been made elsewhere. On the other hand, in limestone country, where underground water usually flows through quite large solution-channels, the oxygen content of the emerging water may be quite high. Minckley (1963) found animals and a reasonable amount of oxygen right into the mouth of the large spring which gives rise to Doe Run, Kentucky, even though the water was also rich in carbon dioxide.

Carbon dioxide in high concentrations is fairly rapidly lost both to the atmosphere and, as we shall see, by interactions with calcium carbonate. It continues, however, to be produced by respiration within stream and river channels, and often more is formed at some points than at others. For instance, Neel (1951) and Minckley (1963) showed that, in small streams in Kentucky, pools and small man-made impoundments reduce the oxygen and increase the carbon dioxide content of the water as it passes through them. Even large rivers may display such features; the great swampy area of the Sudd reduces the oxygen and raises the carbon content of the waters of the White Nile, and the relatively high content of carbonic acid also markedly lowers the pH (Talling, 1958).

Clearly the rate at which such anomalies are restored to a state of equilibrium between the water and the atmosphere depends upon local factors affecting turbulence and the ruffling of the water surface. It has been shown that falls, weirs, and similar structures, which cause entrainment of bubbles, restore lowered oxygen content, and presumably also eliminate supersaturation, more rapidly than does turbulent flow through channels with the same loss of head (Gameson, 1957). We can assume, therefore, that the rougher the stream bed the more rapid is the restoration of equilibrium and the more stable is the environment in respect of dissolved gases.

MAJOR DISSOLVED SALTS

It was stated above that the amount of total dissolved solids in rivers and streams varies with local geography, and several studies of limited areas have illustrated this. For example, Schmitz (1961) gives a map of the Fulda-Eder river-system in Germany which clearly relates the bicarbonate content of the water to the varied geological structure of the drainage basin, and similar data are available from Tennessee (Shoup, 1947, 1948).

Most running waters are 'bicarbonate' waters in the limnological sense, and they therefore show the complicated relationships between pH, CO_2, H_2CO_3, H^+, $CO_3^=$, HCO_3^-, Ca^{++}, and Mg^{++} which are discussed in detail by Ruttner (1963) and Hutchinson (1957). A very useful graphical and algebraical exposition of these relationships is given by Deffeyes (1965). For our purposes it is sufficient to stress only a few points:

1. Rainwater reaching watercourses as run-off from bogs, dense forest litter, and similar substrata tends to have a low pH, because of hydrogen ions produced by dissociation of carbonic acid and the loss of cations by base exchange with the organic matter.

2. Water which has percolated through the soil is also rich in carbon dioxide and similarly tends to be rich in hydrogen ions.

$$H_2O + CO_2 \longrightarrow H_2CO_3 \rightleftharpoons H^+ + HCO_3^-.$$

3. Calcium carbonate, which is a common constituent of many rocks, is almost insoluble in water, but it dissolves fairly readily, as bicarbonate, in carbonic acid, and it neutralizes the soil water where it occurs.

$$CaCO_3 + H_2CO_3 \longrightarrow Ca(HCO_3)_2 \rightleftharpoons Ca^{++} + 2HCO_3^-.$$

4. Calcium bicarbonate in solution is a good buffer-system and thus resists changes in pH, but it remains in solution only in the presence of a certain amount of free carbon dioxide, the so-called *equilibrium* CO_2. Therefore any process which removes carbon dioxide, as does photosynthesis, tends to cause precipitation of calcium carbonate from solution, especially where the bicarbonate is abundant.

It can be understood, therefore, that spring water in limestone regions is often very rich in calcium bicarbonate where it emerges to the surface. As it flows along it loses carbon dioxide to the atmosphere and by photosynthesis, and after some distance this loss becomes loss of equilibrium CO_2, and deposition of calcium carbonate occurs by reversal of the first part of the equation under number 3 above. This process is also aided by the ability of many plants to make direct use of bicarbonate ions.

$$Ca(HCO_3)_2 \longrightarrow CaCO_3 + H_2O + CO_2 \text{ (photosynthesized).}$$

Deposition of calcium carbonate is therefore a common feature of streams in limestone areas, and is a subject to which we shall return (p. 75).

A somewhat analagous situation may occur with iron salts, particularly in boggy and marshy areas where large amounts of organic matter cause reducing conditions (low redox potentials) in subsoil water. These, especially where the pH is low as it usually is in wet non-calcareous soils, reduce ferric iron, and it goes into solution as ferrous bicarbonate; such solutions are often very acid because of excess carbon dioxide and the presence of organic acids. On reaching the open stream, carbon dioxide is lost, the pH rises, the ferrous iron is oxidized, and ferric hydroxide is deposited as a flocculent brown film, often in association with iron bacteria, mainly *Leptothrix ochracea*, which frequently form rust-coloured layers up to a centimetre thick on all fixed objects in the water.

$$4\,Fe(HCO_3)_2 + 2\,H_2O + O_2 \longrightarrow 4\,Fe(OH)_3 + 8\,CO_2.$$

This reaction occurs very slowly in waters which remain acid; so brown bog-waters and the similar dark-coloured waters from tropical rain-forests, of which part of the acidity is caused by organic acids, may retain some ferrous iron in solution. Shadin (1956) records up to 1·5 mg./l. in the River Oka under winter ice, but in most rivers any iron which remains in the water is in particulate ferric form.

Apart, however, from these complications in headwaters, caused by the peculiar oxygen/carbon dioxide relationships of many spring waters, the dissolved content of the major salts is not changed rapidly by internal adjustment. Total concentration varies, as we have seen, with discharge and is generally higher at times of low flow. It also generally increases downstream, as larger and more varied areas of landscape drain into the watercourse. This has been shown to occur in the Great Berg River in South Africa (Harrison and Elsworth, 1958), the Nile (Talling, 1958), and the Ohio (Woods, 1965*a*), and it doubtless occurs in most rivers (Livingstone, 1963). Sometimes such increases occur quite suddenly because of the confluence of a hard-water tributary. For instance, the Rio Manaus raises the electrical conductivity of the water of the Rio Negro by 20 per cent (Gessner, 1960); but more usually there is a moderately steady addition of salts from the drainage basin, and the concentration of ions, particularly of calcium and bicarbonate, rises steadily. Table III, 1 shows a series of analyses of water collected from the River Dee in Wales on one day, and it demonstrates the steady

Locality	km. below source	Conductivity micro-mhos	Ion contents mille-equivalents/l. HCO_3^-	Cl^-	$SO_4^=$	Ca^{++}	Mg^{++}
Llanderfel	8	50	0·23	0·16	0·15	0·26	0·18
Cynwyd	18	54	0·18	0·18	0·16	0·27	0·20
Corwen	22	64	0·36	0·19	0·21	0·44	0·23
Glyndyfrdwy	32	72	0·35	0·18	0·24	0·49	0·21
Llangollen	45	86	0·40	0·20	0·24	0·56	0·22
Pen-y-Bont	59	408	0·92	0·71	2·84	1·26	0·39
Overton	70	488	1·60	0·71	3·36	2·04	0·60
Bangor	80	472	1·57	0·41	3·62	2·07	0·75
Farndon	99	568	2·45	0·44	3·16	3·33	1·29
Figdale	112	625	2·20	1·53	2·52	3·73	1·31

TABLE III, 1. The chemical content of the water in the River Dee, Wales, as determined from a series of samples collected on 4 October 1955 from various points below its source in Llyn Tegid (Bala Lake).

increase in dissolved ions from the source of the river, in Llyn Tegid, to Llangollen. There the river breaks through the mountain mass of the Carboniferous limestone, and, although it receives no major tributaries at that point, it acquires a relatively heavy load of dissolved salts over quite a short distance. Part of this is caused by industrial activities in the valley below Llangollen, which probably account for some of the sulphate and probably an equivalent amount of sodium. But the increases in calcium, magnesium, and bicarbonate are undoubtedly caused by the limestone, and it can be seen that the concentration of these increased steadily as the river flowed out of the limestone area between Pen-y-Bont and Overton on to the flat alluvial plain beyond. This, of course, is full of limestone debris brought down by the river and it continues to supply calcium to the water.

It can also be seen from the table that the river contained fair amounts of chloride and sulphate ions, and, by difference, other cations, almost certainly sodium and potassium. This is the normal state of affairs, as these ions are derived readily from rainfall and the decomposition of rocks. In some parts of the world, particularly in arid areas, running waters may contain much larger amounts of chloride or sulphate. Rivers which are very rich in sulphate occur round the Sea of Azov and in Central Asia (Shadin, 1956), but they are unusual, and such waters never reach the condition of not being fresh water in the ordinary sense of the word. Even where, in the rift valley of Ethiopia, the Waracallo River drains the slightly salt Lake Langanno into the very saline Lake Hora Abgiata the water is fresh enough for cattle to drink. Thus even under these very exceptional conditions running water remains fresh.

Quite ordinary lakes, however, may exert a considerable influence on the dissolved salts of rivers on whose courses they lie. Not only may they concentrate the solution by evaporation in hot dry climates as does Lake Langanno, but biological activity in the lake may have profound effects on the calcium/carbon dioxide complex. The water flowing out of a lake is, under natural circumstances, surface water and so has come under the full influence of the phytoplankton. In hard waters during the summer, photosynthesis by the phytoplankton tends to reduce the calcium content by precipitation of carbonate, and to raise the pH by removal of carbon dioxide, and, under certain conditions, by production of hydroxyl ions. For full details the reader is referred to Ruttner (1963), and, for further discussion of the effects of lakes on the chemistry of the water in rivers below them, to the review by Symons *et al.* (1964).

PLANT NUTRIENT SALTS

The three important ions needed for plant growth are potassium, nitrate, and phosphate, and in addition, trace amounts of a long list of other elements are necessary. The latter have been discussed by Eyster (1964), and they need not concern us further here, since they are presumably only exceptionally in short supply. To my knowledge nobody has ever demonstrated or suggested that any of them is a limiting factor of plant growth in a natural stream. This is, however, in considerable contrast to agricultural findings, so it would be interesting to investigate streams in areas which are known to be short of particular nutrients to see if addition of the missing elements had any effect on the biota.

Similarly, we will not further consider potassium, as it is a common constituent of many minerals and is always present in considerable amounts in all but the most unmineralized waters. At no time has it been supposed that it is a factor limiting plant growth in natural waters. This leaves nitrogen and phosphorus as the two plant-nutrient elements which are most likely to be important.

Nitrogen can exist in saline form as ammonium, nitrite, or nitrate, and of these the last is the most available form for plant growth. It is also the fully oxidized form and so, except under conditions of pollution, it is the normal one which occurs in running water. Phosphorus is more complex since it occurs both as simple ionic orthophosphate and as bound phosphate in soluble and particulate form. In lakes only about 10 per cent or less is present as simple ions, and these are very rapidly taken up by plants (Hutchinson, 1957). It appears, however, that the bound phosphate is continuously released by bacterial action, so in turbulent water the fertility is measurable in terms of total phosphate.

The main sources of nitrate and phosphate in streams are rainfall and the land surface. We have already seen that rain contributes fairly large amounts of nitrate. Some figures given by Voigt (1960) for two rainstorms in Connecticut are, in mg./l. of rainwater collected:

Storm in	May	September
N	0·05	0·07
P	0·01	0·01
K	0·3	0·6

It is known also that drainage from agricultural land produces large amounts of nitrate and phosphate. In Wisconsin it was found that runoff contributed about 7·7 kg. of nitrate nitrogen and 0·38 kg. of phosphorus per hectare per year (Sawyer, 1947), and Owen and Johnson (1966) give similar figures for phosphate draining into streams in southern Ontario. Moreover, the steeper the slope the greater is the contribution; Mackenthun *et al.* (1964) give figures for the annual contribution from arable land in Wisconsin from which the following values have been calculated.

Slope	Nitrogen kg./ha.	Phosphorus kg./ha.
20°	34	1·6
8°	16	0·45

Almost certainly the amounts contributed by non-agricultural landscapes are smaller than this, but even if they are only half as much or less the quantities involved are considerable. It will be recalled that, on average, 1 sq. km. of landscape supports 1·4 km. or less of channel (Chapter I), so 1 ha. supports not more than 140 m. of channel. Thus the amounts of nitrogen and phosphorus entering a headwater stream 1 m. wide are probably considerably in excess of the quantities normally applied as fertilizer even to high-grade agricultural land.

It seems also to be true that much of the nutrient enters streams in particulate form and is then released fairly slowly. Indeed, it has been suggested that nearly all the phosphate reaching streams in agricultural areas does so through bank erosion (Owen and Johnson, 1966). Thus while it may be washed in in large amounts at certain times it is not all swept rapidly downstream and lost. Shadin (1956) reports that ooze collected from floodwaters in the Volga, when soaked for ten days in water from the river, raised the dissolved content of the water as follows:

Ca	from 19·6 to 51·2 mg./l.
Mg	from 6·1 to 51·7 mg./l.
N	from 0·075 to 0·260 mg./l.
P	from 0·0008 to 0·0125 mg./l.

Normally the concentrations of nitrate and phosphate actually in solution in stream or river water are low because the ions are rapidly taken up by plants. Mädler (1961) made an extensive study of the phosphate in streams, and he found that it was soon eliminated from the water. Neel (1951) noted that the concentrations of both nutrients fell as the water passed over riffles, and Minckley (1963) records that in Doe Run, of which the water is exceptionally rich in nutrients from the limestone feeding the spring source, the concentrations fell during periods of low flow.

All these are indications that the nutrients soon enter plants and are then recycled fairly fast in the stream, and similar indications are given by the observation in the Volga that the nitrate content of the water may fall to zero in the summer and rise to well over 1 mg./l. in the winter when plants are not active (Shadin, 1956). In the same way exceptional areas of plant growth, such as the Sudd on the White Nile, act as traps for nutrients (Talling, 1958). We can thus state that, although considerable amounts of nutrients do enter streams, relatively small quantities may actually occur in solution in the water, especially during seasons of active plant, or perhaps bacterial, growth. On the other hand, Buscemi (1966) found that the nitrate and phosphate concentrations increased in the downstream direction in the Palouse River, Idaho. Much, therefore, depends upon local conditions, and clearly the position regarding the amounts of nutrients available and the possibility that they may actually control the growth of primary producers in streams merits further study. It is interesting in this connection to note that Odum (1956) concludes that polluted streams, which are, of course, greatly enriched by nutrients, are possibly the areas of highest primary production on the planet. It is also pertinent to mention here a point to which we shall several times return, namely that running water, simply because it is running, is a richer habitat than still water. This was pointed out long ago by Ruttner (1926; Brehm and Ruttner, 1926), who stressed that current, by preventing the accumulation of a shell of depleted water round an organism, constantly presents fresh material to its surface to replace that used up by metabolism. This concept, of course, applies to plant nutrients as it does to oxygen or any other dissolved substance involved in biological activity.

Finally we have to consider the effects of lakes on nutrient salts. As is well known, all lakes act as fertility traps and become steadily more eutrophic. Indeed, the recent concern about the pollution of lakes throughout the world has made this common knowledge to anyone who reads the newspapers. We can assume, therefore, that lakes on rivers reduce the amount of dissolved nutrients which would otherwise arrive at the reaches below them. It should not, however, be forgotten that much organic matter leaves lakes in the form of plankton and that this also contributes nutrients somewhere downstream. I think that at this stage of our knowledge it is possible only to say that here again is a field open for study. There are, in fact, indications that some lakes may actually increase the nitrate content of the waters downstream. This occurs, for instance, in the Tsimlyansk Reservoir on the River Don, and it has been considered as being possibly caused by nitrogen fixation in the reservoir (Shelomov and Spichak, 1960). Some blue-green algae, particularly *Anabaena* which is a common plankton organism of eutrophic lakes, have long been known to be able to fix molecular nitrogen.

ORGANIC MATTER

All natural surface waters contain dissolved and particulate organic matter, and the amounts are surprisingly high. It is estimated that sea-water contains 0·5–1·2 mg./l. of organic carbon in solution, which is about one hundred times the amount present as plankton and ten times as much as occurs as particulate matter derived from the breakdown of organisms (Wangersky, 1965). In inland waters the quantities are even higher. Einsele (1960) records that the water in Austrian rivers and streams contains about 1–10 mg./l. of organic matter, Maciolek (1966) found an average of 0·67 mg./l. of particulate organic matter in the clear mountain-water of Convict Creek, California, Höll (1955) found 12–15 mg./l. of oxygen demand from potassium permanganate in woodland streams in Germany, and Shadin (1956) records oxygen demands of 5·50–20·15 mg./l. in lowland rivers in the U.S.S.R. To equate figures given as organic matter or as oxygen demand with those given as organic carbon the former should be divided by about, or rather less than, 2; it is not possible to be more precise than this without knowing what classes of organic compound are involved. Undoubtedly some running waters, particularly those draining swamps like the Ob, brown moorland waters, and the soft black waters of some tropical rivers, contain still larger amounts of organic matter.

We know remarkably little about the fate or significance of all this material, although the total quantities involved are clearly very great. Einsele calculates that 32 million kg. a year pass down the River Enns, and that 10 per cent of this is deposited above two dams below Steyr, where about one-tenth is converted into animals in the lakes. The rest presumably flows on into the Danube. Similarly he calculates that about two-thirds of the organic matter in the Traunsee is brought into the lake by the River Traun, and Shadin suggests that many rivers collect more organic matter from outside sources than they produce themselves.

Sometimes, as in the Amazon and the Rio Negro (Gessner, 1961), it seems fairly clear that the oxygen demand of the dissolved and particulate organic matter accounts in large measure for oxygen concentrations of less than 100 per cent saturation. It also seems clear that much organic matter is precipitated by dissolved salts. This effect is particularly obvious in countries such as Scotland and Ireland, where brown peaty tributaries often join hard-water streams and the brown coloration rapidly disappears. Similarly we know that, in sea water at any rate, aeration causes the aggregation of organic matter which is then colonized by bacteria and may form food for organisms (Wangersky, 1965). We can conclude, therefore, that along the course of a river there is probably a steady rain of minute particles of organic matter which can form food for animals. As we shall see in Chapter XXII, this suggestion fits very well with the data of the zoologists.

CONCLUSIONS

We are now in a position to consider, in general terms, the chemical history of a single parcel of water as it flows down various types of stream. This kind of study has been performed in the field several times in Europe by Schmitz (1961), and by others in North America under his influence, and it has produced illuminating results. It is also the basis of the method devised by Odum (1956), which is discussed in Chapter XXII, for measuring the primary production in running water.

Water from a spring will be at the mean annual temperature and probably low in oxygen. In summertime as it flows along it will warm up in the daytime and cool a little at night, and as the days pass this zigzag temperature curve will tend towards an asymptote at the ambient air-temperature. In wintertime the tendency will be downwards but only to 0 °C. even in extreme weather. As it flows the water will gain oxygen, and it will do so particularly during the daytime even though its temperature is rising, as was clearly shown by Tindall and Minckley (1964) in Doe Run, Kentucky. This is, of course, because of daytime photosynthesis.

Water from general run-off or seepage will undergo a similar series of changes, but it will start, probably, with more oxygen and less difference in temperature from that of the air, unless it arises from a glacier or névé in the mountains. In either event it will warm as it descends to lower altitudes, although the reverse may occur in wintertime (Hynes, 1961).

If the spring is from limestone, or some other very calcareous rock, the water will be heavily charged with calcium bicarbonate. It will lose carbon dioxide very rapidly and its pH will rise. After some distance the loss of carbon dioxide to the atmosphere and by photosynthesis will lead to the deposition of calcium carbonate, again aided by photosynthesis (p. 43), and this process will decline steadily as equilibrium is attained. In this well-buffered hard water the pH will not rise above about 8·3 even at times of very active photosynthesis (Entz, 1961), and these changes will occur while the water flows only a short distance—from a few hundred metres to a few kilometres according to the situation.

Water from a non-calcareous spring will similarly lose carbon dioxide, increase its pH, and acquire oxygen in quite a short distance. If the source of the spring is a bog, acid swamp, swampy woodland, or rain forest the water may contain ferrous bicarbonate. As the pH rises and oxygen is acquired this will be deposited as ferric hydroxide, probably over a distance measurable only in tens or hundreds of metres, and this part of the stream will be coated with rust-coloured masses of iron bacteria. Unless the water remains acid little iron will remain in solution. If, as is probable, the water is soft its pH will fluctuate markedly because of photosynthesis, and it may exceed 8·3 in the daytime. Water from such sources is likely to be brownish, because of dissolved organic matter, and to remain so until the stream acquires a richer supply of inorganic ions, usually mainly calcium bicarbonate.

If the water enters a shallow channel with relatively sluggish flow, it will encounter a number of factors which will tend to enhance its daily fluctuation of carbon dioxide, oxygen, and pH. These include photosynthesis of benthic plants which, when they are dense, may cause supersaturation with oxygen in the daytime and correspondingly high levels of pH. Oxygen productions as high as 2·27 g./day/sq. m. of plant bed have been recorded in the English River Ivel (Owens and Edwards, 1961). Correspondingly low levels of pH and oxygen will occur at night not only because of respiration of the rooted plants, but also because of the oxygen demand in the bottom sediments. Edwards (1962; Owens and Edwards, 1963; Edwards and Rolley, 1965) has shown that the oxygen demand of silts and muds in the beds of sluggish streams can be quite considerable, and although many of his figures were derived from organically polluted waters such deposits clearly play an important role in clean streams also.

In all these situations it is probable that, in summertime at any rate, the initial concentrations of nitrate and phosphate will be declining downstream or being maintained at fairly low levels by additions from the land surface. Similarly, the general level of the major ions will be rising by addition from the land and the stream bed. In the water originating from the limestone spring, calcium and bicarbonate ions are not, however, likely to increase, since further addition of these leads to precipitation of calcium carbonate.

Even further downstream, in the large river, the acquisition of the major ions continues, and the diurnal fluctuations become less marked although they may continue to be measurable. Seasonal fluctuations of discharge with their concomitant variations in total dissolved matter, turbidity, oxygen content, etc., tend to replace the more irregular local weather fluctuations in the smaller streams.

We can finish this summary by stressing once again how many of the changes described above are produced by biological processes. A lifeless river would have a very different chemical regime.

CHAPTER IV

Attached algae

The whole subject of aquatic botany has been reviewed and discussed in recent years in the excellent volumes of Gessner's *Hydrobotanik* (1955*b*, 1959, 1963), which are recommended to the reader for a much fuller treatment than is possible here.

It is customary in many works on the ecology of the flora of running water to divide the attached aquatic plants into small species, or 'microphytes', and large species, including flowering plants, which are referred to as 'macrophytes'. There is little justification for this apart from convenience, and the terminology is not without its shortcomings. Many microphytes can produce large and substantial plants, particularly in running water, and some bryophytes, which are usually included with the macrophytes, are quite small.

Here we will consider only the algae, both large and small, including the Cyanophyta but excluding the Charales, which, together with the Bryophyta and Angiospermae, we will reserve for the next chapter.

Algae grow attached to all kinds of solid objects; they also occur as thin films on mud and silt surfaces and, more rarely in running water, as floating or loosely attached thalli in shallow littoral areas out of the current or among dense growths of macrophytes. Larger plants themselves form substrata which are suitable for the growth of smaller species, and all such layers of attached organisms are sometimes referred to as periphyton, or by the German word Aufwuchs, often without its capital A.

The attached community, if we may use that word without any of its ecological overtones, has been discussed, with special reference to running water, by Blum (1956, 1960), Cooke (1956), Round (1964, 1965), and Whitford (1960*b*). It consists primarily of microscopic species, but, particularly in small streams and on rocky substrata, it also contains a number of conspicuous plants. These include several genera of the Rhodophyta (*Lemanea*, *Hildenbrandia*, and *Batrachospermum*), filamentous Chlorophyceae (*Cladophora*, *Ulothrix*, *Oedogonium*, *Stigeoclonium*, and, in quiet places, *Zygnema* and *Spirogyra*), Chrysophyta (*Vaucheria* and *Hydrurus*), and Cyanophyta (*Oscillatoria* and *Phormidium*). At times members of many other genera, such as the diatoms *Diatoma*, *Synedra*, and *Gomphonema*, may be so abundant that they form conspicuous masses of material.

ADAPTATIONS OF STREAM ALGAE

A large number of at least the commoner species of alga appear to be world-wide in distribution, although, as Blum (1956) has stressed, large areas of the earth remain almost uninvestigated, and a great many species are apparently confined to the tropics (Gessner, 1955*b*).

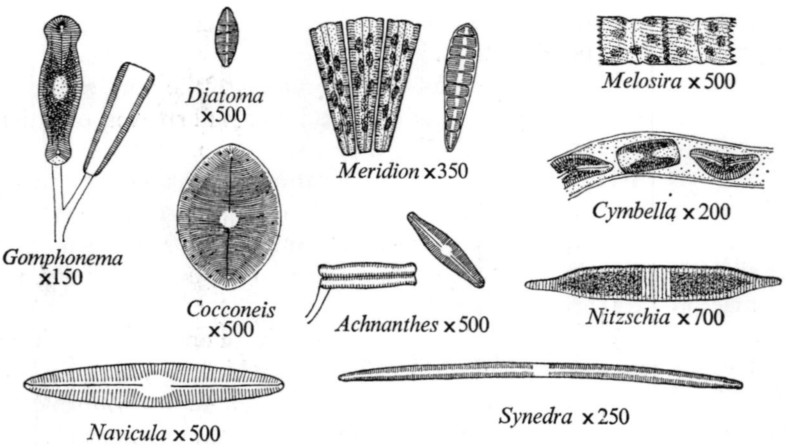

FIG. IV, 1. Diatoms which are common in the attached algal community of streams.

It would seem that even the rare species are very widely distributed and that they are rare primarily because they need rather special conditions for active growth. Several phycologists, e.g. Whitford (1960*b*), have stressed the fact that rare species, or species which are common at only limited times of the year, can often be found as single, often depauperate, cells if the search is sufficiently thorough. Probably, therefore, most species are available at all times; they flourish when conditions become suitable, and many are simple opportunists. We know very little, however, about the means of dispersal of many types of alga, and we cannot at this time explain how species actually get to headwaters, and remain there despite the summation of downstream movement to which they must be subject. This is a problem which is common to all the nonmotile inhabitants of streams, including macrophytes and some groups of animals, e.g. sponges and bryozoans, and it is still not understood.

Algae attached in running water must be firmly anchored if they live on solid surfaces, or able to regain their position if they live on soft substrata. All, or almost all, the diatoms which occur on river silts are motile, and such 'epipelic' (living on mud) genera include *Nitzschia*, *Navicula*, *Caloneis*, *Gyrosigma*, *Surirella*, and *Cymatopleura* (Round, 1964). Also motile are the epipelic filamentous Cyanophyta, *Oscillatoria* and *Phormidium*. 'Epilithic' (living on stones or similar objects) and 'epiphytic' (living on plants) genera are either firmly stuck down by jelly, e.g. *Cocconeis*, a common epiphyte, and *Chamaesiphon*, a small blue-green alga, or stalked, as are many diatoms, e.g. *Cymbella*, *Achnanthes*, and *Gomphonema* (Fig. IV, 1).

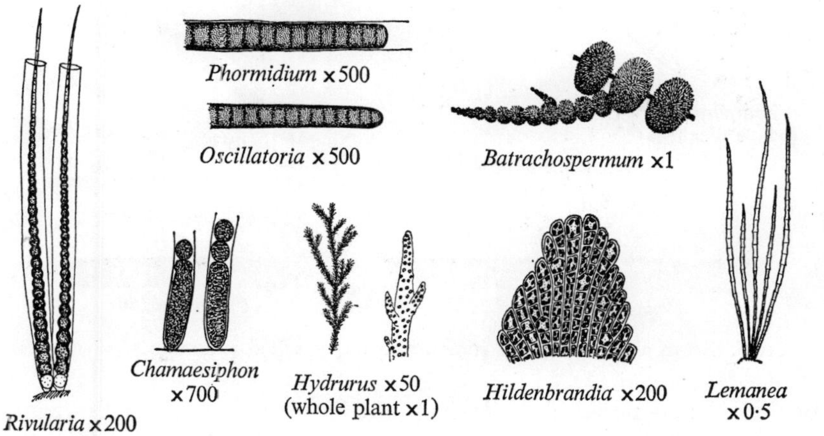

FIG. IV, 2. Red, blue-green, and yellow algae which are common in the attached community of streams.

Often the thallus is flattened and closely applied to the solid surface, where it is, of course, in the boundary layer out of the current. This occurs in the red algae *Hildenbrandia* (Fig. IV, 2) and *Rhodoplax*, and in *Gongrosira* (Fig. IV, 5). Several types of stream alga grow as cushion-like clumps which divert the current over them; such are *Rivularia* (Fig. IV, 2), *Nostoc*, *Schizothrix*, *Vaucheria*, and the tube-dwelling desmid *Oocardium* (Fig. IV, 3). Yet others are attached firmly by rhizoid-like structures at the bases of filaments which trail out as tassels into the current, as in many of the filamentous Chlorophyceae (Fig. IV, 3). In some genera, e.g. *Cladophora*, *Oedogonium*, and *Ulothrix*, the tassels may become very long, and they may exceed a metre in *Cladophora glomerata*. In many genera these basal rhizoid-like structures, or basal plates of cells, persist during unfavourable seasons, when they are very inconspicuous, and new growths spring up from them. This is particularly striking in the genus *Lemanea* (Fig. IV, 2), whose long, conspicuous, greenish-purple tassels reappear at the same point on a rock year after year at the appropriate season.

In addition to the attached algae which occur actually in the current many streams also contain unmodified species which are normal inhabitants of pools and the edges of ponds. They occur in quiet areas

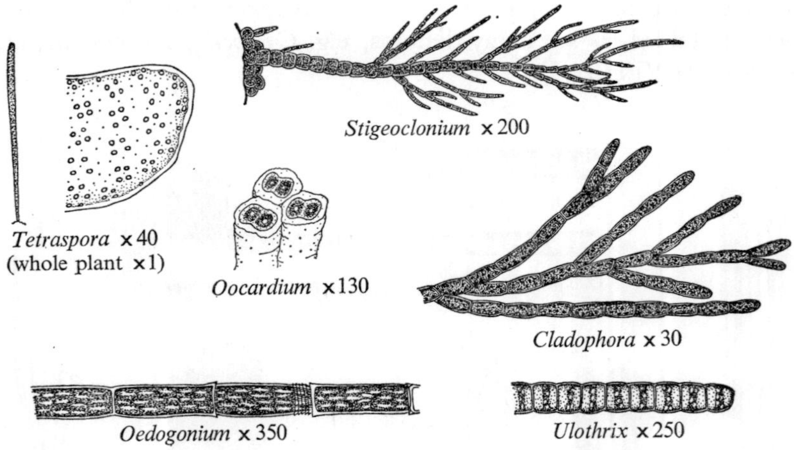

FIG. IV, 3. Green algae which are common in the attached community of streams.

behind obstructions, in vegetation, and in little bays along the banks, and they include filamentous Chlorophyceae such as *Spirogyra*, *Zygnema*, and *Mougeotia*, various Chlorococcales and desmids, and often green flagellates. Such accumulations of pond algae occur primarily at times of low water in warm weather, and they are usually very temporary and sporadic, since slight changes in water level destroy them. They illustrate particularly well how opportunist are many algae, and how quickly they can occupy and exploit favourable situations.

METHODS OF STUDY

Quantitative methods of studying attached stream algae have been reviewed and discussed by several authors (Blum, 1960; Cooke, 1956; Lund and Talling, 1957; Sládečková, 1962); they present rather formidable difficulties.

Round (1964) has devised a method for the epipelic species which involves placing a glass cover-slip on a sample of the substratum in a Petri dish. The algae move to the surface of the deposit after some hours, and attach themselves to the glass where they can be identified and counted. Sampling of solid surfaces or those of plants in a satisfactory way is more difficult. Definite areas of stones can be scraped clean, and Douglas (1958) has devised methods of doing this, by using a brush inside an enclosed area made by pushing a bottomless plastic container firmly against the surface of the stone. A further refinement, which she used on submerged surfaces of immovable rock, was to suck up the dislodged material through the middle of the brush into a sampling jar.

A problem, however, which is presented by all methods which involve scraping or brushing off algae, is that many specimens are broken up and damaged in the process, and that the sample inevitably includes detached particles of the substratum which make counting difficult and unreliable. These methods are, therefore, much better for diatoms, the frustules of which can be chemically cleaned, than they are for the other groups.

An alternative method which should avoid these problems was suggested by Margalef (1949), who advocated coating stones to be sampled with colloidin, which can then be stripped off, carrying the algae with it, and mounted for study. This has often been cited as possibly a good technique, e.g. by Blum (1960), but it does not appear to have been extensively used even by its originator.

Because of the difficulties of working with natural substrata, many workers, observing that almost anything solid left for a while in a stream becomes covered with algae, have taken to using artificial substrata. These have usually been ordinary glass microscope-slides, but many other materials have been tried from time to time, including ground glass, plastic, and photographic film, all of which apparently give rather similar results under a variety of conditions (Sládečková, 1962). Geitler (1927) was one of the earliest to use glass slides in this way, and he noted that the algae which first grew on them were mainly the species which normally grow on mosses. These are the diatoms *Cocconeis* and *Achnanthes*, which are, however, often found also on stones. Some of the species which occur commonly on stream beds do not grow, at least for a very long time, on glass, a point which is stressed by Lund and Talling (1957).

Nevertheless, the glass-slide method has been used very successfully in comparative studies, particularly in connection with pollution and most notably by Butcher (1932 and later papers). More recently it has been widely used in the United States for monitoring water quality, and a special floating apparatus which holds slides vertically in the water has been designed for making comparative studies of floras of diatoms (Patrick *et al.*, 1954). We still, however, know far too little about the development of such artificial populations in relation to the flora elsewhere in the stream. Colonization is rapid at first, and often there are more species present after a short while than there are later on. The numbers appear to be very unstable, especially when they build up to high levels (Butcher, 1947), and many species do not appear at all, or do so only after a very long period (Sládečková, 1962). Lund and Talling (1957) suggest that the primary colonizers are probably species which are not particularly selective about substratum, and they remark, rather drily, that it still remains to be discovered how long such substrata should be left before they begin to acquire floras resembling those of their surroundings. This does not, however, detract from the value of the method in comparative studies, in which it has been very useful. It should also be noted that Gumtow (1955) found no differences between the algal populations on stones on a riffle in the West Gallatin River, Montana, and those on similar stones which had been thoroughly cleaned before replacing them, as long as the latter had been there for a week. However, none of the larger long-lived algae was present in his populations.

Counting and identification of algae is tedious work; the various methods and techniques are discussed by Lund and Talling (1957). Some workers who are interested not so much in the algae themselves as in their capacity as primary producers, have therefore tried to find short cuts. These involve weighing the growths scraped off plastic sheets and extracting and measuring the amount of chlorophyll they contain (Grzenda and Brehmer, 1960), or soaking stones in acetone to extract chlorophyll for measurement (McConnell and Sigler, 1959; Waters, 1961*b*). This last method has shown, as might be expected, that there is more chlorophyll, and so presumably more algae, on unit areas of large stones than of small ones. The critical diameter of stone in the Logan River, Utah, was found to be about 2·5 cm., and this is presumably because of the lesser stability of the smaller stones and their greater tendency to roll.

Blum (1957) is apparently alone in having attempted to apply the transect method of the terrestrial plant ecologists to stream algae. He strung a marked rope across the stream at a riffle, and recorded the presence of all the visible algal species in alternate decimetres of transect. The importance of the various species was then determined by the relative number of decimetres within which the species was present. This method, although it has the disadvantage of being usable only in clear shallow water, undoubtedly gives results fairly closely resembling the actual status of the flora in the stream, and it would seem to be worthy of further use and refinement.

CONTROLLING FACTORS

Our knowledge of the ecological factors which control stream algae is based very largely on field observations, with all the uncertainty which that implies. Nevertheless, it is possible to make general statements some of which may be true, and almost all of which are in need of experimental proof. As stated earlier, many algae seem to be opportunists which appear and flourish when conditions are right for them, and then give way to other species. This enables us at least to guess at their optimum conditions, but as has been pointed out above 'rare' species sometimes appear in great numbers for short periods or in very limited places, so their ecological requirements are probably a complex of conditions about

Epilithic	Epipelic	Epiphytic
SPRINGS		
Meridion circulare	*Meridion circulare*	*Achnanthes minutissima*
Achnanthes	*Fragilaria leptostauron*	*A. lanceolata*
	F. construens	*A. affinis*
	Navicula cryptocephala	*Cocconeis placentula*
	Nitzschia dissipata	*Denticula tenuis*
	Campylodiscus noricus	
STREAMS AND RIVERS		
Hildenbrandia rivularis	*Melosira varians*	*Chamaesiphon*
Lithoderma fluviatilis	*Fragilaria intermedia*	*Oncobyrsa*
Chamaesiphon	*Frustulia*	*Dermocarpa*
Rivularia	*Gyrosigma*	*Rivularia*
Meridion circulare	*Caloneis*	*Aphanochaete*
Diatoma hiemale	*Neidium*	*Chaetophora*
Cocconeis placentula	*Diploneis*	*Oedogonium*
Achnanthes	*Stauroneis*	*Bulbochaete*
Synedra	*Navicula*	*Cocconeis*
Gomphonema	*Amphiprora*	*Achnanthes*
Cladophora	*Amphora*	*Synedra*
Vaucheria	*Cymbella* (motile spp.)	*Cymbella*
Lemanea	*Bacillaria*	*Gomphonema*
	Nitzschia	
	Cymatopleura	
	Surirella	
	Scenedesmus	
	Pediastrum	
	Oscillatoria	
	Spirulina	

TABLE IV, 1. Some common attached algae of various types of habitat in running water. Data from Round (1964).

which we know nothing. Moreover, we do know, from studies made in still water and in cultures, that many algae interact with one another by the production of metabolites which they release into the water. There have been no indications that this occurs in running water, but there is no apparent reason why it should not do so. This subject has been fairly recently reviewed by Hartman (1960), and the possibility of its importance in rivers and streams should not be forgotten.

It is possible to list the algae which normally occur commonly in particular types of habitat, as has been done by Round (1964) from whose data Table IV, 1 is taken, and to show that the species tend to recur in similar habitats. Schmitz (1961) also lists a number of diatoms that have been more or less regularly recorded by various workers from spring streams in several parts of Europe. Such data, coupled with observations on seasonal cycles and altitudinal occurrence and with careful studies of microhabitats and the ways in which species appear and disappear with the changes of season, are the basis of such knowledge as we have of the requirements of stream algae.

Temperature is an obvious ecological factor which should reveal itself in seasonal changes in the flora, but its effects are difficult to disentangle from those of light which is also low in winter. However, there do appear to be species which occur only in the tropics and so presumably need warm water, and there are others which seem to require low temperatures since they occur only in winter at low altitudes but persist for much longer in high streams. Such are the chlorophycean, *Prasiola*, of streams at over 3,600 m. in the Andean tropics where the temperature does not exceed 10 °C. (Gessner, 1955*a*), and the chrysophytan, *Hydrurus foetidus* (Fig. IV, 1), which is common all the year round in high streams in Europe, but is seen only in winter, early spring, and late fall at low altitudes. Probably many of the species which occur in springs are cold-water stenotherms, as are a number of species, e.g. *Phaeosphaera perforata*, *Chlorosaccus fluidus*, and *Meridion circulare*, which Whitford (1960*b*) reports as occurring only at temperatures lower than 15 °C. in North Carolina. At the other end of the scale are several species which are confined to hot springs and do not survive at normal temperatures (p. 399). Others, e.g. the red alga *Hildenbrandia*, occur only in waters that become warm in summer (Findenegg, 1959), and many species, particularly of Chlorophyceae, grow primarily in summertime. Such are species of *Oedogonium* and *Cladophora glomerata* which seems, however, to be limited to temperatures below about 25 °C. (Blum, 1956, 1960). Presumably these algae require higher temperatures for growth even though they persist as reduced structures through the winter. Further study of the floras of high streams in the tropics, which are almost designed as experiments on the effects of temperature, would undoubtedly produce valuable data. For instance, Gessner (1955*a*) found *Hildenbrandia* at 2,000 m. in the headwaters of the Orinoco but not at 3,600 m., and this is a plant which does not occur at low altitudes in the tropics as its upper temperature limit is 20–25 °C. Similarly *Lemanea*, which from observations in temperate latitudes appears to be a cold-water stenotherm, occurs only at high altitudes on Mount Elgon in East Africa (Hynes and Williams, 1962), which indicates fairly clearly that it is limited by temperature rather than by light.

Light. The last example is stressed as it provides a contrast with *Batrachospermum*, another rhodophytan (Fig. IV, 1), which grows in cool seasons, but which unlike *Lemanea* is apparently limited by high light-intensity. It seems to be indifferent to temperature and to water chemistry (Minckley and Tindall, 1963), and Whitford (1960*b*) states that it persists later into the summer in shaded localities and in brown-coloured water than it does elsewhere. During studies on Mount Elgon, when we had been especially commissioned to collect specimens for a taxonomist friend, we could find it only in densely shaded streams, but it was not confined to cold water.

The Rhodophyta of fresh-water are indeed, like their marine relatives, adapted to grow at low light-intensities (Gessner, 1955*b*; Israelson, 1942). *Hildenbrandia* is often found at considerable depths, and in shallow clear water it usually occurs under stones rather than on them (Luther, 1954). We have already seen that *Batrachospermum* appears to be a shade plant, and the same seems to be true of the phaeophytan *Lithoderma* (Fritsch, 1929). On the other hand, while many diatoms appear to be fairly indifferent to light, in that they occur as abundantly in shaded places as in open ones, many Chlorophyceae require a fairly high light intensity. This is indicated not only by their increase in spring and early summer, but by their distribution in stream beds. Warren *et al.* (1960) state that such algae need at least 3·2 per cent of full sunshine to grow in Oregon, and that in midsummer they can obtain this only on open riffles away from the shade of trees in full leaf. It is indeed easy to observe the way in which conspicuous green algae disappear from a stream bed when it enters a deeply shaded reach, and this phenomenon of heavy midsummer shade accounts for the fact that many species are abundant in early summer and fall, and comparatively scarce in high summer. It is odd though that in at least some species, e.g. *Stigeoclonium tenue* and *Ulothrix zonata* (Fig. IV, 3), this summer scarcity occurs even in the absence of trees. Probably it is an effect of temperature, but, as was suggested in Chapter II, the fact that there are species which flourish in the cool water of the early summer and fall and not so well in the summer, probably reflects the conditions under which they evolved before man cleared the streams. A recent study of the flora of hot-spring streams in Yellowstone National Park indicates that the greatest quantities of chlorophyll, protein, and RNA are produced at temperatures ranging from 51 to 56 °C. (Brock and Brock, 1966). This seems to be a fair indication that, in avoiding warmer waters because they are shaded, many stream algae have been driven by evolutionary circumstance to operate at temperatures well below those at which the algal cell is at its most efficient.

Even in the emergent vegetation of large rivers the effect of shade can be observed. In the delta of the Volga the epiphytic growth is greater on the soft low growths of canary grass than it is on the tall reeds and sedges, presumably because of the shade cast by the latter (Gorbunov, 1955).

Light intensity	Net oxygen evolved mg./sq. m./hr.	
lux	8–10 °C.	18–18·5 °C.
25,500	226	361
16,700	200	331
11,400	180	439
7,400	201	182
4,300	121	208
2,150	133	58
1,180	42	59
620	44	—62

TABLE IV, 2. The rate of oxygen production by algal communities in laboratory streams at different temperatures and illuminations. Data simplified from Phinney and McIntyre (1965).

Little experimental work has been done on the effect of light on stream algae, but a notable exception is the study of Phinney and McIntyre (1965) on growths in laboratory channels. They showed that a community kept at just over 18 °C. increased its rate of production of oxygen as light was increased, but that above 11,400 lux, which is considerably less than full sunlight, photosynthesis actually declined. At 8–10 °C. the rate of photosynthesis continued to rise to the highest level of illumination tested; it was, however, much lower than at the warmer temperature. Their data are reproduced in Table IV, 2 which illustrates fairly clearly that this is a community which is efficient only at fairly low temperature and illumination. This, as already suggested, probably reflects its evolutionary history. The one negative figure in the table demonstrates, of course, that, at the lowest light intensity in the higher temperature, respiration exceeded photosynthesis, a further demonstration of lower efficiency in warmer water.

Current is another obvious factor in running water, and there are many algae which occur primarily or exclusively in streams, or in similar places such as stony lake shores where there is continuous water movement. These include *Lemanea*, *Hildenbrandia*, *Audouinella*, several species of *Batrachospermum*, *Hydrurus*, several species of *Cladophora*, *Gongrosira*, *Lithoderma*, and several species of diatom. In the southern United States some algae, e.g. *Oedogonium kurzii*, *Vaucheria ornithocephala*, and *Campsopogon caerulans*, are confined to rapids, and a number of others, e.g. *Stigeoclonium*, *Draparnaldia*, *Chaetophora*, and *Tetraspora*, are confined to rapids in summertime (Whitford, 1960*a*). Indeed, many workers have noted that attached algae are more abundant in faster water, or that certain species occur more abundantly in areas of swift flow (Behning, 1929*b*; Butcher, 1940, 1946; Jones, 1951; Minckley and Tindall, 1963; Kann, 1966). Blum (1960) found that when he mapped the detailed distribution of *Diatoma vulgare* in a small area of the Saline River, Michigan, it occurred only where water was flowing swiftly and was up to about 30 cm. deep; it was absent from sheltered areas downstream of emergent rocks.

There is thus an innate current demand in some species, and different species become dominant at different current speeds. This last point was very nicely demonstrated by Zimmerman (1961*a*, *b*, 1962) who made observations on artificial open-air channels at Zurich, Switzerland. Three channels were supplied with river water flowing at controlled speeds of 5, 20, and 80 cm./sec. A series of experiments was run in which the water was first clean and then variously polluted with sewage, and after each experiment the organisms which had developed were recorded. In the clean water the dominant algae were: at 5 cm./sec. the palmella stage of an unidentified green alga, at 20 cm./sec. *Phormidium subtile*, and at 80 cm./sec. *Hydrurus foetidus*. In water containing 5 per cent of well-purified sewage the important algae were: at 5 cm./sec. *Rhizoclonium*, *Chaetophora*, and *Scenedesmus*, at 20 cm./sec. *Tribonema* and diatoms, and at 80 cm./sec. many Chlorophyceae and diatoms and again *Hydrurus*. The results obtained with heavier loads of pollution do not concern us here, but it is relevant to note that Zimmerman concluded that the effect of the current was often of greater importance than that of the quality of the water.

As was stated earlier (p. 48) Ruttner long ago suggested that current makes water 'physiologically richer' because of its constant renewal of materials in solution near the surfaces of organisms. This is undoubtedly the explanation for the current demand of many species and for the fact that the others need current in warm weather, when dissolved gas contents (CO_2 and O_2) are low. Whitford (1960*a*; Whitford and Schumacher, 1961, 1964) has shown experimentally that many algae grow better, and that others grow only, in running water in the laboratory (Table IV, 3). He found that *Oedogonium kurzii* takes up phosphorus

	Current 15–30 cm./sec.	Aerated still water
Tetraspora lubrica	++	+
Microspora stagnorum	++	+
Stigeoclonium tenue	++	o
Chaetophora incrassata	++	+
C. pisciformis	++	+
Draparnaldia platyzonata	++	+
Oedogonium kurzii	+	o
Mougeotia sp.	++	+
Spirogyra sp.	++	+
Tribonema affine	++	o
Vaucheria ornithocephala	+	o
V. uncinata	++	+
Phaeosphaera perforata	+*	o
Tabellaria fenestrata	++	+
Eunotia pectinalis	++	o
Pinnularia sp.	+	o
Phormidium sp.	++	+
Anabaena variabilis	++	+
Cylindrospermum sp.	+	o
Batrachospermum sirodotia	+	o
Batrachospermum sp.	+	o
Lemanea australis	+*	o

TABLE IV, 3. The effect of current on the growth of algae in the laboratory at 15 °C. + = growth after two weeks, ++ = good growth, o = no growth or death, * = survival for two weeks but no significant growth. Data from Whitford (1960*a*).

(P_{32}) ten times as fast from a current of 18 cm./sec. as it does from still water, and that the rate of uptake continues to rise up to speeds of at least 40 cm./sec. It also respires more rapidly in a current, so its whole metabolism is speeded up. As it is also known that other species of *Oedogonium*, together with *Hydrurus foetidus*, *Cladophora glomerata*, and *Batrachospermum* respire more rapidly the higher the oxygen content of the water, and that they continue this increase of metabolic activity into supersaturated solutions (Gessner and Pannier, 1958), we can conclude that when oxygen *availability* is raised by turbulent flow the metabolism of the plants also is raised and that they will thrive better the faster the flow. It has indeed been experimentally demonstrated, in laboratory channels fed with stream water at different rates of flow, that not only is there some difference in the species which establish themselves as dominants, but that there are differences in the amounts of production. McIntyre (1966*a*, *b*) showed that a channel fed with water at 9 cm./sec., which became dominated by *Stigeoclonium*, *Oedogonium*, and *Tribonema*, produced less material drifting out than a similar channel fed with water from the same source at 38 cm./sec. The dominant plant in the latter was *Synedra*, which grew in dense, brown, felt-like masses as opposed to the green tassels in the slower water. Moreover, the respiratory rate of material from the faster-flowing channel was higher, indicating that its general metabolism was enhanced, and it responded more to movement of the water in the respirometer than did material from the slower channel. This indicates a high degree of adaptation to lotic conditions, and it is significant that the rate of respiration in both communities fell when the oxygen content of the water was lowered. The periphyton of running water would seem to be closely adapted to the particular environment in which it develops.

The substratum is, as we have seen in Table IV, 1, of sufficient importance in the ecology of attached algae to make it possible to refer to epilithic, epiphytic, and epipelic communities; but in running water these distinctions are not very clear. Many workers (e.g. Jones, 1951; Scheele, 1952) have stressed that the differences between epilithic and epiphytic floras are largely statistical and that species are merely more abundant in one than in the other. On the whole, however, the larger species are confined to solid objects rather than to plants, and the slower-growing ones, especially the Rhodophyta, are found only on stable materials such as rocks and large stones.

We know little about the reasons why certain species occur on certain substrata and not on others. Round (1964) stresses, with reference to the epipelic community of running water, that the nature of the sediment itself may be important, and that organic matter and pH may be significant factors. It is also undoubtedly true that different species of macrophyte are very differently colonized by epiphytes, and that even different parts of the same plant offer different qualities of surface for colonization. Some of the latter effect may be merely a reflection of age, young shoots having had less time in which to acquire a flora, but dead leaves are often much more thickly colonized than living ones, and some plants, for example, *Potamogeton pectinatus*, acquire very little periphyton while they are alive.

Similarly, different types of stone in a stream bed often acquire different amounts of epilithic flora, as can sometimes be seen near bridges where alien stones, bricks, and concrete are mingled with the native rock. This is another subject in stream biology which would repay further study.

Sometimes we can guess at the reasons, but we have little real information on which to base our guesses. For instance, *Hildenbrandia rivularis* is not often found in very hard water, and it is suggested that this may be because it is unable to grow on limestone. It is a very slow-growing plant and it may be that it becomes detached from the limestone by solution of the calcium carbonate faster than it can grow to cover the surface (Luther, 1954).

It is also fairly clear that the stability of the substratum is important. We have already seen that larger stones carry more chlorophyll than smaller ones, and careful study by Douglas (1958) of the diatom *Achnanthes* in a small stream in England, showed that when floods cut down the number of specimens by wash-out the effect was much more severe on stones which were rolled by the water than it was on fixed rock. The same effect of scour in reducing numbers of algae has been noted in the Arakawa River in Japan (Koboyasi, 1961*a*).

The alkalinity of the water, or some related parameter such as pH or hardness, has often been considered to exert a considerable influence on algae. Many genera, e.g. *Lemanea*, *Batrachospermum*, and *Stigeoclonium* to cite a few conspicuous forms, occur freely both in soft, acid water and in limestone-spring streams, but others seem to be primarily confined to hard or soft waters. For instance the diatoms *Achnanthes* and *Cocconeis*, some species of *Phormidium*, and *Cladophora glomerata* are found only in alkaline water (Blum, 1960), and the last species is replaced by *Oedogonium kurzii* in the soft-water streams of North Carolina (Whitford, 1960*b*). On the other hand, many, but by no means all, desmids and some other algae, e.g. *O. kurzii* and *Phaeosphaera perforata* (Whitford, 1960*a*), are found only, or primarily, in soft waters.

Butcher (1938), during his studies of growth on glass slides in rivers in southern England, found not only that many fewer specimens grew per unit area in an acid stream (3,200 per sq. mm.) than in an alkaline river (13,200/sq. mm.), but that the dominant forms were different. In the river *Cocconeis*, *Achnanthes*, *Amphora*, *Nitzschia*, *Gomphonema*, *Ulvella*, *Sporotetras*, and *Chamaesiphon* were common, and in the stream most of the flora was composed of *Eunotia*, *Ulvella*, and *Sphaerobotrys*. Moreover, there was a far more marked annual cycle in the stream than in the river, which may have been connected with the chemical differences betwen the waters. Round (1964) also lists many of the same genera when he states that the epipelic community in alkaline running water tends to have as common species *Amphora ovalis*, *Caloneis amphisbaena*, *Navicula cryptocephala*, *N. gregaria*, *N. radiosa*, *Gyrosigma acuminatum*, *Nitzschia sigmoidea*, and *Cymatopleura soleci*. In acid streams the common epipelic species are *Eunotia* spp., *Actinella punctata*, *Frustulia rhomboides*, *Pinnularia* spp., and *Surirella* spp.

Other minerals are presumably also important; many species of diatom are characteristic of saline waters, as are the long, green tubes of *Enteromorpha intestinalis*, and these sometimes appear in salt-polluted streams. There are indications that *Cladophora glomerata* is sensitive to iron salts (Blum, 1957, 1960), and the so-called nutrient salts, potassium, nitrate, and phosphate, and for diatoms silica as well, are undoubtedly also important although we know little about their influence in running water.

Potassium seems to be rarely, if ever, a factor limiting algal growth even in still water, but phosphate and also sometimes nitrate, silica, and even sulphate have been shown to be sometimes in short supply in lakes. We know nothing of any possible influence of relative shortages of potassium or sulphate on algal growth in running water, and it seems unlikely that they would occur except under very exceptional circumstances. It was, however, long ago suggested that flood waters, which bring in, among other things, silica, may encourage the growth of diatoms (Pearsall, 1923), and in at least one instance it has been shown that diatoms can reduce the dissolved silica content of stream water. Müller-Haeckel (1965, 1966) found that 400 m. downstream of a small saline spring in north Germany, which emerged with a constant silica-content of 17–18 mg./l., there was a daily fluctuation of 3–9 mg./l. The low point was in the late afternoon and the high one was at dawn, and the effect was greater in May than it was in October. This she attributed to massive growths of the diatom *Thalassiosira*. However, the same phenomenon was not found in other streams and it may be that the unusual conditions of the saline spring were responsible. This could be by reduction of the flora to one or a very few species with a marked diurnal rhythm. In any event it would be somewhat surprising if dissolved silica were ever a limiting factor in running water as most rocks and soils contain silicates, and it must be constantly renewed in the water near to the diatoms by turbulence. Mokeeva (1964) found no correlation between the silica content of the water and diatom populations in the

River Oka in Russia. In passing it may be noted that the situation in lakes is different because there large masses of water become isolated from the substratum by thermal discontinuities.

Our information on the effects of nitrate and phosphate in running water is also indirect. We have seen (p. 48) that in a uniform reach with no additions along its length these ions tend to decline in amount, and this is undoubtedly because of uptake by plants. It has also long been known that the numbers of algae growing on slides placed in rivers tend to increase downstream, presumably because there are more nutrients, and that the inflow of purified sewage alters the community which develops. During his work on several British rivers Butcher (1932, 1946, 1947) showed that the normal 'slide' community characteristic of the lower reaches is dominated by *Cocconeis*, *Ulvella*, and *Chamaesiphon*, and that it is replaced below sewage outfalls by one dominated by *Gomphonema*, *Nitzschia*, and *Stigeoclonium*. Similarly, as we have seen (p. 63), different algae became dominant in experimental channels when small amounts of sewage effluent were added to the water.

It is also a well-known fact that algae, particularly the conspicuous filamentous Chlorophyceae, thrive below sewage outfalls and also below lakes, ponds, and dams (Stober, 1964; Blum, 1957; Kann, 1966). The increase caused by sewage is usually attributed to increased supplies of nutrients, and that below lakes *may* be caused by nitrogen fixation in the lakes, by bacteria or blue-green algae, or by the slow release of phosphate from littoral deposits during warm weather (Hutchinson, 1957), which is, of course, the time when it is most likely to be in short supply in the river. We do not know if nitrogen fixation occurs in running water, although it may do so, as *Nostoc* and *Rivularia*, species of both of which occur in streams, are known to be able to perform this feat.

Similarly, it has been observed that the amount of algal growth increases at points where rivers flow through areas which are rich in nitrate and phosphate, as has been reported in the Sundays River in South Africa where it crosses the Ecca series of rocks (Oliff *et al.*, 1965). All these observations lead to the conclusion that in many running waters one or both of these ions is in inadequate supply for the maximum growth of attached algae. We stand in need, however, of an experimental approach to this subject, which is now of such importance in relation to pollution. Very few field experiments have been performed on running waters in contrast to the large amount of work which has been done on the fertilization of lakes and ponds.

It has been shown experimentally that the addition of phosphate to the water encourages the growth of *Cladophora glomerata* on rocky shores in the Great Lakes, and that nitrate alone does not do so (Neil and Owen,

1964). Probably this finding would apply to streams also, but to my knowledge the only experiment actually performed on streams was a rather early and uncritical one by Huntsman (1948). He had noticed that in the barren parts of Nova Scotia the numbers of young salmon and other fishes were always greatest below farms, and that this was correlated with increased algal growth. So he tried the effect of leaving piles of ordinary agricultural fertilizer to leach in from the banks of streams, and he found that in fact this did increase the algal and fish populations; but even after two years the effects were apparent for only about 150 m. He also noted that masses of filamentous algae developed in the side arms of the streams, and he concluded that the nutrients were held up in the finely divided bottom materials. We could do with many, more carefully organized, experiments of this type, attempting to separate the effects of the nutrients from one another and to establish the critical concentrations under field conditions.

Scour. One feature of the ecology of attached algae which makes it difficult to study is their instability. During periods of active growth, particularly when the discharge is low and the water is clear, considerable populations are built up, and every solid object in the water may become thickly covered with a brown carpet of diatoms and trailing streamers of filamentous green algae. Silt surfaces also become brown with diatoms and dotted with blackish-green patches of filamentous Cyanophyta. These low-water populations are, however, very unstable and a single storm may wash them away. Butcher (1947) noted that populations in excess of 10,000 cells/sq. mm. rarely persisted for long on his glass slides, and many other workers have commented on the great fluctuations in algal biomass, although the attached algae are less unstable than the very evanescent plankton (Jackson *et al.*, 1964).

Much depends on the algae themselves and on the site on which they are growing. Some, such as the tough tassel-like Rhodophyta and species with closely applied flattened thalli, are normally little affected except by raging floods, and they remain unchanged when filamentous green algae and colonial or stalked diatoms are greatly reduced. Similarly, algae growing on stones which are rolled and rubbed together by the water are much more affected than are those on rocks or moss, as was nicely shown by Douglas in her work on *Achnanthes* to which reference has already been made. Spring floods in areas where there are great amounts of meltwater may reduce algal populations to almost nothing, and anchor and frazil ice also reduce them by abrasion; this must be an annual pattern common to many small streams in areas with a severe winter climate.

Grazing animals are also another factor causing irregular variation. Many invertebrates and some fishes feed on periphyton; Cyanophyta are usually avoided and may even be poisonous (Gajevskaja, 1958), but great quantities of other algae may be consumed quite rapidly. Douglas found that the numbers of the grazing caddis-worm *Agapetus fuscipes* were negatively correlated with the numbers of *Achnanthes* in her stream, and it is often possible to observe cleared patches of stone surface around snails and limpets, which have presumably been caused by grazing.

SEASONAL CHANGES

It will be clear from the discussion on the effects of light and temperature that there are seasonal changes in the flora as a whole. This applies even in springs at constant temperatures in temperate climates, where the winter flora is dominated by diatoms, or sometimes by *Hydrurus*, and the summer flora by Chlorophyceae, Cyanophyta, and Rhodophyta; light is therefore certainly involved (Gessner, 1955*b*). We have virtually no information on seasonal changes in the tropics, but it seems probable that they occur at least where there is an annual rhythm of discharge. In the Nile in Egypt the seasonal flood in the autumn dominates the cycle. As it declines the diatoms *Cocconeis* and *Gomphonema* appear, followed by dense growths of Chlorophyceae, which in June and July are followed by a second growth of diatoms, mostly *Gomphonema* and *Melosira*, with some Cyanophyta. This later growth is swept away by the flood-water, and the cycle is resumed (Abdin, 1948). Doubtless similar cycles occur in many areas nearer the equator, but they may be less complex because of the uniformity of temperature. The post-flood diatom development in the Nile corresponds with the cooler weather of the winter, and it may not occur in a uniformly warm climate.

In contrast many seasonal studies have been made in Europe, Japan, and North America, and long lists of the times of occurrence of various algal species have been published (e.g. Budde, 1928, 1932; Wehrle, 1942; Mack, 1953; Gumtow, 1955; Blum, 1957; Koboyasi, 1961*a*; Bursche, 1962). Usually diatoms dominate the lists of species, and in stony streams and rivers they are also the most abundant forms. Genera which are common in the winter include *Achnanthes*, *Meridion*, *Gomphonema*, *Navicula*, and *Diatoma*. In early spring these are joined by other algae, e.g. *Hydrurus* which may be present all through the winter, *Ulothrix*, and *Phormidium*, and these early cold-water species persist for longer or shorter periods depending upon the temperature regime, or, as already noted for *Batrachospermum*, the amount of shade. During the spring, the winter diatoms decline, although some such as *Cocconeis*, *Synedra*, and *Navicula* persist, and they are joined by others, e.g. *Cymbella* and *Melosira*, and a great variety of Cyanophyta, e.g. *Chamaesiphon* and *Oscillatoria*, and of Chlorophyceae, particularly *Oedogonium*, *Ulvella*, and *Cladophora*. At this season also the Rhodophyta, and some crustose green algae such as *Gongrosira*, which are present all the year, grow and become conspicuous. In soft-water streams desmids are often abundant during this late spring to early summer season. As the water becomes shaded by the opening of the leaves on the trees, this early development of the flora declines, and even in streams which are open to the sky there is often a decline in importance of some Chlorophyceae, presumably because of unfavourable light or temperature conditions. *Cladophora glomerata*, for instance, sheds filaments in June in the Saline River, Michigan (Blum, 1957), but it does not do so in the softer light and lower temperatures of streams in England.

In the early autumn there is often a short recrudescence of growth as the water cools and the leaves fall from the trees, and some of the spring species attain a second maximum. Such are *Diatoma*, *Synedra*, and *Navicula*, while others such as *Cocconeis* and *Melosira* may attain their only maximum at this time (Raabe, 1951). On occasion these autumnal maxima of diatoms may occur earlier during the summer when flood waters lower the temperature, as was observed by Hornung (1959) in the German River Echaz. As the autumn progresses these maxima decline, the summer species disappear or die down to their bases, and the winter diatoms appear. The first of these is *Diatoma vulgare*, or *D. haemale* in small northern and alpine streams, and the former may become very abundant in the late fall, as in the River Ybbs in Austria (Von Mitis, 1938). Another important winter species is *Gomphonema olivaceum*. These two are often considered to form a winter association, but as Blum (1954*b*) has pointed out their seasons do not coincide exactly and they are really in competition. *Diatoma* appears first and forms masses of filaments, and then *Gomphonema* steadily takes over and makes a more massive mat-like growth. This declines in late winter and is replaced by a second growth of *Diatoma*.

In streams which are dominated by meltwater floods and cold winters the cycle may become truncated by physical conditions. In the West Gallatin River, Montana, Gumtow (1955) found that the scouring action of the floods in May and June almost cleaned the stones, and that it was late July before recovery was evident, leading to a build-up of diatoms and other algae, mainly *Tetraspora* and *Oscillatoria*, in August through September. *Navicula* was the most abundant diatom at this time, but it was replaced by *Diatoma* in early winter. During December and January, however, a decline was caused by anchor and frazil ice, and, although the numbers of the diatoms *Navicula*, *Diatoma*, and *Cymbella*, and of *Tetraspora* and *Oscillatoria*, increased a little in the early spring, they were swept away by the meltwater.

On stony or other solid substrata there is thus a fairly clear annual cycle, and in running water the actual nature of the substratum is apparently not of great importance for many species except in respect of stability. We have less information on larger rivers, but it is worth noting that in the quiet-flowing Hampshire Avon and the canal-like River Hull in England Butcher (1938, 1940) found far fewer seasonal differences of growth on glass slides than he did in the swifter-flowing, stony-bottomed rivers Tees, Itchen, and Lark.

COMMUNITY CONCEPTS

Much paper has been used, especially in continental Europe, in attempts to apply to aquatic microphytes the concepts of the plant community developed by the terrestrial plant ecologists. Studies in which various running-water associations are discussed and named include those of Budde (1932), Behre and Wehrle (1942), Panknin (1947), Symoens (1951), and Margalef (1960), and the subject has been reviewed by Blum (1960).

The conclusions derivable from these discussions would seem to be that it is not useful to attempt to apply the ideas and findings of the terrestrial macrophyte ecologists to running-water microphytes. The latter occur as irregular mosaics on stream and river beds and are subject to great seasonal changes. Even though certain groups of species tend to occur together their ratios differ greatly from place to place even in one stream; because of the instability of the habitat the arrangement is probably largely due to chance. Often a 'community' which has been named after two or more species will contain only one of them at a given place and time, and, as there is little evidence of widespread succession or of dependence of one species upon another, the concepts of the 'climax' and the 'sere' are inapplicable. In running water even the epiphytic species are, with few exceptions, not obligatory epiphytes, so many of the ideas behind terrestrial plant sociology are not valid in running water.

On the other hand, succession does occur in some limited situations, and we have seen that there can be competition between species of diatom; the set-up is therefore not without interaction between plants. On the Rhine-falls near Schaffhausen the rhizoids of the abundant mosses are inserted into a layer of the blue-green alga *Plectonema* (Jaag,

1938), which seems to indicate primary colonization by the alga and dependence upon it by the moss; and in crustose calcium-depositing areas, of which more below, the pioneer plants are species of *Schizothrix* (Blum, 1960) or *Gongrosira* (Symoens, 1957; Minckley, 1963), which later are invaded by *Phormidium* and *Audouinella* forming a compact mass on which yet other algae may grow in summertime. Evidence of competition is given by the fact that when a new substratum is provided in a stream it is at first colonized by many species, but the number of these declines as the total population increases (Patrick *et al.*, 1954; Yount, 1956).

Thus, while it is possible to recognize many frequently occurring assemblages of species such as the *Achnanthes–Chaetopeltis–Diatoma–Eunotia* group which grow on glass slides in infertile waters, the *Cocconeis–Ulvella–Chamaesiphon* group which grow similarly in more fertile waters, or the *Diatoma–Meridion–Gomphonema* group typical of stony streams in wintertime, it would seem unwise to attempt to apply to them the clearly defined ideas implicit in the word 'association' as used by the terrestrial ecologists. There are, in effect, few or no such associations in running water, and to use the term in any but its everyday sense is the way to semantics and probably to much more wasted paper.

ZONATION

Zonation of attached algae occurs in respect of altitude, depth, and position on objects in relation to the current. Many studies have shown changes in the abundance of species from end to end of streams, some species being confined to the source regions and others coming in only lower down. Scheele (1952) and Schmitz (1961) give such data for the diatoms of the Fulda River in north Germany and Scheele notes that whereas the upper reaches are inhabited only by running-water species the lower parts contain many which occur also in still water. Similarly Mack (1953) noted a steady increase in the diversity of algae down the length of the Liesingbach in the Vienna Woods. Near the spring only *Gomphonema* was abundant, but lower down it was joined by an increasing variety of diatoms, *Achnanthes*, *Navicula*, and *Cymbella*, and by *Ulothrix*, and finally *Cladophora*, *Vaucheria*, and some *Phormidium* were added to the community.

Zonation within watercourses has also been observed in the Rhodophyta, with *Hildenbrandia*, *Audouinella*, and *Batrachospermum*, or only the last, upstream of a *Lemanea* zone (Budde, 1928; Minckley, 1963). Doubtless many other examples could be cited, and they would not be unexpected in view of the different ecological requirements of the various species, and the known physical and chemical changes in the environment as one proceeds down a stream. It is, for instance, possible that the inability of *Batrachospermum* to photosynthesize bicarbonate ions restricts it in some places to headwater areas where carbon dioxide is always available (Minckley and Tindall, 1963). This restriction would be more effective in hard water than in soft, and it may go some way towards explaining the relatively great abundance of this genus in soft-water streams.

Vertical zonation in running water is much less frequently observed because it takes time to become established and the water level normally fluctuates widely. It has, however, been observed occasionally after long periods of even discharge, and Symoens (1957) described a situation in the Belgian River Meuse in which *Rhizoclonium* grew just above the water level, *Bangia* on the wave-wetted edge, and *Cladophora* below the water surface. Incipient zonations of this kind can quite often be seen by the casual observer, but they are most strikingly developed on floating objects. Behning (1928, 1929*b*) described the way in which boats on the Volga collect a whole community of fouling organisms analogous to those which trouble sea-going ships. *Pleurocapsa* grows above the surface in the region wetted by the waves; reaching to 30–35 cm. below the surface is a zone of *Cladophora* and *Stigeoclonium*, followed by 25–30 cm. covered with diatoms, mostly *Cymbella* and *Gomphonema*. Below this comes a zone of the bryozoans, *Paludicella* and *Plumatella*, and finally the zebra-mussel, *Dreissena*, whose attached shells shelter many other animals. One interesting fact about this peculiar community is that it contains a number of algae, e.g. *Stigeoclonium*, *Cladophora crispata*, and *Audouinella*, which elsewhere in the Volga basin are confined to small streams. Presumably the fact that the floating vessel gives them a firm substratum which is always kept at a suitable depth, and is not subject to scouring by sand at times of high water, accounts for this occurrence outside their normal habitat.

It can also often be seen that algae are arranged on stones and other fixed objects in a definite way in relation to the current. Fritsch (1929) noted long ago that *Phormidium* occurred primarily on the upstream sides of boulders in streams in north Devonshire, England, and Blum (1960) described a similar situation in the Saline River, Michigan (Fig. IV, 4). There *Gomphonema* occurred on the upstream faces of stones with *Diatoma* below it, and their line of overlap extended obliquely upwards along the sides of the stones until, at the downstream end, there was no more *Gomphonema*, and *Diatoma* extended to the tops of the stones. On stones which broke the water surface there was a narrow band of *Ulothrix* on the upstream face above the *Gomphonema*. Similarly,

Lithoderma grows on the upstream sides of stones and not on the downstream faces (Round, 1965). This sort of detailed local distribution makes quantitative sampling very difficult, and it is undoubtedly caused by small differences in local conditions. The upstream face of a stone is the one most likely to receive a disseminule carried by the current, and it also has a thin zone of dead water because of the splitting of the main stream of the current as it passes on each side of the stone. There is also a much larger zone of dead water on the downstream side, and Gessner (1955*b*) was able to show that *Cocconeis* colonized a strip of film wound around a vertical round pole more heavily on the upstream and downstream sides than laterally, where the only dead water was the boundary

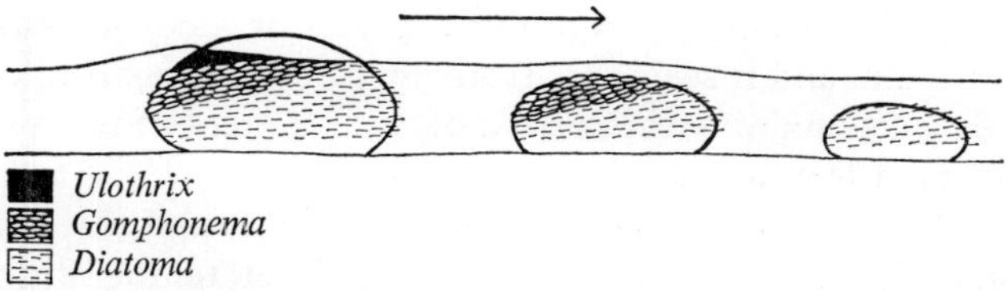

FIG. IV, 4. Diagrammatic illustration of the distribution of algal species on variously submerged stones in the Saline River, Michigan. Redrawn from Blum (1960).

layer. Such observations explain at least in part why some local distributions occur, but not, of course, why some species are largely confined to upstream faces in certain situations and do not also occur on the downstream faces.

TRAVERTINE FORMATION

In hard waters, particularly in headwaters which are fed by limestone springs, deposits of calcium carbonate are often laid down. These may form large solid structures which dam up the stream, producing waterfalls. Or they may even raise the stream so that it stands above the surrounding land like a small man-made aqueduct; Gessner (1959) figures such a stream depicted in a medieval painting. The material of which these structures are formed is often called marl in North America and tuff, or tufa, in Europe. Neither is really suitable as both words have another meaning, and the material is properly called travertine.

Some of the deposition is probably purely chemical and is caused by loss of the equilibrium carbon dioxide necessary to keep calcium bicarbonate in solution, but it is nearly always associated with algae and to a lesser extent with mosses, which, by photosynthesis cause deposition of calcium carbonate around them. The genera involved include

Phormidium, *Gongrosira*, *Gloeocapsa*, *Schizothrix*, *Rivularia*, *Oocardium*, and *Audouinella*, or the chantransia stages of other Rhodophyta, and they occur in various combinations in various places (Lauterborn, 1910; Fritsch, 1949, 1950; Butcher, 1946; Gessner, 1959).

Sometimes the formation of travertine begins at the spring source, but more usually there is a gap of a few tens of metres between the spring and the most upstream deposit (Symoens, 1949; Sourie, 1962). This is doubtless connected with the initially high carbon dioxide content of the water, and Minckley's (1963) study of Doe Run demonstrated how interference with a stream can alter the pattern. The early European settlers of Kentucky constructed two small dams at the head of the valley to operate mills, and now travertine formation begins below them. Previously, as is clear from old deposits, it was laid down much further upstream, and it seems probable that the dam pools have caused the change by maintaining the carbon dioxide content of the water from their deposits of leaves, etc.

Travertine is laid down as crystalline deposits round the cells of the algae (Fig. IV, 5) and it is usually formed only at temperatures above about 14 °C. (Matoničkin and Pavletić, 1961, 1962*a*, *b*). At lower temperatures it may even be eroded, especially in places where the current is swift. The pioneer plants are masses of filaments which at first lie tangled on the substratum, but, as they become encrusted, they tend to align themselves vertically and to produce a surface which goes on growing outward, laying down calcium carbonate and dying off in the lower layers. The desmid *Oocardium* is exceptional as it lives in the mouth of a gelatinous tube which becomes calcified as it grows (Fig. IV, 3), and colonies of this alga produce a cushion-like structure resembling, on a small scale, some of the massive corals.

Sometimes two species occur together and grow at different seasons, each growing up through the other and laying down slightly different types of structure. This produces annual laminae which may build up to large numbers, as occurs for example, in Doe Run, where Minckley found that the spring- and early-summer-growing *Gongrosira* alternated with late-summer-growing *Phormidium*, forming large nodules like water-worn rocks (Fig. IV, 5), and up to over sixty consecutive annual layers at one point. The annual generations of travertine-forming Chironomidae alternating with algal deposition can produce similar layers (p. 138).

As the water passes downstream through a travertine-forming area its chemical content, of course, changes rapidly, and finally it contains insufficient dissolved bicarbonate to continue deposition, which therefore declines. In many hard-water streams, however, small amounts occur

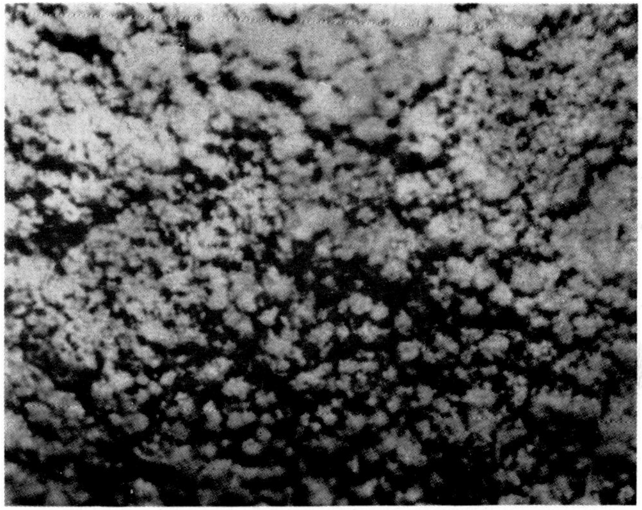

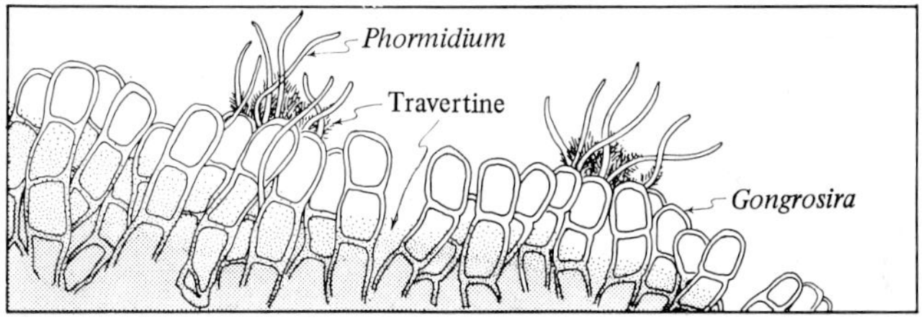

Fig. iv, 5. A cut-open travertine stone showing annual layers, a photograph of the surface of 'living' travertine (× c. 32), and a diagrammatic presentation of the way in which *Gongrosira* and *Phormidium* alternate with one another to produce annual layers. The situation illustrated would be in early summer when *Phormidium* was just starting to grow. From Minckley (1963) with permission.

on stones during the warm low-water period of the summer, and they dissolve again during the winter. In such places some calcium-depositing algae occur more or less permanently. Such a permanent crustose community composed of *Phormidium*, *Schizothrix*, and *Audouinella* is reported from the Saline River, Michigan (Blum, 1957), and it occurs in many calcareous streams in eastern central North America.

Also as one passes through the zone of travertine formation there is often a change of the principal species, which is brought about presumably by the changing levels of pH, carbon dioxide content, etc. In the Bavarian district where *Oocardium* produces the spectacular raised streams on 'growing stone' the upstream areas tend to be dominated by *Rivularia*, *Pleurocapsa*, and *Gongrosira*, with *Oocardium* taking over further downstream (Gessner, 1959). Such successions in importance of different algae are not, however, always observed.

Other algae and mosses grow on the travertine, and in areas of intense deposition they may themselves become quite harsh and calcareous. They do not, however, form solid structures, and their thin coat of calcium carbonate disintegrates into sand-like material. Algae which do this include *Cladophora*, *Vaucheria*, and *Lemanea* and doubtless others, and their fine deposits as well as leaves, sticks, snail shells, and other debris become consolidated by later growth of the smaller algae. The whole bed of the stream therefore tends to become covered by an irregular, spongy, concrete-like deposit, with great numbers of microhabitats for plants and animals (Matoničkin and Pavletić, 1961, 1962*a*, *b*).

CHAPTER V

Higher plants

The remaining photosynthetic plants which occur on the substrata of running water are a mixed assemblage of taxonomic groups comprising the Charales, which are often classified with the green algae, the mosses and liverworts (Bryophyta), a few species of encrusting lichens, and the flowering plants (Angiospermae). These, together with some of the more conspicuous algae, which were dealt with in the last chapter, are the macrophytes of running water. Other groups of plants, such as the ferns and horsetails, have aquatic representatives, but they are of little significance in rivers and streams and so do not concern us here. We know little about the fungi of running water, although they are clearly of considerable importance; they and the bacteria are discussed in Chapter XXII.

For our purposes, the higher plants can be divided into three fairly distinct ecological categories: those that are attached to rocks and other solid objects, those that are 'rooted' into the substratum, and free-floating plants. We have to put the word rooted into inverted commas because this group includes the Charales of which the 'roots' are really rhizoids.

THE ATTACHED PLANTS

Plants which occur always attached to solid objects in streams include the mosses and liverworts, the dark-coloured, very flattened lichens, *Verrucaria* and *Hydrothyria*, and also the two peculiar families of flowering plants, the Podostemaceae and Hydrostychaceae, which occur on waterfalls, particularly in the tropics. They are all essentially inhabitants of fast-flowing water because only there are found suitable sites for attachment.

Bryophytes, particularly mosses, may occur in slower water when suitable sites such as wharfs, piles, and bridges are provided by man.

This is, however, unusual because, in addition to their need for a firm unsilted object to which to attach themselves, mosses are unable to photosynthesize bicarbonate ions (Ruttner, 1963; Gessner, 1959). In contrast, therefore, to most algae and all the flowering plants, they are confined to water where there is an adequate supply of dissolved carbon dioxide. This means that they are particularly favoured in spring regions where the water emerges from the ground rich in dissolved carbon dioxide, but that they decline in importance as this gas is lost to the atmosphere. They are also favoured by turbulent flow, especially where this leads to white water, because this brings in carbon dioxide from the atmosphere. Mosses are indeed the most characteristic vegetation of waterfalls and rocky chutes, and their firm rhizoid hold-fasts and tough tassels are well adapted to resist turbulence. Many species, particularly the dark-coloured ones, are also well adapted to shade.

It is, therefore, sometimes possible to observe variations in the importance of mosses along the length of a stream, as did Entz (1961) in karst spring-fed streams in Hungary. There, below the springs, mosses are important in water containing at least 12 mg./l. of carbon dioxide. Further downstream, at lower carbon dioxide concentrations, other plants and algae take over, but shady reaches, waterfalls, and rapids can cause the reappearance of moss by again raising the carbon dioxide content of the water.

In headwater streams in wet cool climates, as in western Britain (Hynes, 1961) or Yakobi Island, Alaska (Shacklette, 1965), leafy liverworts such as *Nardia* and *Scapania* form wet sponges on the rocks, and they may build small dams in little streams. Shacklette reports that in Alaska they tend to grow over the little pools above the dams, and that such loose carpets often show blow-out holes caused by sudden rises in discharge; the same phenomenon can be observed in Wales and doubtless also elsewhere.

Most reports of liverworts in streams are from soft or acid water, e.g. *Chiloscyphus* from streams along the north shore of Lake Superior (Smith and Moyle, 1944) and *Dermatocarpon* from a high stream flowing from gneiss and schists on the Franco-Spanish border (Nicolau-Guillamet, 1959), and in general *Alicularia* and *Scapania* are recognized as being characteristic of soft waters in Europe (Watson, 1919; Illies, 1952*b*). They rarely seem to be important members of the submerged community, except in very small headwaters, but in several streams on Mount Elgon in East Africa an unidentified member of the Jungermaniales, Anacrogyneae, is the only large plant occurring over long stretches of stream (Hynes and Williams, 1962). It occurs as a flattened, branched thallus spread over many large stones and boulders at altitudes ranging from about 1,300 to 1,500 m. Presumably the altitudinal limits have something to do with temperature, and since submerged stream-dwelling liverworts are always reported from cold or high waters, it seems probable that they are cold-water plants.

Submerged lichens are not very common, but mosses occur in streams all over the world and belong to a great number of genera. Most are confined to small headwater streams, although some, particularly *Fontinalis antipyretica*, occur also in large rivers. This species is particularly eurytopic as it occurs in many kinds of water, and it is sometimes common in organically polluted rivers, where it is apparently able to use ammonium as a nitrogen source (Schwoerbel and Tillimanns, 1964*a*, *b*). Even in large rivers, however, other mosses may appear where conditions are suitable; the famous Rhine-falls are, for instance, thickly covered with *Rhynchostegium* and *Cinclodotus* (Jaag, 1938), which elsewhere in Europe are typical of small streams.

It seems probable that many forms are eliminated by high temperatures, because most records from the tropics are from mountain areas, and the same applies in warm areas such as South Africa (e.g. Harrison and Elsworth, 1958; Oliff, 1960). There are also indications that temperature has an influence at the other end of the scale. In high alpine streams in France, Dorier (1937) found only the genus *Oxyrhynchium*, which lower down is joined by *Crotaneurum*, *Cinclodotus*, and *Fontinalis*; and in Austria the last three genera occur in the stream above the Lunzersee, but in the warmer outflowing stream they are replaced by *Fissidens* (Findenegg, 1959). There are, however, indications that *Fissidens* may be affected by nutrient supplies rather than by, or in addition to, temperature. Lohammer (1954) reports that it occurs primarily in rich water in northern Europe, and it is particularly abundant in Doe Run, Kentucky, which is rich in nitrates and phosphates (Minckley, 1963). It is also a common plant on the brickwork of English canals, of which the water is usually very rich in nutrients.

It also seems that some species are limited by water hardness. In Europe the almost black moss *Fontinalis squamosa* is always reported from very soft water (e.g. Nicolau-Guillamet, 1959; Hynes, 1961), and it is replaced by *F. antipyretica* in harder water. On the other hand, species of *Hypnum* and *Platyhypnidium rusciforme* occur abundantly in both very soft- and very hard-water streams.

We clearly have much to learn about the factors which control the mosses of running water and determine which species occur where, and we need many careful micro-ecological studies such as those made by Matoničkin and Pavletić (1961, 1962*b*) on travertine waterfalls in Yugoslavia. They found that above the zone of deposition, and on newly formed and submerged travertine, *Cinclodotus* was the chief moss. Where the travertine reached to, or near to, the water surface, it was joined by *Platyhypnidium* if the site was well lighted, or by *Crotaneurum* if it was shaded. These masses of travertine and moss grow in a downstream direction and overhang, forming series of little waterfalls in steep places, and *Fissidens* and *Eucladium* occur in the deep shade which is produced under the overhangs. Thus a mosaic of species is formed in water of the same chemical content, and it can be inferred that two important factors are the depth of the water and the amount of illumination.

The depth of the water can indeed be readily seen to be important to mosses in streams everywhere. On rocks which are above normal water level, and so are rarely submerged, there is often a variety of terrestrial mosses and liverworts, especially in shady places. At or near the water surface, where it is at least sometimes exposed to the air, *Platyhypnidium rusciforme* occurs very widely in Europe, and it produces sporophytes freely when it is dry. The truly aquatic mosses such as *Fontinalis* and *Hypnum* occur at lower levels, and they seem rarely to produce spores. Large stones therefore show a fairly clear zonation of mosses and of types of bryophyte reproduction (Stave, 1956).

The only angiosperms which belong to the attached group of higher plants are the very peculiar pan-tropical Podostemaceae and the African and Madagascan Hydrostychaceae, which are both very highly adapted to life in swiftly flowing water (Gessner, 1955*a*, *b*). The genus *Dicraea* occurs as far north as the Khasi Hills in Assam (26° N.) (Hora, 1930), and one species of *Podostemon* extends into Canada. They occur on waterfalls and in torrential reaches of streams, and they are nearly always associated with tumbling white water.

The plants are attached to the rocks by their roots, which are either flattened rhizoid-like structures which enter every irregularity on the surface, broad, flattened, lichen-like structures which spread over the rock, e.g. *Podostemon*, or straplike bands which hang free in the water except where they are attached, e.g. *Dicraea*. These flattened roots are green and photosynthetic, and in some genera, e.g. *Tristicha*, the shoots and leaves are very reduced. In others the leaves are large and cabbage-like, e.g. *Oenone*, or much divided and tassel-like, e.g. *Rhyncholacis*. The last genus superficially resembles *Ranunculus*, and some African species of *Dicraea* are reminiscent of *Myriophyllum*.

These plants grow on the tops of rocks, often attached to the upstream face, with their vegetative parts trailing in the tumbling water or emerging above the water surface only as a wet mass (Fig. v, 1), or they grow in the spray zones of waterfalls. Only their flowering stems rise into the air and can withstand drying, and the flowers are fertilized by insects and often attract great numbers of bees. The seeds are very small, and when they are wetted they rapidly develop a sticky coat from starchy cells on the surface. This attaches them firmly to the rocks, usually on the upstream face with which they first come into contact. They thus have an adaptation which closely parallels that of the eggs of many stream-dwelling insects (p. 139).

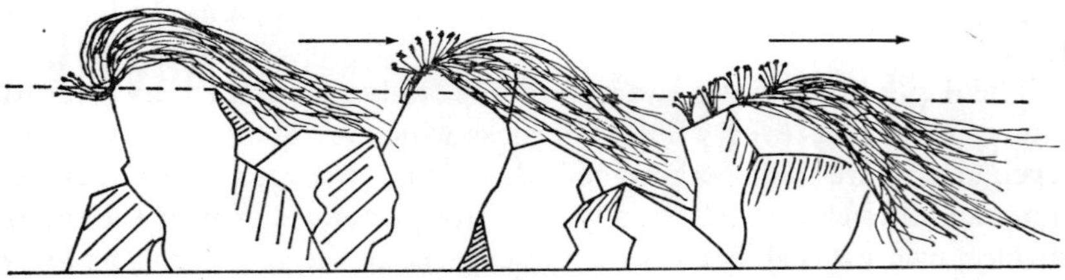

FIG. V, 1. Diagram of the position in which members of the Podostemaceae grow on emergent rocks in fast water. From Gessner (1955*b*) with permission.

Undoubtedly these two very interesting families of flowering plants deserve further study; they are unique in being totally adapted to the running-water habitat, and exceptional in that they always reproduce there by seeding.

PLANTS WHICH ARE 'ROOTED' INTO THE SUBSTRATUM

Plants which are rooted into the substratum include the rest of the angiosperms and the genera *Nitella* and *Chara* of the Charales which are fixed by rhizoids, and they are confined to stream beds into which the roots can penetrate. They do not, therefore, normally occur on rough rocky areas and are thus to some extent limited by current. Some species, however, occur on gravel and rubble bottoms where they are at times subjected to rapid water.

It has been shown that, as with algae, the rate of respiration of aquatic angiosperms increases with the oxygen content of the water, and that the effects of lower oxygen levels are to some extent offset by water movement (Gessner and Pannier, 1958; Owens and Maris, 1964). Similarly, the rate of assimilation is increased by water movement, and some plants which normally occur in streams, such as *Potamogeton densus*, produce a stunted 'hunger form' if they are experimentally grown in entirely still water (Gessner, 1937). It is clearly, therefore, physiologically advantageous for plants to grow in running water, and it is perhaps strange that relatively few species do so, except in very sheltered areas, and that, although some rooted species are *usually* found in rivers and streams, e.g. *Ranunculus fluitans* and *P. densus*, none is *confined* to this habitat. This is in distinct contrast to the algae and mosses, which groups contain many exclusively running-water species.

In fact no 'rooted' plants show any special adaptation to running water, and those that occur there are able to do so primarily because they have tough flexible stems and a creeping growth-habit, either by stolons or rhizomes, with frequent adventitious roots which enable them to hold their position on the stream bed. Plants from running water differ from specimens of the same species in still water in a number of ways (Gessner, 1955*b*). They tend to have smaller leaves and shorter petioles, and they have shorter internodes in both stems and rhizomes. They rarely produce floating leaves (in those genera which normally do so), and they often do not flower or they produce fewer or smaller flowers than plants in still water. This inhibition of growth appears to be a direct effect of the current, as can sometimes be seen in the waterlily, *Nuphar luteum*, in which the upstream and lateral leaves of a clump are often smaller and have shorter petioles than those which they shelter. On the other hand, although the internodes may be short and the leaves small, the shoots of plants growing in rivers, being laid out at length in the light, may become very long.

The reason for these reductions is not known, but it probably results from direct effects on the leaves, and hence on the assimilatory mechanism of the plant, and from this the other effects probably stem. The absence of floating leaves in many situations probably results from the fact that such leaves, like those of land plants, do not assimilate under water, and so are put out of action every time they are submerged by turbulence or rising discharge. The absence or reduction of regular seed-production means, of course, that reproduction is mainly vegetative, so there is little opportunity for the evolution of species, or races, especially adapted to swift running water. Those plants which do produce seeds are in quieter stretches or sheltered places, or they have been exposed to the air by falling water-levels.

The number of species which occur in running water, as opposed to those which occur in quiet bays, side arms, etc., is very limited. Gessner (1955*b*) lists only sixty-one recorded during three intensive studies of rivers and streams in western Europe, and in two of these (East Holstein and Upper Bavaria) only twenty-eight species were found. Each of these plants tends to be most common on a particular type of substratum, and this is, of course, related to the current regime. Butcher (1933) drew up a table showing the dominant plants in various types of

Type of river or stream	Approximate range of total hardness as mg./l. $CaCO_3$	Average pH	Source of river	Type of habitat and approximate range of current velocity				
				Torrential >60 cm./sec.	Non-silted 70–25 cm./sec.	Partly silted 25–10 cm./sec.	Silted <10 cm./sec.	Littoral <10 cm./sec.
Non-calcareous and acid	<20	<6·5	mountains	*Fontinalis squamosa* *Hypnum palustre*	*Ranunculus fluitans* *Myriophyllum alterniflorum*	*Potamogeton polygonifolius*	*Juncus supinus*	none
			hills	*H. palustre* *Platyhypnidium rusciforme*	*R. fluitans* *M. spicatum*	*P. nitens*	*J. supinus* *Callitriche intermedia*	none
			lowland bogs	none	*R. fluitans* *M. alterniflorum*	*P. polygonifolius* *P. alpinus*	*Potamogeton heterophyllus*	none
Slightly calcareous	20–100	7·0–7·5	mountains	*P. rusciforme* *F. antipyretica*	*M. spicatum*	*P. alpinus* *Sparganium simplex*	*Callitriche stagnalis* *Elodea canadensis*	none
Moderately calcareous	100–200	7·0–8·0	mountains	*P. rusciforme* *F. antipyretica*	*R. fluitans* *M. spicatum*	*Sp. simplex* *P. perfoliatus*	*P. crispus* *E. canadensis*	none
			hills	none	*R. fluitans* *M. spicatum*	*Sp. simplex* *Sagittaria sagittifolia* *P. perfoliatus*	*P. pectinatus* *P. crispus*	*Sparganium erectum*
Highly calcareous	>200	7·5–8·0	mountains	*F. antipyretica* *Pl. rusciforme*	*R. fluitans* *Sium erectum*	*P. perfoliatus* *Sp. simplex*	*C. stagnalis* *E. canadensis*	none
			hills	none	*R. pseudofluitans* *S. erectum*	*Sp. simplex* *Hippuris vulgaris*	*C. stagnalis* *P. lucens*	*Sp. erectum*
			lowland bogs	none	*R. fluitans* *S. erectum*	*Sp. simplex* *Sag. sagittifolia*	*P. lucens* *P. pectinatus*	*Sp. erectum* *Glyceria aquatica*

TABLE V, 1. The dominant submerged plants in various types of river in Great Britain. Modified and amplified from Butcher (1933). The littoral category includes only plants which actually stand in moving water.

river in the British Isles, and somewhat similar data have resulted from studies in Germany (Roll, 1938; Schmitz, 1961).

A modified version of Butcher's table, somewhat amplified by my own observations in western Britain, is given as Table v, 1. This is, of course, a considerable simplification, as it takes no account of differences between sizes of stream nor of the great variety of plants which occur along the banks of some rivers. The littoral category should perhaps be called the 'shoal' category, since it includes only plants which stand in the water away from the shore. In such places *Sparganium erectum* is sometimes accompanied by *Glyceria aquatica* in canal-like reaches (Butcher *et al.*, 1931), and *Polygonum amphibium* and *Scirpus*, particularly the large *S. lacustris*, often occupy shoals in less calcareous water, but not to the extent to merit inclusion in a table showing dominant species. It should perhaps also be noted that the word dominant is used here in the sense of predominant, and without any of its ecological implications. In some places also the silted habitat supports great numbers of the common water lilies, *Nuphar luteum* and *Nymphaea alba*, together with *Nymphoides peltata* in continental Europe (Voo and Westhoff, 1961).

In small acid streams the partly silted habitat sometimes supports *Juncus fluitans* and *Lobelia dortmanna* (Naumann, 1925), and in small calcareous streams *Apium nodiflorum*, *Berula angustifolia*, *Nasturtium officinale*, or *Myosotis palustris* are often predominant in this habitat. The last two species, watercress and forget-me-not, flourish in similar habitats in North America to which they have been introduced by man.

We will not here consider the long list of emergent plants which may occur along the banks out of the current, as, unlike the submerged aquatics, they vary much more from one geographical area to another, and they do not belong, strictly speaking, to the running-water habitat.

It emerges from the above discussion that factors controlling rooted plants include current speed, water hardness or some associated parameter such as pH or calcium content, and probably, in view of Butcher's distinction between rivers rising in mountains and those rising in hills, liability to spate.

At very low current speeds the silted bottom is inhabited by still-water species (helophytes), and where the scour of the water keeps rocks and rubble clear of finer deposits only bryophytes or Podostemaceae occur. On coarse gravel some genera with tough stems and stolons and a great development of adventitious roots, e.g. *Ranunculus*, *Myriophyllum*, *Sium*, and *Potamogeton densus*, can maintain a foothold. As the size of the particles declines with decrease in the ruling speed of flow, more fragile stoloniferous plants, e.g. *Callitriche*, can survive together with others with buried rhizomes, e.g. *Potamogeton* spp. and submerged species of *Sparganium*; and on very silted substrata these are joined by others. Species which occur on the shoal habitat combine resistance to current with a need to rise above the water, and they are among the few angiosperms actually standing in the current which regularly produce flowers and fruit.

Water hardness seems to affect some plants but not others. A few, such as *Juncus fluitans*, *Potamogeton polygonifolius*, and *Myriophyllum alterniflorum*, occur only in soft water. There are parallels in the tropics, as in the Amazon basin where the bog-moss *Mayaca longipes* may carpet slow reaches of acid clear-water streams, and the commonest plant in the dark-coloured acid waters of the Rio Negro system is *Tonina fluviatilis* (Gessner, 1959). Other plants, e.g. *Myriophyllum spicatum* and *Nasturtium officinale*, seem to require harder water and a few, e.g. *Ranunculus fluitans*, are apparently indifferent. The Charales belong to the hard-water group, particularly the genus *Chara*; but *Nitella* is also confined to water with a fairly high calcium-content (Gessner, 1959). Either may occur in partly silted or silted areas in many parts of the world.

Much of the above discussion is based upon European species, since most of the studies have been made on that continent, but doubtless the same general pattern obtains elsewhere. There are, for instance, indications of similar distributions involving other species of the same genera in streams along the north shore of Lake Superior (Smith and Moyle, 1944); in South Africa *Scirpus digitatus* seems to replace *Juncus fluitans* as a plant of small acid streams (Harrison and Elsworth, 1958), and in West Africa *Potamogeton octandrum* occurs rooted in rocky and stony substrata (Welman, 1948), and so may be considered to be an ecological equivalent of *Ranunculus fluitans* or *Myriophyllum*.

Two other important factors affecting plants are light and nutrients. Even clear water rapidly reduces the amount of light energy reaching increasing depths, and river water is rarely clear. Angiosperms do not, therefore, grow on river substrata more than a very few metres below the surface, and, in contrast to mosses and some algae, they do not occur in heavily shaded places. For this reason also very turbid tropical rivers, such as the Amazon and many in Africa, are devoid of higher plants. A few plants are, however, less affected than most. In Western Europe *Sparganium simplex* and *Potamogeton pectinatus* can often be seen growing in dappled shade under trees, as can *Vallisneria americana* in North America. In Russia Shadin (1956) reports that *Butomus umbellatus* is less affected by being covered by high water than are *Potamo-*

geton perfoliatus and *P. pectinatus*. All three are abundant in the River Oka in most years, but in the wet summer of 1933 when there was a series of floods, only *Butomus* developed. Possibly this was an effect of light.

We know almost nothing about the influence of nutrients on higher plants in rivers. Carbon dioxide may be a limiting factor in soft waters where no bicarbonate ions are available, because water which is in equilibrium with normal air containing 4·0 ml./l. contains only 0·68 ml./l. in solution—much less therefore than is available to land plants. This, indeed, is one of the chief reasons for many of the peculiarities of submerged aquatic plants which are discussed by Gessner (1955*b*). Similarly, we know little about saline nutrients. Butcher (1933) suggests that they are all obtained through the roots, and that this is why the substratum is of particular importance, but it is now known that aquatic vascular plants absorb the bulk of their nutrient supply through the surfaces of their leaves. It is also known that *Callitriche hamulata*, a common plant in running water, can take up ammonium and phosphate from sewage, but that at nitrogen concentrations above 9–10 mg./l. it does not increase its rate of uptake (Schwoerbel and Tillimans, 1964*a*, *b*). Presumably, therefore, this plant normally operates at lower concentrations, but it has often been established that channels rich in plants reduce the dissolved nitrate content of the water passing through them (e.g. Walter, 1961). It is also known that *Myriophyllum alterniflorum* and *Potamogeton praelongus* concentrate phosphate in their growing tips, and that the concentration is at a maximum during the season of active growth (Caines, 1965). Probably, therefore, phosphate demands fluctuate with the seasons, and they may be at their maximum before or at an early stage in the growing period. Here, once again, we need more investigation under running-water conditions.

There are only two distinct periods in the year for river plants in temperate climates; these are a fairly long active season extending from spring to autumn, and a winter period when some plants shed shoots and leaves and die back to persistent stolons and rhizomes. Many species, however, persist almost unchanged through the winter although their growth is reduced or stopped, and the actual amount of vegetation is reduced because of losses caused by flood damage and other attrition. Butcher (1933; Butcher *et al.*, 1931) distinguished the two groups of plants in English rivers which are listed overleaf.

One could easily add several species which have already been mentioned to both lists, e.g. *Juncus fluitans* to the first and *Nymphaea alba* to the second. Presumably day-length or temperature, alone or in combination, cause the dying down, as they do in land plants, but it should not be forgotten that submerged plants are not normally exposed to freezing temperatures.

Persisting through winter	Dying down for winter
Apium nodiflorum	*Alisma plantago*
Sium erectum	*Nuphar luteum*
Callitriche stagnalis	*Polygonum amphibium*
Elodea canadensis	*Sparganium erectum*
Hippuris vulgaris	*Potamogeton natans*
Myriophyllum spicatum	*P. perfoliatus*
Nasturtium officinale	*P. crispus*
Oenanthe fluviatilis	*P. lucens*
Potamogeton densus	*Sagittaria sagittifolia*
P. interruptus	
Ranunculus fluitans	
R. pseudofluitans	

A striking contrast between the vegetation of well-watered terrestrial habitats and that of river beds is the fact that, on land, the vegetation usually covers 100 per cent of the soil, while it is fairly rare for this to occur in rivers. Gessner (1955*b*) states that in Bavaria only 5 per cent of the area of small rivers and streams studied by one of his pupils had 100 per cent cover, and on 37 per cent of the areas investigated the percentage cover was below 10 per cent. In many places I believe that even smaller amounts would be found, but little study of this kind has been done. One gets the impression from the literature that a considerable amount of plant cover is normal, but this is because the botanists who have been interested in water plants (e.g. Butcher, 1933; Schmitz, 1961) have, somewhat naturally, concentrated on areas where there were reasonable numbers of plants to study.

One reason for our lack of information on this point is that the mapping of plants in water is tedious and difficult, but further use of the ingenious method of remote-controlled photography from a captive balloon, which was developed by Edwards and Brown (1960), could produce much data with relatively little effort.

The chief reason for the open nature of the plant community on river beds is the instability of the plants themselves. When a drifting shoot attaches itself to the bed it causes a local reduction of current, and this leads to deposition, on the downstream side, of material on to which the plant spreads. In time it thus builds up a bolster-like bank of sediment which is held together by its adventitious roots. This diverts the faster water towards the surface and the sides, with the result that the individual hummocks tend to be separated by channels of greater erosive power (Gessner, 1950; Kopecký, 1965). At the same time the increasing height of the mound produces extra turbulence, which ultimately cuts down into the sediment and undermines the whole structure, as is illustrated in Fig. v, 2.

One result of this process is that each mound tends to be formed by

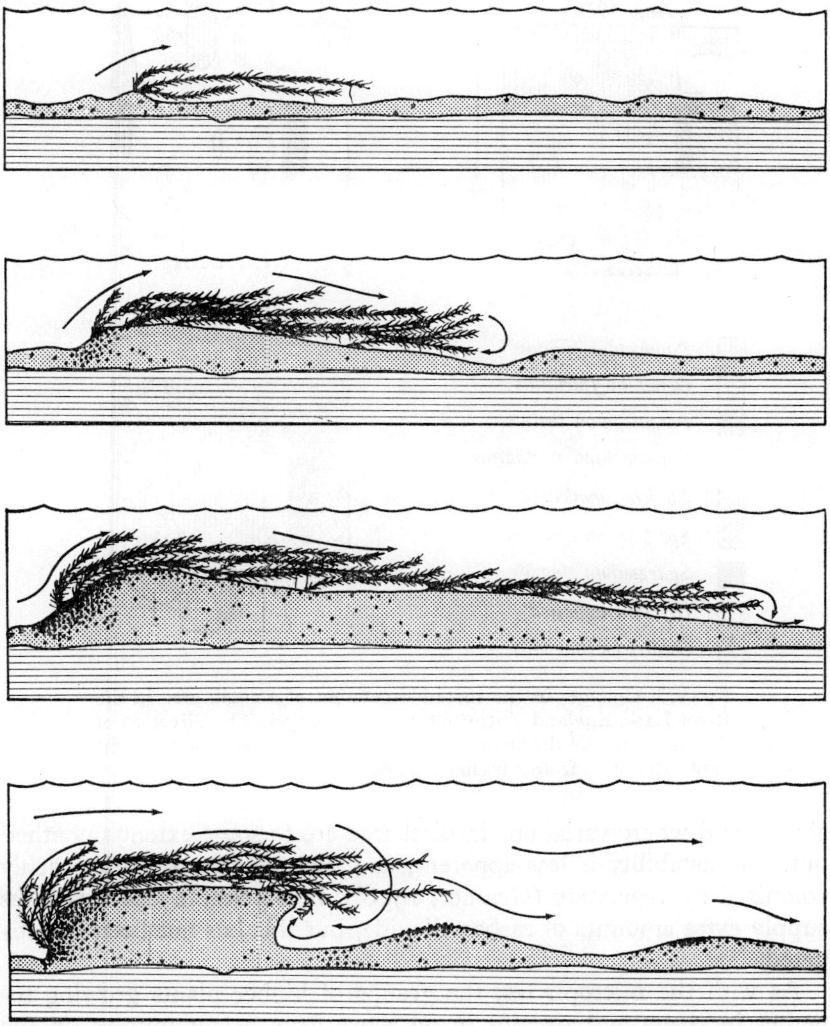

FIG. v, 2. Diagram showing the way in which a rooted plant builds up a bank of sediment, and how it is subsequently broken up by turbulent wash-out. From Minckley (1963) with permission.

only one species, indeed by only one clone, of plant, and to be quite separate from other mounds. Usually very few (less than ten) species are involved at any one place, and their distribution is random and changing. Two long-term studies have been made in Europe (Figs. v, 3 and v, 4), which illustrate very well the shifting nature of the vegetation. Downstream of lakes where there is less sediment to form unstable

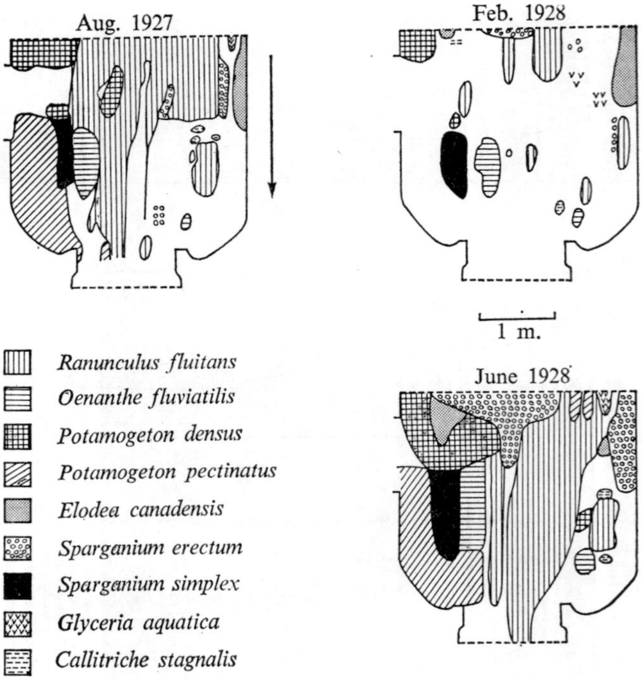

FIG. v, 3. Changes in the rooted vegetation of a small area in the River Lark, England, during a ten-month period. The direction of flow is shown by the arrow; an effluent outfall entered on the right bank. Redrawn from Butcher (1933).

shoals, and where variations in discharge are to some extent smoothed out, the instability is less apparent, and such reaches are often richly colonized by vegetation (Gessner, 1955*b*). It is possible also that lakes supply extra amounts of carbon dioxide, but this has not been investigated.

As with the microphytes, the groups of higher plants growing together in rivers and streams in no sense form a 'community' in the ecological sense, so it is really quite unwarranted to define associations in the way that has been attempted by some European authors. Each

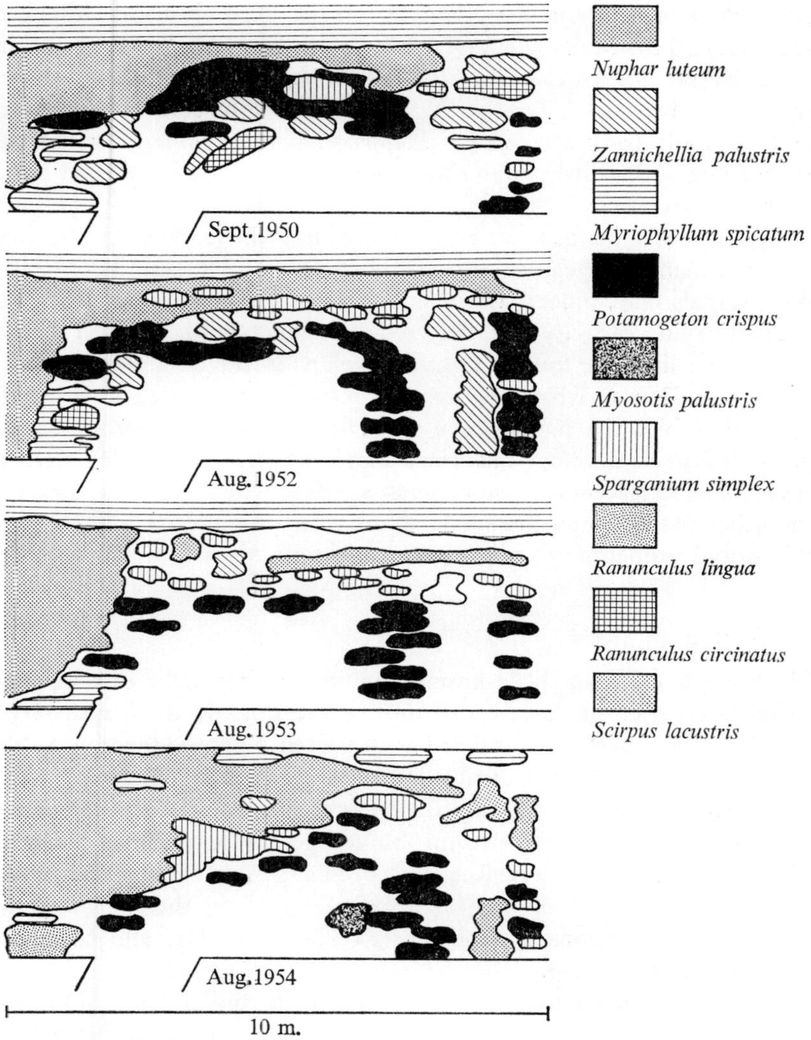

FIG. V, 4. Changes in the vegetation over a 10-m. reach of the River Acher, Germany, during three years. From Gessner (1955*b*) with permission.

species is a separate entity, and the fact that some tend to occur together is a matter of chance. Normally they appear not to interact to any great extent, and colonization of new channels is rapid and is completed in one step. For example, a channel cut from the River Schwarzwasser to some ponds 1·5 km. away was fully colonized by *Glyceria aquatica* and *Elodea canadensis* within a year (Walter, 1961). Just occasionally, however, a limited amount of succession has been observed. Minckley (1963) noted that, in Doe Run, *Myriophyllum heterophyllum* occasionally became established on beds of *Fissidens julianus*, which it then smothered with silt and replaced, and that *Nitella flexilis* colonized the downstream end of a silt bank formed by *Potamogeton diversifolius*. Such small successions are of little importance, and they seem to be all that occur in natural rivers.

On the other hand, successions resembling those in still water may occur in man-made canals, because of their even levels and controlled flow. Canals may, indeed, become so choked with vegetation that their water-carrying capacity is greatly reduced. Narayanayya (1928) reports that only a little *Vallisneria spiralis* develops in the irrigation canals of the Indian Deccan where the current is swift over gravel. This, therefore, resembles the natural non-silted habitat, but, where silt is deposited, *Nitella* develops and traps more. *Vallisneria* then invades the area, to be replaced by *Potamogeton* species, which finally crowd out the other plants when the silt is about 7 cm. deep. In a natural river this would probably not occur because of periodic wash-outs.

FLOATING PLANTS

Floating plants are of little importance in running water in temperate latitudes. Sometimes small amounts of the duckweeds *Lemna minor* and *L. trisulca* become entangled with rooted plants extending to the surface, and they persist during low water; but they do not maintain themselves for long, and they depend upon continual recolonization from sources further upstream. Sometimes also, in deep sluggish lake-outflows, there is a floating mat of emergent sedge or *Phragmites*, continuous with one on the lake, extending along the banks of the river. Such conditions are, however, really lacustrine, and so are of little interest to us here.

In the tropics, on the other hand, where turbid water inhibits the growth of submerged species, floating plants are often important. Many African rivers are lined by floating mats of the papyrus, *Cyperus papyrus*, beneath which oxygen may become depleted (Carter, undated), and Gessner (1959) describes floating meadows of the grass *Paspalum repens* along the Amazon. Another important tropical floating plant is the water hyacinth, *Eichhornia crassipes*, which lines the banks of sluggish rivers in tropical America (Gessner, 1955*b*; Kleerekoper, 1955). The metre-high rosette-like plants float because of the greatly expanded petioles; bunches of dark-coloured roots hang down 80 cm. into the water, and the plant can cover enormous areas. Unfortunately it has been introduced into many parts of the world, and has proved to be a pest, as it interferes with navigation and fishing, blocks channels, and de-oxygenates the underlying water. It can add 10–15 per cent to its wet weight per day, and it spreads very rapidly as it drifts with the wind and can be pushed along by boats (Gay, 1960). Quite recently it has begun to spread in the Nile, whence we shall doubtless hear more about it (Gay and Berry, 1959).

CHAPTER VI

Plankton

At the end of the nineteenth century there was much discussion as to whether or not running water supports a genuine planktonic community. Lauterborn (1893) noted that planktonic rotifers and other small organisms breed in sheltered bays in the Rhine, and that some of them are carried out by currents into the main stream, where he found representatives of several genera, together with diatoms, flagellates, *Stentor*, *Cyclops*, and *Bosmina*. Later, Zacharias (1898) studied several German rivers and recorded long lists of algae, as well as rotifers, cladocerans, and copepods, from the open water. He observed algal blooms in some rivers, e.g. of *Microcystis* in the River Schlei, and he concluded that reproduction must occur in running water and that it is, therefore, meaningful to speak of 'potamoplankton'. He remarked that in small lowland rivers the plankton resembles that of ponds in its general composition, and that in larger rivers it is more like that of lakes, being dominated by diatoms. However, nearly twenty years later Brehm (1911) was stressing that all the organisms found in river plankton occur also in still water, that many of them are of benthic origin, and that those that are true plankters must originate from still waters in the drainage basin. He concluded that the latter are brought in mainly during storms, and that they are only temporary inhabitants of running water. Since then the controversy has smouldered on, but this would seem to have been largely because the various workers were studying different types of river, and, perhaps not surprisingly, their findings differed somewhat.

Many studies in headwater streams have shown that the free water contains representatives of benthic algae, mostly diatoms, washed up from the stream bed. This has been found in North America (Blum, 1956; Clark, 1960; Cushing, 1964; Clifford, 1966*b*) and in many places in Europe, e.g. England (Butcher, 1932), Belgium (Symoens, 1957), Switzerland (Waser and Thomas, 1944), Germany (Wawrik, 1962), Austria (Stundl, 1950), Poland (Starmach, 1938), and Russia (Shadin,

1956). Several workers also report the occasional presence in small streams, mainly during the summer, of truly planktonic diatoms such as *Asterionella*, *Fragilaria*, and *Melosira*, of planktonic rotifers such as *Keratella*, *Notholca*, and *Brachionus*, and of a few cladocerans and copepods (Figs. VI, 1 and VI, 2). These are 'true' plankters in the sense that they are normal constituents of lake and pond plankton, and they are presumably strays from other bodies of water draining into the streams.

A plankton net fished in any small stream will, in addition to micro-organisms, catch specimens of larger, truly benthic, invertebrates, but

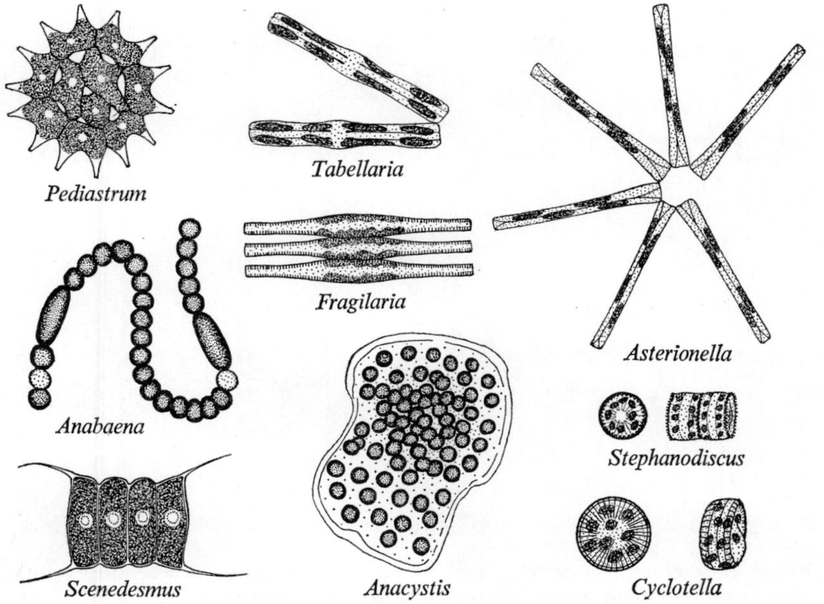

FIG. VI, 1. Common phytoplankters from running water, ×500. Some species of some of these genera are also benthic.

these cannot be considered as plankton; they constitute the drift fauna which we shall be considering in Chapter XIII.

Butcher (1932) found that the numbers of benthic algae in the open water reflect, very largely, the quantity on the stream bed, and there are clear indications from other studies of benthic algae that this is likely to be so. As the populations attached to solid objects increase, the layers become unstable and more liable to wash-out, and this depends, as we have seen, on the nature of the substratum. Spates dislodge algae more readily from loose stones than they do from solid rock or moss, because of abrasion when the stones are rolled. There is also evidence that the varying activity of the different types of benthic alga may influence their availability as potential plankters. Blum (1954*a*) found that the benthic diatom *Nitzschia palea* caused a dense plankton in a small polluted river in Michigan, but that this was much denser in the morning than during the afternoon or at night. Similarly, Müller-Haeckel (1966) observed a diurnal pulse of downstream drift in species of *Navicula*, *Thalassiosira*, *Synedra*, *Nitzschia*, and *Gyrosigma* at different seasons in small streams in Germany. She suggests that it is probably caused by the active production of oxygen and the division of the diatoms during the daylight

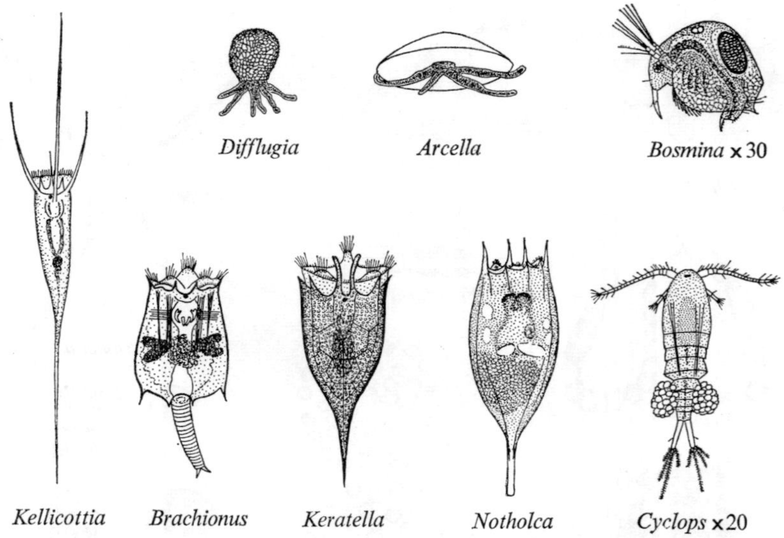

FIG. VI, 2. Common zooplankters from running water, ×70 except where shown otherwise. Some species of some of these genera are also benthic.

hours, which make them, respectively, lighter and more liable to be swept away. The species varied in the times at which their drift was maximal, and clearly quite small differences in physiology, and the ability to move, must greatly influence the presence and abundance of particular species in this peculiar 'drift' plankton of small streams. It is obvious also that the amount of drifting varies seasonally and to some extent with water level. After a long period of low water, when high populations of benthic algae have been built up, a spate will move a large quantity at first, but the numbers being dislodged will fall after a while even if the water level remains high. No direct correlation with discharge or rate of flow can therefore be expected.

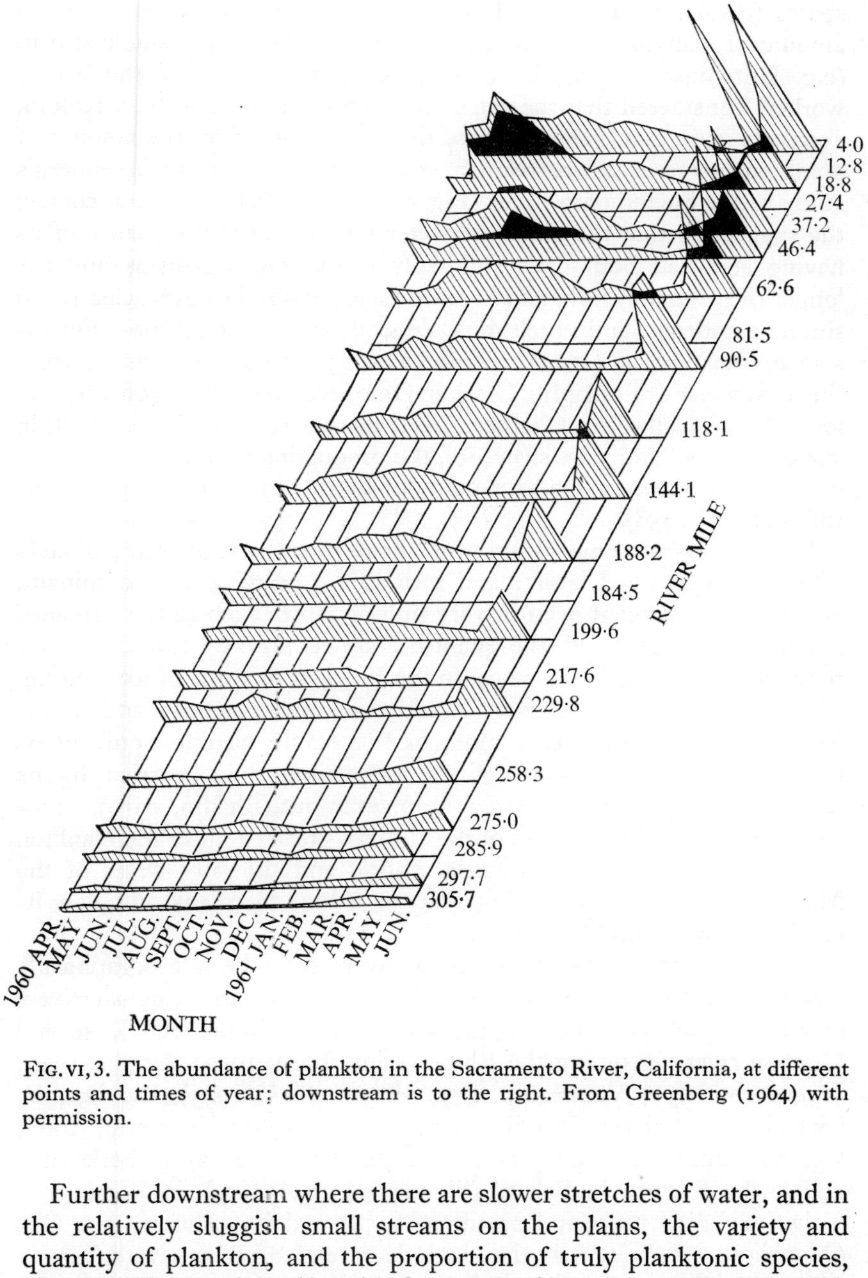

FIG. VI, 3. The abundance of plankton in the Sacramento River, California, at different points and times of year; downstream is to the right. From Greenberg (1964) with permission.

Further downstream where there are slower stretches of water, and in the relatively sluggish small streams on the plains, the variety and quantity of plankton, and the proportion of truly planktonic species, increase. The number of benthic forms may, however, remain quite high in canal-like reaches, and, as in shallower streams, it is increased by spates (Swale, 1964). Several studies have shown an increase in the amount of plankton collected at successive points down a single stream (e.g. Hutchinson, 1939; Sabaneeff, 1952) and many of the earlier workers considered that the age of the water is important (e.g. Kofoid, 1903). It is, in fact, generally true that in any one river the amount of plankton increases downstream, as is very nicely shown by Greenberg's study of the Sacramento River (Fig. VI, 3). Implicit in this is, of course, the suggestion that the older the water the greater is the chance of its having acquired both drift and truly planktonic organisms and the longer the time they have had to reproduce. This is, however, clearly too simple a concept, and much must depend on local conditions. For instance, Lackey *et al.* (1943) recorded over 250 species of organism from the open water of Four Mile Creek in Ohio, even though it is only 64 km. long. As we shall see in Chapter XXII, there is some indication that, in streams of pool and riffle structure, the production of minute organisms in the pools provides food for many of the invertebrates living on the riffles (Illies, 1958).

In larger rivers some plankton is always present, and true plankters often predominate. Diatoms and rotifers are nearly always dominant, and a few are present even in very muddy rivers such as the Missouri (Berner, 1951), although the quantities are reduced by turbidity. This reduction is shown, for instance, in the Volga at the time of ice melting, when the water becomes very muddy (Behning, 1929*a*), and in the Yennisey, which acquires a more or less stable plankton only in its lower reaches, where some of the suspended solids carried by its tempestuous upper waters are able to settle out (Shadin, 1956). Similarly, the turbid flood-waters of the autumn greatly reduce the plankton of the Ganges (Lakshminarayana, 1965), and the silty water of the Missouri was shown by Eddy (1934) to reduce the plankton below its confluence with the Mississippi.

Studies of the plankton of larger rivers have been geographically widespread. In Europe they include some Irish and English rivers (Southern and Gardiner, 1938; Rice, 1938; Swale, 1964), several German rivers including the Rhine (Lauterborn, 1902; Bennin, 1926; Jürgensen, 1935; Vonnegut, 1937; Waser *et al.*, 1943), the Danube (Schallgruber, 1944; Stundl, 1950; Liepolt, 1961; Enăceanu, 1964; Czernin-Chudenitz, 1966) and its tributary the Tisza (Uherkovich, 1965), the Brynica in Poland (Siemińska, 1956), and, in Russia, the Volga (Behning, 1929*a*; Romadińa, 1959) and its tributary the Oka (Monakov, 1964). In Asia they include several Siberian rivers (Greze, 1953; Shadin, 1956; Pirozhnikov and Shulga, 1957), the Ganges (Lakshminarayana, 1965), and the Yangtze (Lemmerman, 1907*b*). In

North America studies have been made on the Californian San Joaquin (Allen, 1921) and Sacramento (Greenberg, 1964), the Ohio (Eddy, 1934; Hartmann, 1965), several parts of the Mississippi and its tributaries (Wiebe, 1927; Reinhard, 1931; Eddy, 1931, 1934; Denham, 1938; Lackey, 1942; Brinley and Katzin, 1942; Berner, 1951), and many other rivers (Williams, 1964, 1966). In Africa most work has been done on the Nile (Abdin, 1948; Rzóska *et al.*, 1955; Klimovicz, 1961), and in South America there have been some studies on the Amazon and the Orinoco (Gessner, 1955*a*, *b*).

From these investigations it is clear that the phytoplankton is always more abundant than the zooplankton, and that diatoms are almost always dominant. The most frequently encountered truly planktonic genera are *Asterionella*, *Tabellaria*, *Fragilaria*, and the disc-shaped forms *Melosira*, *Cyclotella*, *Coscinodiscus*, and *Stephanodiscus* (Fig. VI, 1). Benthic algae which are frequently encountered in large numbers include the genera *Synedra*, *Nitzschia*, *Navicula*, *Diatoma*, and *Surirella*. During the summer, or in permanently warm rivers, these are joined by a variety of truly planktonic Chlorophyceae, such as *Scenedesmus*, *Ankistrodesmus*, and *Pediastrum*, and a variety of flagellates including *Cryptomonas*, *Mallomonas*, *Chlamydomonas*, *Trachelomonas*, *Euglena*, *Synura*, and *Ceratium*. Fairly frequently also Cyanophyta, such as *Gomphosphaeria*, *Aphanizomenon*, *Anacystis*, *Anabaena*, and *Lyngbya* (Fig. VI, 1), occur when the water is warm.

The zooplankton includes protozoans, mainly *Arcella* and *Difflugia*, and sometimes ciliates, which are especially numerous in polluted water; but it is nearly always dominated by truly planktonic rotifers, such as *Keratella*, *Synchaeta*, *Polyarthra*, *Asplanchna*, *Brachionus*, *Kellicottia*, *Trichocera*, *Triarthra*, *Notholca*, *Rattulus*, and *Euchlanis*. Crustaceans, which are so important in still-water plankton, are rarely numerous in the open water of rivers, and those that are found there usually belong to the genera *Cyclops* or *Bosmina* (Fig. VI, 2); *Alona*, *Chydorus*, and *Diaptomus* have, however, been reported fairly often. Other genera of Cladocera seem to be unusual, but several have been found in the Danube (Ponyi, 1962; Enăceanu, 1964), and similar lists have been published for the Siberian rivers Yennisey, Lena, and Ob (Pirozhnikov and Shulga, 1957) and for the Volga (Behning, 1929*a*; Romadina, 1959). It would seem likely, however, that all these, and the occasionally reported *Chaoborus* larvae, are strays from other habitats.

In some rivers certain less generally distributed organisms occur fairly frequently; such are the veliger larvae of the zebra-mussel, *Dreissena*, which is common in a few south-eastern European rivers, and the medusae of *Craspedacusta*, which have been found in rivers in several continents. *Dreissena* larvae occasionally occur in great numbers and are thus biologically significant organisms, but they can occur, of course, only where the adult mussels are present.

We mentioned earlier that plankters have been observed to reproduce themselves *in situ*, and this should not be forgotten when we consider the other varied sources of potamoplankton. Many observers agree that both plant and animal members of the community breed successfully under some conditions, and there is clear evidence of this in the production of blooms of algae. If the flow were entirely laminar, and the channel regular and canal-like, every individual would move steadily downstream; but, as Margalef (1960) has pointed out, the mere fact that the flow is turbulent and gives rise to eddies and backflow, must tend to keep individuals in a river for much longer than the time taken by the main mass of the water to pass. Indeed, Wawrik (1962) has suggested that *Cyclotella meneghiniana*, *Stephanodiscus hantzschii*, and the 'Danube *Melosira*', *M. distans alpigena*, may be truly native to the Danube, surviving and reproducing in eddies, bays, and backwaters, and recolonizing the main stream continuously. It has similarly been suggested that the cladoceran *Bosminopsis deitersi zernowi* and the rotifer *Brachionus bennini* are native to the Volga and its tributaries. But the former has also been found in still waters, and the latter, which is difficult to distinguish from related forms, probably also occurs away from the river (Shadin, 1956).

These are, however, exceptional cases, which do little to modify the general statement that there are no planktonic organisms which are confined to rivers. Most river plankters must originate in still or gently flowing areas, and be constantly, or frequently, supplied to the river, where they may or may not reproduce significantly. This can be seen to be so where peculiar types of water are drained, such as humic swamps which are rich in desmids. These unusual organisms then appear in the river plankton and clearly indicate its source; an example is Van Oye's (1926) study of the Ruki River in the Congo, which he found to contain large numbers of desmids derived from humic lentic waters. Similarly, lakes on the course of a river supply plankton to lower reaches (Galtsoff, 1924; Eddy, 1931) although, as we shall see, its composition soon becomes changed by selective elimination. This happens even when the river is sluggish and deep. Southern and Gardiner (1938) noted that the plankton of the Irish River Shannon resembled, in general composition, that of Lough Derg, into which it flows, but that crustaceans were relatively much scarcer in the river. The plankton there almost certainly originated in the rather similar Lough Ree, 60 km. above the sampling point, where it presumably contained a normal complement of crustaceans; it had then been changed by the river. Sometimes, especially in small rivers and streams, typical lentic plankters occur sporadically some distance downstream of a lake or dam (Pennak, 1943). The reasons for this are examined later in this chapter.

Much true plankton originates actually in the river, in quiet bays and side arms, where it can develop continuously; this was found long ago by Fritsch (1902, 1903, 1905) in the Thames. The composition of such plankton may not be the same as that of the nearby main stream, and this is for two reasons. The first is the selective elimination referred to above, and the second is that the river draws from many sources, some of which may lie far upstream and be very different from the nearby bays and side arms (Ackenheil, 1946). That dead arms, bays, etc., are, in fact, a source of plankton was shown nicely by Des Cilleuls (1928*b*) when he found that the number of planktonic algae increased sharply in the Loire when the river water rose in a spate, and that it then declined as the stock was used up, long before the water level fell. Here we have a clear example of wash-out. Floods, of course, also increase, at any rate at first, the number of small benthic organisms being carried in the water, and there is evidence that some of the truly planktonic forms may lodge for a while and reproduce on aquatic plants or on the river bed (Fritsch, 1905; Blum, 1960). Rice (1938), indeed, as a result of his study on the Thames, contends that this source of plankton may be more important than that from side arms. Probably much depends on local circumstances. Undoubtedly aquatic plants do shelter large numbers of small plants and animals, and these must constantly move out into the water. Probably many animals, such as non-planktonic rotifers and Chydoridae, are normally added to the drift plankton from such sources.

Lakes on the course of a river supply enormous amounts of plankton to lower reaches, as is shown very clearly by figures given by Cushing (1964) from his study of the Montreal River, Saskatchewan (Table VI, 1). From these it can be seen that the series of four lakes increased the concentration of plankton in the river below them at almost all times. The subsequent selective action of running water on the plankton from lakes has been much studied. Even active lake-plankters such as crustaceans can maintain themselves only against flows of up to millimetres per second. Thus if the rate of water movement through a lake exceeds speeds of this order of magnitude, and if the retention time in the lake is short, little plankton will develop. Brook and Woodward (1956) have shown, in a study of four small Scottish lochs, that the rate of water replacement greatly affects the composition of the plankton. In those with short retention times the plankton is reduced in variety and stability, and it consists primarily of species which themselves have a

Upper rapids	17 June	27 June	11 July	25 July	8 Aug.	22 Aug.	3 Sept.
Cyanophyta	1,400	3,500	3,300	3,000	600	1,100	700
Chlorophyceae	3,500	5,400	2,900	3,900	700	1,200	1,000
Bacillariophyceae	9,100	8,100	2,400	1,600	400	600	300
Rotifera	490	270	10	10	10	10	0
Entomostraca	10	0	0	0	0	0	0
Lower rapids	**16 June**	**26 June**	**10 July**	**28 July**	**10 Aug.**	**26 Aug.**	**4 Sept.**
Cyanophyta	8,100	10,900	1,800	3,800	900	4,200	2,600
Chlorophyceae	4,700	4,100	2,000	3,600	800	4,400	2,300
Bacillariophyceae	16,700	11,200	3,900	8,900	800	9,500	4,000
Rotifera	2,080	380	120	280	570	400	480
Entomostraca	110	30	30	10	20	0	10

TABLE VI, 1. The numbers of plankton organisms per litre in samples collected at various dates in 1960 from two stations on the Montreal River, Saskatchewan; based on 50-l. samples poured through bolting cloth of 68 meshes/cm. The upper rapids were above and the lower rapids just below an inter-connected series of four lakes on the river. Data from Cushing (1964).

rapid rate of replacement. This gives some indication of the way in which rivers act selectively on plankton.

It also seems that planktonic crustaceans, at any rate in their older stages, can to some extent avoid lake outflows, so that as the water leaves a lake it tends to be deprived of part of its plankton. Ruttner (1956) states that this applies more to alpine than to lowland lakes in Austria, but Brook and Woodward observed the phenomenon at fairly low altitudes in Scotland, and Chandler (1939) showed that this reduction in the concentration of plankton in water leaving the basin applied not only to crustaceans, but also to other organisms, in the outfalls of two lakes into the Huron River. This phenomenon of total reduction has not been reported from elsewhere and it needs further study, but it seems to be fairly well established that some reduction occurs, primarily among the Crustacea.

Once the water from a lake enters a river the true plankton begins to decrease rapidly. Reif (1939) found that, in the very swift stream arising from a glacial lake in Minnesota, half the plankton had disappeared in 180 m., and Hartman and Himes (1961) showed that 17 km. below Pymatuning reservoir in the Shenango River, Pennsylvania, the phytoplankton was down to 72·6 per cent of the amount leaving the lake, and that the proportion of diatoms had increased. Similar results are reported from Austria (Ruttner, 1956) and Switzerland (Jaag, 1938), and Table VI, 2 shows some of those obtained from below the Lunzer Untersee. It can be seen there that some of the diatoms, particularly

Cyclotella and *Asterionella*, were more persistent than were other algae, and that the rotifers were more persistent than the copepod nauplii. Ruttner also found that during periods of high water the planktonic organisms persisted for greater distances.

The elimination of lake plankton was investigated by Chandler (1937) in a number of rivers in Minnesota, by Reif (1939) and Beach (1960) in

Phytoplankton no./c.c	Distance from lake in metres			
	0	50	700	1,100
Chrysidalis sp.	*250*	*140*	*50*	*0*
% loss		44	80	100
Mallomonas akrolimes	*620*	*140*	*90*	*50*
% loss		78	86	92
Rhodomonas lacustris	*2,290*	*500*	*420*	*230*
% loss		78	82	90
Cryptomonas erosa	*12*	*6*	*5*	*3*
% loss		50	58	75
Cyclotella comensis	*70*	*100*	*110*	*50*
% loss		0	0	29
Synedra delicatissima	*5*	*4*	*2*	*1*
% loss		20	60	80
Asterionella formosa	*8*	*24*	*14*	*8*
% loss		0	0	0
Phytoplankton total	*3,255*	*914*	*691*	*342*
% loss		72	79	90
Zooplankton no./l.				
Rotifera	*15*	*10*	*6*	*7*
% loss		33	60	53
nauplii	*1·7*	*1·0*	*0*	*0*
% loss		4	100	100

TABLE VI, 2. The numbers of plankters collected at various distances below the Lunzer Untersee, Austria, on 4 February 1945. From Ruttner (1956). On 12 April 1945, during a spate, there were no reductions in the numbers over these distances.

Wisconsin, and by Uspenskii (1963) in the Moskva River. Chandler showed that it was not uniform, and that the losses were particularly great where there was aquatic vegetation. He noted, for instance, that 70 per cent was eliminated by 20 m. of thick plant growth in Maple River, and 60 per cent by 15 m. of weeds in Bessey Creek. *Najas* and *Chara* were particularly effective filters, and the plankters adhered to the periphyton growing on all solid objects. For these reasons the rate of elimination increased during the summer as the plants grew up, and it was particularly high in the autumn when fallen leaves accumulated and became coated with sessile algae. Beach observed the same role of plants and vegetable debris in the removal of planktonic rotifers in the Ocqueoc River, Michigan. Lake plankton also settles out along the stream banks and bed, where it becomes stuck, apparently mainly to sessile growths. Beach noted that the spines of *Keratella cochlearis* from the river tended to be shorter than those of specimens from the lakes on its course, and it seems likely that this is because of selective elimination of the specimens with longer spines; these would, one imagines, become more readily entangled.

The results of this kind of elimination are also specifically selective, and Reif was able to show that, in general, Chlorophyceae, desmids, and actively swimming crustaceans were eliminated more rapidly than were diatoms, Cyanophyta, and rotifers. This goes some way towards explaining the composition of river plankton, and it may be that those crustaceans which are frequently recorded from rivers are those which are less likely to be caught up. Perhaps, for example, the rounded shape and small size of *Bosmina* (Fig. VI, 2) favour its survival as compared with the larger and less compact genera which make up the rest of the planktonic Cladocera. There are also indications that many Cladocera are eliminated by silt. These small crustaceans feed more or less unselectively on materials caught by their complex legs, and thus, in turbulent water, they inevitably ingest some silt and small sand grains. This makes them heavy and they are no longer able to maintain their position in the water (Rylov, 1940). Possibly some difference in the feeding mechanism of the Bosminidae favours their survival in rivers. Silt also always reduces the numbers of rotifers, and they are generally less common in silty rivers than they are in clear ones (Williams, 1966).

It is obvious, therefore, that the plankton of a river may vary considerably along its length, both in space, and, in relation to spates, in time. This is shown particularly well by the Nile (Rzóska, 1961). The White Nile originates in Uganda with a rich plankton from Lake Victoria, but it soon loses much of it, particularly because of waterfalls. It acquires more from Lake Albert, which it again to a large extent loses in the 500 km. of papyrus swamp in the Sudd. Yet more is brought in by its tributaries, the Bahr el Gazal and the Sobat, but this declines a little before the river comes under the plankton-promoting influence of the Gebel Aulia dam, from which a rich plankton extends at least to the Egyptian frontier. Presumably there are further losses in the cataracts of Upper Egypt, but these have not been investigated. Here, in this great river, there is time enough for reproduction of those plankters which persist, especially as the temperatures are high. The water normally takes about thirty-five days to flow from Lake Victoria to the head of the dam lake, and then forty days on to Khartoum, so a complex process of selective loss, replenishment, and reproduction results in a great variety of plankton composition along the length of the river.

In temperate climates seasonal effects are important, and Lauterborn (1916, 1917, 1918) long ago observed that the distance travelled along the Rhine by typical plankters from Lake Zurich and Lake Constance varies with the seasons. Also, as in the Nile, reproduction of plankton undoubtedly occurs in large temperate rivers. Behning (1928) calculated that during the summer there is time for several generations of several species of *Brachionus* in the Volga before the headstream water reaches the Black Sea.

In addition to the fairly rapid selection and elimination of species from lakes, the plankton which persists comes under the influence of a number of factors in the river environment which continue to change its composition in various ways. These we can classify under the general headings of water movement, turbidity, temperature, and nutrients.

It was suggested above that falls and cataracts eliminate plankton, and this has been found to be so by several workers (e.g. Des Cilleuls, 1928*a*; Beling, 1929; Seeler, 1932; Fjerdingstad, 1950; Uspenskii, 1963; Hartman, 1965); but even quite steep falls do not eliminate all species. Beling found that some small crustaceans survived passage over the rapids in the rivers Dniepr and Bug, although their numbers were reduced, and Lauterborn (1910) showed that *Fragilaria*, *Cyclotella*, and even intact colonies of *Asterionella* pass unharmed over the 19-m. direct falls on the Rhine near Schaffhausen, although later workers have reported that these falls shatter many diatom colonies (Jaag, 1938). In general, therefore, very swift reaches tend to eliminate true plankters, and it is only in regions of slow flow that they multiply satisfactorily. This is demonstrated particularly well by exceptionally dry years when river discharges become very low and lead to a greater proliferation of plankton than is usual. Schorler (1907) noted that the Elbe became quite green during the very dry summer of 1904, and was particularly rich in species, and Des Cilleuls (1932) observed the same effect in the Loire in 1928 and 1929, although in that relatively shallow river it was clear that much of the reproduction occurred on the river bed, and that the algae then moved up into the water. More recently it has been shown that the diatom *Stephanodiscus hantzschii* reproduces and builds up fairly dense populations in the English River Lee at times of low water, when the temperature and light conditions are suitable. It can, however, do so only when the rate of flow is less than 140 cm./sec., and as soon as the velocity rises above 280 cm./sec. the numbers fall rapidly, presumably because the algae are swept away faster than they can reproduce (Swale, 1964).

It will be clear that the effect of floods on the density of the plankton varies with its origin and with the sampling site. Where, in a shallow stream or river, the plankton is primarily of the drift type, high water will sweep up more individuals, and during periods of rising water the numbers will go up, as has been observed, for example, in the English River Tees (Butcher *et al.*, 1937). The same will apply, in all rivers, to those components of the plankton which originate from other water bodies and are swept into the river. This has been demonstrated, for instance, in the River Sokoto, a tributary of the Niger in West Africa, where Rhizopoda (Holden and Green, 1960), the rotifer *Lecane* (Green, 1960), and the cladoceran *Alona* (Green, 1962) all form a greater proportion of the plankton during the rains than during the dry season. These are presumably all forms originating in other water bodies, although it is possible that some of the rhizopods are from the river bed. In this river the oxygen content falls significantly during high water, and it is known that some species of *Arcella* and *Difflugia*, the two most common genera there, produce gas vacuoles at low oxygen tensions. This would tend to bring them up into the plankton (Green, 1963); but no such explanation can be put forward for the increase in *Alona* and *Lecane*.

In the same way, at a given point below a lake, a spate will increase the quantity of true plankton, because, as we have seen, high water carries lake plankton further before it becomes entangled. On the other hand, where truly planktonic organisms are reproducing in the river in slow reaches along its course, high water will, unless it flushes out side arms and backwaters, reduce them by dilution and wash-out. This is very clear in the lower Nile where, throughout the length of Egypt, the river grows its own plankton without allochthonous additions. This builds up to its highest levels in June and July at the end of the low period and it consists very largely of *Pediastrum*, *Anacystis*, and *Anabaena*; the annual flood then washes most of it away (Abdin, 1948).

The effect of spates therefore varies from place to place and from species to species, and it is impossible to generalize. A very nice parallel situation exists in the coliform bacteria, which are either of soil or faecal origin. In a clean river their numbers per unit volume tend to rise with the water level, since more are washed in from the soil; in a river polluted by sewage, however, their concentration falls because of dilution (Taylor, 1941).

Turbidity is usually an accompaniment of floods, and we have seen earlier that floods are associated with changes in the plankton. Obviously, however, turbidity must also operate in its own right as it interferes with photosynthesis, and Sabaneeff (1956) suggests that it may also interfere with the feeding mechanisms of zooplankters, a point to which we have already alluded. Any river which is always turbid will, therefore, carry little true plankton, and those that are seasonally so, such as the Volga, referred to earlier, or the monsoon rivers of India (Roy, 1955), will carry reduced amounts of plankton at the flood season. It is at present not known how much of the reduction is due to high water and how much to the turbidity. An interesting side effect of the influence of turbidity is that cold weather, by freezing the soil and reducing the amount of erosion, can increase the amount of phytoplankton. This has been observed, for instance, in the Missouri and the Rio Grande in the United States (Williams, 1964).

Temperature is an obvious direct ecological factor, and its importance in river plankton is that as it rises it encourages the development of Chlorophyceae, Cyanophyta, and Cladocera in particular, all of which reach their maximum development in warmer waters. Some diatoms, however, show winter maxima; such are *Diatoma vulgare*, *D. elongatum*, and *Fragilaria* (Blum, 1960; Williams, 1964). Knöpp (1960) measured the potential oxygen production of samples of Rhine plankton brought into the laboratory at different seasons and kept under constant conditions. He concluded that, although light must be a very important factor in plankton development, the temperature must exceed about 12 °C. for really active development of most of the phytoplankton community. This is also indicated by the fact that, in temperate climates, blooms occur only in summertime, and that there is nearly always more plankton during the summer than in winter, as will be clear, for example in Fig. VI, 3. It is, however, difficult to separate the effect of temperature from that of light, as both always increase together. For instance, *Asterionella formosa* reaches a maximum in rivers in early summer, but it does so at the same time in the whole area extending from the Great Lakes to Florida (Williams, 1964). This perhaps indicates that for this species light is more important than temperature.

Blooms are also correlated with the availability of suitable nutrients, and they occur in many widely scattered rivers and are formed by a wide variety of algae (Schmitz, 1954). There is very good reason to suspect that they are often related to human addition of plant nutrients; Mokeeva (1964) observed a correlation between the nitrate content, derived from pollution of the water of the River Oka, and the numbers of Chlorococcales in the plankton. In a large river blooms may be quite localized; Claus (1961) observed this in the Danube near Vienna, but he could find no correlation between the numbers of algae at any one place and any physical or chemical factor that he measured. This is, perhaps, not surprising, since local concentrations of particular nutrients could be rapidly used up in producing a bloom, or, particularly with Cyanophyta which tend to float, be dispersed by turbulence from below, leaving the algal bloom floating on like a raft for a few days.

Human activity can also radically change potamoplankton in other ways, and the most important of these is the construction of impoundments. When dams reduce the rate of flow they encourage the development of true plankters, and, if the volume of impounded water is large enough not to be almost entirely replaced by spates, a permanent lake-community will be maintained (Einsele, 1960). Impoundments also allow suspended silt to settle out, and this encourages phytoplankton, as has been dramatically illustrated by the construction of a series of dams on the previously markedly muddy Missouri River (Neel *et al.*, 1963). There each impoundment tends to build up its own planktonic community, and these differ from place to place, probably because of local conditions. Each now tends to dominate the composition of the potamoplankton below it, which has also been increased. Similarly, the great Gebel Aulia dam on the Nile in the Sudan has an enormous influence on the plankton (Brook and Rzóska, 1954; Prowse and Talling, 1958). There the water is impounded each July, and by September it has backed up for a distance of about 500 km. This leads to a hundred-fold increase of the plankton, and its changing abundance and composition can be seen as the dam is approached from upstream. The water clears, algae, particularly *Melosira* and *Anabaena*, increase in abundance, and great numbers of Cladocera appear. In Russia also this effect of an increase of plankton towards dams, with increasing importance of crustaceans, has been reported (Shadin, 1962).

Such great concentrations of plankton inevitably change the situation further downstream, and the great photosynthetic activity in large impoundments has marked effects upon the chemistry of the water, raising the pH and oxygen content and reducing the hardness of the water. The influence of a large dam is therefore profound and it extends a long way downstream.

The fact that the potamoplankton is under the influence of many climatic factors such as light, temperature, and rainfall, inevitably results in more or less regular seasonal changes. These were noted by early workers in temperate latitudes in both the Old and the New Worlds (Kofoid, 1903, 1908; Lemmerman, 1907*a*).

In small streams and rapid shallow rivers the drift plankton usually reflects the seasonal cycle of the benthic microphytes (Schroeder, 1930;

Butcher, 1932), although it is subject to variations imposed upon it by the occurrence of spates. Clark (1960), however, found little seasonal variation in a canyon section of the Logan River, Utah, and it is possible that the local conditions there prevent much variation in the benthic flora.

Larger or more sluggish temperate rivers on both sides of the Atlantic have been found to show similar cycles. The major rivers studied include the Mississippi (Reinhard, 1931), the Ohio (Brinley and Katzin, 1942; Hartman, 1965), the Illinois (Kofoid, 1903), the Danube (Stundl, 1950; Schallgruber, 1944; Enăceanu, 1964; Czernin-Chudenitz, 1966) and its tributary the Tisza (Uherkovich, 1965), the Volga (Behning, 1929*a*, *b*; Romadina, 1959) and its tributary the Oka, some Siberian rivers (Shadin, 1956), and some German rivers (e.g. Bennin, 1926; Seeler, 1932). Generally speaking the plankton is at a minimum during the winter, and it then consists largely of a few species of diatom and sometimes a few rotifers (Williams, 1966). In spring these are joined by other diatoms, and Chlorophyceae, Cyanophyta, rotifers, and crustaceans occur in increasing numbers and variety as the summer progresses. Maximum numbers are almost always reported during the summer months, and in autumn the amount and the variety of the plankton again declines. Table VI, 3 shows some data from the Danube, and

	1958 April	May	June	July	August	September
Bacillariophyceae	512,930	1,257,340	9,650,000	28,625,000	69,000,000	83,091,800
Chlorophyceae	140	890	860,000	1,575,000	2,210,000	2,322,500
Cyanophyta	2,730	5,050	480,000	750,000	885,100	97,700
Rotifera	65,570	243,200	296,360	200,350	770,000	746,720
Copepoda	25,000	34,000	155,000	100,100	357,000	8,300
Cladocera	600	960	16,160	10,620	39,600	1,160

	1958 *cont.* October	November	December	1959 February	March
Bacillariophyceae	81,020,000	17,200,000	16,590,800	382,000	396,000
Chlorophyceae	4,280,000	180,000	92,400	—	—
Cyanophyta	50,180	12,500	4,200	—	—
Rotifera	52,970	42,270	760	550	2,970
Copepoda	660	—	—	300	2,430
Cladocera	—	—	—	—	—

TABLE VI, 3. The numbers of planktonic organisms found per cubic metre at different seasons in the River Danube in Rumania. Data from Enăceanu (1964).

Fig. VI, 3 shows the seasonal changes in total numbers at various points in the Sacramento River. Small numbers of Chlorophyceae persist throughout the year in some rivers, but Cyanophyta and Cladocera are nearly always confined to the warmer months.

In the River Oka it has been observed that, superimposed on the general pattern outlined above, there is a small peak in numbers of *Asterionella* in the first few days after freeze-up, and another of *Melosira* just before the break-up of the ice (Shadin, 1956). Possibly these are connected with complex interrelations of light, turbidity which is reduced by the freezing of the soil, and temperature, and such small winter peaks may be of more general occurrence than has so far been reported.

In the tropics the composition of the plankton varies between the wet and dry seasons (Van Oye, 1922; Holden and Green, 1960). This is presumably a direct result of the effects of wash-out and wash-in of organisms from various sources, and the build-up of native plankton during periods of low water. In the Nile the flood period, with its attendant turbidity, greatly reduces the plankton each year, but it recovers rapidly, presumably by local reproduction. The first organism to reappear in numbers is *Melosira*, followed by *Anabaena* (Rzóska *et al.*, 1955).

In lakes, and even in small ponds, a characteristic feature of the plankton is its uneven distribution with depth. This has been little studied in rivers, where it seems generally to be assumed that the turbulent flow must result in more or less uniform vertical mixing. In the Volga, however, it has been found that phytoplankton is less abundant near the surface than it is in the deeper water, and that cladocerans and copepods tend to cluster in the deeper water and even to make diurnal vertical journeys as they do in lakes. Rotifers were found to be evenly distributed except for *Brachionus pala* which is an especially powerful swimmer (Behning, 1928; Shadin, 1956). Similar observations are reported from the rivers Oka and Yennisey, and it may be that this is a general phenomenon in deep rivers which remains in need of further study.

In conclusion it can be stated that running water almost always contains free-floating micro-organisms, and that in large rivers or sluggish streams many of these are truly planktonic and may reproduce. Much, however, depends on local conditions, on seasons, and on the incidence of spates, and the whole community is inherently unstable and subject to constant change. In great contrast to the condition in lakes, crustaceans are always unimportant and the animals are represented mainly by rotifers; again in contrast to lakes, the phytoplankton is usually greatly in excess of the zooplankton. Often, however, as was clearly brought out by Williams's (1966) studies of plankton samples from a large number of rivers in North America, when the number of diatoms is high the rotifers are also common. This may be a simple trophic effect or it may be that similar conditions favour both types of organism.

The former, however, seems the most probable as, in some of the rivers he investigated, the ratio of rotifers to phytoplankton increased downstream, indicating that the diatoms had perhaps been heavily grazed. The trophic relationships in the potamoplankton deserve more study, and this could probably be undertaken in artificial experimental systems.

CHAPTER VII

The composition of the benthic invertebrate fauna

The invertebrates which live on, in, or near the substratum of running water include representatives of almost every taxonomical group that occurs in fresh water. There are indeed remarkably few freshwater groups which are not regularly represented in rivers and streams, and those that are absent are very specialized animals, such as the phyllopod crustaceans which are confined to temporary pools. A few others occur only in the open waters of lakes, but it may be noted here that even such specialized occupants of open waters as the cladoceran *Leptodora* and the nematoceran *Chaoborus* are occasionally collected in numbers from running water (Behning, 1928; Minckley, 1963).

In contrast, there are several groups which occur *only* in running water, and many more which reach their maximum development and diversity there. This is undoubtedly a consequence of the permanence of streams as compared with lakes and ponds. Many river systems have been in continuous existence from far back into geological time, whereas lakes persist for relatively short periods and give little opportunity for the development of a purely lacustrine fauna. That they are no less capable than rivers of producing evolutionary change and diversity is shown very well by ancient basins such as those of lakes Baikal (Kozhov, 1963), Ohrid (Stanković, 1960), and Nyasa (Fryer, 1959) which have evolved many endemic groups. But even these are evanescent phenomena; in time all lakes fill with deposits and disappear, and most of their fauna perishes, whereas river faunas have a much better chance of continuity and development. Although rivers may change and evolve they rarely disappear, so they are not evolutionary traps.

It is probably because of this that many of the 'living fossils' of fresh water which are not conspicuously adapted to some highly specialized way of life, such as temporary pools or underground waters, are primarily found in running water. These include various fishes and amphibians, which we shall consider in Chapters XV and XIX, and a variety of arthropods. Among the latter may be mentioned the syncaridan *Anaspides* of Tasmania, the archiperlarian stoneflies of Australasia and South America, and the widespread rhyacophilid caddis-flies (Fig. VII, 1).

Among the invertebrates, several whole families are entirely confined to quite fast-flowing waters. The best known of these are the dipteran Blepharoceridae, Simuliidae, and Deuterophlebiidae, but many other higher taxa reach their maximum development in streams and contain families which are entirely or almost confined to running water. Such

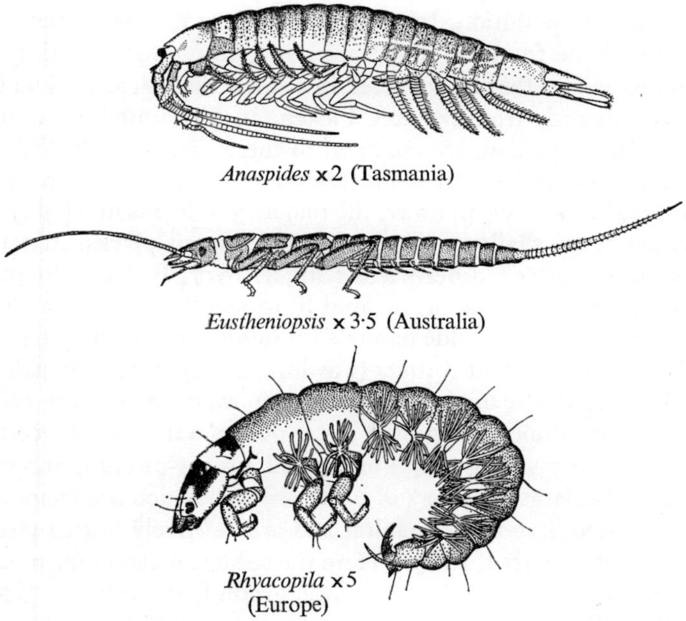

Anaspides x 2 (Tasmania)

Eustheniopsis x 3·5 (Australia)

Rhyacopila x 5 (Europe)

FIG. VII, 1. Survivors of ancient invertebrate stocks which now occur mainly in running water.

are the insect orders Ephemeroptera, Plecoptera, Trichoptera, and Megaloptera, the unionid Pelecypoda, and, in so far as fresh waters are concerned, the crustacean Decapoda and the prosobranch Gastropoda. Ross (1963) suggests that the early evolution of the Trichoptera occurred in cool streams, and he points out that the surviving primitive families are confined to this habitat. It also seems that the Ephemeroptera and Plecoptera have been primarily inhabitants of running water for a very long while. There are similarly many families in other orders of insects, e.g. the odonatan Agrionidae, Cordulegastridae, and Gomphidae, and the coleopteran Elminthidae, Psephenidae, and Hydraenidae, which occur primarily in streams.

It will be noted that almost all the above examples are insects, and this is a reflection of the very great importance of this class in running water. It is very unfortunate, therefore, that taxonomical studies of the young stages of insects have become increasingly unfashionable and neglected in recent times (Wiggins, 1966). Other groups, notably sponges, turbellarians, oligochaetes, leeches, crustaceans, mites, and molluscs are generally less important, although there may be many species or large biomasses in certain habitats.

Micro-organisms have been much less thoroughly studied except in connection with pollution. In areas of slow flow they seem not to differ from those to be found in still water, and Gray (1952) indicated that most of the ciliates that he collected in a small English stream had probably arisen directly from the soil. Picken (1937) pointed out that many types of creeping ciliate are confined to the surfaces of algal filaments, and, since those which can swim freely are round-bodied and rotating and thus liable to be swept away, the majority of forms inevitably occurs only in sheltered shallows. On stones in exposed places, it is usually possible to find attached peritrichs (Stiller, 1957; Badcock, 1949), and, even in mountain streams with rapid flow, sheltered areas, such as the insides of moss tufts, provide habitats for quite normal still-water micro-animals like the bdelloid rotifers (Pawłowski, 1956; Madaliński, 1961). Similarly, under stream beds, where the water percolates relatively slowly between stones and gravel, there occurs a microfauna consisting partly of the very young stages of normal stream-dwelling insects, but containing also large numbers of other animals which are characteristic of this so-called 'hyporheic' habitat and are relatively scarce elsewhere in streams (Schwoerbel, 1961*a*). Here the subterranean fauna meets that of the stream, and larger phreatic (underground) animals like *Niphargus* and *Bathynella* in Europe, and *Asellus stygius* or blind *Crangonyx* species in North America, intermingle with normal stream animals, and may wander up into the overlying stream bed (Minckley, 1961; Maitland, 1966). This peculiar habitat is discussed more fully in Chapter XXI (p. 406).

Under large rivers with coarse sandy beds a special microfauna occurs. This has been studied primarily in the River Oka, a tributary of the Volga. There Neiswestnowa-Shadina (1935) found a specialized but fairly abundant fauna of protozoans, mostly attached, but including the free-living *Loxodes*, several genera of rotifers, occasional Gastrotricha, Turbellaria of the family Otoplaniidae, a group which is known also from marine sands, and oligochaetes of the families Enchytraeidae (*Propappus*) and Aeolosomatidae. All these animals live in the interstitial spaces and attach their eggs to the sand grains, and they may occur in considerable numbers; 108 animals and 41 eggs were recorded per millilitre of sand. Near the shore, where the sand was finer and there was more light, there was a more varied fauna and a flora of diatoms, resembling the micro-associations of still water.

Undoubtedly, as we shall see, the major factor controlling the occurrence of animals is the substratum, and there is a fairly sharp distinction between the types of fauna found on hard and on soft stream beds. Clean, and therefore shifting, sand is a relatively empty zone which is inhabitated primarily by the microfauna mentioned above and by rather few larger animals. The fauna of the hard substrata has its own typical character, and it is here that most of the obviously specialized forms occur; that of the soft substrata is more generally shared with still waters, and it shows much more geographical variety.

One of the most striking features of the faunas of stony streams is their remarkable similarity the world over. Except in zoogeographically isolated areas, such as oceanic islands where many taxa may be altogether absent, the following groups of invertebrates almost invariably provide the major constituents of the fauna.

Parazoa	Spongillidae.
Cnidaria	Hydridea.
Tricladida	Planariidae.
Oligochaeta	Lumbriculidae (with Naididae and Tubificidae in sheltered places).
Gastropoda	Ancylidae and a variety of prosobranch families, e.g. Potamopyrgidae in Australia and Pleuroceridae in North America.
Pelecypoda	Unionidae (or the similar Mutelidae in the Southern Hemisphere) and Sphaeriidae (in sheltered places).
Peracarida	Asellidae, Gammaridae, and Talitridae (except in tropical Africa).
Eucarida	Atyidae and Palaemonidae (in warm regions), crayfishes (Astacidae or Parastacidae), or crabs (Potamonidae and Telphusidae). Crayfishes and crabs seem to be mutually exclusive in most parts of the world.
Plecoptera	All families.
Odonata	Libellulidae, Gomphidae, Cordulegastridae, and Agrionidae (in vegetation), with occasional Coenagrionidae (e.g. *Argia* in North America, and *Pyrrhosoma* and *Ischnura* in Europe).
Ephemeroptera	Many families, but especially Baetidae, Ecdyonuridae, some Leptophlebiidae (flattened genera of which replace Ecdyonuridae in South America and Australia),

	Tricorythidae (except in Europe), Siphlonuridae, Oligoneuriidae, Ephemerellidae, and Caenidae (in sheltered places).
Hemiptera	Corixidae (*Micronecta*) and *Aphelocheirus*.
Megaloptera	Corydalidae (except in Europe).
Trichoptera	Many families but especially Rhyacophilidae, the net-spinning Philopotamidae, Polycentropidae, Hydropsychidae and Stenopsychidae, Leptoceridae, Sericostomatidae, and Helicopsychidae.
Lepidoptera	Pyralidae (except in Europe, and they normally form a small part of the fauna).
Coleoptera	Elminthidae, Psephenidae (except in Europe), Gyrinidae, Hydraenidae, and some Haliplidae and Dytiscidae (Hydroporini).
Diptera	Chironomidae (especially Diamesinae, Orthocladiinae, and Tanytarsini), Simuliidae, Blepharoceridae, Deuterophlebiidae (in Northern Hemisphere mountains), Tipulidae (especially Limoniinae and Pediciinae), Rhagionidae (*Atherix*), and Muscidae (*Limnophora*).

The similarity of fauna extends also to the water-striders living on the surface. In most continents broken water is inhabited by *Rhagovelia*

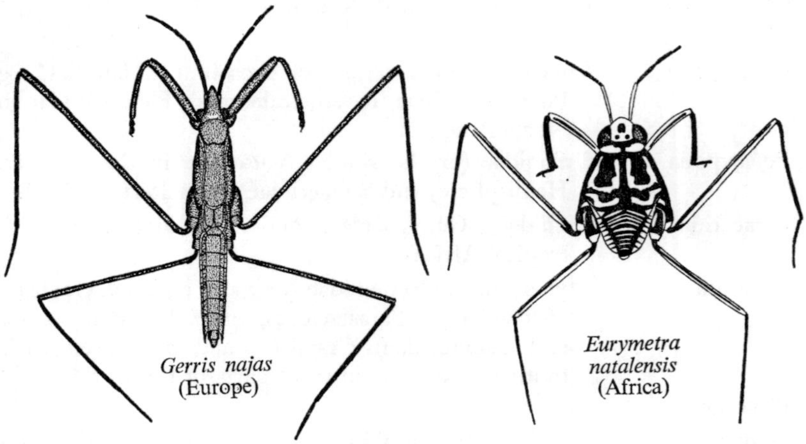

FIG. VII, 2. Two members of the Gerridae which occur primarily on running water, ×3.

although this genus is absent from Europe, and pools harbour large members of the Gerridae. In Europe the common species is *Gerris najas*; in North America it is *G. remigis* (Riley, 1921), in Africa species of *Eurymetra* (Hynes, 1955*a*), and in the neotropics *Cylindrostethus* (Hynes,

1948), and a notable feature about all these gerrids is the rarity of winged specimens (Fig. VII, 2).

The above list stresses the unity of the fauna of stony streams throughout the world, and it also indicates a certain impoverishment of the European fauna, which can undoubtedly be attributed to the Pleistocene glaciations (Thienemann, 1950). There are, however, other regional differences which somewhat alter the character of local faunas, and a few of these may be listed.

1. The very highly specialized bug *Aphelocheirus* is found only in the Old World.

2. The families of Plecoptera which are dominant in the Northern and the Southern Hemispheres are quite different, and it is clear that there has been much parallel evolution. For example, members of the Gripopterygidae in South America occupy the same niches as members of other families in North America and Europe (Fig. VII, 3). (Illies, 1961*a* and 1963). The same appears to apply in Australia (Hynes, 1964*b*).

3. In tropical Africa, probably for historical reasons, the Plecoptera are relatively unimportant members of the fauna, and they are represented by only one species, which, however, is common (Hynes, 1952*b*).

4. Similarly, tropical Africa lacks the Amphipoda and Isopoda which are important constituents of many stream faunas elsewhere. Both these groups and a wider range of Plecoptera occur, however, in South Africa where they form part of an 'old element' in the fauna, and probably represent relicts of a time when the climate of Africa was more temperate (Harrison, 1965*a*).

5. Freshwater crabs, many of which live in stony streams, are confined to warm regions, and the same applies to a lesser extent to decapod shrimps. Thus, for instance, Palaemonidae and Atyidae occur much more generally in warm climates, and it is in the warmer areas that they have penetrated into fast upland streams (Hora, 1923; O'Farrel, 1949; Johnson, 1957; Hynes and Williams, 1962).

6. In certain areas the presence of a peculiar local fauna is reflected in the stony rivers and streams of the region. This applies, for instance, to the River Angara which contains several elements from the endemic fauna of Lake Baikal which it drains (Greze, 1953). Similarly, some of the locally restricted Pontocaspian fauna of south-eastern Europe has penetrated far up rivers which drain into the Black and Caspian Seas (Shadin, 1956; Mordukhay-Boltovskoi, 1964). Thus, *Metamysis*, *Dreissena*, and *Corophium* occur far up the Volga, and they form a distinctly regional element in the fauna.

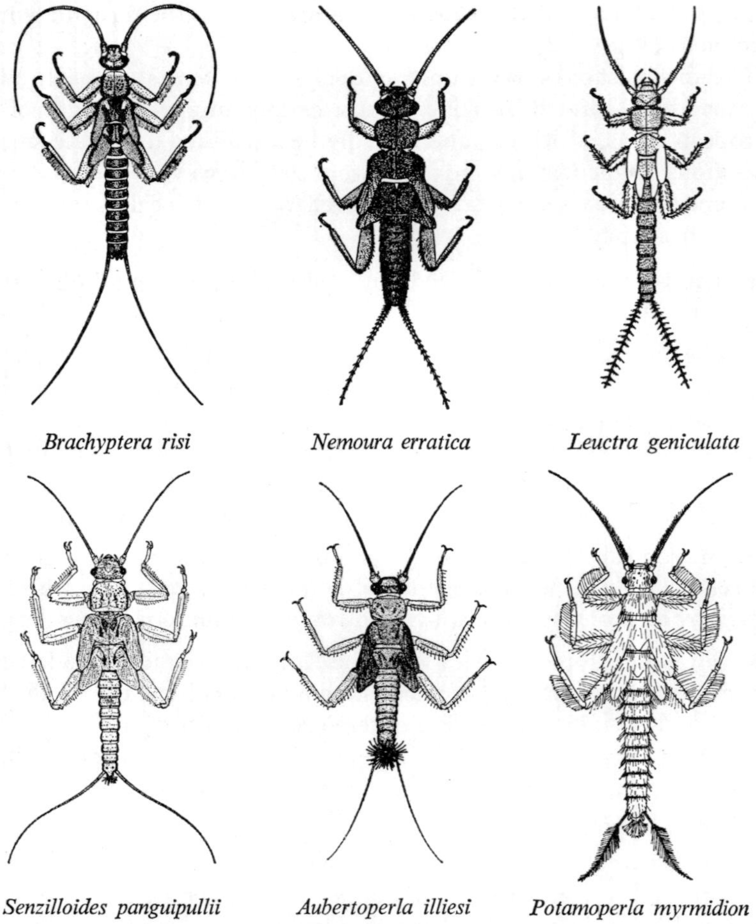

FIG. VII, 3. Comparison between the nymphs of European stoneflies (top row) from three families and their morphological equivalents among the Gripopterygidae of South America (bottom row). From Hynes (1941) and Illies (1963) with permission.

7. In some areas much local evolution seems to have occurred, and as a result some groups have become relatively more varied and abundant than they are elsewhere. This appears to apply to crayfish of the genus *Orconectes* and unionid mussels in the basins of the Mississippi and St. Lawrence, and, although they are fishes, we can include the darters, Percidae, Etheostominae, in this complex. The crayfish and darters are mostly species of hard bottoms, and they form an important part of the stony-bed fauna in eastern North America. Most of the mussels occur

in silty or sandy habitats, but the presence of a wide variety has led to the movement of several species, e.g. of *Lampsilis*, *Lasmigona*, *Anodontoides*, *Strophitus*, and *Alasmidonta*, on to harder bottoms (Clark, 1944). Stony streams in eastern North America tend therefore to be richer in species of crayfish, clams, and small fishes than corresponding streams elsewhere.

The foregoing are, however, quite minor differences which only slightly modify the general statement that the fauna of hard substrata in streams and rivers is remarkably uniform all over the world. Even faunistically isolated areas, like Table Mountain in South Africa (Harrison and Agnew, 1962) and New Zealand (Phillips, 1929; Hirsch, 1958), have faunas which more or less conform to the world pattern. Only on oceanic islands such as Hawaii and other Pacific islands are major portions of the normal fauna missing (Needham and Welsh, 1953; Laird, 1956).

This uniformity is much less evident in the fauna of the softer substrata in the larger rivers. There are still many genera of invertebrates such as *Limnaea*, *Chironomus*, *Tubifex*, and *Limnodrilus* which can be found in rivers in most continents, but the less-rigorous habitat of areas of slower current which allows less-specialized species to occur also permits the local character of the fauna to be dominant. It is easier to recognize a representative collection of animals as being, say, African or North American, than to be sure that it came from a river rather than from a lake. This is certainly not true of a collection from the hard bed of a smaller stream.

It is therefore difficult to generalize, but characteristic creatures of soft riverine substrata are Tubificidae, Chironomidae (Chironominae), burrowing mayflies (Ephemeridae, Potomanthidae, Polymitarcidae) Prosobranchia, Unionidae, and Sphaeriidae; and when plants are present a great variety of organisms may be added. Superimposed on this is the local freshwater fauna, so that, for example, there are great numbers of mussels in the Mississippi basin, crabs in Central Africa, and Caspian crustaceans in the Volga, as well as a wide range of regionally local worms, mayflies, dragonflies, bugs, beetles, and snails, many of which are also to be found in still waters. There are few river 'specialists' in the lower reaches, although examples are provided by some of the large mayflies which make their U-shaped burrows in clay banks. These include *Tortopus* of the Savannah River, Georgia (Scott *et al.*, 1959) and *Polymitarcys* and *Palingenia* in the Volga (Behning, 1928). Similar specialists which are confined to the slower parts of rivers are the sand burrowers which are adapted to an environment of shifting sand and which are discussed in Chapter IX (p. 175). By and large, however, the animals of soft river beds occur both in still and in running water, and they are only facultatively riverine.

A last point in this general consideration of the invertebrate fauna of running water is one which has already been touched upon with reference to decapod crustaceans. In warm climates there is a greater tendency for animals of marine origin to penetrate into fresh waters; this has been discussed by Thienemann (1950) where further details can be found. Clearly this phenomenon increases the diversity of the fauna of the lower reaches of tropical and subtropical rivers and increases still further the local character of their faunas. It is probably more important among fishes than it is among invertebrates, but several groups of crustaceans and molluscs are involved, and species of marine origin can become important in the fauna of running water in the relatively empty environment of oceanic islands. For example, the torrent midge, *Telmatogeton hirtus*, occurs in vast numbers in the rocky streams of Hawaii (Needham and Welsh, 1953). This species is a member of the Clunioninae which is the only primarily marine group of insects, and presumably the lack of normal competitors has allowed it to reoccupy what may have been the original habitat of its ancestors. The volcanic origin of Hawaii makes it unlikely that it is a relict from earlier times which has persisted in its ancestral habitat.

CHAPTER VIII

Anatomical and behavioural adaptations of benthic invertebrates

Current is the most significant characteristic of running water, and it is in their adaptations to constantly flowing water that many stream animals differ from their still-water relatives.

We have seen in Chapter I that where current speeds do not normally exceed about 40 cm./sec. a stream bed is likely to be sand, or even silt at still lower maximum currents of about 20 cm./sec. Where currents frequently exceed about 50 cm./sec. on steep slopes the bed is likely to be stony, and the animals which live there must be able to maintain their position. We have also seen that it is unusual for currents to exceed 200–300 cm./sec. except for short periods and in special places, and that close to the substratum there is a boundary layer 2 or 3 mm. thick in which the current drops rapidly to zero. There are also many regions of dead water on a rough bottom.

It is clear, therefore, that the actual currents experienced by animals on stream beds are very much less than those which can be measured by current meters. This is an important point because, in turbulent flow, the force exerted by the water increases as the square of the current, and, theoretically, at a current of 300 cm./sec. the force would reach 0·5 g./sq. mm. and would be impossible for the small animal to resist (Bournaud, 1963). In order, therefore, to maintain themselves in rapid water benthic animals must be prepared to live in the boundary layer, with suitable anatomical adaptations, or their behaviour must be such as to keep them out of the current altogether in areas of dead water.

At first sight it would appear that truly sessile animals would be particularly successful, but, although sponges and bryozoans do occur, they are rarely abundant and are usually found under stones rather than on them. One of the reasons for this may be that their inability to move prevents their establishing themselves on the tops of stones or leaving them when they become subject to scour or exposure by low water. But current itself may also be a factor. Jewell (1935) states that strong currents do not favour *Ephydatia* and *Spongilla*, although weak ones do, and there are some North American sponges, e.g. *Heteromyenia argyrosperma* and *Carterius tubisperma*, which grow well only in flowing water. Only rarely, however, are sessile animals important in terms of biomass, although they occur frequently, especially in larger rivers (Behning, 1928; Pagast, 1934).

We will consider the various adaptations of mobile stream-dwelling invertebrates under a number of headings, but it should be borne in mind that these are, to a considerable extent, artificial, and that any one species or higher taxon may display several types of adaptation. Also, some of the phenomena to be discussed are not strictly adaptations; they are changes brought about by life in this peculiar environment—changes such as the loss of respiratory structures which are not needed in well-aerated water. We can see a parallel here to the loss of eyes and pigment in subterranean species, which, while being characteristic of the environment, can hardly be described as adaptations to it.

MORPHOLOGICAL ADAPTATIONS

Steinmann (1907, 1908) in his classical studies of the animals of mountain streams was the first to codify a number of anatomical characteristics which he considered to be adaptations to life in rapid waters. His concepts have been much criticized and modified by later workers, notably Popovici-Baznosanu (1928), Hubault (1927), and Nielsen (1951*a*), but many of his main points seem acceptable and they will serve as headings for discussion.

Flattening of the body. Many animals that live on stones are flattened, and this has, of course, the effect of allowing them to live in the relatively still boundary layer. The most striking of these are the water-pennies, the larvae of Psephenidae (Fig. VIII, 9), but there are also examples among the mayflies, e.g. *Epeorus* and *Rhithrogena* (Fig. VIII, 1) (Hora, 1930; Dodds and Hisaw, 1924*a*). Similarly, some gomphid and libellulid dragonflies, e.g. *Ictinus* and *Zygonyx* in India are flattened and limpet-like (Hora, 1928), as are the stony cases of some stream-dwelling Trichoptera, e.g. *Tremma* of Europe (Wesenberg-Lund, 1943) and the sub-family Glossosomatinae of the Rhyacophilidae. But flattening is by no means universal among the animals which live exposed on stones, and it is obvious that very small animals can keep down in the boundary layer without being flattened to any marked degree. Also, of course, flattening can serve other functions and occur for other reasons even among stream-dwellers, and this is why some of Steinmann's critics

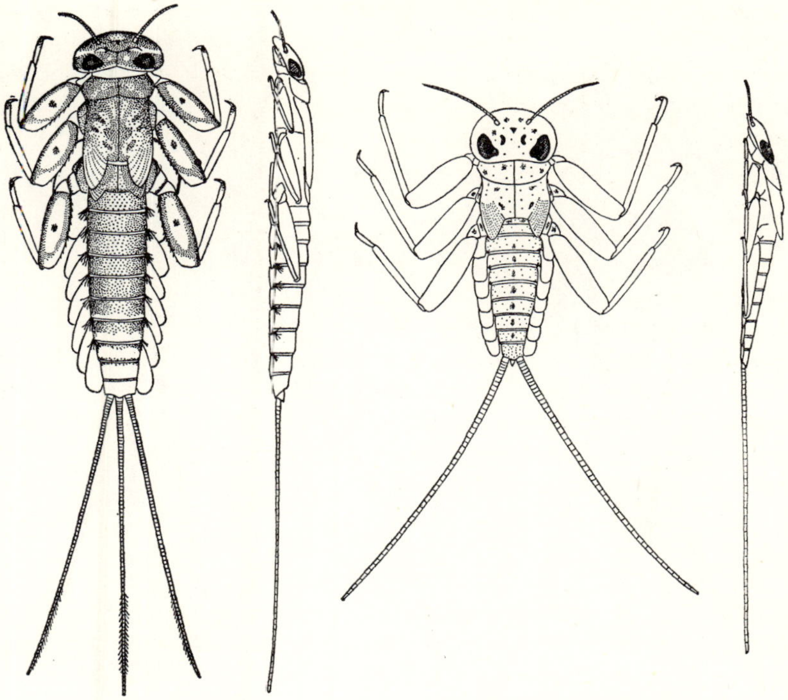

Rhithrogena (Europe) *Epeorus* (N. America)

FIG. VIII, 1. Two flattened mayflies of the family Ecdyonuridae which live exposed on stones. ×5.

expressed doubt about the validity of his suggestion that flattening is an adaptation to current. Some animals that live on sand are very flat (p. 178), and there are many inhabitants of stony streams which are flattened and yet do not live exposed lives on stone surfaces in fast water. These include, among Ephemeroptera, most of the Ecdyonuridae, e.g. *Ecdyonurus*, *Afronurus*, *Heptagenia*, and *Stenonema* in various continents, various Leptophlebiidae in Australia and South America (Figs. VIII, 2 and IX, 9) (Harker, 1954; Illies, 1961*a*), and the peculiar limpet-like *Prosopistoma* of the Old World (Fig. VIII, 3) (Gillies, 1954). The large stoneflies are also flattened, as are some dragonflies, e.g. the libellulid, *Brachythemis* (Fig. VIII, 3) which is common in Central African streams.

These flattened animals are all fairly large, and they live *under* the stones, so it is obvious that their shape allows them to avoid the current altogether. Other flat animals, such as the triclad turbellarians and the glossiphonid leeches, belong to groups which are always flat, and they

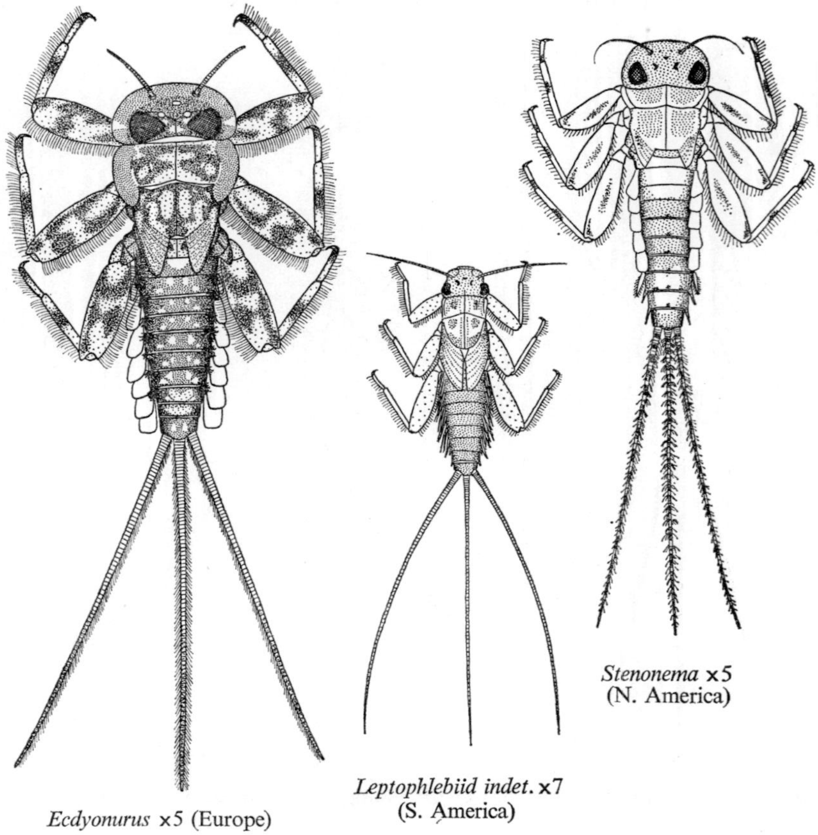

FIG. VIII, 2. Flattened mayflies which live in sheltered places under stones. The flattened Leptophlebiidae fill the niche of the Ecdyonuridae in the Southern Hemisphere.

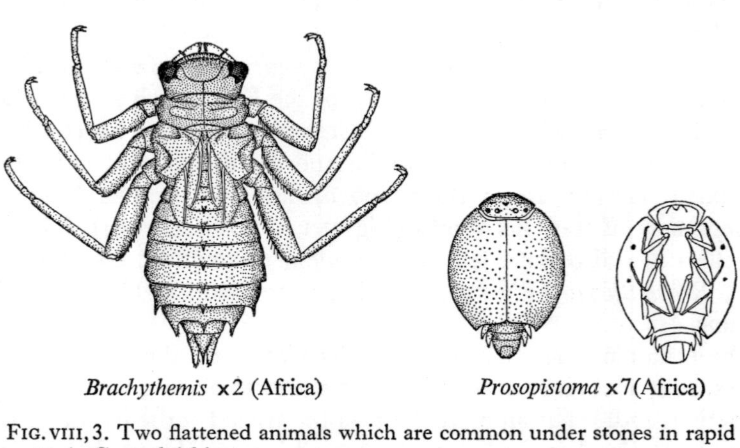

FIG. VIII, 3. Two flattened animals which are common under stones in rapid water in Central Africa.

use their shape to conceal themselves under stones in running water just as they do in other habitats (Fig. VIII, 4).

An interesting additional example is the African tricorythid mayfly *Dicercomyzon*, which must be among the flattest of all mayfly nymphs (Fig. VIII, 10) (Ventner, 1961). It occurs in streams but not in fast water, although it has a well-developed friction-pad (p. 132) and is difficult to dislodge. Its normal habitat is on emergent vegetation and stones near the bank out of the force of the current, and it may be that here the flattening and firm attachment offer protection against predation rather

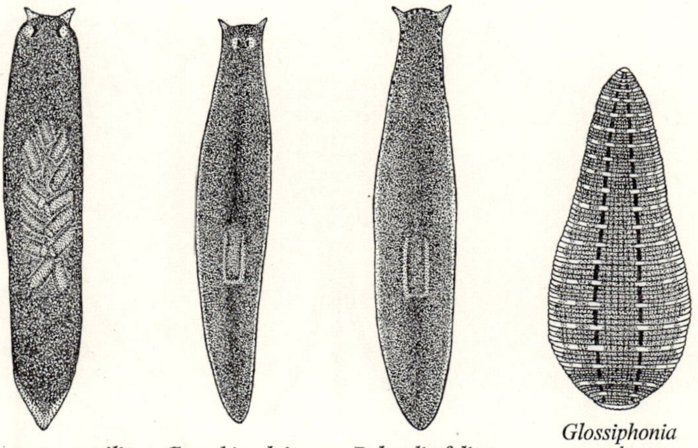

Phagocata gracilis (N. America) — *Crenobia alpina* (Europe) — *Polycelis felina* (Europe) — *Glossiphonia complanata* (Europe)

FIG. VIII, 4. Flatworms and a leech which occur under stones in rapid water.

than against current. Most fishes and other carnivores would find the nymphs difficult to detach and swallow, and possibly the same applies, to a lesser extent as they have no friction-pads, to the European species of *Ecdyonurus* which, at least when well-grown, move towards the banks where they can often be seen lurking on the tops of stones in slack water (Rawlinson, 1939). Flattening is, therefore, by no means a prerogative of animals from swift water, and even among them it occurs for widely differing reasons.

Streamlining. It is well known that a fusiform body offers least resistance to fluids, and an ideally streamlined object is widest at about 36 per cent along its length and tapers to a point at the rear (Bournaud, 1963). This shape is closely approached by many fishes, and one might expect it to be common among stream-dwelling invertebrates. Actually it is rather rare. Almost the only animals that display it to perfection are the baetid mayflies, particularly the genera *Baetis* (Fig. VIII, 5) and *Centroptilum*, and, perhaps significantly, these are usually among the most numerous animals on stony substrata. They are nearly always more abundant than other mayflies, including those which are flattened and

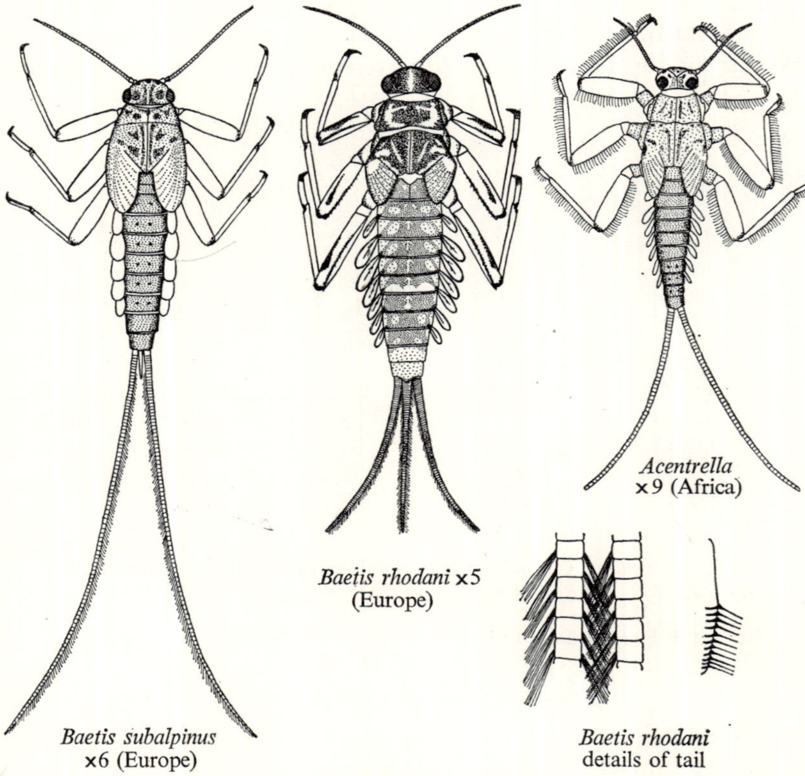

FIG. VIII, 5. Streamlined nymphs of the family Baetidae and details of the tail fan of *Baetis*.

provided with friction discs and thus apparently more highly adapted to life in places exposed to rapid currents (Verrier, 1948).

Baetid nymphs stand high on their legs with their abdomens hanging free in the water, and it can be seen that their tails swing from side to side in the small eddies, and keep the body facing directly into the ever-changing local current (Einsele, 1960). They are also able to swim quite fast in short bursts; the three tails are spread out, and the rows of long hairs which border both sides of the central filament and the inner sides of the outer filaments (cerci) make an efficient fin, which is moved rapidly up and down (Fig. VIII, 5). At rest the tails are relaxed, and they are pushed together by the current so that the bristles are folded (Hubault, 1927). If the current is very swift the nymphs pull their bodies down so that their ventral sides touch the substratum, and in this way they, as it were, duck down into the boundary layer. Some species of *Acentrella* which inhabit fast African streams, where tremendous spates

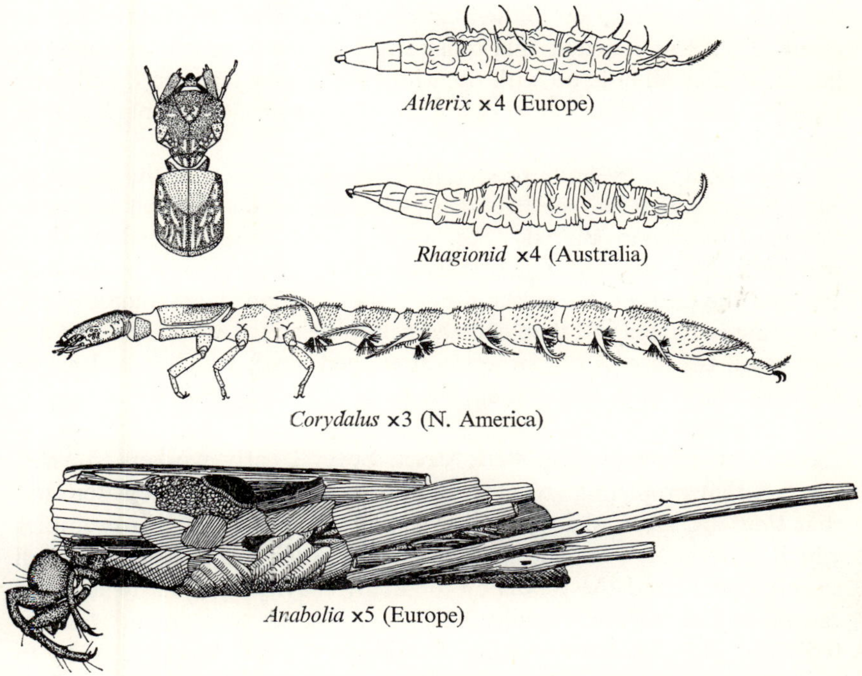

FIG. VIII, 6. A caddis-worm with a long tail to its case and several animals which have lateral projections on their bodies although they live in rapid water.

occur regularly, seem to represent a compromise between the streamlined and flattened forms. They are slightly flattened and have large laterally placed legs, and so are more readily able to crouch down and hold on (Fig. VIII, 5). They seem, however, always to occur with normally streamlined baetids, so the advantage cannot be very great.

Any animal which is limpet-shaped is, of course, to some extent streamlined on its free side, and among the flattened mayflies the streamlining may be quite good, in that they are tapered from front to rear. It is noticeable, however, that this streamlining of limpet-like forms is almost confined to the flattened mayflies, which, like the baetids, have long tails acting as vanes to turn their heads into the current. It is probable that the long cerci of the nymphs of Plecoptera perform the same function, and that the same applies to the long sticks which some caddis-worms, most notably *Anabolia* and *Pycnopsyche*, incorporate into their cases (Nielsen, 1942; Cummins, 1964). *Anabolia* actually lives in quite slow water at the sides of streams, and it is a large and clumsy animal which clambers about on stones and vegetable debris (Fig. VIII, 6). Presumably it is less liable to be swept away by sudden eddy currents if it is automatically swung head-on to the ever-changing directions of water movement. To a lesser extent this must apply to many cased trichopteran larvae, as they all hold on with their legs at the very front end, and their tapering cases are thus free to swing into the line of the current (Fig. VIII, 16).

A less obvious example of streamlining can be seen in the larvae of Simuliidae, which like the Baetidae are among the commonest and most universal of stream animals (Fig. VIII, 15). They are attached by their rear ends, by a method which is discussed below, and their bodies trail out into the current. From the hind to the fore end the diameter of the body first increases and then decreases, but the actual streamlining is not very good because of the large blunt-topped head, which is at least as wide as, and usually wider than, the body.

Reduction of projecting structures. Several early workers stressed the fact that projecting structures increase resistance to the water, and that they are reduced in stream-dwelling animals. The two examples which were cited were the gills of *Baetis*, which are reduced to small simple plates, and the tendency for many mayfly nymphs to lose the central one of their three tails. Neither seems very convincing. Admittedly the gills of stream-dwelling Baetidae are small, but so they are in some fast-swimming, still-water species (Dodds and Hisaw, 1924*a*). And while it is true that many baetid species have short or very short middle filaments, as in the European *Baetis alpinus* and *B. subalpinus* (Fig. VIII, 5), the North American *B. intermedius*, and the Japanese *Baetiella*, other species living in the same habitats retain three. It may be that the loss or shortening of one tail does improve the streamlining; Berner (1950) reports that in Florida a two-tailed species of *Pseudocloeon* lives on the tops of rocks where a three-tailed *Baetis* species lives on the sheltered downstream faces. But several of the flattened genera have only two tails, and these include *Dicercomyzon*, which we have seen does not live in fast water (Fig. VIII, 10), and the ecdyonurid *Bleptus* of Japan which lives in similar places (Uéno, 1931). This matter of two tails versus three therefore remains unproven, and the reduction of the gills may, as we shall see, be concerned more with respiratory physiology than with adaptation to rapid currents.

Obviously no animal with large and numerous projections is likely to be able to exist in exposed situations in the current, but this does not mean that such creatures have no place in streams. The shaggy maggots of Rhagionidae, e.g. *Atherix* (Fig. VIII, 6), are frequent inhabitants of running water in most of the continents; but they live under stones, and possibly their projections aid them to keep station in the slight currents

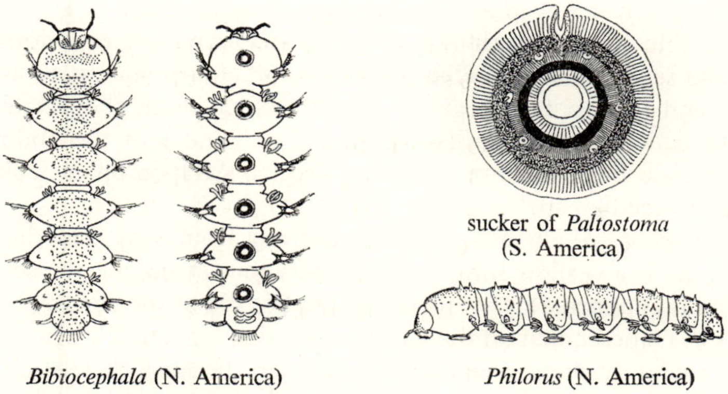

FIG. VIII, 7. Larvae of Blepharoceridae (×4) including the very spiny *Philorus*, and details of the structure of the sucker.

there. The same probably applies to the large lateral projections of the larvae of Corydalidae (Fig. VIII, 6) and Gyrinidae.

Hora (1930) has pointed out that under some conditions a roughened surface may actually reduce the drag of the water, by transferring the shear from the surface of contact between the animal and the water to shear between two layers of water. He notes that many blepharocerid larvae, all of which live in very swift water, are elaborately sculptured or spiny (Fig. VIII, 7), and the surfaces of many stream insects are far from being smooth. Undoubtedly considerations which are important in, say, aircraft or ships cannot be applied directly to these small size-levels.

Suckers. Hydraulic suckers are not uncommon in the animal kingdom; they are a very efficient means of attachment to smooth surfaces, but they are nevertheless rather rare in stream animals. They occur, of course, on leeches, where they are characteristic of the group, and some leeches do live in streams, but they are rarely found in exposed situations. Although they can, under experimental conditions, withstand very strong currents of water, they do not, in contrast to many others creatures living in rapid currents, move upstream against fast water (Dorier and Vaillant, 1954).

Elsewhere in streams, apart from those on parasites or epizoites, such as the Branchiobdellidae and the Temnocephalea, which are stream-dwellers only by proxy, true hydraulic suckers are found among invertebrates only in the Blepharoceridae. The larvae of this highly specialized and almost world-wide family of Nematocera are confined to smooth rocks in very fast water, and the whole segmentation of the body is altered in such a way that it is grouped around a row of six ventral suckers. Thus the head, thorax, and first abdominal segment form the first false segment, then follow four abdominal segments each with a sucker, and the last false segment consists of the sixth abdominal segment to which is more or less fused the remainder of the abdomen. Each division bears a ventro-laterally directed parapod on each side as well as the median sucker (Fig. VIII, 7).

Each sucker, which is apparently derived from a typical dipteran pseudopod, is a shallow funnel-shaped chitinous structure, the soft rim of which is supported by struts which are hooked at the end, and it has a small V-shaped notch on the anterior margin (Fig. VIII, 7). The central area is a slightly dome-shaped piston, which, because the cuticle surrounding it is flexible, can be raised or pushed down by muscles (Hubault, 1927; Hora, 1930). When the sucker is placed in position the soft free edges engage the substratum, and the piston is pushed down. This drives the water out of the anterior notch, which functions as a valve, and negative pressure results when the piston is again raised. Presumably the hooks, being directed inwards, also engage in any roughness on the surface. There is some disagreement among various observers as to how these animals behave in the field; Hora states that they maintain their length across the current and Dorier and Vaillant (1954) that their heads are always downstream. It seems, however, that they have a variety of gaits. They can turn by releasing suckers 1–4, attaching 1 and 2 laterally, and then moving 5 and 6. This series of movements, repeated two or three times, enables them to turn through 180°; and they can walk forward moving either two suckers at a time in the sequence 1 and 4, 2 and 5, 3 and 6, or one sucker at a time starting at the rear. Hubault describes this last motion as like that of a caterpillar. When the sucker of one division of the body is detached, the parapods move forward and engage the substratum thus stabilizing the larva, and if it is moving across the current the downstream parapod acts as a fulcrum. These animals are capable of moving against currents of up to 240 cm./sec. and probably much more (Dorier and Vaillant, 1954).

The absence of true suckers from most stream insects is perhaps surprising, especially as suckers have been evolved in other habitats. The larvae of the psychodid *Maruina* of North America and Asia have a row of eight ventral suckers, so the ability to evolve these structures is present at least in other Diptera. But *Maruina* lives in thin sheets of water trickling down rocks, and it may be that the necessity for really smooth surfaces for the efficient operation of suckers has militated against their development in streams. Hora (1936) has shown that Blepharoceridae are absent from areas in India where the rock surfaces are roughened by deposition of travertine or growths of algae or moss. The

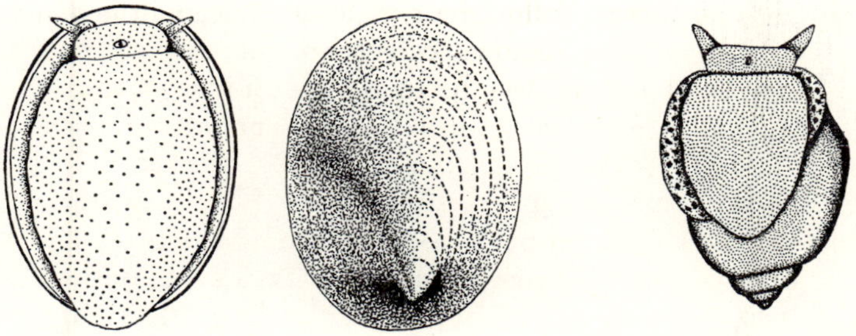

Ancylus fluviatilis x 5 (Europe)

Limnaea pereger x 3 (Europe)

FIG. VIII, 8. Broad-footed snails. *Ancylus* is a typical river limpet and *Limnaea pereger* is common in fairly rapid water.

areas available to them are therefore rather limited, and probably other, simpler, attachment methods are more flexible.

Rather similar to true suckers are the broad feet of gastropods, and these are well shown by the limpet-like Ancylidae which are very characteristic inhabitants of stony substrata almost everywhere (Fig. VIII, 8). Noticeably also the most common normally shaped snail in European streams is the particularly broad-footed *Limnaea pereger* (Fig. VIII, 8), and in Indian torrents the similarly broad-footed genus *Paludomus* occurs among gravel. In this animal the upper parts of the shell are often eroded away so that the large last whorl comes to resemble a pebble (Hora, 1928). Set against this, however, is the experimental finding of Moore (1964) that *Physa propinqua* is less able to maintain its hold in fast water than is *Limnaea palustris*, although the latter has the smaller foot. He suggests that this is caused by differences in behaviour and shape. *Physa* tends to raise the front of its foot in searching movements which *Limnaea* does not do; it is also rather less streamlined. He was further able to show that the ability to hold on was much affected by the substratum, and it has been shown in field studies that the North American prosobranch *Goniobasis* (Fig. VIII, 18) is able to maintain itself in fast water only where it has a solid substratum to which it can attach its foot (Ludwig, 1932). Given a suitable substratum and suitable behaviour patterns stream-dwelling snails can maintain themselves in very fast water, and they can move against currents of over 100 cm./sec. (Dorier and Vaillant, 1954).

Friction-pads and marginal contact. In many insects of the torrential fauna the flattened ventral surface, or some structure round the edge of the animal, is modified in such a way as to make close contact with the substratum. This increases the frictional resistance, and the danger that

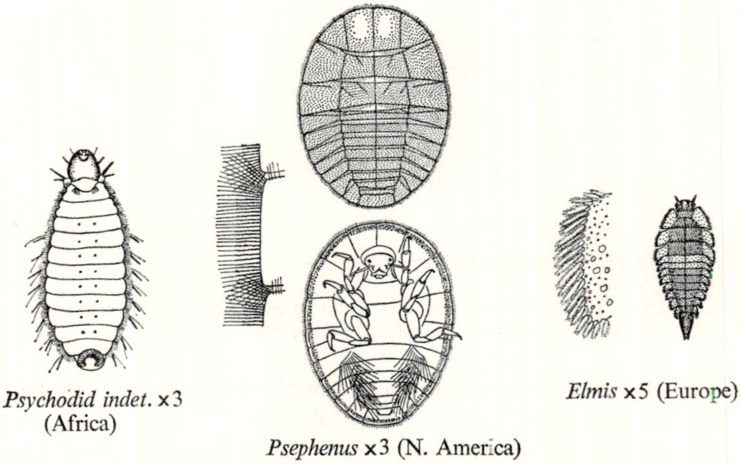

FIG. VIII, 9. Larvae with peripheral bristles which fit the surfaces of stones, and details of the bristles in *Psephenus* and *Elmis*.

the animal may be lifted off by the current is lessened. The simplest example is the soft flexible periostracum round the edges of limpet shells which fits closely to any irregularities in the surfaces (Fig. VIII, 8) Hunter, 1953*b*). Another is seen in *Psephenus* and its relatives, the water-pennies, on which there is a complete peripheral ring of rather complex movable spines which fit the surface and seal off the ventral side of the larva (Fig. VIII, 9) (Hora, 1930). Rather similar spines occur round the edges of the larvae of the European beetle *Elmis*, at least one species of which is often found on stones (Beier, 1948), and on some ceratopogonid and psychodid larvae (Fig. VIII, 9).

Some dragonflies have flattened abdominal sterna with transverse ridges of spines (Fig. VIII, 11), and among the mayflies there are several genera with various modifications of this type. The simplest is found in *Dicercomyzon*, which, as we have seen, is not really an inhabitant of fast water. Here the flattened thoracic sterna are surrounded by a fringe of outwardly directed hairs which make close contact with the substratum (Fig. VIII, 10) (Ventner, 1961). A slightly more elaborate structure of the

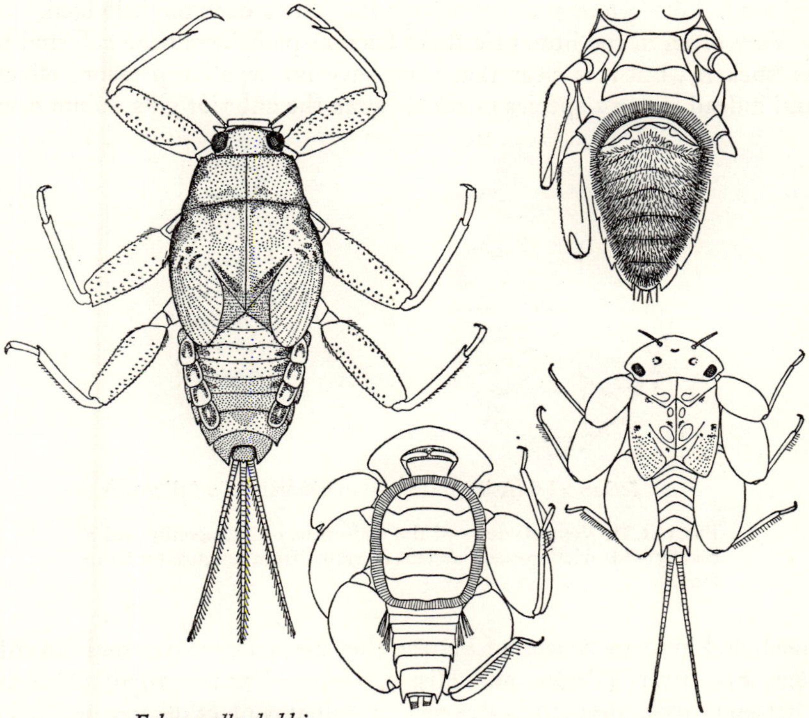

Ephemerella doddsi ×3·5 (N. America)

Dicercomyzon ×8 (Africa)

FIG. VIII, 10. Two mayflies with friction-pads of bristles, that of *Dicercomyzon* being simpler than that of *Ephemerella doddsi*, which resembles seal skin.

same kind is shown by the North American *Ephemerella doddsi*, but here the flat ventral surface also has a felt of backwardly directed hairs (Fig. VIII, 10). In the flattened nymphs of the Ecdyonuridae, and also in some similarly flattened leptophlebiid nymphs in Australia (Harker, 1954), the gills extend out sideways from the abdomen and make contact with the substratum. In the genus *Rhithrogena* and some species of *Epeorus* the front gills are enlarged and turned forwards under the body, thus increasing the area of marginal contact and reducing the possibility of water's flowing in under the nymph (Fig. VIII, 11). In some of these species the hindermost gills are also turned inwards, and in a few North American and Indian species the anterior margins of gills 2–6 bear thickened spiny pads which increase the friction (Hora, 1930). The effectiveness of the inturning of the gills as an attachment mechanism is easily demonstrated if specimens of *Rhithrogena* and some other, unmodified, ecdyonurid are placed together in a pan of water which is then agitated. The *Rhithrogena* are always the last to become dislodged.

Very often in the literature these friction-pads have been referred to as 'suckers', but it is clear that they have no negative pressure effect, and indeed in most species of *Rhithrogena* the anterior gills do not even

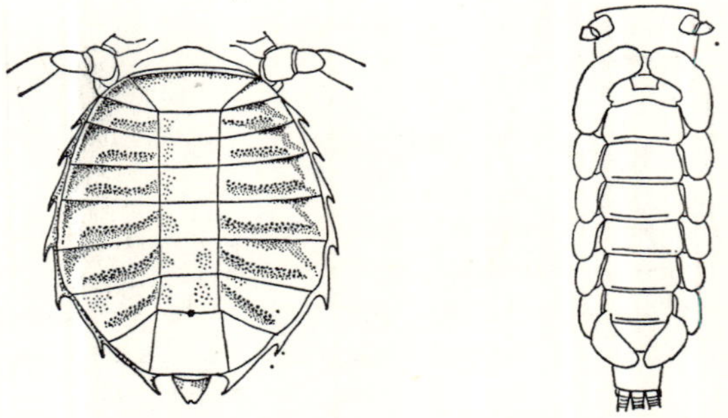

Ictinus ×1·6 (Asia) *Rhithrogena* ×5 (Europe)

FIG. VIII, 11. Ventral views of the abdomens of a dragonfly and a mayfly with friction-pads. *Ictinus* is drawn from a figure by Hora (1930).

meet, and in many species of *Epeorus* they are not even directed inwards (e.g. Fig. VIII, 1) (Dodds and Hisaw, 1924*a*; Verrier, 1956). Also, as Nielsen (1950*b*, 1951*a*) has stressed, marginal contact devices are by no means confined to torrent-dwelling animals. They and friction-pads, however, serve these animals well in their peculiar habitat.

Hooks and grapples. Most stream-dwelling arthropods have well-developed tarsal claws with which they hold on to roughnesses on the surfaces of stones, and the small elminthid beetles rely entirely on their sprawling large-clawed legs for attachment. Other, more unusual, developments of claw-like grapples are found on the posterior prolegs of free-living caddis-worms and the larvae of Corydalidae (Figs. VIII, 6 and VIII, 12). Under normal circumstances the larvae are sprawled on the rocks with both the thoracic and the abdominal claws engaged, and the same method of attachment by circlets of hooks on a forked front proleg and two long posterior prolegs is found in the larvae of Diamesinae which occur on stones in fast water (Fig. VIII, 12) (Thienemann, 1954*a*). In the very characteristically rheophile trichopteran *Rhyacophila*, the posterior claws can be brought forward to meet the base of

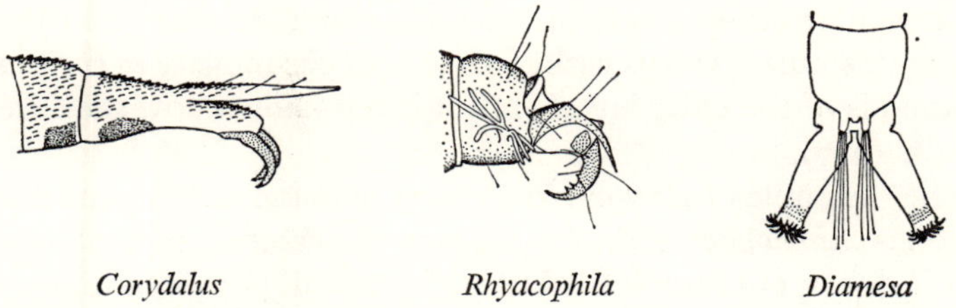

FIG. VIII, 12. Details of the hooks at the rear ends of a number of insect larvae from rapid water.

the proleg, thus forming an efficient gripping mechanism (Fig. VIII, 12) (Nielsen, 1942).

Circlets of hooks on the ends of prolegs are also used by the larvae of Simuliidae and Deuterophlebiidae. In the former they are engaged only in silken mats and are not directly attached to the substratum, so we shall defer consideration of them till later, but in the Deuterophlebiidae

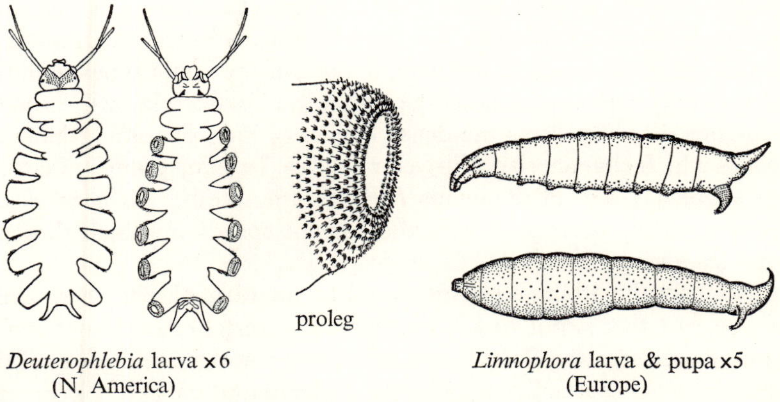

FIG. VIII, 13. Larvae of *Deuterophlebia*, with details of the outwardly directed hooks on the proleg, and the anchor-like larva and pupa of *Limnophora*.

they are used directly on the rock (Fig. VIII, 13). The larvae have lateral lobes on each side of the first seven abdominal segments, on the ends of which are ventrally directed prolegs. Each proleg bears a number of circlets of Z-shaped hooks the free ends of which are broadened and divided into five teeth. By protrusion and retraction of the prolegs the larvae can attain a fairly firm foothold, and they can scuttle about quite fast in currents of over 250 cm./sec. Their normal gait is laterally directed as they alternately loosen the anterior and posterior end of the body and rotate the free end through an arc before re-attaching it (Kennedy, 1958). It is interesting to note that the first instar larvae of the Blepharoceridae have similar prolegs with circlets of hooks in place of the grapple-like parapods of the older larvae, but it is not known how they are used (Hubault, 1927).

Another example of a hook-like process is found in the almost world-wide genus *Limnophora*, in the maggot of which a pair of posterior prolegs is drawn out into fleshy but stiff forwardly directed hooks which function like the arms of an anchor. The structure persists in the puparium, which becomes hooked on to some projection on the stream bed, and leaves the pupa, like the larva, with its head directed downstream (Fig. VIII, 13).

Small size. It will be clear from our consideration of the relatively stagnant boundary layer on stones that very small animals are largely free from difficulties caused by current. Stream animals are not, however, noticeably smaller than their still-water relatives, with two possible exceptions. The world-wide Elminthidae are particularly small beetles, and are certainly smaller than other members of this order which have taken to an aquatic existence; and the hydracarine mites which occur in streams are all fairly small and include none of the larger types found in still water. It may also be noted here that the Hydracarina which occur in streams do not possess the dense fringes of long swimming-hairs on the legs which characterize still-water species. In some groups of closely related species, e.g. in the genus *Lebertia*, the number of hairs on the legs is inversely correlated with the current speeds in which they are found (Schwoerbel, 1961*b*).

Obviously the ability to swim would be no advantage to these small animals, but this is not to say that stream invertebrates do not swim. We have seen that the Baetidae are efficient swimmers, and many of the larger insects, especially among the Ephemeroptera and Plecoptera, have fringes of long hairs on the hind borders of the legs with which they can swim short distances in an ungainly manner to regain their foothold. Really efficient swimming is, however, the prerogative of the streamlined mayflies and, of course, of some fishes.

Silk and sticky secretions. Many stream-dwelling arthropods employ secretions of silk or similar material to attach themselves to stones or other objects exposed to the current. Some, for example several genera of Orthocladiinae (Thienemann, 1954*a*), the torrent midge of Hawaii, *Telmatogeton hirtus* (Needham and Welsh, 1953), and several genera of the Psychomyidae, e.g. *Lype* and *Tinodes*, spin silken tubes on the stone surfaces, in which they spend their lives. Rather similar slimy tubes are made by the amphipod *Corophium curvispinum* (Fig. VIII, 14) in rivers draining into the Caspian area, and some of the aquatic caterpillars, e.g. *Cataclysta* (Fig. VIII, 14) in America and *Aulacodes* in Asia (Hora, 1928,

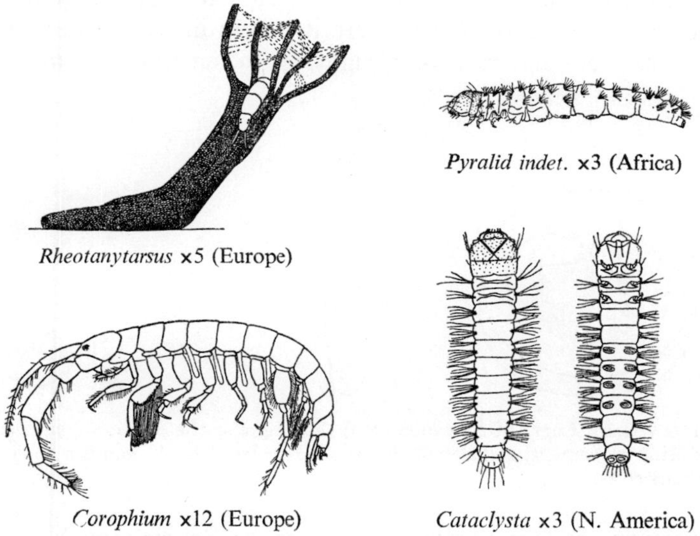

FIG. VIII, 14. Larvae of pyralid moths and *Rheotanytarsus*, and the adult of *Corophium curvispinum*.

1930; Nielsen, 1950*b*; Pruthi, 1928), cover themselves with flat sheets of silk, and they also form their cocoons under this roof.

All stream-dwelling cased caddis-worms attach their cases to solid objects with silk when they pupate, and even those which do not have cases as larvae, e.g. Rhyacophilinae, Polycentropidae, and Philopotamidae, construct rough pupal shelters with small stones and silk. Cased caddis-worms also attach their cases temporarily with silken straps when they moult. They cannot, of course, at that time rely on their claws to hold them, and in many stream-dwellers this habit of attachment becomes more or less permanent. Some Hydroptilidae are almost always found with their cases permanently anchored, as are some Brachycentridae e.g. *Oligoplectrum* (Nielsen, 1943) and *Brachycentrus* (Lloyd,

1921) which, as we shall see, use their legs for other purposes (p. 185) (Fig. x, 3).

In the chironomid tribe Tanytarsini the attached tubes are fairly rigid, being reinforced with particles of solid material, and after running for some distance along the substratum they rise up into the current (Walshe, 1950). The small hydra-shaped tubes of *Rheotanytarsus* are frequent features of stony streams in summertime, and they may form dense patches on the tops of submerged stones (Fig. VIII, 14). Rather similar tubes, but without the hydra-like arms, are formed by the European genus *Lithotanytarsus*. This peculiar animal is known from several areas of calcareous mountains where it occurs in hard-water streams, usually near, but not extending right up to, an underground source

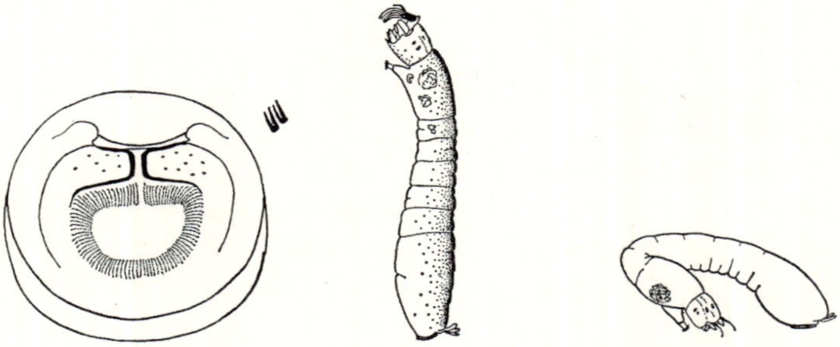

FIG. VIII, 15. Larva of *Simulium* (×6) showing also the position assumed during movement, and details of the rear circlet of hooks and the hooks themselves.

(Thienemann, 1954*a*; Symoens, 1955, 1957; Sourie, 1962). It forms dense masses of tubes on the rocks, and these become encrusted with calcium carbonate forming a rigid but porous travertine. The actual deposition of the calcium is caused by algae which grow on the tubes, but the larvae are confined to areas where this occurs. Each year the tubes erode to some extent in the winter, but new growths occur each summer, and thick laminated deposits may be formed.

The larvae of Simuliidae are among the most expert users of silk (Fig. VIII, 15). Their anatomy, adaptations and habits have been very well described by Grenier (1949). The larva of *Simulium* has very large salivary glands which produce the silk, and with this it makes a tangled mat on the substratum to which it can attach itself by hooks on highly modified prolegs. There are two of these, each of double origin. The anterior one is on the ventral side of the first thoracic segment, and it projects forwards under the head. Its end portion can be retracted by special muscles or erected by blood pressure, and it bears a number of circlets of outwardly directed hooks. When this proleg is placed against the mat of silk and erected the hooks are rolled outwards and engage in the threads. The posterior organ is similar, although so flush with the surface of the larva as to be scarcely recognizable as a proleg, and it forms the rear end of the body below the upwardly turned anus. It also has a complex arrangement of retractor muscles and a disengagement system based on the lever action of an X-shaped dorsally placed sclerite (Fig. VIII, 15). The details of these and their mode of action need not concern us here, where it is necessary only to note that the rear proleg also attaches the larva to its pad of silk by the outward rolling movement of a series of circlets of outwardly directed cleat-like hooks.

Normally the larva is attached by its rear proleg and hangs free in the current, and it moves by bending over so that its head and the openings of the silk glands touch the substratum, where it spins another mat and attaches the front proleg (Fig. VIII, 15). It then moves its rear proleg up to the new pad, releases its front end and repeats the process. In this way it progresses slowly in a leech-like manner. If it is disturbed, or the current becomes very strong, it bends over and attaches its fore proleg to the silken mat. In this way it doubles its attachment to the substratum and brings most of its body into the boundary layer. It also attaches a safety line to the substratum so that if it is dislodged it is carried away on a thread, just like a falling spider, and, like a spider, it can climb back up the thread, by working it between the front proleg and the head, which bears a number of pectinate spines which engage the silk.

These larvae are therefore among the most adapted of all animals to life in rapid running water, and they often occur in very fast currents. The pupae also are housed in silken cases which, although varying a great deal with the species, are generally shoe- or slipper-shaped with the pointed end directed upstream (Fig. VIII, 23). In other Diptera of this habitat the pupae are either glued to the rock, as in the Blepharoceridae and Deuterophlebiidae (Fig. VIII, 23), or they remain in the larval tube, as in the Chironomidae.

All the examples cited so far are arthropods, but one mollusc, the zebra-mussel *Dreissena polymorpha* (Fig. VIII, 17), is almost unique among freshwater clams in retaining its byssus into adult life, and it attaches itself permanently to solid objects. Although it is by no means confined to running water it is remarkably successful there, probably because of the byssus, and it has become abundant on rocks and man-made structures in the rivers in Europe to which it has gained access. It even causes trouble by settling in and blocking water-supply pipes.

The eggs of many stream-dwelling animals are also attached by sticky materials. Those of gastropods, leeches, and many insects are deposited in capsules or jelly masses, which are stuck to solid objects under water or at the water's edge, and such eggs are, of course, firmly fixed from the time of laying. Many insects, however, deposit their eggs on the water surface, and it is among such animals that specialized eggs are found. Some species of Simuliidae scatter their eggs, which merely lodge in the bottom deposits, although once there they tend to stick (Davies *et al.*, 1962). In other groups quite elaborate structures may be found. The eggs of some Plecoptera develop a jelly coat which sticks them firmly to stones, and those of the larger species bear elaborate attachment-plates and are invested in membranes bearing sticky knobs (Hynes, 1941; Brinck, 1949; Knight *et al.*, 1965*a*, *b*). Similar adhesive jelly-coats occur on the eggs of some of the Ephemeroptera which oviposit on the water surface, and in others the eggs extrude long sticky strings or bear conical adhesive structures (Percival and Whitehead, 1928). It seems, however, that in this order there is little correlation between the presence or absence of such adhesive structures, or their elaboration, and the habitat in which the nymphs normally occur (Degrange, 1960). Possibly during their long evolutionary history many members of this ancient and very successful freshwater group have moved back and forth between various types of habitat, and modifications evolved in one have been carried over into others.

Ballast. Most of the stream-dwelling larvae of Trichoptera which build cases do so with stones or large sand grains. This is presumably partly because such materials are the only ones available, but it has the effect of making the insects heavy and less easily swept away. Among the Limnephilidae particularly, quite large stones are used and the cases are

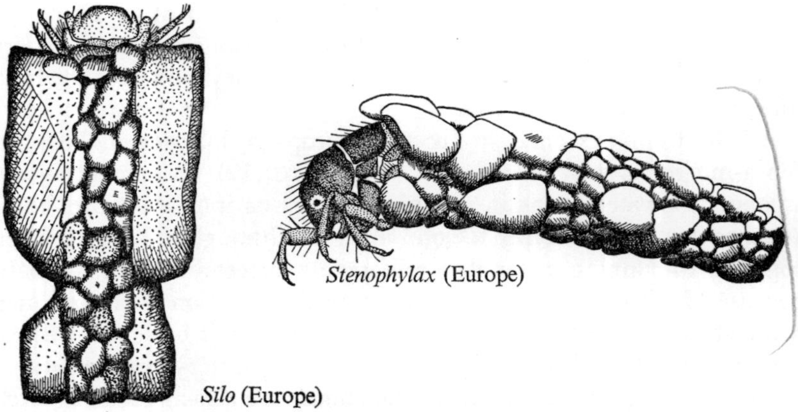

FIG. VIII, 16. Caddis-worms with heavy stone-cases which function as ballast

rough and heavy, e.g. *Stenophylax*, and in the Goerinae the larvae of several genera, namely *Silo*, *Lithax*, and *Goera*, incorporate relatively very large stones into the sides of their cases (Fig. VIII, 16). In fully grown specimens of *Silo* there are usually three pairs of such stones, increasing in size towards the front (Nielsen, 1942), and these animals are so heavy that even if they release their hold in a fast current they merely slide off the stone and fall into sheltered dead water. Webster and Webster (1943) have shown experimentally that larvae of *Goera calcarata*, which were

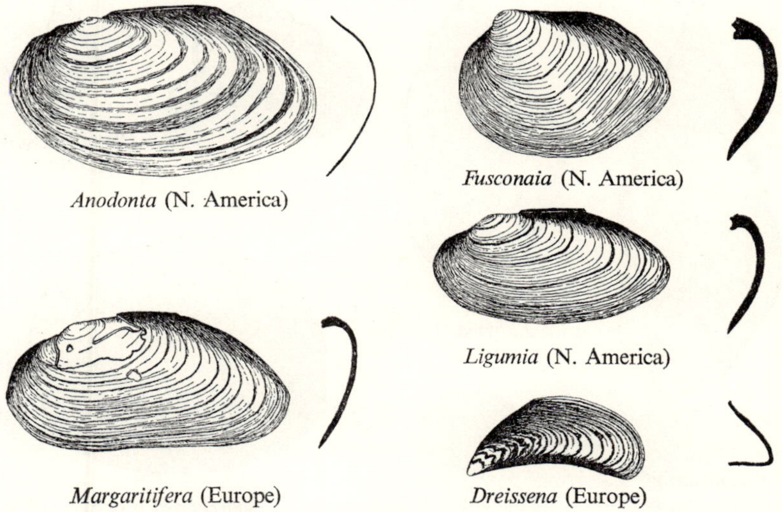

FIG. VIII, 17. The shells of several large bivalves and their thicknesses in mid-line. *Anodonta* lives in quiet water and has a thin shell, as has *Dreissena* which is attached by a byssus. The others occur among gravel and stones in faster water and have more massive shells. Note the erosion of the umbo of *Margaritifera* which is typical of this species from soft water. (*Dreissena* ×1, others ×$\frac{1}{3}$.)

forced to construct new cases, made significantly heavier ones in running than in still water, and it may be that in all these genera larger stones are incorporated in faster water. This has not been investigated.

Among molluscs also it is noticeable that running-water species are often massive as opposed to still-water forms. Species of *Fusconaia* from stony substrata in North America are often very thick-shelled, and generally speaking all the Unionidae from fast water are much heavier than genera such as *Anodonta* from soft substrata (Fig. VIII, 17). This is also evident in Europe where the heavy-shelled *Unio* species inhabit harder bottoms than the thin-shelled *Anodonta* species, and it is noticeable that even the pearl mussel *Margaritifera*, which inhabits hard substrata in the Northern Hemisphere, has a fairly massive shell despite the fact that it is usually found in non-calcareous water.

Stream-dwelling Prosobranchia also tend to have massive shells, as is shown for example by the common North American genera *Goniobasis* and *Pleurocera*; but the situation here is not simple as this does not apply to very small snails. The genus *Potamopyrgus* of southern latitudes, one species of which is now widespread in Europe (Thienemann, 1950), often inhabits quite rapid streams and is fairly thin-shelled (Fig. VIII, 18).

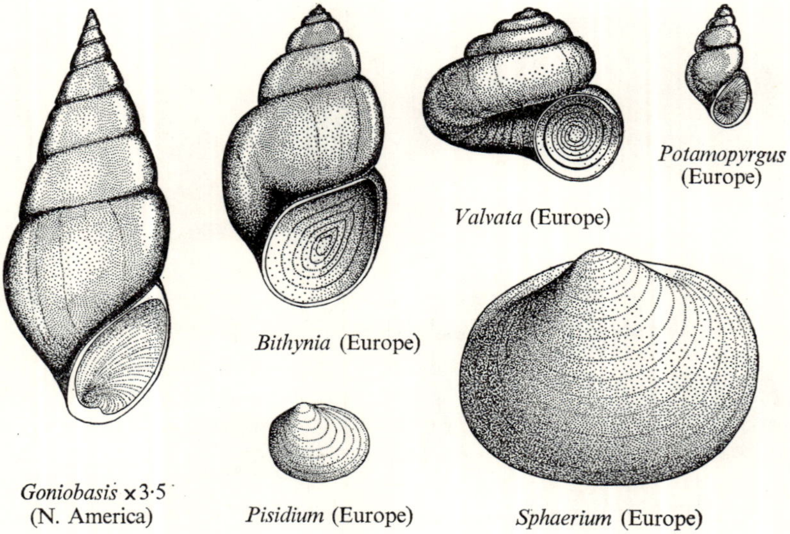

FIG. VIII, 18. Prosobranch Gastropoda and Sphaeriidae which occur in streams and rivers. *Goniobasis* has a heavy shell and lives in rapid water; *Bithynia*, *Valvata*, and *Sphaerium* are thinner shelled and occur more usually in shelter, but the two small genera, *Potamopyrgus* and *Pisidium*, often occur in rapid water. (×5 except where shown.)

When it is disturbed it releases its hold and falls down into the gravel, and it may be that the shell thickness in the larger molluscs is as much concerned with protection against crushing by moving stones as it is with ballast. Presumably the danger of being crushed during a spate is less for the smaller species as they can more readily fall into crevices in the underlying deposit. This would appear to apply also to the thin-shelled but small Sphaeriidae, particularly *Pisidium* and some species of *Sphaerium* (Fig. VIII, 18).

An interesting observation which seems to relate shell form to the speed of the current is that of Goodrich (1937) who found that species of *Pleurocera* in streams in Alabama became more obese, i.e. wider in relation to length, in the downstream direction. He found that the degree of obesity was similar in headwaters and in small tributaries, and that where, in one river, there was a rapid near the mouth the snails resembled those of the headwaters in being relatively less obese. These facts remain unexplained, but they would appear to be in some way related to current, and it is possible that the more elongate shape of the specimens from headwaters and rapids is of benefit in faster water. Perhaps, as with the tails of Ephemeroptera and the sticks on the cases of *Anabolia*, the longer tapering shell serves to head the snail into the current. This explanation cannot, however, account for the observation that the obesity of some North American clams decreases towards the headwaters, e.g. *Fusconaia flava* (Fig. VIII, 17), nor for the fact that in several genera of Unionidae, e.g. *Quadrula metaneura*, and in the prosobranch *Io* of the Tennessee River system, the sculpturing of the shell becomes more and more elaborate in the downstream direction (Adams, 1915; Ortman, 1920; Goodrich, 1941). Doubtless though water movement must be in some way involved, and it would seem that this phenomenon is worthy of further study.

Another example of a relationship between shell shape and current speed is found in the common European *Ancylus fluviatilis* (Starmühlner, 1953). It can be seen over much of that continent, and possibly the same applies to other Ancylidae elsewhere, that specimens living in fast water are higher and more steeply conical than those from slower stretches. This is, of course, in direct contrast to what has been said about streamlining earlier in this chapter, and it likewise remains unexplained. Possibly the higher cone is more resistant to crushing by moving stones than the lower one, but this is mere speculation. A parallel example is provided by the true limpets of the genus *Patella* in the intertidal waters of North Atlantic coasts. In these animals also the shells of individuals living on exposed areas are taller than are those of specimens on more sheltered rocks.

Life in vegetation. Most of the animals which live in vegetation in streams show no particular morphological adaptations which distinguish them from similar animals from still water. In dense growths of plants and mosses they are completely sheltered from the current, and in such places a varied microfauna can be found. Some of the larger animals, however, do possess anatomical characteristics which can be correlated with their habitat. Thus dragonfly nymphs, which live *on* plants rather than *in* them, e.g. Agrionidae (Fig. X, 9) and some Libellulidae, have long legs and stout claws. The European species of the stonefly *Taeniopteryx*, which in contrast to their North American relatives all live among plants, have long tergal spines, which presumably hold them in place and which are absent from the North American species (Berthélemy, 1966). A similar development of dorsal spinous projections is found in some South American Gripopterygidae (Fig. VIII, 19) (Illies, 1961*a*, 1963), and it is perhaps significant that the common European mayfly *Ephemerella ignita*, which is a moss-dweller (Hynes, 1961), is distinguished from its benthic relatives in that continent by the presence of dorsal spine-like processes on the abdomen. We can conclude, therefore, that in at least

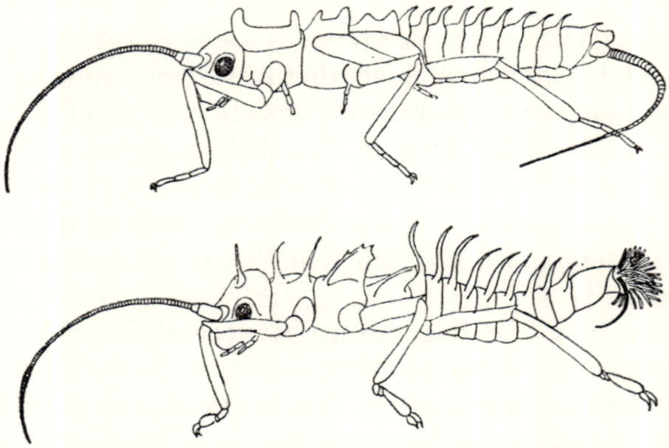

FIG. VIII, 19. Lateral views of *Taeniopteryx* from Europe (above) and *Araucanioperla* (?) from South America (*below*), showing similar developments of dorsal spines in these two unrelated plant-dwelling stoneflies. From Illies (1961*a*) with permission.

some insects dorsal processes aid in fixation among the plants of running water.

Reduction in the powers of flight. We have seen in the last chapter that the water-striders of streams tend to be apterous, and it will be appreciated that the ability to fly is of less advantage to stream-dwelling insects than it is to, say, pond species. Indeed, a tendency to disperse widely may be a positive disadvantage because streams in different valleys are often separated by considerable ecological barriers, and the chances of a species' reaching another suitable stream by air would appear often to be less than that of reaching it by water by way of confluences. It is, therefore, not surprising to find that, in many groups, some species, though by no means all, show reduced powers of flight.

Many stoneflies have reduced wings, and few of them fly very well; in some species both long-winged and short-winged specimens occur

(Aubert, 1945). This may be regarded as a sort of double insurance which, as in the water-striders where occasional fully winged specimens occur, allows for dispersal as well as ensuring that many individuals remain near to suitable breeding sites. A few stoneflies, e.g. *Scopura* of Japan and *Andiperla* of South America, are quite apterous, and in many species only the males are flightless. This keeps them near the water so that they are available for females when they emerge, and in all these species the females are clumsy fliers, and so are not able to fly much further than is needed for oviposition. It has also been observed that the length of the wings decreases with increasing altitude in many species of stonefly (Kühtreiber, 1934; Hynes, 1941; Brinck, 1949). This is also probably correlated with the need for individuals to remain near their stream of origin and not to be swept away by the higher winds on the mountains.

Among beetles, stream- or spring-dwelling species in several genera of Dytiscidae, e.g. *Hydroporus*, *Deronectes*, *Oreodytes*, and *Platambus*, and Palpicornia, e.g. *Anacaena* and *Hydraena*, have reduced flight apparatus (Jackson, 1956*b*, 1958); and among the Elminthidae, while some species fly well and come to light at night, others are unable to fly and have lost their hind wings (Young, 1954).

Even among the Ephemeroptera, which are all winged and capable of flight, some running-water species are unable to move far from the water because they have no time to do so. All adult mayflies are short-lived, but many will survive for a few days in the sub-imago stage during cool weather (Lyman, 1944). Some running-water forms, however, live for only very short periods and perform only one flight. In Central Africa the females of *Prosopistoma* never shed their sub-imaginal skin, and they perform a single dawn flight, lasting less than an hour, during which they mate. They have tiny legs and, like an aircraft with no undercarriage, they can only crash-land on to the water to oviposit (Gillies, 1954). Much the same applies to the North American Polymitarcidae *Ephoron* and *Tortopus*, and in the family Oligoneuriidae the adults shed their sub-imaginal skins in flight, a moult which does not include the wings, and they also do not land between the time of emergence and falling exhausted after oviposition on to the water (Edmunds, 1956). Similarly the Deuterophlebiidae, although they fly fast, live only for about two hours, and so are effectively confined to their stream of origin (Kennedy, 1958). It can be seen, therefore, that these insects, although quite capable of flight, are as effectively restricted to their stream of origin as they would be if they could not fly at all.

As with several of these adaptations it can be clearly seen why the loss of ability to disperse is advantageous in the running-water habitat, but it is equally clear that this particular modification is not exclusive to stream dwellers even among the aquatic insects. For example, a whole tribe of the Limnephilidae, the Baicalinini which are endemic to Lake Baikal, are flightless (Kozhov, 1963), and some lake-shore stonefly species have produced local flightless populations in exposed lakes, e.g. *Capnia atra* in southern Sweden (Brinck, 1949). The reason behind these changes to flightlessness is the same as that for running water—the disadvantage to the species of the loss of specimens from a restricted habitat.

Almost all the modifications of stream animals discussed above should be interpreted in this light. They are adaptations which enable animals successfully to inhabit rivers and streams, but similar modifications of almost all the types considered here are to be found, for the same or for different reasons, in quite other environments.

BEHAVIOURAL ADAPTATIONS

Several authors have stressed that successful adaptation to life in running water involves both behaviour and morphology (e.g. Nielsen, 1950*b*; Einsele, 1960). This is self evident since no morphological adaptation will function successfully unless it is properly used, and in many respects behaviour is perhaps the more important, although it is really impossible to disentangle the two aspects. Much of what follows has, therefore, morphological undertones.

Many common stream creatures, e.g. flatworms, annelids, crustaceans, and a great number of the insects, are quite ordinary representatives of their groups, and they persist in running water simply because they avoid the current by living under stones or in the dead water behind obstructions. Their 'adaptation' consists therefore in their cryptic habits, and in their avoidance of the current. The same applies to those larger members of the fauna, already mentioned, which are to some extent flattened and so able to creep under stones. This is true even of the relatively very large freshwater crabs and of torrent-dwelling Natantia, which are cryptic in their habits (Hora, 1923).

Avoidance of the current can be conveniently studied in *Gammarus* species, as they are often abundant enough to reveal their habits by their distribution. In small streams, or in larger ones where the current is slow, they are normally fairly evenly distributed in areas where they can shelter under stones. Being poor swimmers, and without any attachment mechanisms, they can only scuttle from shelter to shelter. In larger streams, where the current is markedly swifter in the middle than along the sides, movement between stones involves exposure to swifter water, so the centres of such streams have relatively less dense populations than their edges. This can often be observed very simply in the field, as can the fact that, at times of low water when the current in the centre is relatively slack, the animals tend to spread out from the bank, as it then becomes possible for them to move to stones which were previously inaccessible (Hynes, 1954). The same sort of behaviour pattern is doubtless to be found in all the animals which live in the slower water along the sides of streams. They select areas where the current is tolerable, and they move further out or back into shelter as the flow varies. This applies, for instance, to many genera of mayflies, e.g. *Leptophlebia*, *Habrophlebia*, *Caenis*, and *Baetisca* (Pleskot, 1953; Leonard and Leonard, 1962), and to snails (Moore, 1964).

Animals which actually burrow down into the substratum are less

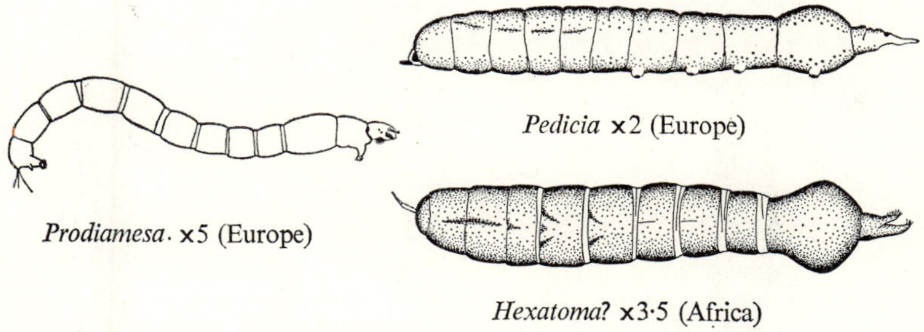

FIG. VIII, 20. Larvae of Diptera which burrow in the substratum. *Prodiamesa* is found in silty sand; the two members of the Tipulidae burrow in gravel and have inflatable rear ends which anchor them while they push forwards.

affected by changes in current, and all that they require in the way of a behaviour pattern is to remain buried. Many animals such as the annelids and some larvae of Diptera, have this habit as a birthright, in that it is typical of all members of their group; they are also long and narrow as befits burrowing animals (Fig. VIII, 20). But several other groups of insects have taken to the same way of life, and they have also become long and narrow. These include, among the stoneflies, the Leuctridae, Capniidae, and Chloroperlidae of the Northern Hemisphere (Fig. VIII, 21) (Hynes, 1941) and many Gripopterygidae of the Southern Hemisphere (Fig. VII, 3). Several genera and species of mayflies exhibit the same habit and shape. The best known of these are *Baetis pumilus* of Europe, which is much longer and narrower than its streamlined relatives on the tops of stones, and has shorter legs (Fig. VIII, 21) (Macan, 1957*a*), and the genus *Paraleptophlebia* of the Northern Hemisphere (Leonard and Leonard, 1962); there are doubtless many others. *Habroleptoides* is particularly modified in this respect, being very long and narrow and having the legs inserted laterally, the most efficient position for moving through narrow crevices between stones (Pleskot, 1953). This lateral insertion of the legs is also, to a lesser extent, to be seen in the stonefly families listed above. Many such animals, despite the fact that they are never exposed to current, are, nevertheless, strictly confined to swift stream habitats, because only there can they find large crevices between stones which are continuously provided with a gentle flow of

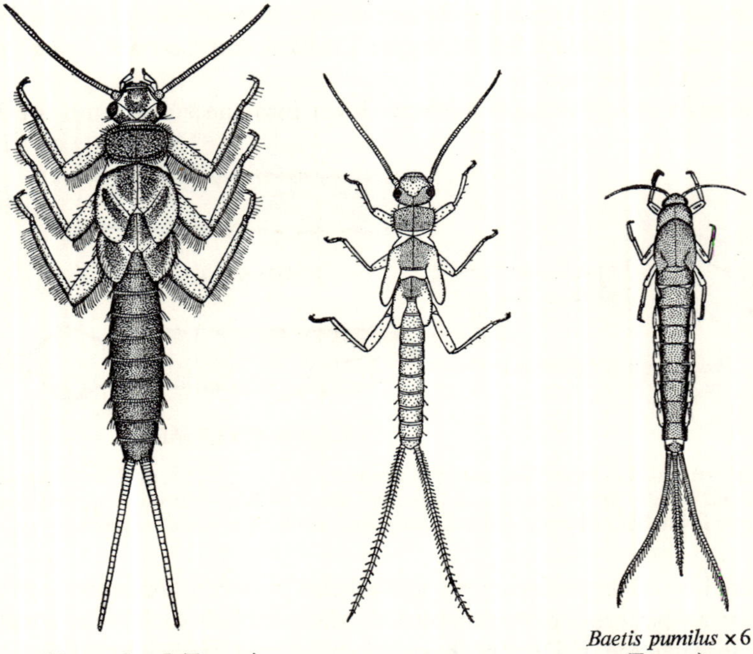

Chloroperla ×5 (Europe) *Leuctra* ×5 (Europe) *Baetis pumilus* ×6 (Europe)

FIG. VIII, 21. Elongate stonefly and mayfly nymphs which burrow in gravel.

well-oxygenated water. In fine sediments there is much less seepage of water to renew the oxygen supply.

Even among the highly modified dwellers on the exposed surfaces it is clear that there is much selection of habitat. Wu (1931) records that she found *Simulium* larvae only in current speeds between 17 and 84 cm./sec., and Ambühl (1959) has shown experimentally that many invertebrates tend to select different current speeds even within the relatively low range of up to 18 cm./sec. Similarly, Phillipson (1956) has demonstrated very elegantly that the common European *Simulium ornatum* always chooses a current of between 50 and 120 cm./sec., and that most of its larvae settle in the narrower range of 80–90 cm./sec. Experimental studies on the maximum current that can be resisted by species in artificial channels have not, however, always given very informative results, although the list of these published by Dittmar (1955*a*) does place a number of common European species in the order in which one would expect to find them. He found that *Liponeura* and *Simulium* were very resistant, at over 240 cm./sec., followed by a group resisting 100–30 cm./sec. which includes many well-known dwellers on the tops of stones (*Euorthocladius*, *Epeorus*, *Rhyacophila*, *Ancylus*, and *Brachycentrus*). Below this, resisting up to between 80 and 100 cm./sec., came a mixed group of animals whose habitat is either on, or under, stones in fast water (*Lithax*, *Micrasema*, *Rhithrogena*, *Agapetus*, *Dugesia*, *Elmis*, *Dinocras*, *Baetis*, *Silo*, and *Glossosoma*), followed by a list, resisting 48–77 cm./sec., of more cryptic animals (*Torleya*, *Amphinemura*, *Atractides*, *Isoperla*, *Ecdyonurus*, *Protonemura*, *Apatania*, and *Limnaea*). This appears to fit very satisfactorily in general terms with our earlier discussion of morphological adaptations, but, in the same year, Dorier and Vaillant (1955) published a similar, and more extensive, experimental

	Dittmar, 1955*a*	Dorier and Vaillant, 1955		
	Mean max. speeds resisted	Max. resisted	Max. moved against	Max. observed in
Liponeura[1]	>300	>240	>240	220
Simulium	280	240	117	114
Epeorus[1]	124	>240	230	109
Rhyacophila	122	200	100	125
Ancylus[1]	118	240	109	24
Lithax	99	132[2]	52[2]	24[2]
Rhithrogena[1]	96	182	125	—
Agapetus[1]	94	89	36	36
Dugesia[1]	93	125	48	<10
Elmis	91	216	—	—
Dinocras[1]	86	240	182	—
Baetis	84	177	102	30
Ecdyonurus[1]	57	154	99	—
Protonemura	53	198	147	145
Limnaea[1]	48	202	117	14

1. Indicates that the same species was studied. 2. Figures for *Goera*, a close relative of *Lithax*.

TABLE VIII,1. Data extracted from Dittmar (1955*a*) and Dorier and Vaillant (1955) showing, in cm./sec., the maximum current speeds resisted, moved against, and inhabited in nature by a number of European stream animals. Dittmar's results are the mean maxima of 10 trials with 5–10 animals on a stone in an experimental channel. Dorier and Vaillant's maximum resisted is for at least 1 min. in an artificial channel.

study which included several of the same genera and at least nine of the same species. Unfortunately, as can be seen from Table VIII, 1, the agreement between the two sets of figures is by no means good, and this, perhaps, confirms the point made above that this experimental approach is not very useful. Almost certainly animals under laboratory conditions are subject to many influences, such as exposure to light and unsuitable substratum, which disturb their behaviour, and it is perhaps not surprising that they respond in a peculiar manner. Indeed, it is difficult to draw any firm conclusions from the extensive set of figures given by Dorier and Vaillant except that the most resistant included the most typically rheophile animals (Blepharoceridae, *Simulium*, certain Orthocladiinae, Ecdyonuridae, Baetidae other than *Cloeon*, Plecoptera, caseless Trichoptera, and some Gastropoda); but it also included leeches. Also, in general, these were the animals which moved against the swiftest currents.

Probably much more informative data could be obtained by careful observation in the field, and this has been admirably done for some European Trichoptera by Scott (1958). He studied the detailed distribution of larvae in an English stream, which was peculiarly suitable for the investigation in that most of the stones were more or less rectangular blocks and it was possible to distinguish clearly between their upper, upstream, downstream, lateral, and lower surfaces. Some heavily cased species, of *Stenophylax* and *Odontocerum*, were always under stones in areas where the surface current was less than 20 cm./sec. These species feed on the detritus which collects in such situations. *Glossosoma* and *Ecclisopteryx*, which scrape algae off stones, were commonest at current speeds of 20–40 cm./sec. and occurred on surfaces other than the lower ones; and for the relatively common *Glossosoma* it was possible to demonstrate that, as the current increased, the larvae occurred less abundantly on the exposed upper surfaces, and more frequently on the more sheltered vertical surfaces. They were also more common on medium-sized stones than they were on smaller ones, which Scott suggests was because of the greater stability of the larger stones. He also found that *Glossosoma* and *Silo* moved into faster water before attaching their cases for pupation, and he suggested that this may be an adaptation which preserves the pupae from desiccation during summer droughts. At surface current speeds of over 40 cm./sec. *Hydropsyche* and *Rhyacophila* occurred, but only under stones, and in this zone there was a rich fauna of the small insects on which these larvae feed. It is also possible that *Hydropysche* needs fairly fast water in which to make its net (p. 188), and that only in fast water does it find sufficient flow beneath the stones. Interestingly also, *Rhyacophila*, although usually under stones, occurred abundantly on the upper surfaces of large stable stones in fast water (40–50 cm./sec.) where moss provided shelter.

It is clear therefore that in the field it is possible to demonstrate very definite selection of particular current conditions by these insects, and to show that it is at least to some extent related to food, a factor which is missing from laboratory experiments, as are probably several others of similar importance. As we have already seen with reference to snails (p. 132) one of these is undoubtedly the nature of the substratum. It has been shown experimentally that the stonefly *Perlesta* selects particles of larger size (4–16 mm.) when it is offered a range of types of substratum in an artificial stream, and that this reaction is complicated by silting, which makes finer particle sizes even less attractive although having little or no effect on the response to larger particles (Lauff and Cummins, 1964). Similarly, Scott suggests that *Glossosoma* does not occur in water slower than 20 cm./sec. because in slow water small mineral particles collect among the algae on the stones and interfere with feeding; this also is a factor which is likely to be absent from laboratory experiments.

Before leaving this subject it should perhaps be once again emphasized that the current speeds mentioned in the last few paragraphs are those of the main mass of the water, or of its surface; they are not the speeds to which the animals are actually exposed in the boundary layer, or in the shelter under stones or in moss. On these we have virtually no information, and it is difficult to see how it can be obtained, especially as the current must differ greatly at points only millimetres apart.

Behaviour in relation to current can also vary with different stages in the life history. As we have already seen, at least some caddis-worms move into faster water to pupate, and *Ecdyonurus* moves towards the banks, and so into slower water, as it grows. The same applies to several other genera of Ephemeroptera, e.g. *Oligoneuriella*, *Baetis*, *Epeorus*, and *Ephemerella*, which all move to the banks when they are ready to emerge (Verrier, 1956), and doubtless this is also true of many other insects. These are, however, changes in behaviour which are restricted to a limited phase in the life history and are understandable as being probably concerned with emergence. More fundamental and long-term changes in behaviour pattern were demonstrated by Beauchamp (1933, 1935, 1937) in the common European flatworm *Crenobia alpina* (Fig. VIII, 4). This cold-water species moves upstream when it is warm or sexually mature and ready to breed, and downstream when it is spent and hungry. So this group of reactions results in a series of migrations which ensure that the species remains in cold water, that it breeds in the headwaters, and that it does most of its feeding further downstream where the worms are less crowded and suitable food is correspondingly more plentiful.

Probably other complex patterns of adaptive behaviour remain to be investigated. There are, for instance, indications that several mayflies migrate up or downstream or into or out of deeper water at certain phases of their life histories (Rawlinson, 1939; Harker, 1953; Macan, 1957*b*; Leonard and Leonard, 1962).

There are also indications that individuals of the same species from different habitats react differently to current. Dorier and Vaillant (1955) found that specimens of the flatworm *Dendrocoelum* and of the dragonfly *Sympetrum* which had been collected from still water were less able to withstand current than were those from streams. It would appear, therefore, that at least some of the behaviour pattern may be learned, and this introduces a further factor into an already complex subject.

Some of the animals which are usually found under stones react negatively to light, and this normally keeps them out of the current during the daytime. Mayflies and stoneflies have, however, been reported by several workers to move on to the upper surfaces of stones at night (Hubault, 1927; Lyman, 1945; Dorier and Vaillant, 1955; Verrier, 1956), and Beauchamp (1933) observed that *Crenobia alpina* is more active in the evening than during the day. This is an aspect of the behaviour of stream animals which would obviously repay further study in relation not only to current, but also to feeding habits. In fact, though, several of the animals which are generally reported as being *found* under stones can often be *observed* clambering about on the upper surfaces, but they are readily disturbed by movements of the observer and they rapidly seek shelter. This applies, for instance, to many flattened mayflies and to at least some of the large stoneflies, and to see them one has to approach the water with the same care that is needed to see trout. It is possible that the sudden illumination of a stream bed at night with a flashlight reveals animals which are not so readily seen in the daytime simply because they normally hide when they see the observer approach. Somebody with a penchant for watching animals, equipped with a suitable hide and Polaroid glasses to remove the reflection of the water surface, could well clarify these observations which are of considerable interest. Undoubtedly though, many stream animals do avoid strong sunlight, and move into the shade at times when the sun is high. Some Baetidae can be seen to do this in shallow streams, and the same applies, particularly in the tropics, to many water-striders, which are even more readily observed (Hynes, 1955*a*). On the other hand *Simulium* larvae move towards light, although this tendency is subservient to their movement into a suitable current speed (Grenier, 1949).

We have seen above that some stream animals, most notably *Crenobia alpina*, undertake regular migrations upstream and downstream and that many invertebrates can move against quite swift water. It is also evident that most animals normally face into the current, so that any move they make tends to take them upstream. This is some compensation for the fact that any involuntary move, or loss of hold on the substratum, inevitably carries them downstream. This downstream displacement of animals is an important feature of stream biology, and has come to be known as drift. We shall be discussing it in Chapter XIII as it has, in recent years, attracted a great deal of attention, but it often seems to be overlooked that there is a lot of active movement in the other direction. If this were not so the headwaters would soon become devoid of animals. Among the insects much of this upstream compensatory movement can be performed by flying adults, but this cannot account for the maintenance of position of the fauna as a whole, and even some of the insects are very poor fliers.

Upstream movement of invertebrates actually in the water is not easy to observe, but there are many suggestions in the literature that it occurs, and this is another aspect of invertebrate behaviour which would repay further study. There are indications that *Dugesia dorotocephala*, a North American ecological equivalent of *Crenobia alpina*, also moves upstream during the summer, and the same applies to the snail *Physa integra* (Noel, 1954). The little prosobranch *Potamopyrgus jenkinsi*, which has become so widespread in Europe, must have spread upstream to achieve its present distribution, and it has in fact been observed to do so in an active manner (Adam, 1942; Heywood and Edwards, 1962). Even quite large and clumsy snails apparently undertake upstream movements. *Campeloma decisum* shows a strong tendency to move against a current, and large aggregations of this species at the downstream ends of rapids in streams in Michigan have been interpreted as resulting from the inability of specimens moving upstream to surmount the barrier of swifter water (Bovbjerg, 1952*b*, 1964). Gastropods can cover quite large distances despite their leisurely pace; Shelford (1937) estimates that they move about a mile (1·6 km.) a year upstream when recolonizing areas after unusual droughts, and Hynes (1960) records a movement of at least 2·4 km. in a year by *Limnaea pereger*, following its elimination by pollution.

Crustaceans also readily travel upstream. The mitten-crab, *Eriocheir sinensis* (Fig. VIII, 22), which breeds in brackish water, regularly travels 750 km. up the Yangtze Kiang, and it has been found 1,400 km. from the sea. Since its introduction into Europe it has been collected as far inland as Czechoslovakia (Hoestlandt, 1948), and some crayfish make regular migrations both upstream and downstream, e.g. *Pacifastacus klamathensis* in Oregon (Black, 1963). Movements of such large animals are fairly easy to follow, especially when it is known, as with *Eriocheir*, from where they originate. With smaller crustaceans it is more difficult, but even so there are clear indications that amphipods move upstream at times. Noel (1954) reports the probability of an upstream movement of *Gammarus fasciatus* during the summer in a spring stream in New Mexico, and Minckley (1964) observed mass upstream movements of *G. bousfieldi* in Kentucky, and was able to follow one individual for half an hour during which time it moved 5 m. Similarly in Europe, it is certain that *Gammarus pulex* moves upstream. When this species was introduced into a small stream in the Isle of Man it appeared in numbers

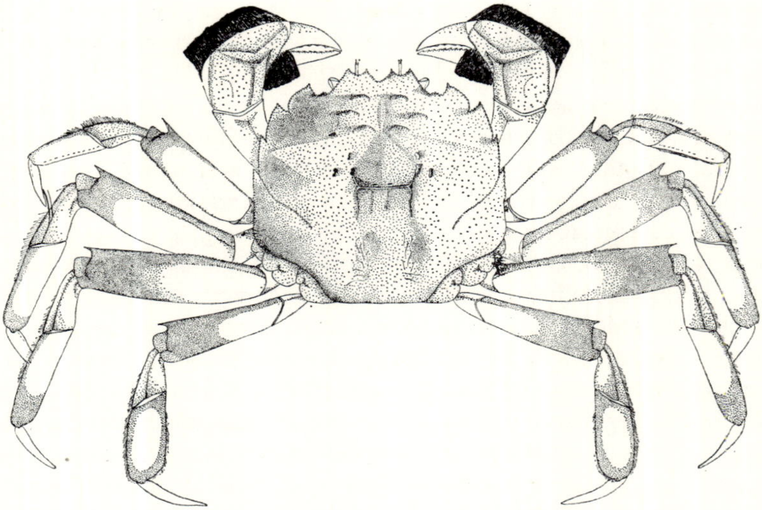

FIG. VIII, 22. The mitten-crab *Eriocheir sinensis* which migrates long distances upstream. $\times \frac{1}{2}$. From *Scientific Publication of the Freshwater Biological Association* 19, (1960) with permission.

in the headwaters within four years of becoming established, and had thus moved up 3 km. Even the tiny mites are able to undertake migrations against the current; Schwoerbel (1959) found that *Lebertia tuberosa* tends to concentrate near the cool sources of streams in the Black Forest at times when the air temperature is high, and this clearly can result only from upstream migration.

Insects also move against the current, and, despite the difficulties of making observations, there are several reports of upstream movement of young stages. In Manitoba, when temporary streams fill with snow water during the spring melt, the mayfly *Leptophlebia cupida* moves upstream out of lakes and rivers and travels up to at least a mile (1·6 km.) at an estimated rate of 200 m./day (Neave, 1930). The same phenomenon has been seen in Ontario, and in Europe the small burrowing species *Baetis pumilus* (Fig. VIII, 21) has been observed to travel upstream at least 2 km. into tributaries of the River Allier after an exceptional drought (Verrier, 1953). Similarly in Germany, Engelhardt (1951) was able to be certain that the net-spinning Trichoptera had recolonized a previously dry stream by upstream movement of the larvae.

There is thus already a substantial body of evidence that invertebrates can and do move upstream by active migration against the direction of flow, and this behaviour is, of course, an important adaptation enabling them to live in running water. Doubtless further study will show that this pattern is almost universal, and it would seem necessary that this should be so for the maintenance of the fauna in this peculiar habitat.

Insects which have actively flying adults have a further method of regaining ground lost by drifting, and this may be one of the reasons why they so often dominate the fauna of small streams. Several authors have reported the upstream flight of mayflies and stoneflies (e.g. Muttkowski, 1929; Müller, 1954*f*), and it has been assumed that it must also occur in *Simulium* (Davies *et al.*, 1962). Roos (1957) made a special study of this phenomenon on a river in central Sweden, over which he placed nets to catch insects moving both upstream and downstream. He found that flight was predominantly upstream among the Trichoptera, Ephemeroptera, Plecoptera, and females of *Simulium*, and that this was particularly true of individuals which were ready to oviposit. He was able to observe that many individuals, particularly among the Trichoptera which are more readily followed than the others, moved long distances upstream, even passing over small lakes. They thus carry their eggs far upstream.

It has even been established that some non-flying adult insects move fairly consistently upstream. Thomas (1966) observed the emergence of *Capnia atra* on to snow-covered ground from a stream in northern Sweden and noted that they walked about 50 m. at right angles to the stream to the edge of a wood. They continued on the same course for a further 50–100 m. and then, mostly, turned through 90° and headed in an upstream direction, even though the stream was no longer visible to them. His extensive observations leave no doubt that most specimens orientated themselves in this way, so that is another, albeit probably minor, way in which species maintain their position in streams. It appears that *Allocapnia pygmaea* behaves in a similar manner in North America, so the phenomenon may be quite general in winter stoneflies and possibly in spring species also. The latter would, however, be much more difficult to observe in the absence of snow, and this behaviour may be an adaptation peculiar to species which emerge in very cold weather. It is difficult to imagine that any individual can walk as far as it could have flown, but at that season it is often too cold for the insects to fly.

EMERGENCE OF ADULT INSECTS

Some stream-dwelling insects are faced with particular problems when the time comes for the adults to emerge. For many the difficulties are not greater than they are for other aquatic insects, and nymphs and pupae either crawl out of the water and moult, e.g. many Ephemeroptera and Trichoptera, all Plecoptera and some Diptera, or rise to the surface and emerge there, e.g. some Ephemeroptera, particularly Baetidae, and many Chironomidae. These are 'traditional' methods of emergence and are not greatly disturbed by the current.

Even insects which pupate in cases fixed to the substratum face no particular problems as long as the pupa can free itself. It is well known, for instance, that the pupa, or more strictly speaking the pharate adult, of the Trichoptera has large and peculiarly crossed mandibles. These are attached by tendons to muscles inside the adult head, and when the time comes for emergence the contraction of the adductor muscles increases the distance between the pointed tips, which are inserted into the silken mesh of the cocoon and case. In this way the silk is torn and the pupa can wriggle free and swim ashore (Fig. VIII, 23) (Hubault, 1927). The larva of the moth *Aulacodes*, which also pupates in a closed cell under the canopy of silk spun by the larva, leaves a slit at the top of its shelter through which the pupa can escape into the water (Pruthi, 1928). Problems, however, arise where the pupa is itself fixed to the substratum, or is unable to free itself, as when the pupal case faces into a fast current. This occurs, for instance, in the caddis-worm *Oligoplectrum maculatum*, which, as we shall see, attaches its case with the head end directed into fast water. The pupa has thus to emerge against the current. The last-stage larva, however, builds a circular flange round the mouth of the tube, so that the whole structure comes to resemble a small trumpet (Fig. X, 3), and it is suggested that this, by deflecting the water, creates a small dead space immediately in front of the case into which the pupa can emerge (Nielsen, 1943, 1950*b*).

The pupae of the Blepharoceridae are firmly attached to the rock by sticky chitinous secretions at the sides of the abdomen, and those of the Deuterophlebiidae are similarly stuck down. Pupae of the Simuliidae are also immobile, so in all these three groups the adult fly has to emerge under the water (Fig. VIII, 23). Nothing is known of the details of the emergence of Deuterophlebiidae, but in the other two families the head

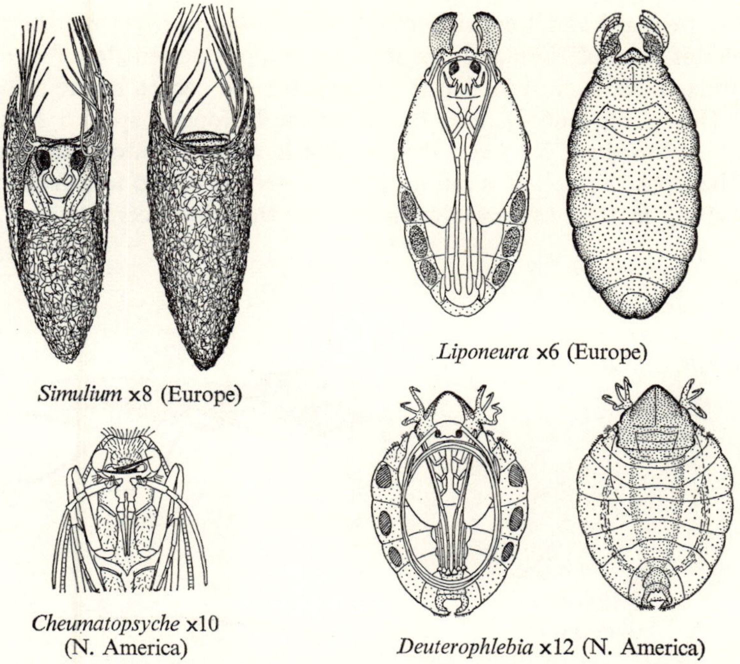

FIG. VIII, 23. The large crossed mandibles of a caddis-fly pupa and the pupae of three families of Diptera which are firmly attached to stream beds and from which the flies emerge under water. The dark patches at the sides of *Liponeura* and *Deuterophlebia* are the points of attachment.

of the pupa is directed downstream, and, just before emergence, gas accumulates between the pupal and adult cuticles. The fly thus emerges in a bubble which rises rapidly to the surface. *Simulium* is said to fly off at once (Wu, 1931), while Blepharoceridae are swept to the bank where they crawl out (Hubault, 1927). It seems very probable that this method of emergence in a bubble is also used by the Deuterophlebiidae, as it is difficult to imagine how these delicate little flies could otherwise attain the air unwetted and undamaged.

OVIPOSITION BY AERIAL ADULTS

Oviposition in running water presents a special problem to insects as it is often difficult for the adult to get to the area where the young stages will live. Nevertheless, despite the lack of any apparent morphological adaptation to aid them, the females of many Trichoptera and of several species of Baetidae creep into the water and lay their eggs on the undersides of stones (Fig. VIII, 24) (Nielsen, 1942; Verrier, 1956; Macan,

1961*a*; Leonard and Leonard, 1962). Brickenstein (1955) reports that females of *Neureclipsis* may swim at times, ovipositing females of Hydropsychidae have been recorded from considerable depths in the Mississippi (Fremling, 1960*b*), and the common European species *Hydropsyche angustipennis* has been seen to dive in and swim down at least 9 cm. (Badcock, 1953*c*). It is interesting to note that baetid females found ovipositing in this way are always without their long cerci, which are

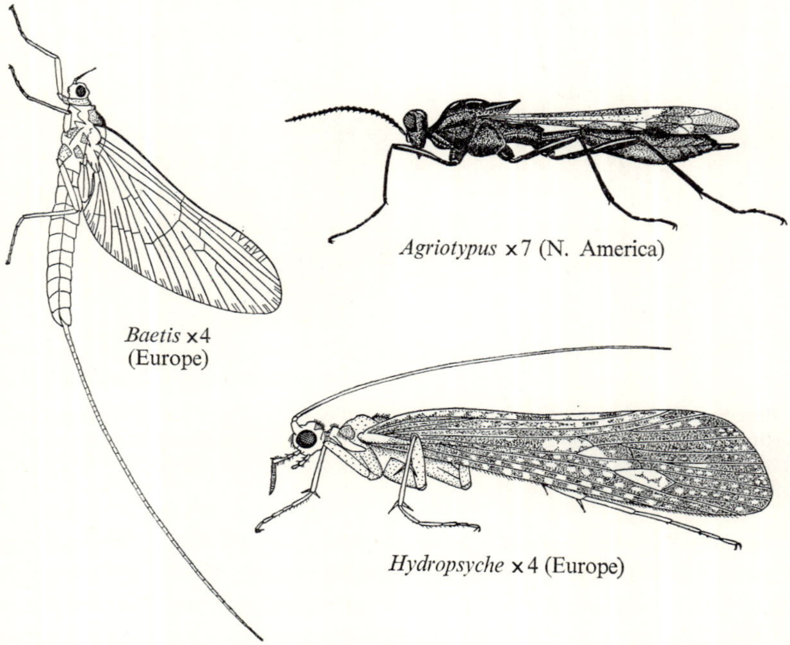

FIG. VIII, 24. Adult insects which creep down into quite rapid water to oviposit.

shed on entering the water. The long tails, which are important in stabilizing the rather clumsy flight of these insects, would be a considerable hindrance in fast water. It seems likely also that at least some of the Lepidoptera oviposit directly on the substratum, since the Indian pyralid *Aulacodes* dives readily into the water (Hora, 1930). Perhaps even more unexpected is the fact that the adult females of the very small and frail ichneumonid *Agriotypus*, which parasitizes the larvae of Goerinae, crawl successfully under the water in search of their hosts (Fig. VIII, 24) (Fisher, 1932).

All these insects become surrounded by a film of air adhering to their bodies, and it is remarkable that, despite this added hindrance of the elastic film of the air/water interface, they maintain their hold and arrive at their destination. *Agriotypus* has not only to find larvae of *Silo* or *Goera*, but it must select among them specimens that are about to pupate or have already done so. Presumably therefore, a considerable amount of time has to be spent under water by each female before she has laid all her eggs, each on a different host.

Most insects, however, avoid actually entering the water, and this applies even to some Trichoptera. Thus *Crunoecia*, whose larvae live at the water's edge, lays its eggs there, but *Oligoplectrum*, whose larvae live on the exposed tops of stones, also oviposits at the edge (Nielsen, 1942, 1943). Presumably the females of the latter genus could not reach a point very close to the larval habitat, and the task of getting there has to be left to the young larvae.

Oviposition at the water's edge, with subsequent migration of the larvae to their permanent habitats, is in fact a common expedient, and is found in the Limnephilidae (Wiggins, 1966), most Diptera, e.g. Chironomidae, most species of *Simulium* (Peterson and Wolfe, 1958; Davies *et al.*, 1962) and Blepharoceridae (Hubault, 1927; Giudicelli, 1964), and in a few mayflies, e.g. *Habroleptoides* (Pleskot, 1953; Macan, 1961*a*).

Other insects deposit their eggs at the water surface; the eggs then sink and, as we have seen, stick to the substratum. Oviposition is achieved in one of three ways, the simplest of which is practised by the large stoneflies. These extrude a mass of eggs which adheres to the underside of the tip of the abdomen. The female then runs, or swims, out on to the water surface; the sticky substance holding the eggs together dissolves and they fall separately to the bottom (Hynes, 1941; Brinck, 1949). A further development of this method is for the female to fly over the water with her egg-mass and either drop it or to glide down to the surface and touch the egg-mass down on to the water. This 'dive-bombing' method is used by many small stoneflies (Hynes, 1941; Minshall and Minshall, 1966) and by a number of mayflies (Needham *et al.*, 1935; Verrier, 1956), and it is particularly insects with this method of laying that can so often be seen to oviposit on the shiny tops of automobiles or on wet asphalted roads, which they presumably mistake for water surfaces. In the small stoneflies, especially the Nemouridae, and possibly in other groups also, the sticky secretion which holds the eggs together has the property of dissolving, like camphor, almost explosively. The eggs are thus shot vigorously apart and scattered over a considerable area (Hynes, 1941).

Yet other insects, e.g. many Ecdyonuridae (Leonard and Leonard, 1962), the dragonfly *Cordulegaster*, and some Simuliidae (Davies *et al.*, 1962), deposit their eggs one or a few at a time by flying over the water and dipping the tip of the abdomen at intervals. Unlike the dive-bombing of whole clumps of eggs this is not done on areas with the swiftest and most turbulent flow, presumably because of the risk of entrainment and loss of the female. When all the eggs are dropped in one mass her fate is of no consequence to the species, and, indeed, it can often be observed that dive-bombing females may become trapped on the water surface in swift places and swept downstream.

Finally a few insects avoid the water altogether and lay their eggs in masses on branches, leaves, and bridges often some considerable distance above the water. This seems to be the habit in some Trichoptera, in all the Megaloptera, and in *Atherix* (Muttkowski, 1929; Hamilton, 1940; Wesenberg Lund, 1943; Kaiser, 1961). The larvae then hatch and fall into the water and disperse. It is perhaps strange that this method of oviposition is not more widespread, as almost all streams, unless man has cleared them, are overhung by vegetation. It is also odd that, in at least the Megaloptera and *Atherix*, the eggs are often laid in large clusters contributed by many females. This seems to be very much the proverbial single egg-basket, and one would have expected that evolution would have selected wider dispersal. Unlike the dive-bombers there would seem to be no advantage to the female in laying all her eggs in one place, and even less for several, often many, specimens to lay in close proximity.

CHAPTER IX

Further adaptations of benthic invertebrates

PHYSIOLOGICAL CONSIDERATIONS

We have already noted in Chapters IV and V Ruttner's suggestion that moving water is physiologically richer than still water and that this causes plants to be metabolically more active in running water. The same effect can be observed in animals, and several laboratory studies have shown that arthropods from running water consume more oxygen than similar species from still water, even when both are anaesthetized to eliminate any differences caused only by activity (Table IX, 1) (Fox *et al.*, 1934, 1937; Walshe, 1948). It is therefore clear that, in general, the metabolic rate of stream animals is higher, at a given temperature, than that of similar species in still water; and it has even been found that metabolic rate varies with the habitat within a single species. Fox and Simmonds (1933) demonstrated that the respiratory rate of the common European *Asellus aquaticus* from a fast stream was about one and a half times that of specimens from slow streams.

Species	Habitat	Oxygen consumption cu. mm./g./hr.
Ephemerella ignita	stream	950
Caenis sp.	pond	290
Ecdyonurus venosus	under stones in swift stream	604
Ephemera danica	burrowing in sand in swift stream	370
Ephemera vulgata	burrowing in mud in a pond	278

TABLE IX, 1. The oxygen consumption at 10 °C. of urethane-anaesthetized nymphs of various European mayflies. Data from Fox *et al.* (1934).

In general it seems to be true that whereas the respiratory rate of still-water animals is of the so-called 'independent' type, i.e. it remains more or less constant as the oxygen content falls and begins to decline only at quite low oxygen levels, that of stream-dwelling animals tends to be 'dependent' at least at low and medium concentrations. That is to say that it falls as the oxygen content falls, or remains constant only at the higher oxygen levels but begins to fall when the oxygen content is still quite high. At least one *Baetis* species remains dependent even above 100 per cent air-saturation at 10 °C. (Fox *et al.*, 1937), and it is probable that this applies to some other species also; it definitely is true of some algae and higher plants (Gessner and Pannier, 1958). There are, however, exceptions to the general rule that stream animals have dependent respiration. Berg (1952), during his study of two European species of ancylid limpet, found that the situation is the reverse of what he expected. *Ancylus fluviatilis*, which is a species of streams and rocky lake shores, has an independent respiratory rate, while *Acroloxus lacustris* from ponds, sheltered lake shores, and large sluggish rivers is dependent. This stands as a warning against generalizing from too few data obtained from too narrow a range of animals.

Other warnings are also in order when attempting to assess the ecological significance of laboratory experimental work. Many, probably most, animals from temperate climates display a fair amount of physiological adjustment to temperature, and, as a consequence, their oxygen consumption at any one temperature varies with the seasons. This is true of the limpets cited above (Berg *et al.*, 1958), and the amount of adjustment to temperature may be different for individuals of different sizes. This has recently been shown to occur in the European stonefly *Perla marginata*, of which medium-sized nymphs adjust their metabolic rate better to higher temperatures than do large or small specimens; and Istenič (1963) shows that this is very neatly related to the life cycle of the species, which takes three years to mature. The little ones hatch in the autumn, when the temperature is already falling; they grow during the winter, and by summertime they are medium-sized and can continue their activity and growth despite the higher temperatures, to which they are then able to regulate. The same applies during their second summer, and by their second autumn, when they are entering the large category, temperatures are falling and they no longer need the regulatory power, since they will have grown to full size and emerged before temperatures again rise. The unwary experimenter, therefore, looking for large specimens with which to work, or, alternatively, collecting his experimental material only in the summer when the water is warm and pleasant for fieldwork, could obtain a quite erroneous picture of the metabolic regulatory ability of this species.

It should also be pointed out that, as in experiments on the resistance to current which were discussed in the last chapter, factors other than oxygen content and temperature can affect metabolic rate as measured in the laboratory. Wautier and Pattée (1955) showed that the oxygen consumption of nymphs of *Ephemera danica* dropped by 65 per cent when they were allowed to dig in sand, their normal habitat, and that that of *Ecdyonurus venosus* fell by 30–40 per cent when they were allowed to sit on stones rather than on a smooth glass surface. These reductions were of the same order as those produced by anaesthetizing with urethane, and they are a clear indication that animals may be disturbed and respire unusually when they are in the unfamiliar, and probably uncomfortable, environment of laboratory glassware. Eriksen (1964) has further shown, for the American species *Ephemera simulans*, that the respiratory rate falls to a minimum when the grains of the material in which the nymphs are allowed to burrow are of a particular size, indicating that they become quiescent only when they are in the correct habitat. Similarly, *Hexagenia limbata* respires less when it is allowed to shelter in a glass tube which somewhat resembles its natural burrow than it does when it is free in a respiratory chamber. *Siphlonurus armatus*, which is a swimming species and is thus less affected by the substratum, showed no such changes in respiratory rate when given sand or stones. Thus each species can be expected to react differently.

Another factor affecting respiration is current, and this almost certainly operates through rapid replacement of the water in the immediate vicinity of the animal. It has been studied experimentally in several Ephemeroptera, Trichoptera, and Plecoptera, and the results are of considerable interest. Ambühl (1959, 1961, 1962; Jaag and Ambühl, 1964) showed that the respiratory rate of *Rhyacophila*, *Baetis*, and *Rhithrogena* is directly related to the current speed, and that faster currents decrease the lowest levels of oxygen concentration at which the insects can survive (Figs. IX, 1 and IX, 2). This last effect is also known to occur in the stoneflies *Acroneuria* and *Pteronarcys* (Knight and Gaufin, 1963, 1964), and in the caddis-worms *Wormaldia* and *Hydropsyche* (Philipson, 1954), and it effectively restricts all these animals to running water. It is not, however, universal, as can be seen in Fig. IX, 2, e.g. *Anabolia* and *Ephemerella*, and it occurs probably only in the most rheophile organisms. To the above obligately running-water animals we can add *Agrion*, in which the respiratory rate falls to only 50–60 per cent of normal when it is kept in still water, and this is insufficient for survival (Zahner, 1959). *Simulium* also has an apparent requirement for current (Wu, 1931), and, even though some species of the genus can survive in quite low oxygen tensions, none can tolerate still water for

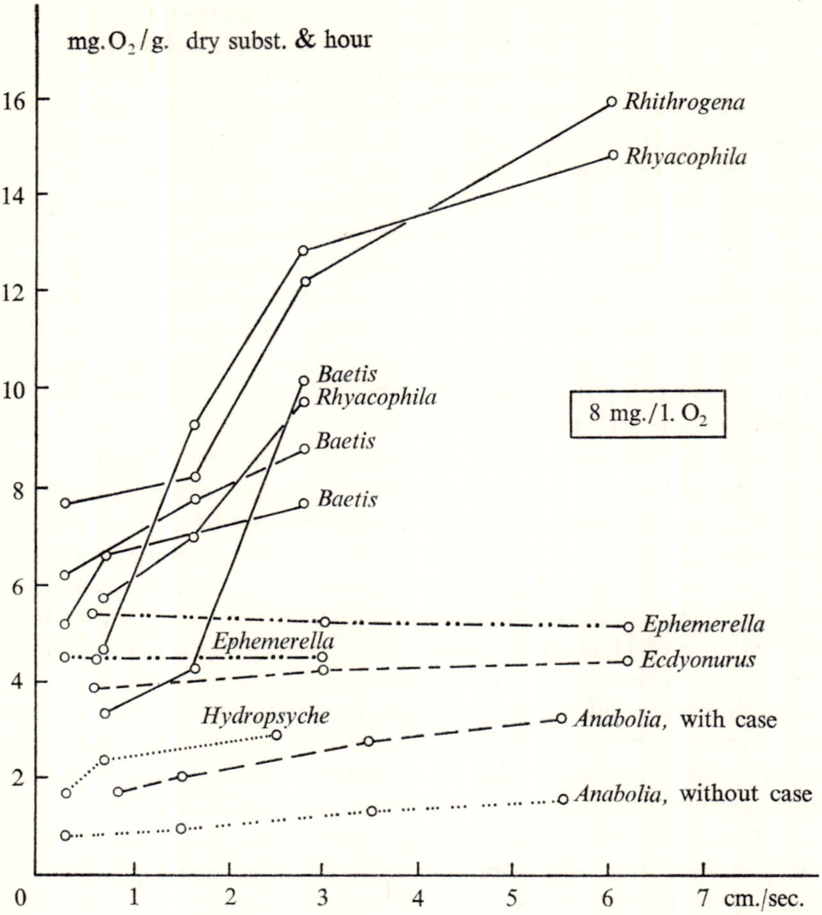

FIG. IX, 1. Graphs showing the effect of current speed on the oxygen consumption of some freshwater animals; measured at 18 °C. in water containing 8 mg. oxygen/l. From Jaag and Ambühl (1964) with permission.

long. It would seem that, like *Agrion*, their respiratory functions are so tied to water movement that even 100 per cent saturation with air is inadequate in still water.

Not all stream-dwelling animals are, however, so much affected by current. Ambühl found that in *Ecdyonurus venosus* the effect of current was apparent only at fairly low oxygen concentrations, and that in *Ephemerella* it was not apparent at all (Fig. IX, 1). This then is an important way in which species may differ in their dependence upon environmental features.

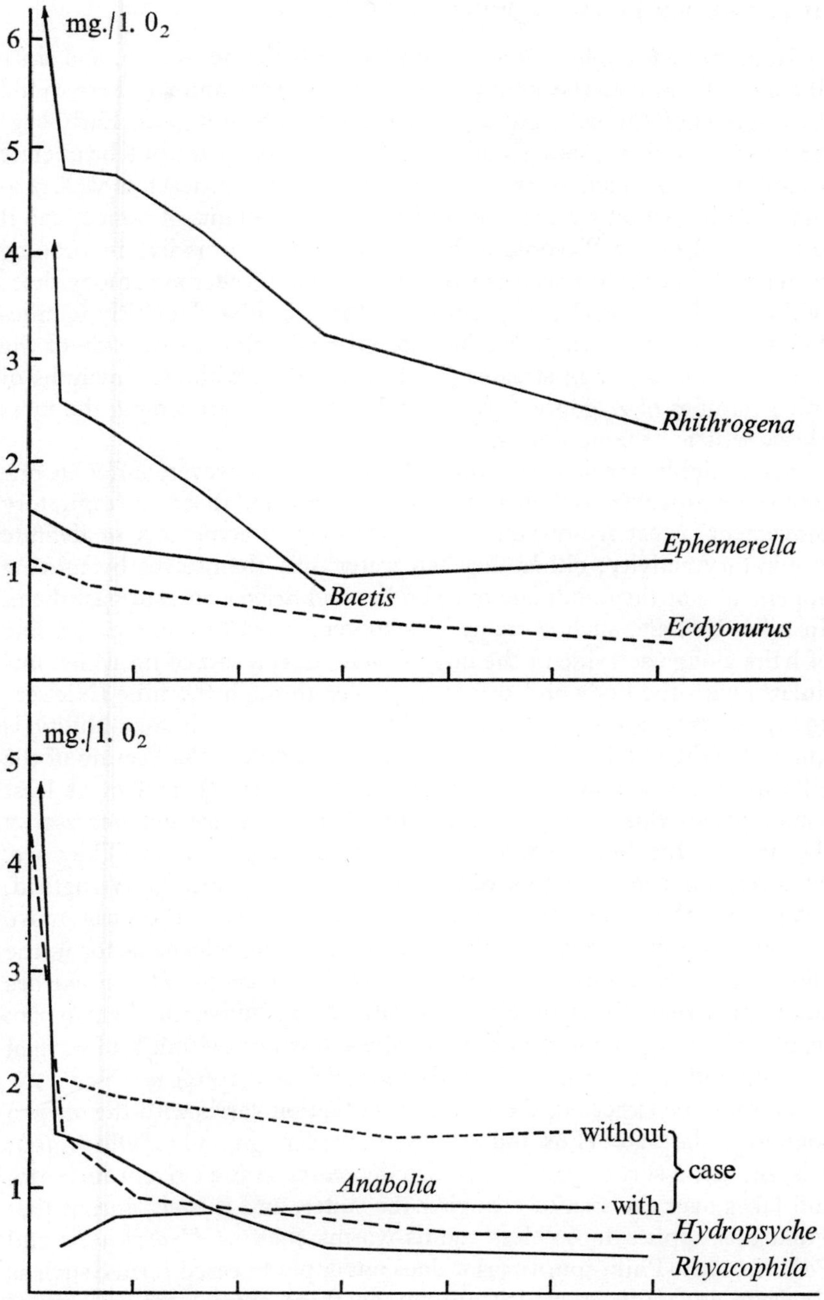

FIG. IX, 2. Oxygen concentrations which are just too low for the survival of experimental animals at different current speeds. From Jaag and Ambühl (1964) with permission.

It might be thought that as running water is rich in oxygen, and since it constantly renews the supply on the surface of the animal, there would be a tendency towards reduction of the respiratory organs. Early suggestions that this is so (Dodds and Hisaw, 1924*b*) do not bear close examination. In fact, stream-dwelling animals are at least as well provided with respiratory structures as are the inhabitants of ponds, and it can be said of the Plecoptera that *all* the gilled forms live in running water and that the few species which occur in still water are among those without gills. Similarly, the nymphs of the primitive damselfly families Polythoridae and Epallagidae have tubular gills down each side of the abdomen and occur in streams (Lieftinck, 1962), while the nymphs of other families of Zygoptera, which lack these gills, are among the most characteristic of pond animals.

Some highly specialized stream-dwellers do, however, differ in one important aspect from their still-water relatives, and that is in respiratory movement. Most freshwater arthropods display some sort of definite respiratory activity; either they fan water over themselves by moving appendages or they undulate their bodies and drive a current past them. In tube-dwellers, such as many Trichoptera and Chironomidae, a line of hairs along each side of the body functions as a sort of fin which undulates with the body and drives the water through the tube (Nielsen, 1942). If they are deprived of their case their respiratory ability is impaired (Figs. IX, 1 and IX, 2). In the Ephemeroptera the beating of the gills produces a respiratory current (Eastham, 1932), and in at least some species this is their primary function; they are not necessarily the area of the body where all gaseous exchange occurs. They are, however, important in this respect in burrowing species (Wingfield, 1939), and their size, and the amount of water that they can move through the tube at a given rate of beating, is a significant factor in the ability of the nymphs to tolerate low oxygen tensions. *Hexagenia* has larger gills than those of *Ephemera* (Fig. IX, 6), and with them it can regulate its oxygen consumption at values down to 0·8 ml./l. of oxygen as compared with 1·2 ml./l. in *Ephemera* (Eriksen, 1964).

As would be expected, the speed of ventilation varies with the oxygen content of the water (Fox and Sidney, 1953; Knight and Gaufin, 1964); it is, in fact, a sort of panting. It also decreases as the current increases and takes over the task of changing the water, and it is significant that while this applies to caseless caddis-worms such as *Hydropsyche* and *Polycentropus* (Philipson, 1954) it does not apply to cased forms, such as *Anabolia* and *Halesus*, which are among those animals whose rate of oxygen consumption is not affected by current (Fig. IX, 1) (Ambühl, 1959). It will be clear why this should be so when it is considered that they are able to produce their own current, and that the cases protect their bodies from outside water movement. But some of the most highly adapted of stream-dwellers have lost this ability to ventilate, or like the stoneflies they do it only when they are in respiratory distress (Gaufin and Gaufin, 1961; Knight and Gaufin, 1966); such animals have, therefore, come to rely entirely on the current. This applies to several of the caseless Trichoptera, e.g. *Rhyacophila* and the Philopotamidae, several Ephemeroptera, e.g. *Baetis* and those genera which have their gills incorporated into a friction device, and the dipteran families Blepharoceridae and Simuliidae. These animals are therefore confined to running water by their respiratory physiology. This point is well illustrated by some of the experiments of Ambühl (Fig. IX, 2), who refers to these species as rheostenic (confined to streams) as opposed to merely rheophile species which usually occur in streams but are able to live in still water for at least short periods.

In all the animals we have been considering to this point the respiration is directly through the body surface, and this applies also to pulmonate snails, which, like those from the deeper parts of lakes, have their mantle cavities full of water and which, like the latter, are not able to make frequent journeys to the surface (Hunter, 1953*a*). The rate of gas exchange therefore tends to be proportional not to the weight of the animal, but to its surface area (Istenič, 1963). This puts an extra respiratory strain on larger individuals, even though the oxygen consumption per milligram of body weight usually declines with increasing size (Knight and Gaufin, 1966). Although the decrease in surface/volume ratio can be to some extent offset, as it is in some stoneflies and caddis-worms, by an increase in the number of gill filaments as the animal grows (Hynes, 1941; Nielson, 1942), it seems to be an important factor controlling life histories. Warm water dissolves less oxygen than cold water, so while a small specimen may be quite all right at higher temperatures a large one may be in respiratory distress. Pleskot (1953) suggests that this simple relationship controls the life history of the mayfly *Habroleptoides*. The small nymphs hatch during the summer and are well able to respire in the warm water, but the remaining large specimens of the previous generation are eliminated by the rising temperature in the spring. Hilmy (1962) has indeed shown that the oxygen consumption in this genus is particularly high in the last instar, almost as high as it is in very young specimens, so the older specimens are subjected to the double strain of increased respiratory rate and a decreasing ratio of surface area to weight. It seems probable that this sort of consideration applies to some stoneflies also, e.g. *Brachyptera* (Hynes, 1961), and the relationship between surface area, temperature, and oxygen may also explain the observation by Steffan (1963, 1964) that, in each genus or sub-genus of Elminthidae in Central Europe, the adults of the species which occur only in cold headwaters are larger, and heavier, than those of species which extend further downstream. The respiration of these beetles depends, as we shall see, on surface area, so they are quite comparable to the larval and nymphal stages of other insects.

A great many stream animals are more stenothermic, that is restricted to a narrower, in this instance cold, temperature range than their still-water relatives. This has been shown experimentally to apply to mayfly nymphs (Whitney, 1939) and it can often be inferred from field observations. Thus, for example, most species of stonefly in Austria are confined to waters where the summer temperature does not exceed 15 °C., and very few occur in waters which regularly exceed 25 °C. (Pomeisl, 1961). It seems very probable, however, that the intolerance of higher temperatures is at least as much concerned with respiration and the lower availability of oxygen in warmer water, as it is with temperature itself. This is reflected in the finding of Knight and Gaufin (1964) that a decrease of only 5·6 °C. from 15·6 °C. decreases by a factor of 2·4 the concentration of oxygen which is lethal to *Pteronarcys*, and in the discovery, which most stream biologists soon make for themselves if they do not learn it from someone else, that it is much easier to transport stream animals alive in moist moss than it is in water. Under these conditions they have immediate access to atmospheric air, and, within limits, the temperature becomes less important. It can be inferred from this that animals which can make direct use of atmospheric air would tend to be among the most resistant to high temperature. A nice example is given by Cole (1957), who records his surprise that an adult of *Stenelmis crenata*, from a stream in New York, survived for over a year at room temperature in a closed vial containing a little water. As we shall soon see, the elminthid beetles are able to respire both in air and in water, and this specimen had access to air in the vial and so was not killed by the temperature. It might well have died had it been unable to reach the air even if the water had remained saturated with oxygen.

A further point about temperature is that very large numbers of the stream-dwelling invertebrates remain active in very cold water. Indeed, the winter, even under freezing conditions, is a period of great activity and growth in temperate streams, in contrast to the general shut-down which occurs on the land (Hynes, 1963). This ability to continue active life at very low temperatures is carried out of the water on to the land surface by at least a few insects. Both in Europe and North America, the adults of some stoneflies regularly emerge on to snow and ice, and they mate and oviposit while the weather remains very cold. Adults of

Allocapnia are unaffected by freezing temperatures, and they become active as soon as the sun shines (Frison, 1929). It may be significant that all the winter stoneflies are black, and are thus able to absorb the maximum amount of radiant heat, but it should be noted that black is by no means an unusual colour in this order of insects and that it is to be found in many summer-emerging species also. However, *all* the winter species are black.

Lastly, in relation to the physiological effects of water movement, it seems very probable that, as with oxygen and plant nutrient salts, other substances in solution are more readily available than they are in still water. This may explain the fact that whereas, in general, molluscs are absent from, or very scarce in, soft-water lakes and ponds this is not so marked in streams. Indeed, we have already noted that the pearl mussel *Margaritifera* may occur in very soft waters, and some other molluscs are often to be found in soft-water streams. Possibly the constant change of water surrounding the animal makes the fewer calcium ions more readily available, and so, effectively, more concentrated. Here also may lie the explanation of the odd fact that, in lentic waters, a large volume of water to some extent compensates for low calcium content. Macan (1963) reports that *Planorbis carinatus*, although generally a typical hard-water species, occurs commonly in the English lake Windermere which contains only 5 mg./l. of calcium. A similar example is the snail *Myxas glutinosa* which occurs in the same lake and in the even softer Welsh lake, Llyn Tegid, but elsewhere in Britain is known only from hard water. Possibly the almost continual movement and wave action in these lakes keeps the water flowing over the snails, and provides the sort of physiological environment that is found in a stream. The same consideration almost certainly applies to other ions, and an example is provided by the northern Atlantic brackish-water amphipod *Gammarus duebeni*, which in some parts of its western European range occurs and breeds in freshwater streams. That it does not do so more widely is almost certainly due, at least in part, to competition, but this is an aspect which does not concern us here. It is readily maintained and bred in the laboratory in brackish water, but it will not breed, although it lives and grows, in fresh water—even when this is taken from streams in which actively reproducing populations of the species occur—unless sodium chloride is added or the water is running (Hynes, 1954). This indicates that in still water there is a shortage of sodium or chloride ions, or perhaps merely of osmotic pressure, and that this is corrected by water movement. Subsequent, unpublished, investigations showed that this amphipod breeds successfully even if suitable fresh water is only stirred round in a jar without any addition of ions to the system.

It is thus undoubtedly true that, at least as regards many dissolved substances, flow increases the effective physiological concentration in the water, and that, for animals as well as plants, running water is a richer environment than still water of the same chemical content.

SPECIAL RESPIRATORY MECHANISMS

Life in running water presents no special problems to groups of aquatic origin which have respiratory systems adapted to working under the water. Indeed, as we have just seen, it is a particularly satisfactory environment in this respect. For groups of aerial origin, however, difficulties arise when they cannot, because of the danger of being swept away, rise to the water surface to renew their air supply. Pulmonate snails, as we have seen, evade this problem by filling the mantle cavity with water, which thus comes to function, with the rest of the body surface, as a gill; and most insects, in their larval or nymphal stages, have closed, or effectively closed, tracheal systems, often supplemented by tracheal gills, and they rely on cutaneous respiration. This is, of course, possible for aquatic insects in a way that it would not be on land, because there is no problem of water loss; and it does not make osmotic regulation any more difficult than it is for all other freshwater creatures. Adult insects, however, for reasons which are not understood, do not seem to have evolved such simple solutions to this difficulty, and pupae which are fixed to the substratum or are otherwise immobile are faced with a special problem. Particularly in small streams, the water level may fall considerably during quite short periods and expose them to the atmosphere. It is imperative, therefore, either that pupation should occur in areas which are not likely to become dry—and we have seen that this occurs in at least some Trichoptera—or that the pupa be able to respire successfully both in air and in water.

Generally speaking, although there are exceptions among the Lepidoptera, Plecoptera, and Diptera but only, so far as is known, in still water, adult insects in only two orders are fully aquatic. These are the Cryptocerata and the Coleoptera, and they normally respire at the water surface. Many stream-dwellers in these two groups continue this practice, e.g. the Corixidae, Nepidae, Belostomatidae, and Dytiscidae, but it confines them to relatively slow water. Some families, however, most notably the Aphelocheiridae, Elminthidae, and Hydraenidae, have moved into swift water, and surface respiration is no longer possible for them. The first two have overcome the problem by the evolution of plastron respiration (Thorpe, 1950); respiration in the Hydraenidae remains uninvestigated.

All bubbles of air carried under aerated water function as gills when they are in contact with a respiratory surface, because, as the oxygen in the bubble is used up, more comes out of solution from the surrounding water to replace it. This is because the nitrogen in the bubble dissolves very slowly in the nitrogen-saturated water, and the invasion of oxygen restores the equilibrium. The bubble therefore lasts a long while—as long as some nitrogen remains—and it supplies the animal with much more oxygen than was originally present. Ultimately, however, the nitrogen is lost and the bubble has to be replaced, and this entails a

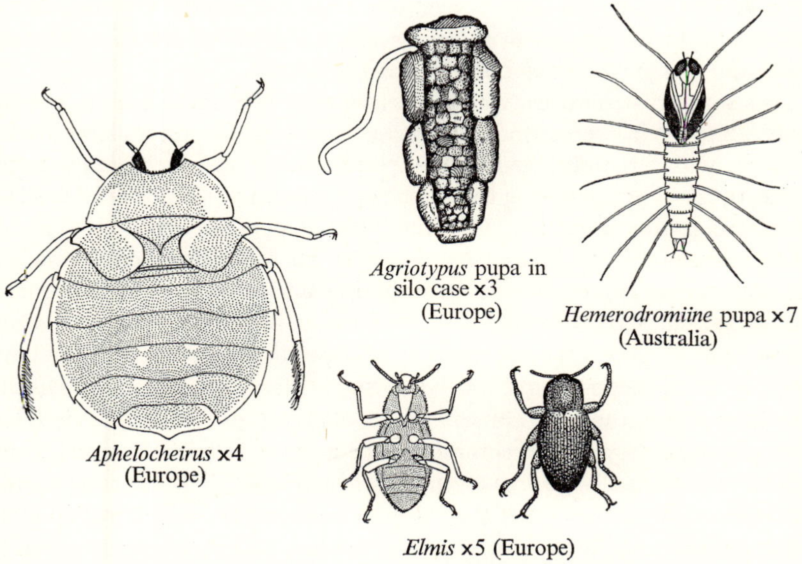

FIG. IX, 3. Insects which use plastron respiration. The stippled area on the under side of *Elmis* indicates the extent of the plastron; in *Aphelocheirus* it covers most of the body.

visit to the surface. The plastron uses this principle, but it eliminates the surface visit.

Some area on the body which includes the spiracles is covered with a dense felt of unwettable hairs (Fig. IX, 3); in *Aphelocheirus* there are about $2\frac{1}{2}$ million/sq. mm., each 5–6 μ long and bent over at the tip. This felt holds a thinly spread layer of air from which the nitrogen is unable to dissolve because the bubble is uncollapsible, and it forms a constant source of respiratory oxygen. It will, of course, collapse under pressure, but this requires four or five atmospheres and represents a considerable depth of water, and *Aphelocheirus* has special 'depth-detectors' which inform the insect of the pressure on the system. These are sensory bristles set among the plastron hairs which are depressed if the bubble becomes too thin. Similar, but less efficient, plastrons, in that the hairs are more widely spaced, occur on Elminthidae (Fig. IX, 3), and these beetles are, like *Aphelocheirus*, freed from any need of contact with the air, although, of course, the mechanism works quite well in air and is in no way damaged by being dried. Elminthidae can occasionally be seen to recondition the plastron by extruding a bladder-like bubble from the first thoracic spiracle, and so presumably speeding up gaseous exchange (Beier, 1948). This practice cannot, however, reduce the total amount of gas available to them, because, even if some nitrogen is lost from the extruded bubble, the slight negative pressure caused by the elastic unwettability of the plastron brings more out of solution in the surrounding water when the bubble is withdrawn.

Anatomical structures which permit fixed or immobile pupae to respire either in water or in air have been termed spiracular gills by Hinton, who has published several papers on them (1947, 1953, 1958, 1962, 1964). They occur on the pupae of Simuliidae, Deuterophlebiidae, Blepharoceridae (Fig. VIII, 23), some Tipulidae, some Empidae (Hemerodromiinae) (Fig. IX, 3), and *Psephenoides*, an Asian psephenid which is almost unique among beetles in pupating under water. A functionally rather similar structure forms part of the cocoon of the ichneumonid *Agriotypus* (Fig. IX, 3) (Fisher, 1932; Thorpe, 1950). Although they have been quite independently evolved, spiracular gills are all basically similar in that they all consist of an elongate or branched air-filled cuticular structure which communicates at its base with a spiracle, and which maintains a large area of air/water interface. They are constructed to various patterns, so that they do not collapse or become completely wetted if the outside pressure rises somewhat, and they become open to the air if they are dried out. The gills of the common European *Simulium ornatum* remain functional up to excess pressures of 50 cm. of mercury, equivalent to about 7 m. of water, which would seem to be ample provision for sudden spates. In other species, however, the resistance to increased pressure is lower, and there appears to be a relationship between the resistance of the gills and the width of the streams in which the species normally occur. Stream width is, of course, more or less directly related to the maximum depth of flood-water and hence to the pressure which the pupae may at times have to endure. For further details the reader is referred to Hinton's very interesting papers.

Spiracular gills vary in position, being prothoracic in most of the Diptera, although the Hemerodromiinae bear them also on seven abdominal segments, and *Psephenoides* bears them in association with the 2nd–7th abdominal spiracles. In *Agriotypus* the structure is basically quite different, although similar in function, being an air-filled tubular outpushing of silken felt which projects from the case of the trichopteran host and communicates internally with the dry cocoon chamber (Fig. IX, 3).

It will be appreciated that, under water, these gills function in fundamentally the same manner as does a plastron, and that they give a constant supply of well-oxygenated air to the spiracle with which they communicate. Should the pupa be exposed to the air, however, and this is likely to happen to all except perhaps *Agriotypus* which is presumably no more exposed to risk of desiccation than is its host, the spiracle can continue to function in the normal way. Moreover, because the respiratory area is then, as in terrestrial insects, reduced to a single spiracle suitably protected by a felt of hairs or some similar device, excessive loss of water is prevented. Spiracular gills thus combine a large surface for gaseous exchange under water with a small protected area for respiration in air where loss of water is important, and they are a beautiful example of a dual-purpose organ.

THE WATER SURFACE

Many animals live on the surface of still water and form the so-called neuston community, although it is not a community in any ecological sense. It includes such creatures as the water-striders, Gerridae, and the whirligig beetles, Gyrinidae, with their peculiar paddle-like legs and their divided eyes (Figs. VII, 2 and IX, 4), and both of these groups occur commonly on running water, although not where it is swift. Likewise other insects, such as Collembola and the little *Microvelia*, which inhabit the edges of ponds and lakes especially where there is emergent vegetation, occur in similar situations on rivers, and also under stones at the sides of quite swift water. These insects do not, however, differ in any marked way from their still-water relatives except possibly in behaviour. Presumably those species which live near rapids do not normally venture out on to them, and it is interesting that at least two species have developed quick-return mechanisms based upon chemical propulsion. It seems almost certain that others remain to be discovered.

The common European *Velia caprai*, which lurks in sheltered places at the sides of streams, can, like all the Veliidae, run on the water surface, but it is also capable of moving quite fast while keeping its legs still, and it can travel up to a quarter of a metre in this way. The motion is achieved by releasing, from the backwardly directed rostrum (Fig. IX, 4), a secretion which, like the glue holding some stonefly eggs together

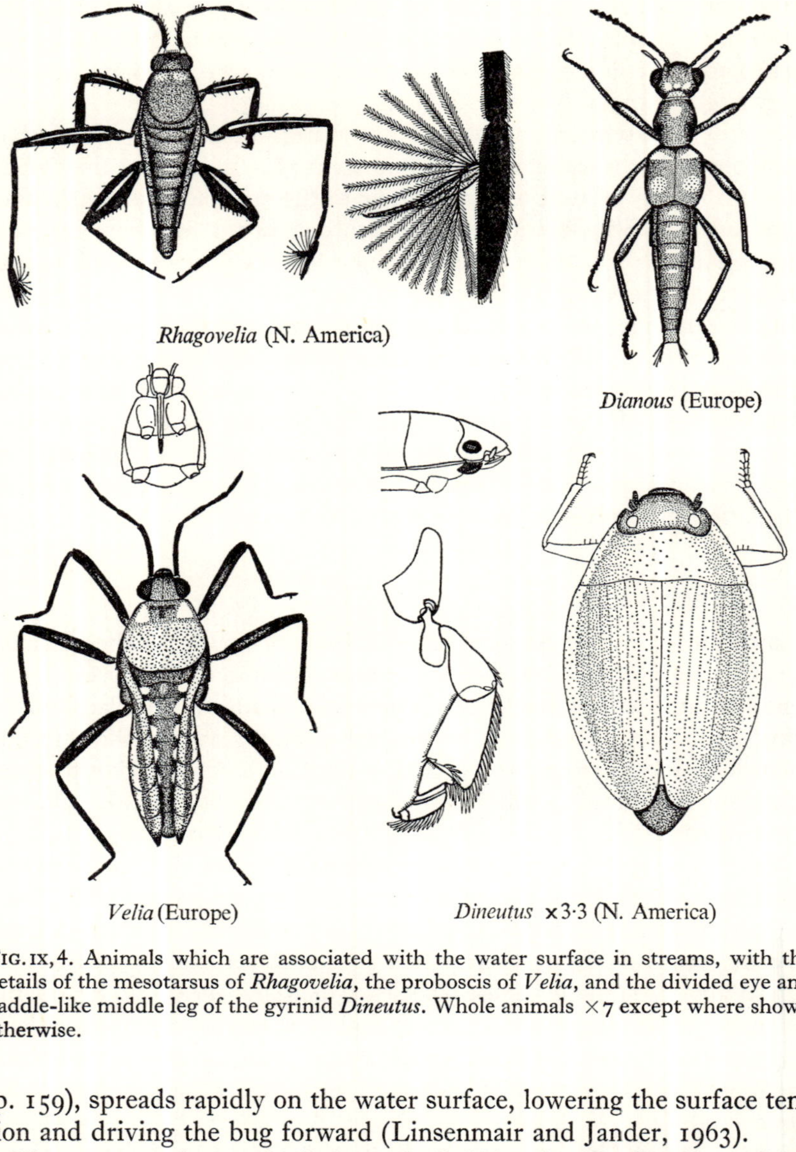

FIG. IX, 4. Animals which are associated with the water surface in streams, with the details of the mesotarsus of *Rhagovelia*, the proboscis of *Velia*, and the divided eye and paddle-like middle leg of the gyrinid *Dineutus*. Whole animals ×7 except where shown otherwise.

(p. 159), spreads rapidly on the water surface, lowering the surface tension and driving the bug forward (Linsenmair and Jander, 1963).

More spectacular are the little staphylinid beetles, the Steninae, which hunt at the water's edge and so risk falling in at times. Some of these, in the genus *Stenus*, are primarily pond forms, but in western Europe *Dianous coerulescens* is common along the banks of stony streams and it has a quite remarkable ability to zoom across the water surface to safety (Fig. IX, 4). This little metallic bluish-green beetle, which has an orange spot on each elytron, is only about 1 cm. long, yet it can move across the water at speeds of up to 60 or 70 cm./sec., and it can keep this up for several metres. It heads away from the light, and so quickly regains the bank, and this rapid movement, which is many times faster than it can run, is caused by the release of a surface-tension-lowering secretion from glands at the tip of the abdomen (Jenkins, 1960; Linsenmair and Jander, 1963). It is easy to demonstrate, by repeatedly making a specimen swim in this way, that it soon runs out of fuel, so presumably this method of movement is physiologically costly, but it is a very efficient life-saving device for an animal that lives beside swift water.

These are, however, mechanisms for *avoiding* swift water. One genus of water-strider, *Rhagovelia*, has taken to swift water as its normal habitat, and it is to be found particularly at the tail ends of riffles where it dashes about on the broken surface seeking drowning insects. Only the very young stages are confined to smooth water at the edge, and they move out into the turbulent areas as they grow (Hynes, 1948). In these insects the middle tarsus has a most interesting modification which gives an unusual grip on the surface film (Coker *et al.*, 1936). The terminal segment is deeply cleft, and at the base of the cleft arise the two highly modified, normally terminal, claws and a complexly branched bristle. The bristle, which can be folded like a fan into the cleft, has a number of plumose rays which can be spread down into the water, where, supported by one of the claws, they form a sort of oar-blade. These structures, which are shown in Fig. IX, 4, enable the insect to move very fast over rapidly flowing water where other water-striders are quickly swept away.

SAND AND SILT

Soft shifting sand and light silt are particularly difficult substrata for benthic animals, as it is not possible to make tunnels in them and they tend to envelop and overwhelm small organisms. Mixtures of the two, on the other hand, and consolidated silt, form more rigid substrata which can be burrowed into or walked upon, and they present no special problems. Indeed, specimens of Tubificidae and Ceratopogonidae have been found at considerable depths (200 and 20 cm. respectively) in river mud even though most of the fauna occurs in the top 5 cm. (Ford, 1962).

We have seen earlier (p. 114) that soft sand is inhabited by a number of specialized micro-organisms. Many of these are firmly attached to the sand grains, such as the stalked protozoan *Clathrulina* and a great variety of rotifers. The mobile forms, such as the turbellarian *Otoplana* and the oligochaetes *Potamodrilus* and *Aelosoma*, possess special protrusible sticky cells with which they can anchor themselves at times, and even the nematodes which are found in this particularly unstable habitat can attach themselves to sand grains by their hind ends (Neiswestnova-Shadina, 1935, 1937; Shadin, 1956).

Larger animals are scarce and very few are adapted to life in pure sand. This is perhaps not surprising as, in view of the reasons for its occurrence, shifting sand is poor in foodstuffs. There are, however, a few specialists in this type of environment among the mayflies (Fig. IX, 5).

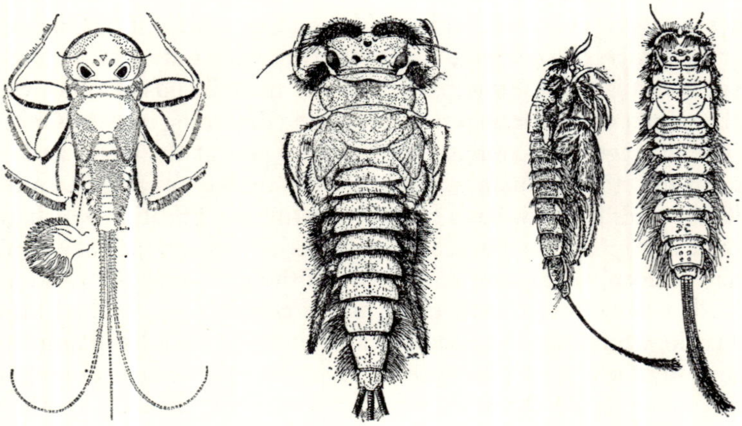

Anepeorus (N.America) *Dolania* (N. America) *Behningia* (Europe)

FIG. IX, 5. Mayfly nymphs which burrow in sand in rivers. *Anepeorus* from the Illinois Natural History Survey (Burks, 1953) and the Behningiidae, *Dolania*, and *Behningia*, from Edmunds and Traver (1959) with permission. Note the ventral position of the gills and the extreme hairiness.

These include the rare family Behningiidae, which are known from eastern Europe, Siberia, and south-eastern North America. The nymphs, which burrow in the sand, are very hairy, and this supposedly keeps the sand away from their bodies; and the gills instead of being in the usual dorsal or lateral position are ventral and boxed in on each side by the legs (Edmunds and Traver, 1959). The one sand-dwelling genus of the Ecdyonuridae, *Anepeorus*, similarly has ventral gills (Burks, 1953), and this would seem to be a device which keeps them free of sand.

It appears, indeed, that a covering of long fine hairs is a not uncommon means of keeping the body free of fine particles and preventing smothering. Most of the burrowing mayflies are markedly hairy (Fig. IX, 6), and their gills, which are held over their bodies in contrast to those of most mayfly nymphs, cause an axial current down the length of the body. By this means the fragile burrow is not disturbed and such particles as are dislodged are kept away from the surfaces of the feather-like gills (Eastham, 1932, 1937, 1939). Similarly among the stoneflies the nymphs of species which live in silty habitats are markedly hairy. For example, the hairiest stonefly nymphs in Britain are those of *Leuctra nigra* which occurs only in silty streams, and the next most hairy is *L. geniculata*

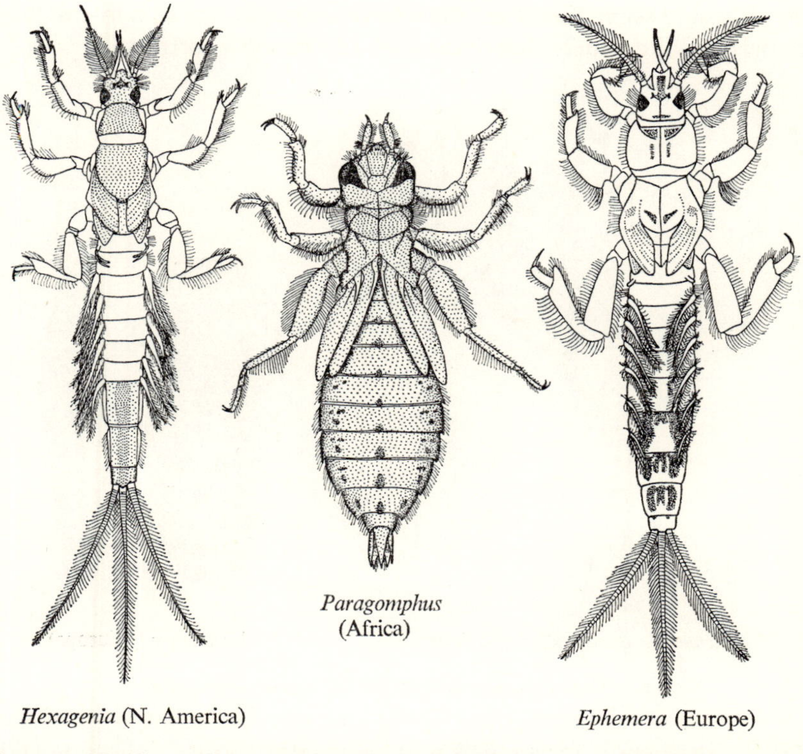

FIG. IX, 6. Mayflies and a gomphid dragonfly which burrow in soft silty substrata. All ×3.

(Fig. VII, 3) which burrows deep down into stony substrata where silt collects (Hynes, 1941). Other species in the genus which live in clean gravel are much less hairy (Fig. VIII, 21).

Another method of dealing with the respiratory problem caused by submersion in soft particulate substrata is open to the nymphs of Anisoptera, which, as is well known, respire through the anus. Several genera of the Gomphidae occur on soft substrata, where they lie buried, often with only the elevated, almost frog-like, eyes and the tip of the tail protruding, and in some genera the last abdominal segment is elongated, forming a respiratory siphon which protrudes above the substratum (Hora, 1928; Corbet, 1956*b*, 1957*b*). It is also worth noting that the Gomphidae, which are the most successful in exploiting this habitat, have also much the most hairy nymphs among the dragonflies (Fig. IX, 6). Presumably there are other advantages, such as the preservation of freedom of movement, to be gained from keeping the deposit from closely investing the body, in addition to those involved in respiration. It would seem, therefore, that extreme hairiness in stream-dwelling animals is quite widely associated with life in soft substrata.

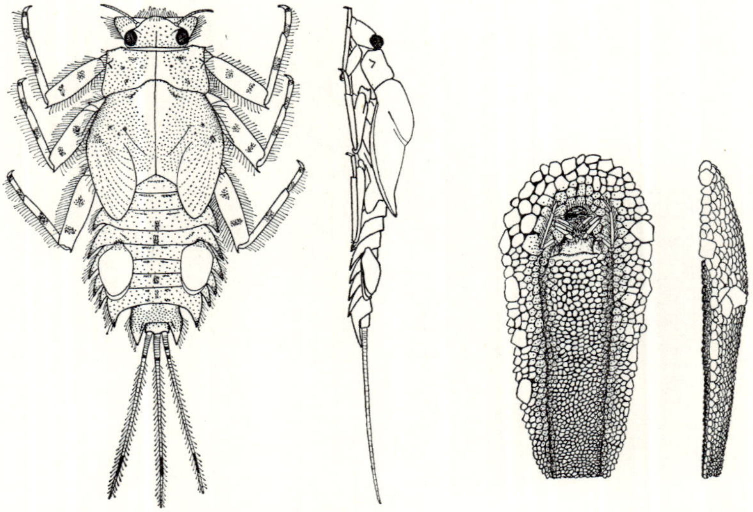

Ephemerella hecuba ×3·5 (N. America) *Molanna* ×2 (Europe)

FIG. IX, 7. Very flattened animals which live on silty substrata.

It is also possible for animals to inhabit soft deposits if they are very flattened and thus prevented from sinking down among the particles. This is seen in the caddis-worm *Molanna*, which is a common inhabitant of sandy lake shores and of fairly stable sandy substrata of rivers in the Northern Hemisphere (Fig. IX, 7). Flattening is also characteristic of some African Gomphidae and Libellulidae (Corbet, 1956*a*, 1957*a*) and of a few mayflies, e.g. *Ephemerella hecuba* which lives in sheltered silty localities (Fig. IX, 7) (Nielsen, 1950*b*).

In the family Caenidae, many of which inhabit silty substrata, it has been suggested that the dorsal position of the gills, and more particularly the way that they beat, enables these insects to live in silty places (Eastham, 1932). As has already been stated, the beating of the gills draws a current of water over the bodies of most mayflies, bringing it in from the front and usually the sides also. In the Caenidae the first gill is reduced, and the second forms a sort of operculum over the others which lie superimposed like the pages of a book (Fig. IX, 8). When they beat they cause a current not along the length of the nymph, but from side to side. Thus, if the animal rests on the silt with one side slightly raised, it can draw water in from that side without also disturbing and drawing in the silt. Implicit in this idea is, of course, the assumption that in these nymphs the gills really are important in the uptake of oxygen, and as we have seen this is certainly not true of all mayfly nymphs,

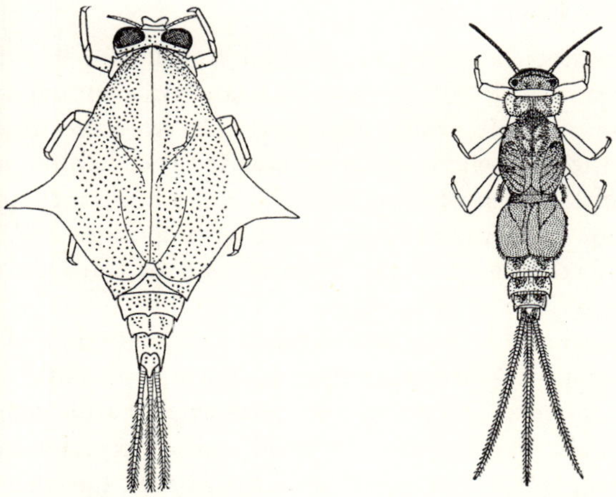

Baetisca ×5 (N. America) *Caenis* ×5 (Europe)

FIG. IX, 8. Nymphs of mayflies which occur in silty places in streams.

although it is of some. Here it seems reasonable to assume that it is, as, apart from the operculum which acts only as a baffle and perhaps as a shield against damage, the gills are very thin and fragile. The exact significance of these structures would repay further study. We may also note here that the gills of *Ephemerella hecuba* are dorsally placed (Fig. IX, 7), and, while this is in no way remarkable for a member of the Ephemerellidae, it does ensure that they can maintain a respiratory current without becoming clogged with silt. In the nymphs of most families of mayflies the normal lateral position of the gills would militate against their being able to live on silt surfaces even if they did become flattened.

The genus *Baetisca* also inhabits silty areas at the edges of streams, and the gills of its peculiar nymphs are quite covered over by the enormous dome-like pronotum (Fig. IX, 8). This may be connected with the silty habitat, but it occurs also in *Prosopistoma* (Fig. VIII, 3), which lives in stony places. *Prosopistoma*, however, unlike *Baetisca*, swims with its gills, so the covering there is probably connected with its propulsion mechanism. It is therefore, like flattening, probably a modification which has arisen in different genera for entirely different reasons. It would seem that the lateral projections on *Baetisca* are devices which prevent it from sinking into the silt and so are analagous to the projections on the abdomen of *Ephemerella hecuba* and the flange of sand grains round the case of *Molanna* (Fig. IX, 7).

PECULIAR ASSOCIATIONS

Associations between animals, in symbiotic or host–parasite relationships, are widespread in all types of habitat, and running water is no exception. For example, the oligochaete *Chaetogaster limnei*, which occurs on aquatic pulmonate snails in most types of habitat, also occurs on ancylid limpets in streams (Geldiay, 1956; Maitland, 1965*b*), and various chironomid larvae are associated with sponges and bryozoans on which they feed (Pagast, 1934; Codreanu, 1939; Kaiser, 1947), as are the larvae of Sisyridae with sponges.

There is also what might be described as a normal range of protozoan and helminth parasites of invertebrates. Normal parasitic insects are however, very scarce as compared with terrestrial or still-water habitats, and this is probably a reflection of the difficulties experienced by adult females in seeking out and ovipositing on their hosts. For example, as we have seen (p. 159), *Agriotypus* has to crawl about under the water in a laborious manner in order to oviposit in goerine larvae. Nevertheless, although it is not characteristic of the groups to which they belong, a few truly aquatic insects have taken to parasitism, and an interesting example is the orthocladiine *Symbocladius*, which attacks the nymphs of Ecdyonuridae in the Northern Hemisphere (Claassen, 1922; Codreanu, 1939; Fontaine, J., 1964). The newly hatched larva climbs on to the host and lodges under the wing pads, where it envelops itself in a bag of salivary secretion and feeds on the blood of the mayfly, which ultimately dies.

Parasitic insects are, however, very unusual, and of more general interest in the context of running water is the association with other invertebrates of many types of chironomid. Larvae of *Parachironomus* and *Eukiefferiella* are often found in association with gastropods, the latter frequently occurring under the mantles of limpets (Geldiay, 1956; Kaiser, 1947). Other genera have been found in the pupal cocoons of Simuliidae and on the legs of crabs (Fig. IX, 9), and one species of *Dactylocladius* is associated with blepharocerid larvae in New Zealand.

It will be appreciated that all these larvae benefit in that their hosts, by providing shelter in exposed places, allow them to exploit habitats which would not otherwise be available to them. A rather similar relationship seems to occur with the larvae of *Plecopteracoluthus*, which occur in gelatinous tubes on the nymphs of some North American perlid stoneflies (Steffan, 1965*a*), and *Hydrobaenus ephemerae*, which occur attached to the nymphs of some European species of *Ephemera* and apparently derive food from the respiratory current of their host (Fontaine, J., 1964).

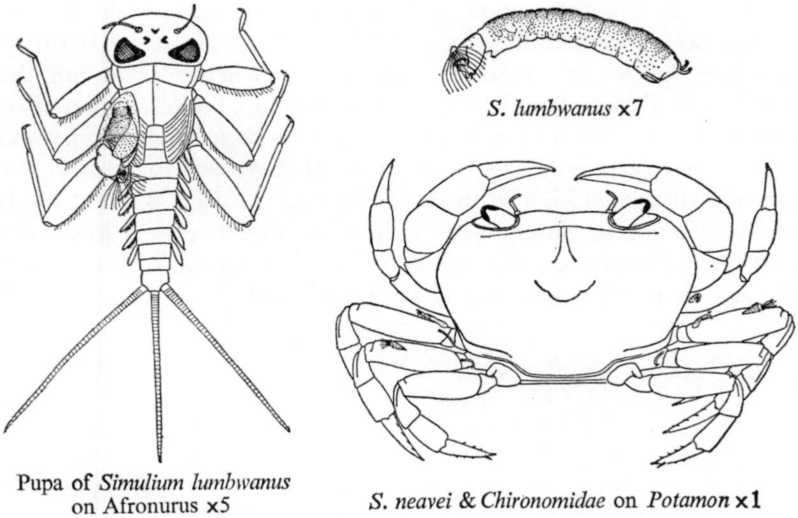

FIG. IX, 9. Phoretic associations between *Simulium* species and mayflies and crabs which are found in streams in Central Africa. Chironomid larvae can also be seen on the crab. Note that in phoretic *Simulium* larvae the circlet of hooks is sub-terminal and the eyes are very small: cf. Fig. VIII, 15.

Another interesting, and as yet unexplained, association is that of several species of *Simulium* with mayfly nymphs or crabs (Fig. IX, 9). So far this phenomenon is only reported from tropical Africa, the Himalayas, and Turkestan (Fontaine, J., 1964). *Baetis* and *Afronurus* have been found carrying larvae or pupae on their backs, and larvae have been found on the ventral surface of *Elassoneuria* (Marlier, 1950; Van Someren and McMahon, 1950; Berner, 1954; Corbet, 1960*a*). In each instance the *Simulium* species appears to occur only in association with a particular genus of mayfly. Similarly, crabs of the genus *Potamon* harbour particular species of *Simulium*, one of which, *S. neavei*, bites man and is a vector of onchocerciasis. This species lives and pupates on the carapace or legs of the crabs, and is found nowhere else, and a related species, *S. woodi*, is reported to occur only in the exhalent passages of the respiratory system (McMahon, 1952; Lewis, 1960, 1961; Lewis *et al.*, 1960). It seems that the *Simulium* derive no food from their hosts, and the reason for the association is quite unknown, although it appears to be obligatory. It is possible that the greater mobility of the host may save the larvae from dessication at times of low water (Corbet, 1961), but other species of *Similium* occur successfully in the same streams, and are indeed far more numerous. It has also been suggested that these species, which in contrast to most African rapid-water species have long pupal gills, may, by this association, be able to inhabit swifter water. This, however, seems most unlikely in view of the fact that in other continents long-gilled species are common in fast water. It is also intriguing that phoretic Simuliidae should be known only from a limited region of the earth, where the habit of using other animals as a substratum would appear to have arisen at least twice, and that they seem to occur only in clear cold streams at high altitudes (Fontaine, J., 1964). Mayfly nymphs and *Simulium* occur closely together in all continents except Antarctica, as do large decapod crustaceans and *Simulium*, so the opportunity to form associations has not been lacking elsewhere.

CHAPTER X

Feeding mechanisms and food of benthic invertebrates

FEEDING MECHANISMS

Running water presents no particular feeding problems to the majority of invertebrates, which eat just as they do in other habitats, but it does offer some extra possibilities. These are the algae which grow on areas of silt-free substratum, and the more or less steady supply of small particles and organisms carried by the current.

Scraping of algal-covered surfaces is, of course, a normal way of feeding in many groups of animals, and stream-dwelling snails, such as *Ancylus*, feed in the traditional way with their radulae. It is, however, not very usual in insects, and it occurs primarily in species which live on the stony substrata of streams. The stonefly *Brachyptera* has a brush of bristles on the stout galea, which in other stoneflies is a weak structure, and it uses this to scrape encrusting algae from stones (Fig. x, 1) (Hynes, 1941); and some ecdyonurid mayfly nymphs have elaborate modifications of the mouth parts which enable them to scrape stones in swift water. The nymphs of *Ecdyonurus* and *Rhithrogena* feed with their heads facing

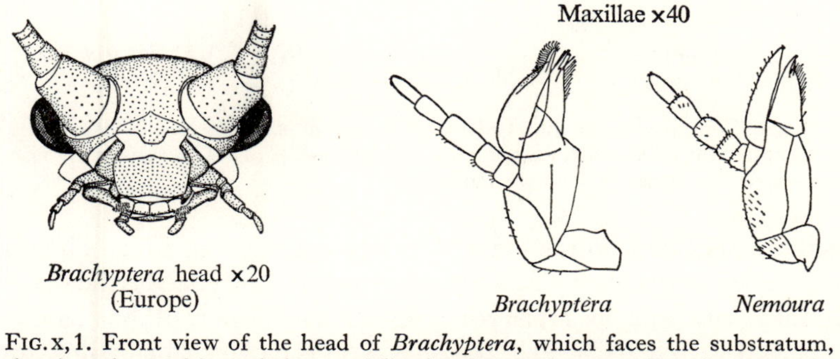

FIG. x, 1. Front view of the head of *Brachyptera*, which faces the substratum, showing the position of the specialized galea; and comparison between the maxilla of this genus and that of an unspecialized stonefly.

upstream (Fig. x, 2); the broad, hairy, or very soft, labrum is pressed against the stone thus protecting the mouth parts from the current, and the labial palps, and in *Rhithrogena* the maxillary palps also, which bear stout bristles on the lower surface, brush off the algae in the slack water behind the labrum and between the large mandibles (Strenger, 1953). Some trichopteran larvae, e.g. *Ecclisopteryx*, *Apatania*, and *Glossosoma* (Nielsen, 1942; Scott, 1958), feed on attached algae, and, as we have seen, *Glossosoma* occurs only where its scraping mechanism is not impeded by fine sand.

The organic drift in the water is exploited in many ways by various arthropods. The simplest method is the development of fringes of hairs

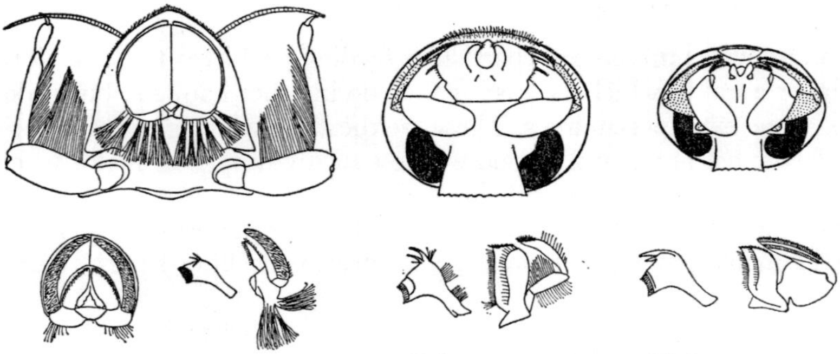

FIG. x, 2. Ventral view of the heads of three mayfly nymphs and details of the mandibles and maxillae in ventral view. The fore-legs of *Oligoneuriella* are also shown, as is a dorsal view of the labium.

which are held up into the current to trap particles for food. These are found on some mayfly nymphs, caddis-worms, and blackfly larvae.

In the African mayfly *Tricorythus*, the outside borders of the mandibles are fringed with hairs, which, together with the hairy labial palps, form a funnel with its base directed towards the mouth (Fig. x, 3) (Barnard, 1932). In other mayflies, e.g. *Isonychia* in North America and *Murphyella* in South America, the forelegs carry long fringes of hairs, and these are held out like a net to trap food particles that are brought by the current (Fig. x, 3).

It would seem likely that *Oligoneuriella* (Fig. x, 2) and *Elassoneuria*, which also has a fringe of hairs on the fore-legs (Berner, 1954), similarly catch particles in this way, and then clean off the hairs by passing them between the thickly haired labial and maxillary palps (Fig. x, 2). Pinet (1962), however, claims that *Oligoneuriella* brushes algae from stones with the fringes of hairs. Also, it would appear that *Corophium curvispinum*, which lives in a tube, strains the water through the dense fringe of hairs on its second pereiopods (Fig. VIII, 14). This would be aided by the respiratory current produced by its pleopods. A similar system is used by some Trichoptera, e.g. *Brachycentrus* in North America (Lloyd, 1921) and *Oligoplectrum* in Europe (Fig. X, 3) (Nielsen, 1943). These

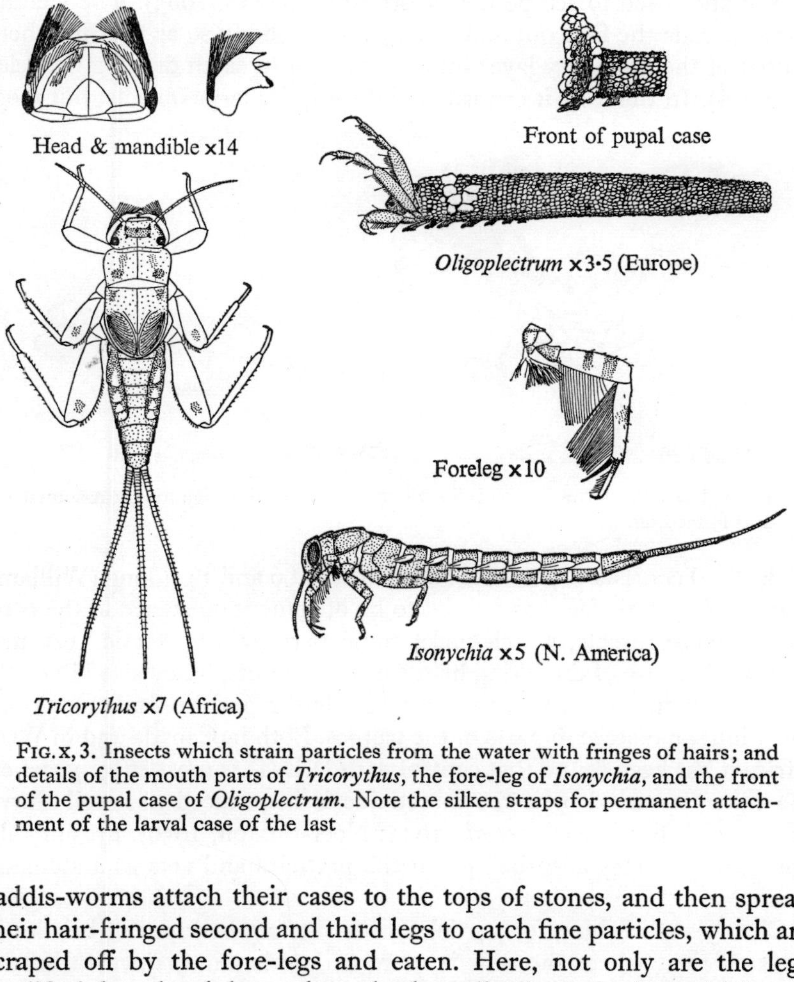

FIG. X, 3. Insects which strain particles from the water with fringes of hairs; and details of the mouth parts of *Tricorythus*, the fore-leg of *Isonychia*, and the front of the pupal case of *Oligoplectrum*. Note the silken straps for permanent attachment of the larval case of the last

caddis-worms attach their cases to the tops of stones, and then spread their hair-fringed second and third legs to catch fine particles, which are scraped off by the fore-legs and eaten. Here, not only are the legs modified, but the abdomen bears backwardly directed spines which prevent the pressure of the water on the legs from pushing the larva backwards into the case. Clemens (1917) calculated that in Cascadilla Creek, New York, enough particulate matter was passing downstream to fill the gut of *Isonychia* eight times in twelve hours, so this passive method of feeding can be regarded as quite efficient.

Simuliid larvae, apart from a few primitive genera and the first instar of *Prosimulium* (Sommerman *et al.*, 1955; Davies, 1960), have a pair of large structures on the head which carry a large folding fan and a smaller accessory fan. The anatomy and function of these are well described by Grenier (1949), and they have perhaps evolved from groups of comb-like bristles used to scrape the substratum (Davies, 1965). The feeding larva spreads the fans out and often twists its body so as to bring them up out of the boundary layer into the current to catch drifting particles (Fig. x, 4). In this way it can catch and eat quite large objects, but most

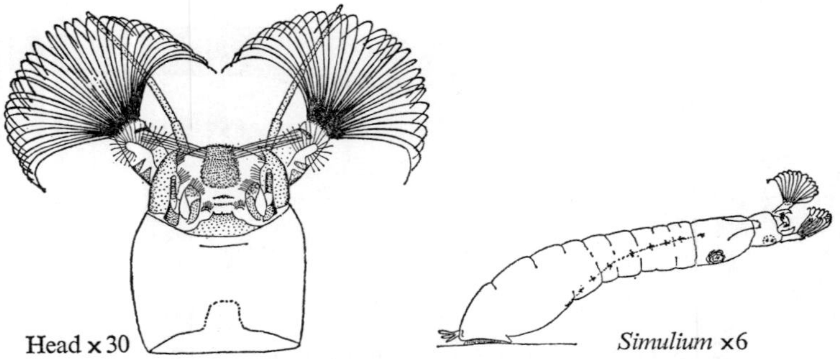

FIG. x, 4. Head of the larva of *Simulium* and a position commonly assumed during feeding.

of the food consists of particles between about 6 and 14 μ long (Williams *et al.*, 1961). This fact has proved to be of some importance in the control of these insects, which make some parts of the Arctic very unpleasant because of the biting habits of their adults. They also kill cattle and poultry in some temperate areas (Anderson and Voskuil, 1963) and carry human disease in parts of the tropics. Both in Canada and in West Africa it has been found that control with D.D.T. is most effective when the water is turbid or if the insecticide is introduced in a soil slurry (Fredeen, 1962; Fredeen *et al.*, 1953; Noel-Buxton, 1956). Presumably the toxin becomes adsorbed on to the particles and acts as a stomach poison.

Clearly such fishing for particles with outstretched fans of bristles is most efficient where the flow is more or less laminar, since turbulent eddies tend to remove particles from the fans. It can indeed frequently be seen in the field that *Simulium* larvae congregate in places where the flow is smooth, and that they are less abundant in, or absent from, stretches of rough turbulent water (Peterson and Wolfe, 1958).

More indirect methods of straining the passing water for particles are used by some of the larvae of Diptera and Trichoptera. Those of *Rheotanytarsus* occur on solid objects where they make tubes of silt and silk which project up into the water (Fig. VIII, 14) (Walshe, 1950, 1951). The top of the tube bears a number of outwardly splayed arms, between which the larva spins strings of saliva to catch passing particles. At intervals it eats and replaces the network section by section, and it also obtains particles to add to the tube or for the construction of new arms from among those which become stuck to the strings. Fishing in this manner occurs only in the summer, and the larva lies quiescent in its tube during the winter.

More elaborate structures are produced by many trichopteran larvae of the families Psychomyiidae, Hydropsychidae, Polycentropidae, and Philopotamidae, all of which spin silken nets from salivary secretions. The simplest of these are the elongate sausage-shaped structures of the Philopotamidae, which are composed of many superimposed layers of rectangular silken meshwork. In *Wormaldia* the individual threads produced by the salivary glands are 1·0 μ thick and the rectangular holes between them are $8\frac{1}{2}$–45 μ wide and 6–$7\frac{1}{2}$ μ high. Several layers thus make a dense felt which will catch tiny particles (Nielsen, 1942). Most workers report that the nets, which are fixed among stones with their widely open front facing the current, have a small hole at the downstream end, and it has been suggested that this indicates that the whole structure is a modification of the normal caddis-worm case. Sattler (1962), however, who made a detailed study of the nets of a species of *Chimarra* which lives in madicolous habitats in South America, could find no terminal hole. He therefore suggests that the simple-tube theory may not be correct. It is possible, though, that the peculiar habitat of this particular species may account for the absence of the hole. It is unique in that its nets are built in thin sheets of water, and they fill with water and often hang like dripping sausage-skins from a vertical substratum. The larvae have a modified, almost broomhead-like, labrum with which they clean tiny food particles from the nets.

More widely open funnel-like tubes of felt-like silk are spun by the Polycentropidae, whose larvae are carnivorous and eat small animals which are swept into the net. They live in a silken tube which opens into the rear of the net, and they rush out, like spiders, to attack their prey. The nets are held open partly by the current and partly by threads attached to nearby objects, and they vary somewhat from genus to genus. In *Polycentropus* they are, to use Wesenberg-Lund's (1911) terminology, shaped liked swallows' nests. In *Neureclipsis*, of which the net and its construction have been beautifully described and figured by Brickenstein

(1955), they are trumpet-shaped; the terminal tube may be turned right round so that the larva sits with its tail upstream, and the trumpets may be as much as 20 cm. long (Fig. x, 5). In *Plectrocnemia* the nets are rather untidy sack-like, structures and they are often quite large, 5–10 cm. long and 4–5 cm. wide, with the living tube opening somewhere towards the downstream end. The genera also vary in that some, e.g. *Neureclipsis*, spin nets only during the summer, while others, e.g. *Plectrocnemia*, make nets at all seasons. They also vary in their reaction to current. Most will not attempt to construct nets in still water, and some construct them more readily at certain current speeds than at others (Edington, 1965). *Polycentropus*, however, will spin a snare-like structure in still water as do some related genera which are regular inhabitants of lakes and ponds, e.g. *Holocentropus* (Philipson, 1954).

The Hydropsychidae are the master-spinners among freshwater invertebrates, and they also make nets only in running water, and in temperate climates only during the summer. They are world-wide in distribution, and are usually an important element in the fauna of running water (Fig. x, 7). They can even reach pest proportions either by slowing down the rate of flow in water conduits with their nets, as is reported from Japan (Uéno, 1952; Tsuda, 1961), or by occurring as clouds of adults which are attracted to lights and which may cause allergic reactions in humans as, for example, along the Mississippi (Fremling, 1960*b*).

The nets are spun on any solid substratum, natural or man-made, and they are usually fairly evenly spaced where they are common (Fig. x, 5). This may be because the larvae stridulate, thus warning others of their presence and so preventing the nets from being constructed where they would interfere with one another (Edington, 1965). Each larval dwelling consists basically of a silken living-tube with a large vestibule, one side of which is usually set at right angles to the current and is composed of netting. The living tube and the vestibule incorporate a number of foreign bodies so that the sides and top of the net area are supported by pieces of stone, twig, etc. (Fig. x, 6). The nets have been described by several workers (Wesenberg-Lund, 1911; Noyes, 1914; Krawany, 1930; Philipson, 1953; Kaiser, 1965) and particularly by Sattler (1955, 1958, 1963*b*). They are more or less circular, and have in the central and lower portions a remarkably regular rectangular mesh made of double threads. Sattler has shown that the thread is the product of the two salivary glands secreting simultaneously, and that the net is made by figure-of-eight movements of the head of larvae. This produces a very regular mesh which in *Hydropsyche* is about $340 \times 170 \mu$ and in *Diplectrona* a little smaller, $190 \times 130 \mu$; these dimensions may be related to the sizes

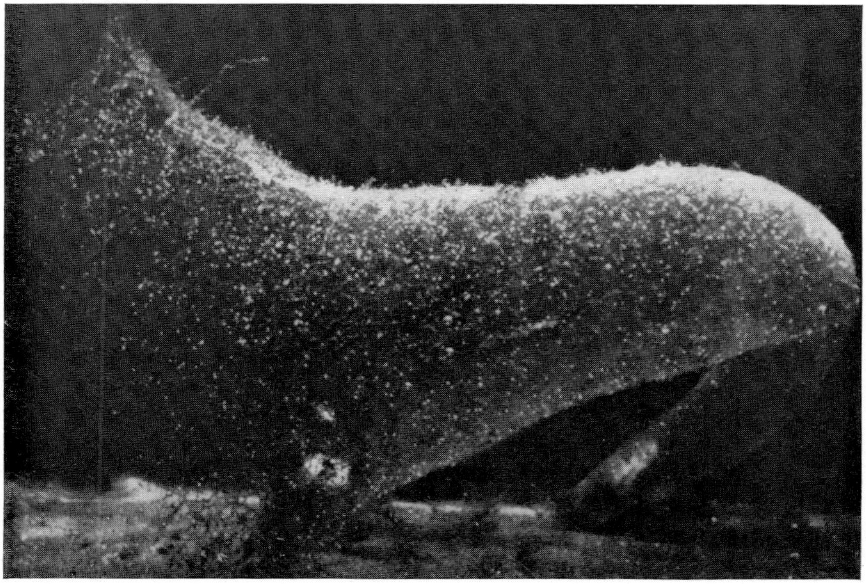

FIG. X, 5. Photographs of nets of *Hydropsyche* as seen in the field (*above*) and of *Neureclipsis* (*below*) spun in the laboratory. From Edington (1965) and Brickenstein (1955) with permission.

of certain of the mouth parts of the larva, although there is some disagreement about the way in which the mesh-size is determined (Kaiser, 1965; Sattler, 1965). This method of spinning leaves the top lateral areas of the net without a meshwork, and this is filled in irregularly after the main part has been spun. The function of the net is to catch the

FIG. X, 6. Cases and nets of *Diplectrona* and *Hydropsyche*. From Sattler (1963*b*) and Philipson (1953) with permission. The case of *Hydropsyche* is drawn from a specimen which was built on the side of a jar of stirred water and the mesh of the net is shown schematically.

small animals and pieces of vegetation on which the larva feeds, and even where it is, rather rarely, not set at right angles to the current it acts as a snare. Krawany (1930) noted that under any one stone the nets may be set at various angles, and he therefore questioned their function as drift catchers; but he had overlooked the fact that flow is turbulent, and that under a stone it is not uniform in direction. Where nets are spun in the

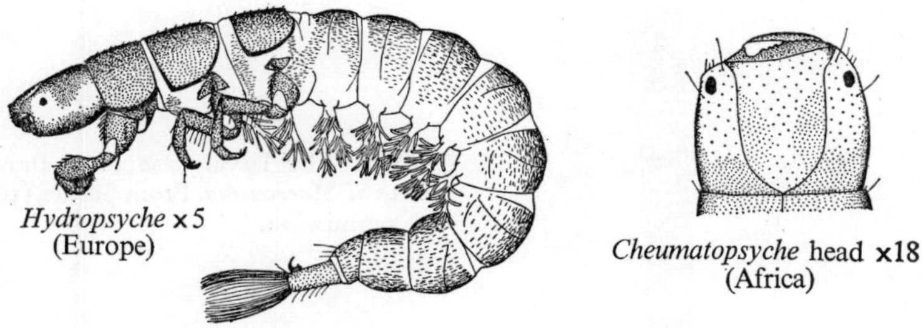

FIG. X, 7. The larva of *Hydropsyche*, a typical member of the Hydropsychidae, and the head of an African species of *Cheumatopsyche* with an asymmetrical clypeus.

open it can be observed that all of them are set at right angles to the local current (Fig. x, 5), and Kaiser (1965) has shown that the direction of the current is sensed by special setae on the top of the head of the larva.

In most of the species the head is quite bilaterally symmetrical, but in some African species of *Cheumatopsyche* the clypeus is peculiarly notched on the left side (Fig. x, 7). The function of this modification is quite unknown and it would seem to be worthy of study.

A very interesting variation of the hydropsychid net has been described from the Amazon region (Sattler, 1963*a*; Sattler and Kracht,

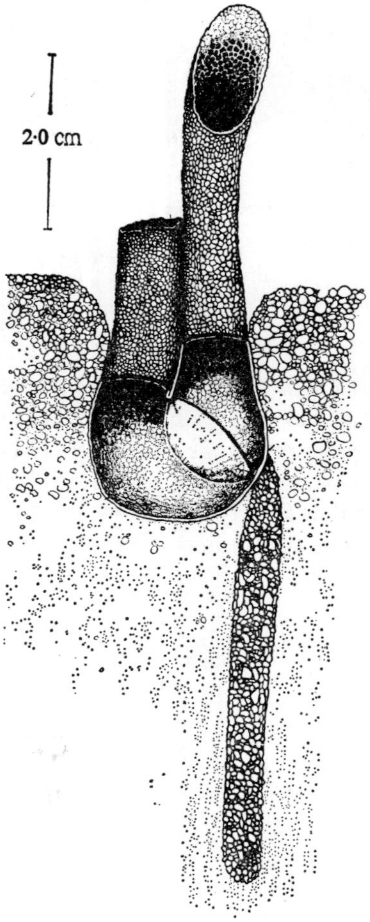

FIG. x, 8. The larval case of a Brazilian species of *Macronema*. From Sattler (1963*a*) with permission.

1963). This is made by a species of *Macronema*, and it occurs in sandy-bedded woodland streams in a structure which is quite deeply embedded in the substratum (Fig. x, 8). The entrance consists of a sand and silk funnel, which closely resembles a ship's ventilator and rises about 3 cm. from the stream bed. At the bottom, a centimetre or more below the surface, is the net, in front of which opens the even more deeply buried living tube; and behind the net a tube with a horizontal opening rises to just above the substratum. Since the more or less vertical opening of the ventilator faces upstream, the whole structure functions like a pitot tube, using the pressure caused by the flow to drive water through the net. Moreover, the fact that the opening of the inflow is set well above the substratum ensures that only the flowing water is filtered and that the

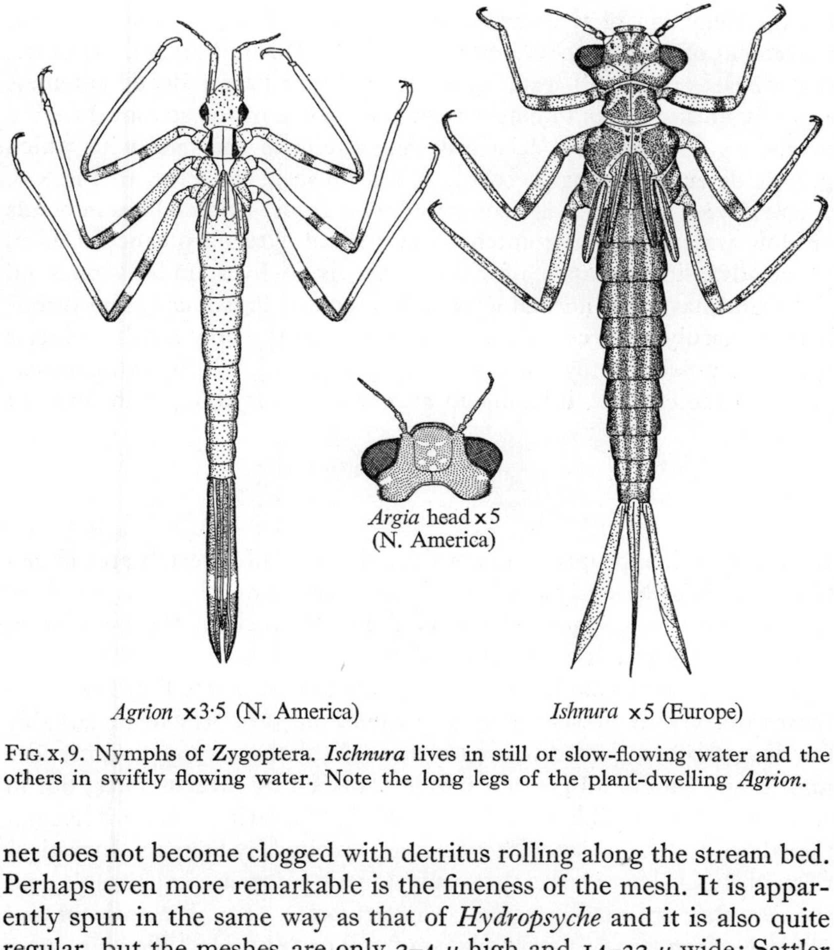

FIG. X, 9. Nymphs of Zygoptera. *Ischnura* lives in still or slow-flowing water and the others in swiftly flowing water. Note the long legs of the plant-dwelling *Agrion*.

net does not become clogged with detritus rolling along the stream bed. Perhaps even more remarkable is the fineness of the mesh. It is apparently spun in the same way as that of *Hydropsyche* and it is also quite regular, but the meshes are only 2–4 μ high and 14–32 μ wide; Sattler calculates that each net has about two million holes. The larva must therefore feed on very small particles, which it brushes off with its hairy fore-legs. Apparently the normal food is fine detritus, but this statement is based only on gut examinations. With a net of such very fine mesh it seems likely that bacteria are important food items, and we know from studies made with membrane filters that stream waters contain many thousands of bacteria per millilitre (p. 437). It is probable that many of the filter-feeders in running water actually feed to a large extent on bacteria.

It was stated earlier that running water presents no particular problems for traditional feeding methods, but this was a slight over-statement. Some animals, most notably dragonfly nymphs, rely on movement for the detection of their prey, and turbulent flow, of course, causes movement of inanimate and inedible objects. It is noticeable, therefore, that whereas most still-water dragonfly nymphs have reduced antennae and rely on their eyes to detect their prey, some of the stream dwelling forms, e.g. *Agrion*, have relatively long antennae and tarsi with which prey is detected (Richard, 1960). It can be seen, however, in Fig. x, 9, where *Agrion* and *Argia* are compared with *Ishnura* which lives in ponds or slow water, that long antennae and small eyes do not occur in all stream-dwelling nymphs of Zygoptera. *Argia* lives under stones on riffles and has quite normal eyes and antennae. Presumably the turbulence is hardly apparent beneath the stones so the only moving objects there are possible prey. In contrast, *Agrion*, which lives on plants exposed to the current, inhabits an area where everything is in constant motion.

FOOD

It may seem inappropriate to consider the food of invertebrates in any biotope as it is obviously as varied as the invertebrates themselves. There are, however, a few general points about the food of the benthos in running water which are worth making.

Although many running waters support higher plants they are by no means universally present, and even where they are present remarkably few invertebrates feed directly on them (Whitehead, 1935). On land a high proportion of all green leaves is damaged by invertebrates, but in running water the nibbled leaf is something of a rarity, and most live out their lives and fall away without being exploited by animals. Investigations in Russia have shown that many aquatic plants are unpalatable, and the few that are readily eaten by invertebrates, e.g. *Potamogeton* and *Ceratophyllum*, are primarily plants of still or very slowly flowing water (Gajevskaja, 1958). Mosses, on the other hand, are directly eaten by quite a range of arthropods, and particular species in several insect groups eat large amounts, e.g. some Nemouridae and Trichoptera (Hynes, 1941; Chapman and Demory, 1963). There is, however, no indication that any of these arthropods is entirely dependent upon moss as a source of food; they all seem able to eat other things.

Several early workers, e.g. Hora (1928), assumed that encrusting algae must form the base of the food nexe, since, in many streams, they are the only plants which occur in quantity, and there are, as we have seen, several genera which are specialized algal scrapers. Some, particularly gastropods and the trichopteran Glossosomatinae, are often quite abundant in suitable places and are then clearly important in the ecosystem (Brook, 1955; Warren *et al.*, 1960). Indeed, it has been shown that *Agapetus fuscipes* can exercise direct control on the diatom *Achnanthes* (Douglas, 1958). Clearly any animal which lives habitually on the upper surfaces of stones has little choice but to feed on algae, but many such animals are remarkably unselective and they eat everything that they scrape off. For instance, *Ancylus* living in polluted but well-oxygenated water can often be observed to keep small areas free of sewage fungus, and, perhaps surprisingly, many species of Baetidae and Ecdyonuridae eat a great deal of detritus.

We have already discussed the particle feeders and the possibility that at least some of them exploit the drifting bacteria of the water, as well as drifting animals, plants, and detritus. Indeed it has been shown experimentally that *Simulium* larvae will develop from egg to adult on pure suspensions of bacteria at about the same concentrations as occur in natural waters (Fredeen, 1960, 1964). Here again, though, it is worth stressing the point that most particle feeders are very unselective, and that they have considerable flexibility in their diets. *Simulium* is capable of scraping food from the stones as well as filtering it from the water (Anderson and Dicke, 1960), and the predaceous net-spinning Trichoptera can hunt their prey, and many are quite omnivorous.

Many running-water invertebrates feed to a large extent on dead leaves of both terrestrial and aquatic origin, and some insects feed habitually on water-soaked wood. These include many caddis-worms of the family Limnephilidae (Lloyd, 1921; Slack, 1936; Nielsen, 1942), some Elminthidae (Young, 1954; Warren *et al.*, 1960), the large stonefly *Pteronarcys* (Muttkowski and Smith, 1929), and species of *Polypedilum* and *Brilla* among the Chironomidae (Zvereva, quoted by Shadin, 1956). Important consumers of dead leaves include a wide range of filipalpian stoneflies (Hynes, 1941) and the amphipods (Steusloff, 1943; Hynes, 1954). Crayfishes and crabs also feed on such material after they have become well grown (Warren *et al.*, 1960; Williams, 1961, 1962), and they often represent a considerable proportion of the total biomass.

Plant debris breaks down into more or less amorphous detritus which lodges in the stream bed, where it becomes mixed with small mineral particles and diatom frustules. This mixture, often together with pieces of still recognizable leaf-tissue, is the main foodstuff of a very wide range of invertebrates (Bengtsson, 1924; Wissmeyer, 1926; Slack, 1936; Moon, 1939; Hynes, 1941, 1963; Jones, 1949*b*, 1950; Brown, 1961; Warren *et al.*, 1960; Chapman and Demory, 1963). These include species in most orders of the insects, most of the smaller Malacostraca, most oligochaetes, and some molluscs. Among the latter all the pelecypods are particle feeding, but it is less generally known that some prosobranchs consume large amounts of detritus, e.g. *Valvata* and *Bithynia*, and that *Viviparus*, which is often exceedingly abundant in rivers, feeds on detritus not only with its radula but also by ciliary action. It has a

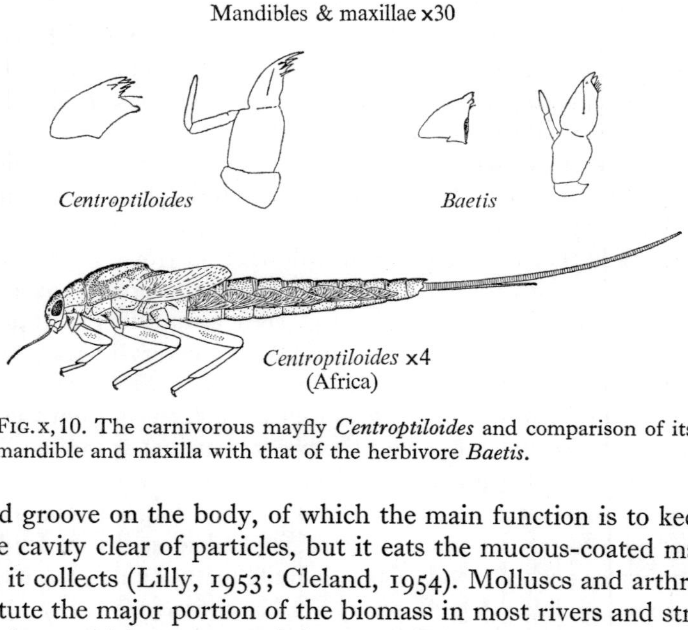

FIG. X, 10. The carnivorous mayfly *Centroptiloides* and comparison of its mandible and maxilla with that of the herbivore *Baetis*.

ciliated groove on the body, of which the main function is to keep the mantle cavity clear of particles, but it eats the mucous-coated material which it collects (Lilly, 1953; Cleland, 1954). Molluscs and arthropods constitute the major portion of the biomass in most rivers and streams, so detritus, which so many of them eat, is a very important source of energy to the community.

Ivlev (1945) suggests that detrital material is made available to animals through the activity of micro-organisms, as he was able to rear chironomid larvae on filter paper together with a suitable bacterial culture. Another suggestion that readily digestible micro-organisms are perhaps an important source of nourishment is provided by the fact that detritus-feeding mayfly nymphs pass their food very rapidly through their guts (Brown, 1961). One would expect the food to pass particularly slowly if relatively stable and indigestible substances such as cellulose were being attacked. This whole matter of the role of detritus in the energy relations of the ecosystem is an aspect of stream ecology which is in great need of further study, and it is one to which we will return in Chapter XXII.

There are indications that the material is passed many times through the guts of animals. For instance, it is easy to demonstrate that newly hatched *Gammarus* survive much better in the laboratory if they are allowed to feed on the faeces of adults than if they are given only the autumn-shed leaves which serve very well as food for the adults (Hynes, 1954). Similarly, Minckley (1963) believes that in Doe Run, Kentucky, the very abundant species *Asellus bivittatus* is almost entirely dependent upon the faeces of the equally abundant *Gammarus minus*. A most interesting fact in this connection is the observation by Pleskot (1953) that *Habroleptoides modesta* and *Habrophlebia lauta*, two European members of the Leptophlebiidae which feed on detritus, eat their own faeces directly from the anus, a habit which closely resembles that of the rabbit. This they do by pulling up the abdomen to the mouth parts with a foreleg. It is possible that this habit of passing food at least twice through the gut occurs in other detritus feeders also, since it would seem likely to increase the efficiency of digestion of refractory material.

Little needs to be said about carnivorous species, although it should be pointed out that there are some unexpected ones. For instance, it seems to be generally assumed that all mayfly nymphs feed on vegetable matter (Degrange, 1960), but several species, e.g. *Ephemerella doddsi* in Western North America (Muttkowski and Smith, 1929; Warren *et al.*, 1960) and *Centroptiloides bifasciata* in Africa are entirely carnivorous (Fig. x, 10) (Agnew, 1962), and other species, including even flattened Ecdyonuridae (Crisp, 1956), are occasionally reported as having eaten animals. Edmunds (1957) lists a number of genera which he believes, on morphological grounds, may be carnivorous.

Among carnivores, as in most habitats, there is a fair amount of selection. For example, the one large species of *Perla* in Britain certainly selects *Simulium* and chironomid larvae, and possibly also *Baetis*, in preference to other animals (Mackereth, 1957); and some groups, most notably amphipods, mites, and elminthid beetles, appear to be largely avoided by invertebrate predators, even where they are abundant (Jones, 1949*b*).

CHAPTER XI

Factors controlling benthic invertebrates

It will be clear from the considerations of the last three chapters that many factors regulate the occurrence and detailed distribution of stream-dwelling invertebrates. The most important of these are current speed, temperature, including the effects of altitude and season, the substratum, including vegetation, and dissolved substances. Other important factors are, liability to drought and to floods, food, competition between species, shade, and, of course, zoogeography. No animal can occur naturally in a region to which, for some historical reason, it has not gained access, even if suitable *loci* occur there. That present distributions are often restricted by historical accident, is well illustrated by the many examples of stream and river species which have been carried beyond their original distributional areas by man, and which have become successful, even important, members of the fauna in new areas. Among these are the American crayfish *Cambarus affinis* and the Australasian snail *Potamopyrgus jenkinsi* in Europe (Thienemann, 1950), and the Asiatic clam *Corbicula manillensis* in North America (Sinclair, 1963; Sinclair and Isom, 1963). On a less grand scale the recent introduction of the amphipod *Gammarus pulex* into Ireland is likely to produce far-reaching effects on the invertebrate communities of streams there.

Many of the factors listed above have been used by various authors (e.g. Ricker, 1934; Berg *et al.*, 1948) as bases for the classification of streams and rivers, or for reaches of them, and these classifications give a fairly clear indication, for a given region, of the species which are likely to occur in particular places.

Obviously many of these factors are interrelated—current, for example, very largely controls the substratum—but it is convenient, if somewhat artificial, to discuss the findings of field work under several separate heads.

CURRENT SPEED

We have seen that many invertebrates have an inherent need for current, either because they rely on it for feeding purposes or because their respiratory requirements demand it. These are the typical rheostenic species, and many workers have found that particular species are confined to fairly definite ranges of current speed as measured in the field (e.g. Ambühl, 1959; Phillipson, 1956; Pomeisl, 1953*a*; Scott, 1958). These are all, however, results obtained by measurements obtained at one time, and, as we know that current varies with discharge, no absolute values of current requirement or tolerance are obtainable from most field studies. Indeed, Phillipson and also Zahar (1951) have shown that *Simulium* larvae change their positions on the stream bed as the current varies and that they go into shelter during spates. They are thus rather rarely swept far away from their normal habitats.

Occasionally, where because of special circumstances the current varies rather little, fairly precise differences of requirement between species can be obtained. Moretti and Gianotti (1962) studied the distribution of two caddis-worms, *Agapetus fuscipes* and *Silo nigricornis*, in the very uniform and constant spring-fed streams of Fonti del Clitunno in Italy; and they showed that, at current speeds ranging from 7·6 to 38 cm./sec., *Agapetus* is definitely more rheophile than *Silo*. As current increased *Agapetus* became more and *Silo* less common, although both occurred, often on the same stone, over the entire range studied. The very slight differences between the preferences of these two species showed up particularly clearly in the positioning of the pupae, which are, of course, free of any complications caused by feeding requirements. The percentage distributions of the pupae on the various faces of the stones were as follows.

	Agapetus	*Silo*
Upstream	27·32	29·46
Upper	24·04	7·14
Downstream	15·84	14·28
Lower	4·91	16·07
Lateral	13·93	16·51

The relative scarcity of *Silo* on the exposed upper surfaces and their relative abundance on the lower, sheltered, surfaces indicate their preference for slower water. Further simple studies of the same kind in favoured locations would undoubtedly reveal similar slight differences between species.

There are even indications that specimens of the same species collected from different current speeds tend to differ. We have already seen that the common European limpet *Ancylus fluviatilis* is taller in swift than in slow water (p. 143). Somewhat similarly, the common snail of the Northern Hemisphere *Limnaea pereger* is smaller and thicker-shelled in fast water than it is elsewhere (Starmühlner, 1953), and in New Zealand it has been observed that the peculiarly coiled caddis-worm *Helicopsyche* (Fig. XI, 1) is larger in swifter reaches (Allen, 1951). It would be interesting to establish whether these differences are caused by the selection of particular microhabitats by certain types of individual, or whether they

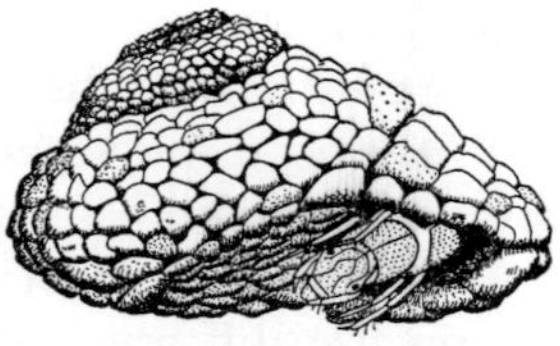

Helicopsyche ×7
(N. America)

FIG. XI, 1. The larva of *Helicopsyche,* an almost world-wide genus of Trichoptera.

are the result of the action of the microhabitat on the development of the specimens that happen to be there.

Persistently very rapid current may make life intolerable for almost all species. For instance, very few invertebrates are to be found in the mid-stream area of the Danube near Vienna, although many occur in the gravel beds of lateral parts of the river (Stundl, 1950). At the other extreme, stagnant or very slow areas in rivers which at times flow swiftly are often without much fauna. This is because silt collects during periods of low discharge, and the conditions become unsuitable for riverine animals; but the periodic wash-outs caused by spates render the environment too unstable for lacustrine species. This has been demonstrated very nicely by Mauch (1963) in his study of the fauna of the River Moselle in Germany.

Many workers have found that within even fairly uniform groups of animals, e.g. *Simulium* (Grenier, 1949), different species have different current preferences. The result is that on rough substrata, where there is considerable local variation in current speed over quite short distances, careful study reveals a mosaic distribution of animals. When Cianficconi and Riatti (1957) placed artificial cubic stones in the Apennine River Potenza and watched how they were colonized, they found that many species occurred only on certain parts. Mites never colonized the exposed upstream faces, and the little caddis-worm *Hydroptila*, the net-spinner *Polycentropus*, and the beetle *Elmis* avoided the upper face as well. Linduska (1942) was able to describe the mayfly fauna of a typical boulder in Rattlesnake Creek, Montana, defining just where each of a long list of species would be found when the current was flowing at a certain speed. On a 50 cm. boulder in water with a surface speed of 240 cm./sec. he postulated that only *Ironopsis* sp. and two species of *Baetis* would be found on top, although all three might also occur elsewhere. Half-way down the sides would be *Iron longimanus* and *Ephemerella doddsi*, both of which might reach the top on the sheltered downstream side. A third *Baetis* species would be found anywhere where the current was less than 120 cm./sec. and *Rhithrogena doddsi* would occur only where the current was less than 60 cm./sec. The other species, *Cinygmula*, *Rhithrogena virilis*, *Ephemerella inermis*, and *E. yosomite*, would be found only right under the stone, and the last two only where debris had collected. It will be appreciated, therefore, that only the most careful study of the positions of animals in the field can reveal the details of their current preferences. Linduska also makes the point that some species, for example, *Leptophlebia adoptiva*, are very fragile, and never leave the shelter down among the stones, they may thus occur in what appears to be very swift water, but they never expose themselves to it. The same, of course, applies to all the species whose normal habitat is down among the gravel in the stream bed. This fauna is discussed further in Chapter XXI (p. 406).

It is, therefore, almost impossible from field studies to define the current requirements of individual species in exact numerical terms. Many species need the current to be close by but cannot stand being actually in it; we might consider the analogy of modern man and his need of an electrical power supply. Nevertheless it is possible to show in general terms that, in any one stream, as the current, measured either on the surface, in mid-water or as near as possible to the substratum, changes from place to place at a given rate of discharge so the fauna changes. For instance, on rocky substrata in streams in the Tatra mountains *Baetis carpathica* replaces *Rhithrogena semicolorata* as the current increases, and *Diura* and *Ameletus* occur only in slow-flowing water (Kamler, 1960, 1962, 1966). In lowland streams on the Flemish Plain the importance of *Gammarus* in the fauna of rooted plants declines and that of Chironomidae increases with increasing current (Albrecht, 1953*b*), and in a similar stream in Holstein Orthocladiinae were found to be characteristic of swifter places than were Tanytarsini (Nietzke, 1937). In soft-water streams in Wales certain mayflies decline in importance and stoneflies replace them as the current increases (Jones, 1943). In streams in northern England different species of net-spinning caddis-worm are found in fast-flowing and slow reaches (Edington, 1965), and in New Zealand the importance of the caddis-worm *Pycnocentrodes* decreases and that of parnid beetles increases as the current declines (Allen, 1959). Many other studies elsewhere have produced similar results, and from what we now know about stream animals these are to be expected.

Similarly, it is known that whole communities change, either wholly or in part, as the current, measured at one time or water level, varies from place to place. Results of this kind can be found in reports of stream surveys from many parts of the world, and the same general rule applies to tropical Africa (Marlier, 1954) or tropical America (Fittkau, 1964) as to, for example, temperate Oklahoma (Hornuff, 1957). Indeed, Zimmerman (1961*a*, *b*) has shown experimentally that identical artificial channels supplied with water from the same river, but flowing at different speeds, develop very different living communities.

One can conclude, therefore, that current speed is a factor of major importance in running water, and that it controls the occurrence and abundance of species and hence the whole structure of the animal community. Its mode of action is, however, complex; it is very variable in time and over very short distances, and it is almost impossible to quantify except in general terms.

TEMPERATURE

Temperature is, of course, intimately related to latitude, altitude, season, and, in spring-fed or lake-fed streams, to the distance from the source. It is, however, possible in any geographical region, to distinguish warm from cool streams, even though, in temperate latitudes, this terminology applies only to the summer. Moreover, as we have seen, spring-fed and shaded streams which are cool in summer are actually warmer in winter than the streams which are 'summer-warm', to introduce once again the more precise terminology of the German limnologists.

Many field studies of individual groups of animals show that there is a clear relationship between the altitude and the occurrence of certain species. This is particularly true of studies on Simuliidae, e.g. those of Grenier (1949) and Zahar (1951). Zahar, for instance, found that in south-east Scotland *Simulium latipes* and *S. monticola* occur only above about 200 m., while *S. variegatum* and *S. reptans* are found only below this altitude. *S. ornatum*, on the other hand, was recorded at all altitudes, and Grenier also found this species, and also *S. aureum*, to have very wide altitudinal range in France. Both these 'species' are now known, however, to be species groups, so further taxonomic study has thrown some doubt upon their apparently very wide tolerance of conditions (Davies, 1966). If a species does appear to have a very wide tolerance it may be taken as an indication that its taxonomy would perhaps bear further study.

Among leeches also, certain species appear to occur only at high altitudes, e.g. *Trocheta bykowskii* and *Erpobdella monostriata* in Poland (Pawłowski, 1948, 1959*b*), and others at low ones, e.g. *Trocheta subviridis*, *Erpobdella octoculata*, *Glossiphonia complanata*, and *Helobdella stagnalis* there and in Denmark and Britain (Madsen, 1963; Mann, 1959). Such findings are generally interpreted as resulting from differences in temperature tolerance, and it is quite usual to find that some species seem to be cold stenotherms, some warm stenotherms, or at least apparently needing warm water at some times of year, while others appear, at any rate on the basis of present taxonomy, to be indifferent.

The effects of temperature itself, as distinct from mere altitude, are more clearly revealed by comparative studies of high-level glacier torrents and similar streams arising from névés, springs, or lakes, which are of course, relatively summer-warm. Dorier (1937) studied several such streams in the French Alps and found that below the glaciers *Diamesa* was almost the only inhabitant. Lower down it was joined by other chironomids and some mayflies and stoneflies, and the fauna became richer in variety the further downstream he collected. In the other streams, in contrast, the fauna was relatively rich right up to the top. Some of these differences may, of course, be caused by factors other than temperature. For instance glacier meltwater is cloudy and so discourages the growth of algae and moss, but many normal streams with no moss have a rich fauna, so it would seem that very low temperatures persisting throughout the year eliminate many invertebrates. This subject is discussed further in Chapter XXI (p. 402).

We have already seen that the oxygen requirements of certain species suggest that they must be confined to temperatures below certain values. Such suggestions are borne out by field studies; for example, as already stated (p. 168), very few stoneflies occur in Austrian streams in which the temperature rises at times to more than 25 °C.; in the Tatra Mountains of Poland, the ratio of Plecoptera to Ephemeroptera, both as regards numbers of individuals and numbers of species, decreases as the annual range of temperature increases (Kamler, 1965, 1966), and similar results have been obtained in Canada (Fig. XI, 2).

Zahner (1959) showed that the damsel-fly *Agrion virgo* replaces *A. splendens* along reaches in which summer temperatures attain only about 16–18 °C. *A. splendens* is a little less susceptible to lack of oxygen, and it occurs in water up to 24 °C.; it is thus usually downstream of *A. virgo*. It would seem that, in the range of overlap, competition between the two species occurs, but this is a point to which we shall return. Presumably in other instances, where closely allied species occupy adjacent reaches of streams, similar considerations must apply. Many observations of such vicarious but contiguous distributions have been made, e.g. on Ecdyonuridae in Austria (Pleskot, 1951) and on various groups of insects in the Pyrenees (Berthélemy, 1966), and they almost certainly reflect the direct, or indirect, effect of temperature coupled with interspecific competition.

In temperate climates the temperature varies with the season in most

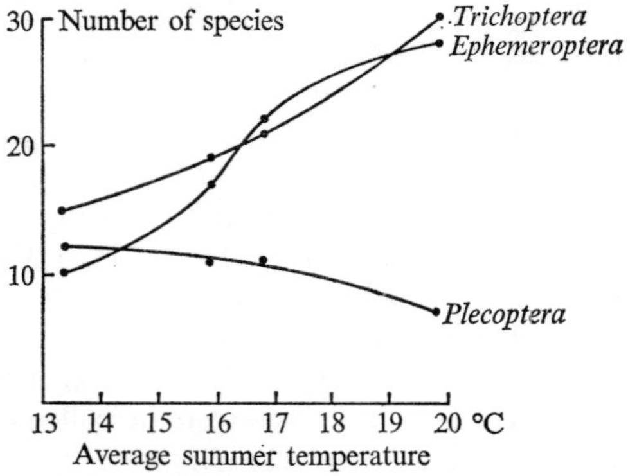

FIG. XI, 2. The numbers of species of three insect orders recorded as emerging adults at four points with different average summer temperatures along a stream in Algonquin Park, Ontario. Redrawn from Sprules (1947).

streams, and the extent of this variation and the rate at which it occurs are as important as the highest level it attains. Studies on *Simulium* have shown that upland species normally produce only one generation a year, and that most lowland species have two or even three. On the other hand, species such as *S. ornatum*, which occur over a wide altitudinal range, seem able to vary the number of generations to suit local circumstances (Grenier, 1949; Zahar, 1951). This type of relationship to temperature may be more important than the actual tolerance by the individual, but it is not easily investigated in the laboratory. Angelier *et al.* (1963) who studied the mites of the River Céret in southern France, suggest that the species which are able to produce two or three generations in a year tend to dominate in the warmer waters, while those with only one are dominant in colder places. They stress, therefore, that knowledge of the life history is important for an understanding of the ecology of any species.

The seasonal swing of temperature also controls the rhythm of the life histories of many animals. Low winter temperatures may slow down growth rates, so the longer the winter the less time remains in which growth can occur. As a result, the higher the latitude or the altitude the later is the emergence time of many stream insects; a fact which has been very well shown for the stoneflies of Sweden and the Pyrenees (Brinck, 1949; Berthélemy, 1966). If the remaining summer season is too short for a species to complete its life history it will not occur. This concept has been used by Ide (1935) to explain the distribution of mayfly species in an Ontario stream. He found that the number of species increased, and the length of the emergence period of each species decreased, in the downstream direction. Where, as in the genera *Leptophlebia* and *Ephemerella*, there were several closely related species, their flight periods were separated in time, and the early species extended further upstream than the later ones. Most of these facts can be explained by reference to the maximum temperature attained and the increase in rate of warming downstream. This is not, however, the whole explanation. Some mayflies, including two species studied by Ide, occur only in the headwaters. Presumably they are unable to tolerate the warmer water and cannot complete their life histories further downstream in the relatively short period intervening between the inhibiting cold of winter and too high a temperature.

A similar explanation is given for the absence of *Heptagenia lateralis* from stony streams in northern England which attain summer temperatures above 18 °C. (Macan, 1960*a*). This species grows little in winter, and it has to do much of its growing during the spring since it hatches in the autumn from eggs laid in early summer. It seems likely, therefore, that warmer streams heat up too fast to allow it to complete its nymphal development before the temperature becomes too high. The eggs can survive warmer water as is shown by the occurrence of the nymphs on lake shores in which the water markedly exceeds 18 °C. in summer. Lakes, however, warm up far more slowly than the streams, and they thus allow time for nymphal development.

As would be expected from the above considerations, differences of the emergence dates of insects at any one point on a river or stream in different years can often be explained in terms of water temperatures during the preceding months (Ide, 1940). One difficulty, however, in making any generalizations on this matter of temperature and occurrence of species is the well-established fact that species do not always react in the same way in different parts of their geographical range. The reasons for this are unknown, but the phenomenon is undoubted and running water is not without its examples. Thus, for instance, several species of Trichoptera appear to be far more cold stenothermic in northern and western Europe than they are in Rumania (Botosaneanu, 1960), and, conversely, the common European limpet *Ancylus fluviatilis* occurs almost everywhere on hard substrata in western Europe, whereas in Transcarpathia in southern Russia it is reported to occur only where the water temperature does not exceed 10 °C. (Ivlev and Ivassik, 1961). An analysis of the reasons for this kind of anomaly would be a valuable contribution to ecology.

A classical but still somewhat puzzling example concerns the stream Tricladida of western Europe. Thienemann (1912) long ago showed that, in the Sauerland of western Germany, *Crenobia alpina* occurs only in the upper parts of streams where the temperature does not exceed 13 °C. At lower points *Polycelis felina* takes over, and it persists until the maximum temperatures reach 15–16 °C., when it is itself replaced by *Dugesia gonocephala*. This pattern is repeated, more or less, in Wales (Carpenter, 1928*a*), although there *D. gonocephala* is absent; and in some parts of northern Germany *P. felina* is missing and *C. alpina* and *D. gonocephala* come into contact (Illies, 1952*b*). At least the upper two species are known to migrate downstream during the winter, and both can occur at sea level in small streams in western Britain, but it would appear from the field observations that they are cold-water stenotherms. This cannot, however, be the whole explanation as *C. alpina* has been found in alpine pools at 22 °C., and many individuals were, moreover, sexually mature, a state which in streams is attained only in winter at much lower temperatures (Steinböck, 1942). It has been established that stream specimens can survive for short periods at temperatures higher than those in which they normally occur (Schlieper and Bläsing, 1952), so the whole question of the occurrence of these species in relation to temperature stands in need of re-examination. This is particularly so because *C. alpina* is usually cited in European limnological literature as the classical example of a cold-water stenotherm. Perhaps it, like the Rumanian Trichoptera and the Transcarpathian *Ancylus*, varies greatly from place to place, or it may be that several other factors are involved. Edington (1966), for instance, suggests that the *pattern* of temperature change, i.e. whether localities warm up early or late in the season, may be as important as the *actual* temperatures attained; we have seen above that this is true for *Heptagenia lateralis*. Obviously we need more thorough work on temperature relationships.

Another example comes from the New World, where Chandler (1966) showed that, in Indiana, the triclad *Phagocata gracilis* is confined to spring areas, almost certainly by its cold stenothermy, and that *Dugesia tigrina* is very definitely a warm-water worm which becomes inactive at temperatures below about 11 °C. Two other species, *Cura foremani* and *D. dorotocephala*, are eurythermal but apparently less tolerant of high temperatures than *D. tigrina*; so we have here a group of species comparable to those in Europe. Chandler did not, however, find them all in the same watercourse so other factors are definitely involved. He suggested that fast flow may exclude *Cura foremani* and that slight pollution may eliminate *D. dorotocephala*.

Nevertheless, temperature is a very important factor in stream ecology, and this is shown particularly well where it changes suddenly along the course of a river. In the Vienna Woods the hot springs of Baden quite drastically change the fauna of the River Schwechat, eliminating all stoneflies and limpets and substituting large numbers of snails (Starmühlner, 1961). In Yellowstone National Park, Wyoming, the inflow of hot springs to the Firehole River makes similar changes in the fauna, some species being found only in the warm or the cool reaches and some being apparently indifferent (Armitage, 1961). Man-made changes produce similar results. The Norris Dam of the Tennessee Valley Authority on the Clinch River has, because the water flows from below the thermocline, produced many miles of cold water in a formerly warm stream, and this has completely altered the fauna (Tarzwell, 1939); and a power station outfall on the Delaware River produces marked reductions in the fauna where it raises temperatures above about 32 °C. (Coutant, 1962).

Another factor connected with temperature is ice. In severe climates rivers and streams may freeze over for long periods, but this, as we have seen, only rarely leads to de-oxygenation, and it indeed protects the water from further cooling. Surface ice does not usually come into direct contact with invertebrates, although when it does freeze them in it apparently kills them (Brown *et al.*, 1953). Anchor and frazil ice, however, form near to the animals and they have been credited with the destruction of invertebrate fauna. Dr. J. Hanson of the University of Massachusetts informed me that a severe cold spell one year in early winter, before the icing over of streams, almost eliminated the genus *Allocapnia* from the area round Amherst, and Gaufin (1959) reports that the severe winter of 1948–9 in Utah 'completely wiped out thousands of organisms' in the Provo River and he attributed this to anchor ice. Possibly the damage is the result of scouring of the stream bed as the ice forms and detaches in a daily rhythm. In contrast, other workers have not found that anchor ice damages invertebrates. They do become caught in it, and when it drifts away they drift with it, but the presence of ice does not always increase the numbers of drifting animals (Logan, 1963). The ice, of course, unlike surface ice, does not become very cold, as it is immersed in water at 0 °C., and Benson (1955) found that when he collected anchor ice from Pigeon River, Michigan, all the frozen-in specimens of a considerable number of species of invertebrates were alive.

It would seem, therefore, that the effects of underwater ice may not always be as severe as might be expected, but that this is a subject which is in need of more study. Very probably, however, the fact that it forms *on* the substratum, and not *in* it, protects much of the fauna, which has merely to move down a very short distance to avoid it. The hyporheic habitat is a refuge from both low and high temperatures as it is from swift currents. In this respect stream animals which live on a coarse substratum are very well provided with a shelter from the elements which is always close to hand, and this is a considerable compensation for the many rigours which stream life presents.

THE SUBSTRATUM

Almost any thorough field study of stream-dwelling invertebrates reveals that certain species are confined to fairly well-defined types of substratum, and that others are at least more abundant on one type than they are on others. Thus, among molluscs, *Ancylus* occurs only on solid surfaces no matter how fast or slow the rate of flow (Geldiay, 1956). *Goniobasis*, on the other hand, requires a solid surface only if the current is swift; in slow water it occurs also on silt and sand (Ludwig, 1932). Here we can presume that we understand the reasons; the flat, limpet-like *Ancylus* (Fig. VIII, 8) can move around only on rock and stone, and the long narrow, but heavy, *Goniobasis* (Fig. VIII, 18) needs a solid surface only when it is necessary for it to hold on. Similarly, it was shown long ago that the common *Ephemera danica* of Europe occurs primarily in places where the range of particle size is 0·05–3 mm., and that it becomes scarcer in places where the main bulk of the deposit is outside this size range (Percival and Whitehead, 1926). Presumably here the reason has something to do with the ability of the nymphs to burrow and to maintain their burrows open. Other examples are the occurrence of the mayfly *Coloburiscus humeralis* in New Zealand streams only where the bed is composed of stable, loosely packed gravel and its absence from areas subject to scour (Wisely, 1962), and the preference of nymphs of *Paragomphus* in South Africa for areas of coarse sand because fine sand enters the rectum and interferes with the respiratory apparatus (Keetch and Moran, 1966). It is also fairly understandable that sessile animals such as sponges should occur only on solid surfaces; on soft and occasionally shifting substrata they would be overwhelmed. Thus the bryozoan *Urnatella gracilis* occurs under a wide variety of conditions in the Ohio River, but always on solid objects like sticks, shells, or stones (Weise, 1961), and it has adopted the same habitat in the lower reaches of the Danube, into which it has been introduced by man (Elian and Prunescu-Arion, 1964).

Often, however, it is by no means clear why certain species apparently prefer particular substrata. In Britain the two very similar stoneflies, *Dinocras cephalotes* and *Perla bipunctata*, share out certain river beds between them, the former occurring primarily on stable stony areas (Hynes, 1941). The reasons for this are unknown, although it may have something to do with colour, as the fairly uniform, darkly coloured *Dinocras* is less conspicuous on the generally darker-coloured stable substrata, while the strongly patterned *Perla* tends to merge with the more varied colours of the shifting stones. In the Middle West of North America the crayfish *Orconectes propinquus* is found only on hard substrata (Bovbjerg, 1952*a*), and there is no apparent reason for this, especially as very similar species occur on soft substrata. Moreover, it has been shown that it occurs primarily on shale and sandstone and is replaced, in Ohio, by *O. rusticus* in limestone areas (Rhoades, 1962). This occurs irrespective of the calcium content of the water, and it also remains quite unexplained. In Rattlesnake Creek, Montana, Linduska (1942) found that the very fragile mayfly *Leptophlebia adoptiva* was always associated with large rocks no matter how fast or slow the current where these occurred. Our lack of understanding of why many such associations with particular types of substratum exist is emphasized by genera in which each species occurs on a different type with no particularly associated morphological specialization. A fine example of this is the genus *Ephemerella* of which many rather similar species occur in central North America, and each has a special fairly well-defined stream habitat, ranging, for the various species, from swift gravel riffles, through shelter under stones and debris, to vegetation and silt (Leonard and Leonard, 1962). Another example is the genus *Xenochironomus* in which one species *X. taenionotus* burrows only in vertical clay banks in streams in Florida, while *X. rogersi* burrows only in horizontal beds of clay (Beck, 1965).

The result of these preferences is that as the type of substratum varies from place to place so does the fauna. This fact has been established by many workers in several continents: for example in all parts of Europe (Behning, 1924; Percival and Whitehead, 1930; Schräder, 1932; Vonnegut, 1937; Berg *et al.*, 1948; Jones, 1949*a*; Marlier, 1951; Albrecht,

1953*b*; Illies, 1958; Einsele, 1960; Sowa, 1961; Kamler, 1966; Winkler, 1963; and Thorup, 1966), throughout the U.S.S.R. (Shadin, 1956), in North America (Shelford, 1937; Shockley, 1949; Sprules, 1947; O'Connell and Campbell, 1953), and in tropical and temperate Africa (Marlier, 1954; Harrison and Elsworth, 1958; Allanson, 1961); and the same is true of small streams as of large rivers like the Volga.

In general it can be stated that the larger the stones, and hence the more complex the substratum, the more diverse is the invertebrate fauna. Sand is a relatively poor habitat with few specimens of few species, apart from its microfauna, but silty sand is richer, and muddy substrata may be very rich in biomass although not in variety of species. Table XI, 1 shows that different substrata support different proportions

	Rubble	Gravel	Sand	Muck
Ephemeroptera	35·5	4·6	9·3	20·3
Trichoptera	7·0	1·7	1·7	3·8
Plecoptera	4·1	2·1	0·7	0
Chironomidae	38·2	67·6	83·9	74·8
Simuliidae	10·8	21·4	0·9	0
Miscellaneous	4·4	2·5	3·5	1·0
Ratio of total numbers emerging to total numbers emerging from sand	4·6 in rapids 3·3 in pools	2·1	1·0	1·8

TABLE XI, 1. The percentage composition of adults of various groups of insects emerging into traps set over different types of substratum in streams in Algonquin Park, Ontario, and comparison of the total numbers emerging from the different types of substratum. Data from Sprules (1947).

and numbers of different types of insect, and Table XI, 2 from the classical study of Percival and Whitehead (1929) emphasizes this point still further. Sprules (1947) suggests that, at least in so far as stony substrata are concerned, the factor which controls the local abundance of the fauna is the space available for colonization. He compared the numbers of insects emerging into traps from four areas, three of which were covered with stones embedded in gravel and one of which had layers of rocks piled upon one another. Although the first three differed widely in other respects (width, depth, rate of flow, and temperature) and had different faunal compositions, the numbers of insects obtained were very similar —5,393, 7,175, 5,440. From the fourth place, however, 31,505 insects emerged into the same area of trap; i.e. five to six times as many. It seems very probable that sometimes this matter of the availability of sheltered crevices is the main factor but it is clearly not always so as can be seen in Table XI, 2. Percival and Whitehead actually found fewer

	Loose stones	Stones embedded in bottom	Small stones and gravel	*Cladophora* on stones	Loose moss	Thick moss	*Potamogeton* on stones
Ephemeroptera							
Baetis	14·62	5·23	1·90	1·60	4·60	1·74	1·90
Rhithrogena	12·04	—	7·70	0·20	0·11	+	—
Ecdyonurus	6·70	4·07	0·60	0·20	0·30	+	—
Ephemerella	0·80	8·56	—	9·40	11·36	4·82	4·50
Caenis	0·48	—	1·90	1·05	1·57	0·42	—
Plecoptera							
Leuctra	1·90	1·66	—	0·30	+	+	—
Isoperla	1·28	0·17	0·60	+	0·27	0·32	0·20
Amphinemura	1·41	0·17	—	+	0·47	0·17	—
Perla	0·10	—	—	—	0·13	—	—
Chloroperla	0·41	—	—	—	—	—	—
Trichoptera							
Polycentropus	1·20	2·96	—	1·15	+	0·30	1·50
Hydropsyche	0·78	—	—	0·80	0·20	4·50	1·60
Glossosoma	2·98	3·73	3·20	+	—	—	—
Agapetus	20·10	20·96	18·20	2·00	0·21	—	—
Rhyacophila	0·74	0·13	—	1·30	0·40	0·25	+
Psychomyia	0·46	3·48	0·60	3·20	—	+	—
Hydroptila	—	0·26	—	5·40	1·74	0·19	—
Ithytrichia	—	0·13	—	0·13	—	0·46	—
Crunoecia	1·05	1·00	1·30	0·16	2·32	0·10	—
Coleoptera							
Elminthidae	3·10	7·76	34·00	8·10	6·16	4·20	—
Diptera							
Chironomidae	17·00	11·43	5·20	39·80	54·00	40·90	42·00
Simulium	3·03	—	—	0·10	1·30	1·54	1·30
Mollusca							
Limnaea	—	3·76	0·60	1·60	0·34	3·00	17·30
Ancylus	—	11·06	2·60	3·20	0·97	—	0·40
Oligochaeta							
Naididae	0·80	1·50	7·70	12·20	3·57	33·00	—
Arachnida							
Hydracarina	1·92	1·56	0·66	4·30	3·30	3·00	11·00
Crustacea							
Gammarus	1·30	—	0·60	+	1·23	1·02	4·50
Average total per square metre	3,316	4,600	3,375	44,383	79,782	431,941	243,972

TABLE XI, 2. The average percentage composition of the principal elements of the fauna on various types of stream bed in rivers in north-eastern England, based on eighty samples. Data from Percival and Whitehead (1929). + = less than 0·1 per cent.

animals on loose stones than on embedded ones. It would seem likely that an additional factor is stability, and that Percival and Whitehead's loose stones were more subject to movement and wash-out than were Sprules's piles of rocks.

On stony substrata the presence of silt reduces and changes the fauna. The Missouri contains fewer species of mussel than the Mississippi (Van der Schalie and Van der Schalie, 1950) because it is more silty, and Sprules (1941, 1947) was able to observe this effect on insects when beavers built a dam across a stream in Ontario, on which he was quantitatively trapping insects. The dam raised the water level about 40 cm.

and caused the deposition of sandy silt. This reduced the total number of insects emerging, especially of Ephemeroptera, Plecoptera, and Trichoptera, and it increased the proportion of Chironomidae. In Russian rivers also, particularly the Oka and rivers in the Caucasus and Central Asia, it has been found that the deposition of silt or sand on stony substrata reduces their fauna, even when the deposits remain for only part of the year. In the Oka for example, the presence of *Spongilla* on stones can be taken as an indication that the area is always free of silt (Shadin, 1956). Similarly, continual deposition of fine clayey silt on to muddy substrata in the Vaal River, South Africa, has been considered to be responsible for the poverty of fauna there (Harrison *et al.*, 1963), and deposits of iron hydroxide in spring-fed streams reduce the fauna as compared with similar streams with no deposits (Thorup, 1966).

In silty or sandy areas any solid objects which are present become rapidly and often densely populated by lithophile animals (Bick *et al.*, 1953; Scott, D. C., 1958), and the more shelter such objects provide the denser is the colonization. Cianficconi and Riatti (1957) found that when they placed grooved artificial stones with the ridges across the current they acquired more animals than they did when the ridges were parallel to the flow and the grooves thus more exposed.

It should also be noted that to some extent the nature of the solid objects in a stream affects the animals which colonize it. We have seen that shelter is important and that more animals occur on irregular stones than on smooth ones, but some animals occur only on wood. This applies to certain beetle larvae, e.g. *Macronychus* (Young, 1954), and we have seen earlier that various animals feed on submerged wood (p. 193) and so would be expected to occur primarily there. In streams in the Eastern Congo Marlier (1954) was able to distinguish quite clearly between the faunas of stones and wood occurring together in swift water. On the stones there was a wide variety of mayflies and caddis-worms as well as stoneflies and other insects; on the wood were larvae of Dryopidae, only three genera of mayflies, *Tricorythus*, *Elassoneuria*, and *Adenophlebia*, and the nemertine *Polymorphanisus*. Possibly this reflects feeding habits, but this can be true to only a limited extent as *Tricorythus* is a filter feeder (p. 184) and *Polymorphanisus* eats mayfly nymphs. Similarly, in the Russian rivers Ural and Northern Dvina, wood has been found to be an important substratum for normally lithophile species in areas where other fixed substrata are rare. Here also special wood-eating forms such as *Polypedilum* are found together with some which feed on other things (Shadin, 1956).

The result of all these factors is that, in general, the fauna of clean stony runs is richer than that of silty reaches and pools both in number of species and in total biomass. This is well brought out by comparison of riffles and pools in a single stream as shown in Table XI, 3. But even such apparently clear-cut results as these need to be interpreted with caution. As we shall see in Chapter XII, the complex mosaic-like pattern of stream beds makes reliable quantitative information extremely difficult to obtain, and even within one fairly uniform type of substratum, such

	Riffle	Pool
Baetidae	32·5	23·5
Chironomidae	13·1	39·9
Parnidae	16·5	13·2
Ecdyonuridae	12·3	3·9
Perlidae	0·6	4·2
Aeschnidae	+	0·3
Hydropsychidae	9·6	0·4
Sialidae	0·5	0
Leptoceridae	2·4	1·9
Psychomyiidae	2·6	1·1
Ancylidae	0	0·2
Ephemeridae	0	0·6
Ceratopogonidae	2·3	3·5
Libellulidae	0·1	0·3
Hydracarina	2·7	2·6
Limnaeidae	+	+
Simuliidae	1·3	0
Others	9·6	4·4
No. of samples	21	33
Av. no. per sq. m.	1,003	652
Av. mg. dry wt. per sq. m.	625	348

TABLE XI, 3. The average percentage composition by number of the faunas of riffles and pools in the Black River, Missouri, during the summer, as shown by the seventeen most common groups of invertebrate. Data from O'Connell and Campbell (1953). + = less than 0·1 per cent.

as the stony riffle, other factors may alter the detailed distribution pattern of the invertebrates.

This has been very well demonstrated in a recent and excellent study by Egglishaw (1964) on a stream in Scotland. He sampled a single riffle by a standardized netting technique, and he not only counted the animals retained by a mesh of twelve threads per cm., but he also measured the volume of vegetable detritus caught. The results of one of his series of samples are shown in Table XI, 4, from which it is clear that many species increased in number as the amount of detritus increased. This applied throughout the year, and for the few sets of samples to which it did not apply there seemed always to be a reasonable explanation, such as too little variation between the samples or a recent spate which could have rearranged things. It will be noted from the table that

Sample no.	1	2	3	4	5	6	7	8	9	10
Plecoptera										
Leuctra inermis	3	8	11	34	24	34	32	32	47	49
Isoperla grammatica	3	4	6	7	4	10	20	14	6	27
Chloroperla torrentium	14	10	6	10	11	16	11	15	14	12
Amphinemura sulcicollis	5	3	3	6	11	21	19	11	19	20
Other Plecoptera	—	2	—	3	2	1	4	9	5	10
Ephemeroptera										
Baetis spp. (mostly *rhodani*)	19	6	5	22	30	65	48	52	40	51
Rhithrogena sp.	17	8	11	15	16	14	28	33	40	38
Caenis (?) *rivulorum*	1	4	1	5	2	3	1	3	3	3
Other Ephemeroptera	—	—	—	1	4	—	3	—	9	11
Trichoptera										
Limnephilidae	3	3	2	1	2	4	6	2	5	4
Hydropsychidae	2	—	—	3	—	2	2	1	1	4
Rhyacophila sp.	1	1	1	2	3	1	—	4	—	—
Other Trichoptera	—	1	1	—	3	—	—	—	10	—
Coleoptera										
Esolus parallelopipedus	3	4	3	3	7	2	16	2	8	3
Other Coleoptera	—	2	1	1	2	4	11	4	8	3
Diptera										
Chironomidae	4	18	11	10	31	39	51	14	73	51
Other Diptera	—	1	—	—	3	8	5	—	3	6
Hydracarina	1	2	—	—	1	4	1	2	2	3
Other Groups	—	1	—	1	1	1	1	2	3	—
Total	76	78	62	124	157	229	259	200	296	295
Plant material (ml.)	1	1	2	2	4	7	7	8	10	12

TABLE XI, 4. The numbers of various animals in ten samples taken in April 1962 from Shelligan Burn, Perthshire, Scotland, by a standardized netting technique. Data from Egglishaw (1964).

the abundance of at least one fairly common species, *Chloroperla torrentium*, was not correlated with plant detritus; the same applied to *Simulium* in samples other than those shown in the table. *C. torrentium* is a carnivore and *Simulium* a filter-feeder, so it is perhaps not surprising that they are unaffected by plant detritus. In contrast most of the animals which were clearly correlated with the detritus actually feed on it, e.g. *Leuctra*, *Amphinemura*, *Baetis*, *Rhithrogena*, and probably many Chironomidae. Availability of food is clearly therefore important, but it may not be the whole story, at least not directly, because *Isoperla*, which also increased in abundance with increasing detritus, is a carnivore.

Egglishaw further conducted experiments in which trays containing uniform bricks and varying amounts of sterilized plant-detritus were left in the stream bed for three weeks to become colonized. Here again similar results were obtained, thus proving that the correlation between detritus and the abundance of certain species was caused by active seeking-out by the animals and not by any physical washing-in of both to the same area. Moreover, when he substituted equal amounts of shredded rubber, which is food for no animal but which might be expected to provide similar purely physical conditions, there was no correlation between the number of animals that moved in and the presence or amounts of rubber. The plant detritus therefore acts as an attractant, and this still further complicates the influence of the substratum on the invertebrates.

A somewhat different observation of the effect of detritus on stream faunas is that of Oliff and King (1964), who found that, in the Mooi River in South Africa, 'natural pollution', caused by the accumulation of fallen leaves, led to a great increase of naidid worms and a reduction in numbers of Ephemeroptera. Clearly, therefore, both the amounts and the nature of the deposits are important.

The presence of vegetation also greatly affects the fauna. This has been known for decades, and Table XI, 2 shows that Percival and Whitehead long ago found that there are more animals in moss, rooted plants, and filamentous algae than there are on stones, and that certain species or groups are confined to, or more abundant in, one or the other habitat. More recent studies have shown the same thing, and Minckley's (1963) thorough study of a spring stream in Kentucky also shows that the presence of vegetation affects the fauna of nearby bare areas (Table XI, 5). His two upper stations were densely covered with beds of the moss *Fissidens* which housed enormous populations of the isopod *Asellus bivittatus* and the amphipod *Gammarus minus*. These were so abundant that they, as it were, spilled over on to the few bare stony riffles and formed a large proportion of the, much less abundant, fauna there. Further downstream at station III, where the *Fissidens* was rare, the isopods and amphipods in the riffles were much scarcer, and their place was taken by Ephemeroptera, and Diptera, mostly Chironomidae. Minckley's study also showed that there were considerable differences between different types of plant. *Nasturtium*, *Myriophyllum*, and *Myosotis* contained relatively, and absolutely, fewer *Asellus* than did *Fissidens*, and, presumably because of the silt they collect round their roots, many more oligochaetes. Other, similar, differences can be seen in

Station no.	I				II				III			
Miles from source	0·1				1·9				3·1			
	Fissidens beds	Bare riffles	*Nasturtium* beds	Shifting sand	*Fissidens* beds	Bare riffles	*Myriophyllum* beds	Silty to sandy pools	Travertine riffles	Rubble riffles	*Myosotis* beds	Silty to sandy pools
Turbellaria	2	4	+	1	5	4	+	+	+	+	+	+
Oligochaeta	1	1	27	5	1	+	25	27	1	4	42	40
Isopoda	68	53	8	33	44	25	28	5	2	4	1	1
Amphipoda	24	26	25	12	21	34	21	7	1	2	10	7
Ephemeroptera	+	+	+		2	5	2	+	45	70	2	1
Trichoptera	2	8	3	10	18	22	1	+	13	3	+	+
Diptera	3	6	24	18	6	4	5	25	36	13	27	30
Gastropoda	+	1	10	13	2	+	17	32	1	1	10	9
Pelecypoda	+	+	3		+	+	+	2		+	3	2
Mean no. animals/sq. m.	87,600	2,700	9,000	1,860	102,000	7,550	34,700	4,640	1,530	1,540	9,960	5,830

TABLE XI, 5. The percentage composition (to nearest whole number) and mean number of animals per square metre, on various types of substratum at three stations in Doe Run, Kentucky. Data recalculated from tables in Minckley (1963) and based on twelve months of sampling. Only those groups of animals are included which formed at least 2 per cent of the total number of animals in one of the habitats. + = <0·5 per cent.

the table; for instance molluscs, mostly *Goniobasis*, were much commoner in the loose growth of the higher plants than they were in the moss. It is also clear that all types of plant were more heavily colonized than the non-vegetated areas of substratum. This is a quite general finding, as is the fact that communities of species found in plants differ from those on bare areas (e.g. Albrecht, 1953*b*; Angelier *et al.*, 1963*a*; Kamler, 1962; Greze, 1953; Hynes, 1961; Winkler, 1963).

Within each type of plant, such as moss, coarse-leaved and fine-leaved angiosperm, etc., there are also differences, and the reasons for these are often far from clear. Greze (1953) studied the extensive weed-beds in the lower reaches of the River Angara in Siberia, and found quite large differences between the populations of invertebrates inhabiting different species. Even quite similar plants, such as three species of *Potamogeton* (Table XI, 6), differ quite considerably, and much work

	Per square metre	Per kg. wet weight of plant
Potamogeton perfoliatus	3,604 (19·93)	4,004 (23·26)
P. lucens	2,555 (14·02)	1,277 (7·01)
P. pectinatus	57 (0·04)	162 (0·12)
Limnanthemum nymphaeoides	392 (0·69)	196 (0·34)
Myriophyllum verticillatum	44 (0·02)	44 (0·02)
Hydrilla verticillata	293 (0·46)	234 (0·35)

TABLE XI, 6. The numbers, and in brackets weights to nearest 0·01 g., of animals in different species of plant in the River Angara. Data from Greze (1953).

needs to be done before we can begin to understand why this should be. Harrod (1964), who investigated the fauna of four species of plant in the River Test, England, concluded that some differences were caused by the different amounts of shelter provided by the plants. But this is clearly not the whole explanation, because if it were the various plants would differ mainly in total fauna and not in faunal composition, except for the rare species which feed on only one plant. She, however, found that *Hydra* was absent from *Veronica beccabunga* and *Carex*, although present on *Ranunculus* and *Callitriche*, that *Stylaria* was rare on *Ranunculus* and *Callitriche* and was relatively scarce on *Carex* and *Veronica*. These distributions have clearly nothing to do with food derived directly from the plants. Similar correlations are known from still water (e.g. Krecker, 1939), and we have at present no reasonable explanation for them.

DISSOLVED SUBSTANCES

Dissolved substances in stream water can for our purposes here be loosely divided into four groups, namely oxygen, salinity, acidity, and hardness, although, as usual with any attempt to categorize ecological factors, some of these, e.g. oxygen and acidity, and acidity and hardness, are often linked together.

Oxygen is rarely a factor in the ecology of invertebrates in clean rivers, because as we have seen it hardly ever drops to low levels. In polluted waters, on the other hand, lack of oxygen can be of very great importance, but this is outside the scope of this book, and the reader is referred to Hynes (1960) for further information. In unpolluted rivers and streams oxygen falls to very low levels under only two conditions, continuous ice-cover for long periods under rather special conditions and excessive autumnal leaf-fall into pools in almost dry streams. The first condition appears to be almost unstudied from the viewpoint of invertebrate ecology, although it is known to have some influence on fishes in northern Siberia (p. 319). In most places, however, it seems that sufficient open water remains to allow replenishment of the relatively small amounts of oxygen consumed at zero temperatures. It will be recalled that rivers and streams do not begin to freeze right over until the whole mass of the water has reached at least 0 °C. Indeed, in some Russian rivers the minimum level of oxygen concentration is not attained until the spring, when rising temperatures permit decay of the plant debris resulting from the previous summer's growth (Greze, 1953). We can take it then that freezing over and any consequent lack of oxygen are of little importance to invertebrates.

Similarly the effects of rotting leaves and the production of 'black water' described from certain areas in the North American Middle West (Schneller, 1955; Larimore *et al.*, 1959) apply only to intermittent streams and will be considered in Chapter XXI. They have no general significance in stream ecology.

In rivers and sluggish streams, however, there are often places, such as stagnant bays and dense stands of emergent vegetation, where, as in ponds, oxygen may fall to low levels in summertime; and where, as in the Suwannee River, Florida, large amounts of groundwater enter, the oxygen content may be low at all times (Beck, 1965). Here, of course, only animals of the normal pond-fauna can survive, and this adds yet another factor contributing to the mosaic-like pattern of river faunas. A nicely worked example of the way this effect operates is the study by Mann (1961*a*, *b*), who investigated the respiration of certain western

European leeches in relation to their ecology. Some species, e.g. *Piscicola geometra* and *Erpobdella octoculata*, show dependent respiration and require well-aerated water for survival; *Piscicola*, in particular, has such a high rate of respiration that it can survive only in running water or in open lakes. Others, on the other hand, e.g. *Glossiphonia complanata, Helobdella stagnalis*, and *Erpobdella testacea*, have independent respiration and so can survive in stagnant places (Fig. XI, 3). The result is that

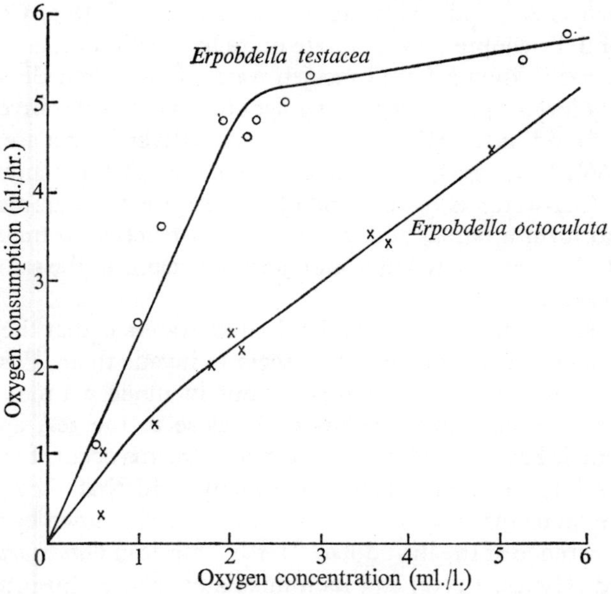

FIG. XI, 3. The oxygen consumption in summer at different oxygen concentrations of two species of the leech *Erpobdella* at 20 °C. Results expressed as consumption per gram. From Mann (1961*a*) with permission.

in the genus *Erpobdella* two, very similar, species share out the river habitat between them, *E. testacea* living in swampy places and *E. octoculata* in the flowing water. In fact, the situation in regard to these two species is even more complicated and more nicely tailored to the situation. *E. octoculata* lives for more than a year, and thus some specimens are large in late summer and have to face the lowest oxygen concentration in this condition, but in *E. testacea* the difficult period of the year is faced only by young specimens with a large surface to volume ratio, and this is of further help to the species (p. 167). Incidentally, it is interesting to note that in winter, when no oxygen problems arise, *E. testacea* displays dependent oxygen consumption just like its sister species. Perhaps there is some physiological premium on independent respiration which is paid only when necessary.

Salinity in running water has been little studied except with regard to pollution, but slightly saline natural streams do occur in some parts of the world. Many freshwater animals are unable to tolerate much salinity and are thus doubtless eliminated by salts (Beadle, 1957). On the other hand many aquatic invertebrates, particularly among Oligochaeta, prosobranch Gastropoda, Crustacea, Odonata, Hemiptera, Coleoptera, Diptera, and Trichoptera can tolerate quite high salinities, and even one mayfly species is known from brackish water (Berner and Sloan, 1954). The salinity levels up to which a long list of running-water invertebrates have been found in the salt-polluted German River Werra are given by Heuss (1966). Many are quite high and are measurable in many parts per thousand, (sea-water is 35 per mille). There are thus ample animals available to form a fauna, and it would be instructive to make a comparative study of naturally saline and normal streams in the same general geographical area.

One interesting feature about inland saline waters is that the ordinary brackish-water coastal animals never seem to invade them. Possibly this is because they usually lie far inland; but in modern times man has many times created saline streams quite close to the sea, and yet no coastal animals have moved into them. Such streams occur, for example, in western England and northern Germany, and that they can form excellent habitats for at least some coastal animals is shown by the widespread occurrence of the introduced North American *Gammarus tigrinus* in England (Hynes, 1955*c*) and its immediately successful introduction into the River Werra (Schmitz, 1960). It may be that other examples of this kind will occur during the next few decades, as man moves animals about by means of modern transport which is fast enough to permit them to survive the journey.

Acidity. Acid waters are almost by definition poor in calcium unless the acidity is man-made or caused by carbon dioxide in spring water and thus very transient. It is, therefore, very difficult, from field studies, to distinguish the effects of high concentrations of hydrogen ions from those of generally low base-content. Acid waters have, however, often been reported to be without species which are known elsewhere to occur in very soft, but neutral, waters, so it would appear that pH has some direct influence. For example, Albrecht (1953*b*) could find no *Gammarus* in the spring region of a small Flemish stream although it was abundant further downstream. This she attributed to the low pH and to the fact that iron hydroxide was being deposited there. Similarly the absence of *Crenobia alpina* from peaty streams in Wales, of *Centroptiloides* from certain South African streams and of some species of chironomid from streams in the English New Forest have all been attributed to low pH (Carpenter, 1928*a*, *b*; Agnew, 1962; Hall, 1951). Many other such examples occur in the literature, but almost all of them lack any proof that acidity is actually the controlling factor. Occasionally though the circumstantial evidence is fairly convincing. In 1955 a volcanic eruption showered the Kamchatka River with ash; this led to high acidity in the river and many chironomids and molluscs disappeared (Kurenkov, 1957). In the Amazon region molluscs are scarce in the very soft, acid, black waters, although, as we shall see, a few genera have overcome the physiological problems involved. In some areas, however, where for local reasons the pH is raised without any marked increase in water hardness, ordinary snails occur. Sioli (1963) found that where the Rio Arapuins joins the Rio Tapajós the pH is raised from 4·5 to 6·4–6·6 without any marked increase in calcium, and quite suddenly at this point a rich molluscan fauna appears. Possibly therefore, molluscs are more controlled by shell erosion than they are by calcium shortage.

Acid streams often have rich invertebrate faunas (Jewell, 1922; Winkler, 1956; Beck, 1965), and many stream animals apparently thrive at low pH levels. Such are, for example the common European stonefly *Nemurella picteti*, which is often found in peaty moorland headwaters (Pomeisl, 1961) together with species of *Leuctra*; and several genera of dragonflies occur at pH values as low as 4·4–4·8 in streams in Tennessee (Wright and Shoup, 1945). Iron-depositing streams have a low pH and ferrous spring streams have often been reported as supporting few species (e.g. Dittmar, 1955*a*; Thorup, 1966). But even in such situations some invertebrates are often common despite the fact that iron hydroxide is deposited on them (Hynes, 1953). It seems probable that the presence of the deposit and the probably low oxygen-content of the water caused by the oxidation of ferrous ions are at least as important in eliminating species as is the low pH. The importance of acidity as such therefore remains to be substantiated at least in so far as many invertebrates are concerned.

Hardness. Water hardness, on the other hand appears definitely to be of some importance, although it should be stated at the outset that even this often remains to be proved. Moreover, in naturally occurring hard waters not only are calcium and magnesium present in relatively high concentrations but it is nearly always true that other ions, e.g. chloride, sulphate, and sodium, are also more abundant than they are in soft waters. Hard water has, therefore, not only more calcium than soft water but nearly always more of several other ions, and also a higher osmotic pressure.

It is perhaps appropriate, therefore, first to consider how water hardness might affect animals. Very few freshwater animals are completely watertight, so most of them have problems with osmotic pressure, which they overcome by various combinations of three physiological devices (Beadle, 1957). These are absorption of salts from the water, the production of urine which is hypotonic to the general body fluids, and the lowering of the osmotic pressure of those fluids and hence a lessening of the tendency of the water to enter. Clearly the first and third of these devices are easier to operate when the outside osmotic pressure is not very low. To live in very soft water therefore demands very efficient osmo-regulation, and animals such as the mitten-crab *Eriocheir* and the river-crab *Potamon* which have not even got as far as producing hypotonic urine would not be expected in very soft water, and they do not in fact occur there.

Many Malacostraca and most Mollusca secrete shells of calcium carbonate, so, on *a priori* grounds, one would expect them to be confined to waters where these two ions are readily obtained. It is indeed generally true that amphipods, isopods, crayfishes, snails, and bivalves are more common in hard than in soft waters, as has been reported by several workers (Clarke and Berg, 1959; Reynoldson, 1961; Slack, 1955; Shoup, 1943).

Calcium not only affects the efficiency of osmo-regulation but it has also been shown, for groups as far apart as flatworms and fishes, that both it and magnesium affect the rate of respiration. Both ions raise respiratory rates at low temperatures and lower them at high temperatures, to which they also increase resistance (Schlieper *et al.*, 1952). It can therefore be expected that water hardness has some important effects, but results from field studies are conflicting and often baffling.

Many groups of insects, e.g. Simuliidae (Grenier, 1949) and most Ephemeroptera (Illies, 1952*b*), seem to be quite indifferent to water hardness, and the same applies to the common *Spongilla lacustris* (Jewell, 1939). On the other hand the Malacostraca and Mollusca listed above, and also leeches (Mann, 1955) appear to be definitely scarcer in softer waters. Nevertheless, representatives of these apparently hardness-sensitive groups are sometimes found abundantly in very soft streams and many such findings are quite unexplained. For instance, European species of *Gammarus* have often been reported from soft waters, and, although it has been 'proved' experimentally and theoretically that it is not possible for *G. pulex* to reharden its cuticle after a moult when the water contains less than 5 mg./l. of calcium, thriving populations of this species have several times been found in waters which contain less than this (Hynes, 1954). One must, therefore, as I have elsewhere stressed in this book, always interpret laboratory experimental results with caution. But the reverse also applies, and a nice example comes from our present area of consideration. Dittmar (1955*a*) after investigating a soft-water stream in the German Sauerland suggests that the absence from his collections of *Polycelis felina* was because the water was too soft. A few years earlier Illies (1952*b*) had suggested that this species was absent from a tributary of the River Weser, not very far away, because the water was too hard! In fact both these gentlemen must have been wrong; *P. felina* is quite indifferent to water hardness and it occurs in both very soft and very hard waters in Britain.

Some molluscs are often found in surprisingly soft water. A few of them, such as the Ancylidae and the *Ampullaria* of the black, and very soft, waters of some tributaries of the Amazon have, as it were, avoided the issue by reducing their shells to only the horny conchiolin layer and so eliminating their special need for calcium (Sioli, 1963, 1964; Fittkau, 1964), but others, such as *Ancylus fluviatilis* and *Limnaea pereger*, both of which are common in Europe, and *Physa pumilia* in Florida (Beck, 1965), occur quite often in very soft streams. They are, however, much less common there than in harder waters; they are small and relatively thin-shelled, and *A. fluviatilis* seems not to tolerate less than 2 mg./l. of calcium (Maitland, 1965*b*).

Perhaps the most surprising of the molluscs in this respect is the mussel *Margaritifera margaritifera* which has the reputation on both sides of the Atlantic of occurring only in soft water (Steusloff, 1939; Murphy, 1942). Yet it is apparently so lavish with calcium that it alone among the freshwater bivalves supported a medieval pearl-fishing industry, and it produced at least one pearl which was big enough to find a place in the crown of the kings of England. In fact it is not confined to soft water, and in hard water it produces a form which was long thought to be a different species, but it is certainly most common in soft waters and it often occurs where there is very little calcium. The umbo of its shell is indeed often dissolved away and replaced by white nacre (Fig. VIII, 17). Here the secret seems to be merely longevity; it lives and grows for over a century, so its annual uptake of calcium must be minute (Hendelberg, 1960; Björk, 1962).

Within some other groups there appear to be definite preferences for certain water hardnesses, but sometimes this has really nothing to do with hardness as such. For instance, mosses tend not to grow in very calcareous places because of their need for free carbon dioxide, therefore moss-dwelling species are absent (Dittmar, 1953). Such waters also deposit travertine and this is the only habitat of certain species, such as *Lithotanytarsus* (p. 138), some other species of Chironomidae and the elminthid beetle *Riolus* (Geijskes, 1935; Beier, 1948). Travertine is also a poor substratum for molluscs as it is difficult for them to crawl over, and it overwhelms sessile animals such as sponges (Jewell, 1939).

There remain, however, many unexplained and apparently valid observations which indicate that within certain taxonomic groups some species are calciphile and others are calcifuge. This has been found, for example, among Chironomidae in streams in the English New Forest (Hall, 1960), among stoneflies in Europe (Illies, 1952*b*), within the mayfly genus *Heptagenia* in Britain (Macan, 1957*a*, 1961*a*), among the Elminthidae in France (Berthélemy, 1966), and in the mite genus *Sperchon* in Germany (Schwoerbel, 1961*b*). It should, however, be added that sometimes some of the conclusions derived from such observations do not survive further study. For instance, Illies found that the *Nemoura* species occurring in the soft-water tributaries of the River Mölle was *N. cinerea* and that in calcareous tributaries it was replaced by *N. cambrica*. In Britain, however, the former species is common in many hard-water streams and the latter in very soft water, particularly in Wales, the Roman province of Cambria, after which it is named.

On the other hand several workers have suggested that the European stoneflies *Nemurella picteti* and *Capnia bifrons* are soft-water and hard-water species respectively, and the same applies to the mayflies *Heptagenia lateralis* and *H. sulphurea*. So far all observations seem to support these conclusions, but there are no reasonable suggestions as to why they should be true.

We can conclude therefore that water hardness is a controlling factor in the ecology of at least some stream invertebrates, but that very often we have little idea as to how it operates and that we need much more critically collected and assessed data.

LIABILITY TO DROUGHT

As we have seen earlier the fixed pupae of many insects have special structures which enable them to survive if the water-level falls and exposes them to the air. Others, on the contrary, such as *Glossosoma* and *Silo*, apparently attempt to avoid this by moving into deeper water to pupate (Scott, 1958). Be that as it may, it is certain that *Agapetus*, which is a close relative of *Glossosoma*, is frequently exposed to dryness when it is fairly near to emergence in the early summer in western Europe, and this does not seem to interfere with the successful appearance of the flies.

It would appear though that the ability to withstand drought is by no means universal among stream invertebrates, and that many of them take few steps to avoid its consequences. Mauch (1963) observed that when the level of the Moselle became very low during the dry summer of 1959 large numbers of invertebrates were stranded by the receding water. Many *Gammarus* were dried out and killed, and the population of *Heptagenia sulphurea* was markedly reduced. Mussels retired into the substratum, but made no attempt to follow the retreating water's edge; together with the prosobranch snails, they merely closed their shells and waited. In this condition they were able to survive for a few days, but many were killed.

Other workers studying smaller streams which dried out during exceptionally rainless years have found that certain species were eliminated (Engelhardt, 1951; Hynes, 1941, 1958, 1961; Kamler, 1966; Larimore *et al.*, 1959; Sprules, 1947). The conclusions to be drawn appear to be that, with some exceptions, species which are in an active state during the dry period are killed. Droughts usually occur in summertime, so it is the late-emerging insects, such as *Leuctra decepta* in eastern North America and *L. fusca* in Europe, which are eliminated. Species which are in the egg stage during the drought survive as long as the water reappears before they are due to hatch. It seems that there is no mechanism to delay hatching, so that if the dry period persists for too long the larvae either fail to hatch or they die soon afterwards (Hynes, 1958, 1961). It would, however, seem likely that this does not apply to the inhabitants of temporary streams in very dry and uncertain climates, because if it did no fauna would survive the occasional year when the streams fail to flow at all.

The exceptions, that is those invertebrates which can endure drought even in an active state, include at least some flatworms, oligochaetes, harpacticoid copepods, Elminthidae and their larvae, some chironomid larvae, and Hydracarina. Presumably they creep down into the hyporheic habitat and there find sufficient moisture, or even water, to allow them to survive (Beier, 1948; Hynes, 1958). There are, indeed, a few species, such as the leech *Dina lineata*, which are found only in places which dry up more or less regularly (Mann, 1959); it is possible that only in such places can they compete with others like them but without the ability to avoid the consequences of drought.

There appear also to be some exceptions to the apparent ability of eggs to remain alive out of water. Among these is the common European mayfly *Rhithrogena semicolorata*, which was observed to have been eliminated by ten weeks of drought which ended in time to have allowed some eggs to hatch under water (Hynes, 1958). Probably many such species exist.

It is clear therefore that dry periods, whether they are a normal or exceptional occurrence, exert a considerable influence on the fauna. It is generally true of streams in temperate climates that if they dry up they do so in the summer, and that they do not contain late-flying insects nor many higher crustaceans and molluscs. But this does not mean that such streams are short of inhabitants. It is well established that the eggs of many insects, for example *Prosimulium gibsoni* and *Simulium damnosum*, can survive dry for many months (Anderson and Dicke, 1960; Crisp, 1956), and intermittent streams often have rich and varied faunas, as we shall see in Chapter XXI.

Drought often only affects part of a stream, of which the headwaters continue to flow, and then, when the water returns, the formerly dry reach is often rapidly recolonized from areas which remained wet. We shall be considering the phenomenon of downstream drift in Chapter XIII, and we have already seen that upstream movement also occurs (p. 153). More interesting from an ecological standpoint is the fact that often, after a disaster such as an unusual drought, species which were not there before appear in the affected area. Sprules (1947) found that after a stream which he was observing in Algonquin Park dried up in one year, it produced so many Chironomidae that he caught as many insects emerging from his area of study as he did in the following year when no drought occurred and the full complement of late-emerging Plecoptera, Ephemeroptera, and Trichoptera was present. And when a reach of the Afon Hirnant in Wales refilled after a long and unprecedented drought, several species of dytiscid beetle which had not been there before appeared for a while, and the stonefly *Nemoura cinerea* colonized the stream and passed through one generation. It had not been there before and it did not reappear (Hynes, 1958, 1961). *Simulium ornatum* is a similar opportunist species which can move rapidly into a disaster area, pass through one generation and disappear again, as was observed by one of my students working on a similar Welsh stream which dried up.

SPATES

It has often been observed that periods of high water reduce the invertebrate fauna in streams. An early study was that of Moffett (1936) of a number of localities in Utah. Some of his data are given here in Table XI,7. South Willow Creek suffered from a cloudburst on 7 August 1934; this produced the first major flood in fifty years and completely altered the stream bed. On 20 August Moffett could find no animals at all, but by late September recovery had begun, and this continued into November with some loss during the winter. He sampled two other streams for comparison, City Creek, which had not flooded for four years, and Mill Creek, which had not flooded for many years, and his

Stream	South Willow Creek			City Creek	Mill Creek
Date	22–3 Sept. 1934	11 Nov. 1934	2 Mar. 1934	17 & 26 Oct. 1934	27 Oct. 1934
Chironomidae	120	282	158	36	137
Simuliidae	17	690	433	5	1
Tipulidae	P	1	P	1	—
Ephemeroptera	8	18	166	172	194
Trichoptera	2	10	5	23	1,108
Other Diptera	2	5	4	19	43
Turbellaria	2	4	2	45	62
Oligochaeta	P	5	7	4	46
Coleoptera	1	3	4	7	8
Nematoda	P	1	P	P	1
Plecoptera	P	21	24	69	35
Gastropoda	—	—	—	1	37
Pisidium	—	—	—	—	2
Total no.	152	1,040	803	382	1,673
g./sq. ft	0·356	1·742	1·97	1·14	3·44
No. of samples	28	14	14	15	10

TABLE XI, 7. The average number per square foot of animals taken from three streams in Utah with a Surber sampler. Data from Moffett (1936). P = present.

data from these are also shown in the table. Probably these streams were not strictly comparable, but the data do indicate that the faunal composition was different from that in the recovering stages in South Willow Creek, and that these were initially dominated by Chironomidae and Simuliidae which have short life cycles. The more normal animals such as Ephemeroptera, Trichoptera, and Plecoptera reappeared only slowly, probably from the headwaters.

Similar disastrous effects of spates are reported from all over the world. In Austria a storm on 11 May 1951 increased the flow of the Mauerbach in the Vienna Woods by 15,000 times and wiped out the fauna (Pomeisl, 1953*a*); an October flood greatly reduced the fauna in Cascadilla Creek, New York (Mottley *et al.*, 1939), and in Wales July floods cut down the fauna of the River Towy from between 613 and 1,091 animals per square foot to 445, and this was further reduced to only 48, mostly the little beetle *Esolus*, by more high water in August

(Jones, 1951). The reduction may not be, and probably usually is not, uniform. Allen (1951, 1959) showed how the effect varies from place to place in the Horokiwi River, New Zealand. A small spate in June 1958 disturbed only the centre of the river, leaving only 31–105 animals per square foot, whereas at the sides the population remained high at 1,374–2,115 animals per square foot. In the River Endrick in Scotland, Maitland (1964) found that winter floods reduced the fauna quite markedly, and that in sandy areas, on which the effect was greater than on stony ones, only burrowing animals, such as Tubificidae, Chironominae, and *Pisidium*, were able to survive the winter. On the other hand the floods did sweep down specimens of some species normally found only on stony areas, so that they were collected on the sand. The effect may thus be different in different reaches and may actually lead to some increases. Minckley (1963) found that a severe flood, emanating mostly from the large spring at the head of Doe Run, reduced the fauna near the spring but increased it in areas further downstream, presumably by the simple process of washing animals out of their original habitats and depositing them lower down the channel.

Floods may also have nothing to do with rainfall, and yet have the same effect. Sprules (1947), in his study of insects emerging into traps on a stream in Algonquin Park, found that a seven-day flood produced by the rupture of a beaver dam in May 1940 resulted in his catching only about half as many adults during that year as he had caught in 1939. It was also clear that the Plecoptera, Ephemeroptera, and Chironomidae had been more severely affected than the Trichoptera and Simuliidae, and this differential effect has been observed in several investigations. Minckley noted that in Doe Run *Gammarus* was much less affected than *Asellus*, and Jones that *Baetis* was more reduced than were *Ephemerella* and *Leuctra*. Even within some species the effect may vary with size; Harker (1953) found that large specimens of *Ecdyonurus torrentis* were frequently crushed and mutilated by rolling stones during spates in a small English stream, but that small specimens survived. The effect of a spate can therefore be different at different seasons.

One result of these phenomena is that streams which are more liable to spates have less abundant and less varied faunas than others. This has been established by studies in Sweden (Badcock, 1953*b*), Bavaria (Engelhardt, 1951), Wales (Jones, 1948), Transcarpathia (Ivlev and Ivassik, 1961), and North America (Nevin, 1936), and it is doubtless a universal phenomenon. An interesting consequence of this is that, in general, small tributaries, being less exposed to the effects of storms covering limited areas, are richer than the larger streams into which they flow, a point which was nicely demonstrated by the Transcarpathian studies

(Table XI, 8). Another consequence is that where man clears the forest and increases the intensity of run-off, he, at the same time, cuts down the variety and abundance of the stream fauna. This point was well brought out by Badcock (1953*b*) in her comparison of streams in a wooded area of Sweden with very similar streams in the now relatively bare landscape of North Wales.

Rivers	No. of stations	No. of samples	Biomass, g./sq. m.	No. per sq. m.
Teresva	10	46	3·266	329
Tributaries	87	264	4·994	557
Borzhava	6	16	3·459	424
Tributaries	8	20	6·250	697
Latoritza	6	6	1·800	170
Tributaries	31	31	4·643	436

TABLE XI, 8. The average amounts of invertebrate fauna in three small rivers and their tributaries in Transcarpathia. Data from Ivlev and Ivassik (1961).

A further effect of flooding is that in areas where spates are seasonally regular there tends to be a corresponding seasonal change in the density of the fauna. Gaufin (1959) found that in the Provo River, Utah, a mountain trout-stream with spring run-off of meltwater, he always found fewest animals in the period April to June. Some of this reduction he attributed to the difficulties of sampling during high water and some to the emergence of early species of Plecoptera, but much he concludes was because of losses caused by wash-out.

In South Africa a number of studies have also demonstrated a relationship between the flood season and the amount of invertebrate fauna. In Cape Province, an area of winter rainfall, the Great Berg River suffers much disturbance of its bed, and consequently the fauna is at its lowest level, in the winter (Harrison, 1958; Harrison and Elsworth, 1958). In the Transvaal, however, rainfall occurs mainly in the summer, and rivers like the Jukskei Crocodile, the Tugela, and its tributary the Buffalo have richer faunas in the winter than in the summer (Allanson, 1961; Oliff, 1960, 1963). This effect can, however, vary from place to place and be altered by special local circumstances. Harrison and Elsworth found that the fauna of stony reaches in the Great Berg River was far less affected than that on other types of substratum except where the floods scoured the stony areas with abrasive sand. Presumably elsewhere the rough stony areas provide adequate shelter from fast water. In the Vaal River the presence of a dam which backs up the water for 64 km. actually reverses the effect of high water because of the abundance, in the summertime flood season, of the larvae of *Cheumatopsyche* and

Amphipsyche (Chutter, 1963). These larvae feed on the plankton which develops in the dam lake during the summer, and at the same time the impoundment removes the abrasive silt and sand by settlement and so protects the river downstream from at least one consequence of high discharge. It also evens out the fluctuations in flow. It seems likely, therefore, that the two most important factors in spates are abrasion and fluctuation, and that a high rate of flow itself, as long as it does not go on all the while as in the Austrian Danube (p. 198), is not of great importance; temporary shelter can always be found.

All the above examples are, however, somewhat complicated by seasonal changes of temperature. In the tropics many areas have seasonal changes of rainfall with little change of temperature, and it is fortunate that the invertebrate fauna of one tropical stream has been sampled throughout the year. Van Someren (1952) studied the Sagana River at three points on Mount Kenya, almost on the equator, and some of his numerical results are shown in Table XI, 9. There are two rainy seasons

Altitude in feet	7,500	5,865	5,600	
Jan.	137	247	78	
Feb.	185	164	152	
Mar.	124	183	110	
Apr.	55	111	207	long rainy season
May	41	36	28	
June	19	59	15	
July	53	40	33	
Aug.	46	132	77	
Sept.	175	48	112	
Oct.	130	110	110	
Nov.	93	31	35	short rainy season
Dec.	112	91	95	

TABLE XI, 9. The average number of invertebrates per square foot at three altitudes in the Sagana River, Kenya, in 1949, based on three samples per month from each station. Data from Van Someren (1952).

which produce high and fluctuating discharges down the steep valley; the table shows that such periods produce a definite reduction in the fauna which, however, recovers fairly rapidly, since breeding seems to be continuous in tropical streams. He also sampled a number of other streams on Mount Kenya and found more animals per square foot in the Naro Moru River (177 at 6,300 ft. and 232 at 8,000 ft.) than in the Burgeret River (50) and the Sirimon (47) both at 7,000 ft., and the average of all his samples in 1949 from the Sagana River was 97. The Naro Moru River is very exceptional for a tropical stream in that it is glacier-fed and so has a very much more regular discharge than the others; this possibly accounts for its richer fauna.

FOOD

The availability of food is an obvious factor controlling the occurrence and abundance of species. We have already touched upon this subject at several points, e.g. when discussing the relationship between the distribution of insects and detritus in stony substrata (p. 211), the distribution of some Trichoptera in relation to current (p. 184), the distribution of animals on plants (p. 213), and the special fauna of submerged wood (p. 210). We could add here the often noted association between filamentous algae and chironomid larvae (e.g. Von Mitis, 1938).

Generally speaking species occur, or are common, only where their food is readily available, but it should not be forgotten that few running-water invertebrates are very specialized in their diets. For instance, many insects can eat either algae scraped from stones, or detritus, and diets may change somewhat with season according to the availability of algae (Chapman and Demory, 1963).

This can lead to some apparent anomalies. For example, in the River Polenz, a tributary of the Elbe, Albrecht and Bursche (1957) found that the number of algae decreased downstream over a reach of 6 km. although the lower reaches were better lighted, and so presumably should have produced more growth. However, at the same time the population of invertebrates increased and it is likely that the effect had been produced by increased grazing. It would, though, appear most unlikely that such a situation would be a stable one unless some other factor were involved, and it seems probable that detritus also increased downstream, thus allowing larger numbers of invertebrates to live there and to produce a marked effect on the algal population without depriving themselves of food.

Occasionally this effect may work the other way, and large developments of algae may overtake the fauna and eliminate much of it. Harker (1953) reports that during her study of the mayflies in a small English stream a sudden growth of Cyanophyta eliminated several species, although she did not discover how this came about. When similar effects of diatoms on *Simulium* larvae have been noticed it has been reported that they grow over and smother the larvae (Sommerman *et al.*, 1955).

SHADE

A few investigations have indicated that there is a definite correlation between shade and the occurrence or abundance of certain species along the lengths of variously shaded streams. Thorup (1966) made a careful study of a spring stream in Denmark, the Rold Kilde, and he found that *Baetis rhodani* was more abundant in unshaded areas than under trees, and that the same seemed to apply to *Agapetus*, *Ecclisopteryx*, *Helodes*, and *Ancylus*, all of which feed to a large extent upon attached algae. The net-spinning caddis-worm *Wormaldia* showed, however, an opposite correlation. Beier (1948) has also noted that the larvae of Elminthidae appear to be reduced in numbers by shade, and this may also be correlated with food.

Similarly Hughes (1966*a*, *b*), during his study of streams in Natal, found that, although many species seem to be indifferent to shade, the mayflies, *Centroptilum sudafricanum*, *Tricorythus*, *Afronurus*, *Adenophlebia*, and *Castanophlebia*, were more abundant in shady places than in exposed ones. Other mayflies, however, e.g. *Baetis harrisoni*, *Centroptilum excisum*, and *Euthraulus*, and the stonefly *Neoperla*, and the trichopterans *Cheumatopsyche* and *Hydropsyche*, were more abundant in open places. Note that here the two species of *Centroptilum* were found to react differently.

Hughes discusses the possibility of indirect action through algal growth, temperature, oviposition habits of adults, and organic debris from the trees giving the shade, but whether it acts directly or not it seems that shade can be a factor in the ecology of stream invertebrates and that it merits more study than it has yet received.

OVIPOSITION HABITS

It is well established that many, indeed probably most, insects lay eggs only in places where the young can thrive. This subject has been discussed, with special reference to the aquatic habitat, by Macan (1961*a*). Adult rheophile insects are attracted to streams, and usually this attraction is of the ovipositing female; but the damse-flies are a special case, since the male and female oviposit in tandem and the males defend territories for the purpose. The females are attracted to the males, and in this way the eggs of *Agrion*, for example, are always laid in running water (Zahner, 1960).

Sometimes though it appears that the special requirements of ovipositing females may limit the occurrence of the young stages to a range narrower than that which they could successfully occupy. In Scotland

Simulium larvae tend to be scarce in heavily wooded streams except in open places (Zahar, 1951). This may be either because the larvae tend to move towards the light or because of a failure of the females to oviposit in shaded reaches. The latter explanation seems the more likely as we know that the phototropism of the larvae is subservient to their rheotropism (p. 148). Failure to oviposit in shaded reaches has also been suggested by Hartland-Rowe (1964) to account for the absence of *Rhithrogena* from a small tributary of Gorge Creek in Alberta, and Pinet (1962) believes that the patchy distribution of *Oligoneuriella rhenana* in European rivers results from very special requirements of the females for oviposition. The adult stage, therefore, although often very brief, may exert a considerable influence on the detailed distribution of the young stages of at least a few species, and it is possible that quite a number of anomalous distributions may be accounted for in this way.

It has also been shown that some species of the snails *Bulinus* and *Biomphalaria* fail to lay eggs satisfactorily in turbid stream water in Rhodesia although the snails themselves live quite well in it (Harrison and Farina, 1965). It would seem that this would limit them to certain localities or their breeding to certain seasons.

OTHER BIOTIC FACTORS

In any community of living organisms the usual factors of predation, disease, and inter-specific competition operate, and there are some interesting examples of biotic interactions from running water.

In Belgium it has been reported that planted coniferous woodland composed of *Picea* and *Thuja* exerts a deleterious effect on the fauna of streams flowing through it (Huet, 1951), and pinewoods in Moravia have been similarly accused (Zelinka, 1950). It would seem that the effect of riparian vegetation on the invertebrate fauna of streams would repay further study, and that there may be many other examples of certain trees controlling to some extent the composition of the aquatic fauna which is continually exposed to their litter. This point is discussed further in Chapter XXII.

Catastrophic epidemic disease is not often noted in wild populations of invertebrates except where they have attained pest proportions under unusual conditions created by man. In the nineteenth century, however, three species of European crayfish were attacked by a disease of unknown origin, which appeared in France in 1876 and spread rapidly to western Siberia (Vivier, 1965). The large edible species, *Astacus fluviatilis*, was almost wiped out, and it still remains a rather rare animal. Its smaller relative, *A. pallipes*, was less affected, but Duffield (1933) who made a long series of observations on this species in the River Ock, a tributary of the Thames, found that it suffered a marked diminution in numbers at intervals of about thirteen years. He also observed that mortalities appeared to spread upstream. The crayfish apparently persisted in the side streams from whence the main stream was, each time, recolonized. In France it had been noted earlier that the initial rate of spread upstream, in any one watercourse, varied between 2 and 60 km. a year, and there was some evidence that the disease persisted in lakes (Vivier, 1965).

Macan (1962, 1963) describes a very nice example of how a very small change in a stream can, through the agency of one species, exert a large influence on the general biota. He had for some years been studying a small stream in the English Lake District, and subsequently piped water was taken to a small hamlet near its source. As a result the inhabitants of the hamlet no longer had to pump all the water to their houses, and they became more lavish in their use of it. This in turn led to the overloading of a septic tank, which previously had worked adequately, and small amounts of sewage found their way into the stream. The enrichment of the stream encouraged a large increase in the population of the flatworm *Polycelis felina* and, at first, of many of the species of insects. However, the numbers of *Polycelis* continued to rise, and then several species of mayfly were adversely affected. *Ecdyonurus torrentis* disappeared altogether, and *Rhithrogena semicolorata* and *Baetis rhodani* declined markedly in abundance. Macan notes that these are all species which live on the tops of stones, while others such as *B. pumilus* and the stoneflies, which live under stones, were unaffected. He suggests that the reason was that *Polycelis*, which is a carnivore and traps its prey with mucus strings laid down as it moves over the stones, was selectively eliminating the species which came into contact with its traps. Thus a minor chemical effect which, possibly itself indirectly, had affected one species had produced far-reaching effects on others which, in its absence, might well have benefited from the slight enrichment of the environment.

Another example of the influence of predation is the finding of Straškraba (1965) of the sudden diminution, as compared with upstream reaches, of *Gammarus fossarum* below a weir in a stream in Czechoslovakia. The weir forms the upstream boundary to trout and it would seem that here we have a clear instance of direct predatory control. Obviously this is an important feature of life in all habitats, but it is not always possible to observe its effects so directly.

There are also some fairly well-established examples of competitive exclusion by closely related species. We have already seen that the two European species of *Agrion* inhabit reaches of streams with rather different temperature regimes and that this can be related to some extent to their respiratory requirements (p. 201). It so happens that their optimum temperature ranges do not overlap much, so they must compete with one another only over limited areas (Zahner, 1959). In another dragonfly genus, however, there is clearer evidence of competition. Normally in Europe *Cordulegaster bidentatus* inhabits springs and *C. annulatus* the lower reaches of small streams (Geijskes, 1935), but where, as in parts of northern Germany, *C. annulatus* is absent, *C. bidentatus* may extend far downstream (Illies, 1952*b*). Presumably in other areas it can compete with its rival only in the cold waters of springs.

The British Isles, which because of the effects of the Pleistocene ice ages have a rather limited fauna, provide many examples of species having a wider ecological range where they are free of competitors. There *Astacus pallipes* is the only crayfish, and it occurs in all types of river as long as the conditions are suitable for crayfish. On the continent of Europe it is confined to small streams, and other species occupy the larger ones and the rivers. In Wales the white flatworm *Phagocata vitta* is often found in small streams, whereas in Sweden it is confined to groundwaters (Dahm, 1949). Very probably this is because of the absence from Britain of *Dugesia gonocephala*, a species which appears to compete with it on the continent. On the British mainland *Gammarus pulex* occurs in all suitable waters from high-level streams in the mountains to large rivers. Across the Channel the high cold streams are occupied by *G. fossarum*, and the low warm ones by *G. roeseli*, neither of which occurs in Britain. And on the other British Isles where even *G. pulex* is absent, or was until recently, many streams are occupied by *G. duebeni* which elsewhere, except in a few special areas, is confined to brackish water. Presumably it is unable to compete with marine species on the one side and freshwater ones on the other, although it can live and breed in sea water and in at least some types of fresh water (p. 169) (Hynes, 1954).

A last example concerns the insects. In England and Ireland the large Perlodid stonefly *Diura bicaudata*, which incidentally is of particular interest as being one of the few aquatic insects to occur right around the Northern Hemisphere, is confined to high-altitude streams or to the shores of lakes. Low-level streams and rivers are occupied by the rather similar *Perlodes microcephala*. In the Isle of Man, which lies in the middle of the Irish Sea, *P. microcephala* is absent and *D. bicaudata* occurs in streams right down to sea-level (Hynes, 1952*a*).

We can conclude, therefore, that many aquatic species are able to occupy a wider range of conditions than those in which they are normally found, but that they are prevented from doing so by biotic factors. Anomalous distributions such as those mentioned on p. 204 with respect to temperature may therefore very possibly be caused by peculiar biotic circumstances, and this is especially likely to be true where the species is at the edge of its geographical range and so possibly beyond that of a normal competitor.

THE PROXIMITY OF SUITABLE HABITATS

Although it may be considered strange to consider the close proximity of suitable habitats as a factor controlling the occurrence of aquatic invertebrates, and despite the fact that a suggestion made many years ago that it is one has been described as irrational by Brinck (1949), it does nevertheless often lead to unexpected occurrences in unusual places. Moreover, quite large numbers of individuals may be involved and give the impression that the species is well established.

We have already seen (p. 213) that Minckley found that in places where beds of moss were densely populated by *Gammarus* and *Asellus* these animals spilled over on to the riffles, on which, elsewhere in the stream, they did not occur. We know from the experiments of Moon (1940), who studied the colonization of trays of substratum laid down on the beds of lakes and streams, that stream invertebrates move around actively, so it is not surprising that close to large populations considerable numbers can be found which are, as it were, out of place. Mankind himself offers many parallel examples, and a superficial observer of our species might well conclude that we were capable of living out our lives in water or on rugged mountain slopes.

Both Hynes (1941) and Sprules (1947) have noted this proximity effect on the detailed distribution of stream insects, and it is one which should always be considered as a possibility when a population is found which appears to be in a habitat unusual for the species.

We may conclude, therefore, from the discussion in this chapter that the general considerations of animal ecology apply to stream invertebrates in full measure. And it seems appropriate here to restate the three ecological principles of Thienemann (1954*b*), one of the earliest students of stream invertebrates, if only because they appear to have been ignored by the English-speaking world. Two of them were first propounded in 1920, yet they are repeatedly rediscovered and put forward as new ideas, and one can frequently overhear German scientists at international meetings expressing shocked surprise that they are apparently almost unknown outside Central Europe. They are:

1. The greater the diversity of the conditions in a locality the larger is the number of species which make up the biotic community.

2. The more the conditions in a locality deviate from normal, and hence from the normal optima of most species, the smaller is the number of species which occur there and the greater the number of individuals of each of the species which do occur.

3. The longer a locality has been in the same condition the richer is its biotic community and the more stable it is.

The reader will appreciate that these three simple statements incorporate or imply many of the concepts that are included in this chapter.

CHAPTER XII

Quantitative study of benthic invertebrates

By their very nature river beds are difficult to sample accurately; they are not only often hard and full of boulders and large stones, but the current always interferes to some extent with the successful operation of apparatus. Moreover, the varied distribution of flow and eddy causes equally varied types of substratum, which, as we have already seen, harbour different communities of animals. To assess populations in terms of numbers or biomass per unit area therefore presents formidable, perhaps insurmountable, difficulties. Wurtz (1960) goes so far as to conclude that 'to sample, randomly, a population composed of burrowing, sedentary and peripatetic species, as well as solitary and gregarious species, and have the results mean anything quantitatively, is a highly dubious procedure'. This is certainly true, but dubious or not it is essential for a full understanding of the ecology of any environment to have some measure of the abundance, or at best the relative abundance, of the species, and many workers have attempted to obtain this. It is possible to assess the findings statistically, and methods for doing so have been evolved with special reference to stream-bed samples (Harris, 1957); many workers (e.g. Longhurst, 1959) have, after careful discussion of the validity of results obtained, concluded that, if several samples are collected, a fair picture of the abundance of the commoner species is obtained. We shall return to this point of the validity of sampling later on.

SAMPLING METHODS

The various types of apparatus and method which have been evolved for quantitative sampling of the benthos in running water have been discussed by Albrecht (1959), Macan (1958*b*), and Cummins (1962), to whose comprehensive papers the reader is referred for more detail than is given here. In particular this chapter owes much to the excellent review of Albrecht which is, unfortunately, not readily available in many parts of the English-speaking world.

In some favoured locations it is possible to use apparatus which was designed for still water, such as the Ekman and Petersen grabs, which work properly only in soft sediments. In passing it may be noted that these instruments are often referred to as 'dredges'. This is, of course, an incorrect use of the word, which is derived from the verb to drag, and means an instrument which is pulled across or through the substratum. Grabs, on the other hand, bite into the bottom from above and do not move across it. It seems unfortunate not to retain this clear and useful verbal distinction between two entirely different processes. However, to return to the apparatus, it is usually difficult to operate a grab from a boat in a river, as the current tilts it sideways, and often the only way is to allow the boat to drift with the current with the grab just above the bottom, and then to drop it suddenly so that it bites in vertically (Schräder, 1932). The same applies to the various lacustrine coring devices which can be used on soft substrata. Indeed, working in deep running water presents very great difficulties, which could perhaps be solved by scuba-diving. At present most quantitative devices can be operated only where the water is shallow enough for wading, although some of the grabs can be used on poles. Such are the pole-mounted Ekman grab used by Berg (1938) in the sub-littoral regions of Lake Esrom and a somewhat similar, but slightly more effective, grab designed by Allan (1952) which is capable of working in small gravel, although even it is prone to jam on small stones and thus fail to close.

Shallow-water samplers are of various kinds. Some workers have used grabs which are pushed by hand into the substratum and so bite out a definite area (Ford and Hall, 1958; Minckley, 1963). Others have used corers in this way, and an interesting addition to this technique is the use of very low-temperature fluids to freeze the sample *in situ* before the core is withdrawn, thus retaining it in the sampler without disturbance (Shapiro, 1958; Efford, 1960).

A whole family of devices has been designed which are pushed down into the substratum and then pushed or pulled forward a fixed distance like a shovel to take up a definite area of substratum (Macan, 1958*b*; Dittmar, 1955*b*; Allen, 1940, 1951; Kamler and Riedel, 1960*b*; Hynes, 1961). Such dredge-like devices, if they are heavy enough, can also be used on lines in fairly deep water (Usinger and Needham, 1956); but all, in contrast to grabs and corers, incorporate some sort of netting to allow water to pass through as the sample is taken; otherwise animals and some of the substratum are lost by displacement. All must of course, be moved in an upstream direction to prevent, as far as possible, loss of displaced animals in the outflowing current.

Another family of devices is perhaps best described as box or drum samplers. They consist of square or cylindrical boxes open at top and bottom which are pushed into the substratum. In some of the circular ones the bottom is jaggedly toothed and the top is fitted with lateral handles so that the sampler may be rotated while being pressed in, and thus cut its way downward (Wilding, 1940), a modification which is particularly useful on rough stony bottoms. In all of them a definite area of substratum is isolated, and it can then be dealt with in a number of ways. Some investigators have pushed a metal plate through the substratum under the box and then lifted the whole area out (Kamler and Riedel, 1960*b*), but this can, of course, be done only with a small sampler. Others have baled out the contents (Whitley, 1962). Others have removed and washed the stones and then stirred up the remaining substratum and fished the enclosed water repeatedly with a net until no further animals could be collected (Needham, 1934; Idyll, 1943); or else, after stirring, have slipped in a meshed box, which fills the sampler and can then be closed and withdrawn taking with it all stirred-up material (Davidson and Wilding, 1943). A few have made use of the current to wash out the stirred-up animals. This is done by opening a port in the box on the upstream side and allowing the water to flow through into a net, taking all suspended matter with it (Neill, 1938; Waters and Knapp, 1961). The last method works, however, only where there is sufficient current to carry the most active animals, and, of course, all box samplers must be higher than the depth of the water, which restricts their use to shallow places.

Similar box samplers can be used for sampling plant beds, and Gerking (1957) has used this method to separate samples of plants from those of the underlying substratum. He used a large box, cut off and removed the enclosed plants, and then took a bottom sample with an Ekman grab inside the box. This can, of course, be done only on a reasonably soft substratum, but it is on such materials that rooted plants normally grow in rivers.

A logical extension of the ideas behind the current-operated box-sampler is the well-known Surber (1937) sampler. This widely used instrument has become for the stream-invertebrate worker a symbol of office almost equivalent to the butterfly net of the entomologist. Instead of physically enclosing the area to be sampled, this device merely marks it out with a frame, usually with sides to prevent too much eddying, upstream of a net, and the operator picks up stones, etc., from the area and scrubs them in the mouth of the net. He then stirs up the whole area a few times and allows all the debris to be swept into the net by the current. Similar apparatus has been devised in Europe (Müller, 1958; Badcock, 1953*b*), but it is difficult to use in very cold weather or where the current is slack (Leonard, 1939).

A somewhat comparable method which has been quite widely used in Europe is to lift up and clean off a definite number of stones into the mouth of a net. If the stones are retained and their areas measured, by drawing round them on a piece of paper with a pencil, it is possible to relate the catch to a definite area of stream bed (Schräder, 1932; Albrecht and Bursche, 1957). Such a method, however, cannot collect animals which are at all deeply buried in the substratum.

It will be appreciated that nearly all these shallow-water samplers incorporate some sort of netting, and that in most of them this cannot be too fine or the water passes too slowly. As Jónasson (1955, 1958) has so clearly demonstrated, the size of the mesh can be very important in life-history and production studies, so it is often important to use a very fine mesh. Because a fine mesh impedes the flow and causes eddies it can only be used in shovel-like devices, or in box samplers where the operator himself wields a net in the enclosed water, and for good quantitative work such methods appear to be the most suitable. In practice, however, neither shovels nor boxes work very well on rough bottoms. Shovels get stuck on, or ride over, large stones, and they greatly disturb the substratum causing animals to move away, and it is difficult to get them to bite in very deeply. Box samplers are very difficult to embed far into rough bottoms and they tend to leak round the edges; and in swift water the very act of lowering them on to the substratum causes downward eddies and the loss of some of the more active creatures.

Alternative methods of quantitative sampling are to set out trays of substratum in the river bed, and then to lift them complete with the fauna that has moved in after a fairly long interval (Moon, 1935; Wene and Wickliff, 1940; Egglishaw, 1964), or to put in entirely artificial stones (Britt, 1955; Cianficconi and Riatti, 1957), layers of hardboard or some such material bolted together (Hester and Dendy, 1962), or wire-mesh boxes filled with brushwood or some other bottom material (Scott, D. C., 1958). Such methods, however, are unlikely to provide any measure of the population in the river bed as a whole, although they are a source of specimens, and under some circumstances they are the only way to obtain large numbers of certain species. Artificial substrata have also proved to be very useful in the comparative censusing of black-fly larvae which seem to be attracted to white surfaces or strips of polyethylene sheet. (Wolfe and Peterson, 1958; Williams and Obeng, 1962.)

If it is merely required to have some numerical assessment of the differences between areas, much simpler methods can be used, and they are often more satisfactory. They include, for example, general collecting with a hand net for a fixed time or kicking up the substratum in a standardized way in front of the mouth of a net. Such techniques were employed by Mauch (1963) in his study of the Moselle, and they are frequently used by workers interested in assessing the effects of pollution. Morgan and Egglishaw (1965) have recently studied the validity of this type of method and, as can be seen from Table XII, 1, it gives very

	Sample number I	II	III	IV	V	VI
Plecoptera						
Amphinemura	7	26	4	13	3	19
Protonemura	7	2	5	4	0	1
Leuctra	6	25	19	16	4	6
Brachyptera	12	0	0	3	0	4
Isoperla	1	2	0	1	0	1
Ephemeroptera						
Rhithrogena	8	18	7	15	3	9
Ecdyonurus	0	0	8	7	4	3
Baetis	24	23	19	17	9	21
Trichoptera	3	1	3	3	0	2
Diptera						
Simuliidae	19	0	0	1	0	2
Chironomidae	3	1	3	2	1	1
Other intervertebrates	2	3	2	1	0	2
Total	92	101	70	83	24	71

TABLE XII, 1. The numbers of invertebrates collected on a stony substratum in Allt Leathan, Scotland, in January 1960 in each of six samples resulting from three standardized kickings-up of the substratum in an area 24 × 46 cm. upstream of a net. Data condensed from Morgan and Egglishaw (1965).

consistent results. Only one of their samples, number V, was greatly different from the others, and only those organisms which are known to aggregate, e.g. *Brachyptera* and Simuliidae, were markedly different between samples. They also demonstrated that such samples collect the number of animals which more thorough samplers reveal to be present on a fairly definite area of stream bed, so it might be possible to quantify this technique, which can be used on all sorts of substratum. In using it the flat-ended net is held firmly on the substratum facing upstream, and the bed is disturbed just upstream of it by rolling over stones and gravel in an upstream direction with a booted foot for a definite distance. This is done several times, and Morgan and Egglishaw found that after three or four such kickings-up of the same area few additional specimens or species are taken. Some animals, e.g. Ancylidae, Hydroptilidae, and *Coryneura* which are firmly attached to stones or embedded in crevices, tend to be missed by this technique. As we shall see, however, all methods are to some extent selective, and the simplicity and flexibility of this one indicates that it merits further study to make it more truly quantitative. It is also possible to record the abundance of the various organisms obtained in such ways on an arbitrary logarithmic scale, and so to save a great deal of laborious sorting work. Such records are, however, only of limited value in the study of the fundamental problems of stream biology, and so have little place in this book. They have been reviewed, with special reference to the study of pollution, by Bick (1963) and Hynes (1964*a*).

In some places a very large proportion of the fauna is composed of insects, and it is, at least theoretically, possible to obtain some idea of the population by trapping the adults as they emerge from a definite area of stream bed. This idea was used by Ide (1940) and Sprules (1947) in their studies on streams in Ontario, but it is clear, from the fact that about twice as many insects were obtained when the traps were emptied every two hours as compared with daily emptyings, that large numbers of insects escape collection. Mundie (1956) has discussed the design of traps which may overcome some of the inherent problems, but Macan (1964) has shown that different designs of trap produce somewhat different results and that at least one species of stonefly tends to avoid them altogether. It is thus fairly clear that traps for adults, while useful in a comparative way are not likely to lead to firm estimates of the total population of even emerging insects.

The sampling of moss fauna presents fewer difficulties, other than the tedium of sorting. The moss can be scraped from a definite area with little danger of losing the inhabitants, because of their cryptic habits and efficient hold-fast mechanisms (Minckley, 1963; Kamler and Riedel, 1960*b*). Alternatively, a definite volume may be collected by plucking tufts at random (Frost, 1942), and the volume collected can then be related to a definite area of substratum as calculated from the average thickness of the growth (Hynes, 1961).

The actual collection of samples is, however, often only part of the problem. All types of sampling device, except emergence traps, collect a fair amount of fine debris as well as animals, and many collect a large quantity of substratum as well. An efficient means of extracting the animals is therefore essential. Hand-sorting in flat pans is possible, but it is so time-consuming that it limits the number of samples that can be dealt with. It is also not particularly efficient as it is easy to overlook specimens covered by or attached to other objects.

Two principal methods of overcoming this problem have been tried in various forms. The first of these is elutriation, and the simplest technique is merely to place the sample in a trough, with baffles, along which a current of water flows and washes out the animals and lighter debris into a sieve (Moon, 1935). A more sophisticated development of this idea is to place the sample in a vertical tube up which a current of water or a stream of air-bubbles passes. This brings up animals and light debris towards the surface where they can be drawn off through a lateral pipe (Lauff *et al.*, 1961). The same effect can be obtained even more simply by using large amounts of water and a shallow dish, which is operated in the same manner as in gold-panning except that, of course, here the heavy material is discarded and the light is retained, usually in a sieve.

These methods can be applied to both live and preserved samples; in many respects they work better with the latter, as dead animals are less liable to remain attached to heavy objects. Moreover, preservatives like formalin tend to cause animals to leave their retreats; for example, caddis-worms usually leave their stony cases and are thus more readily floated out. Even so, some heavy animals such as large molluscs and some caddis-worms remain in the debris, which has to be inspected before being discarded.

It is possible under some circumstances to cause living samples to sort themselves, at least in part. Britt (1955) found that a dilute solution of alcohol and hydrochloric acid causes animals to release their hold on stones, and Dittmar (1951) suggests the use of an intermittent electric current to chase animals out of debris. The current passes mainly between particles so the animals tend to move out into the free water where the electric field is less intense. J. D. Bayless of the North Carolina Wildlife Research Commission has, in a mimeographed report, advocated a similar method for use in the field as an aid to hand sorting.

Animals can also often be induced to walk or swim out of particularly difficult material. Williams (1960) suggests that they can be extracted from freshwater soils by covering the sample with clean sand under water. After twenty-four hours the great majority will have burrowed upwards into the sand from which they are easily washed out. Similarly, animals can be separated from large volumes of vegetable debris, which does not lend itself to any method of separation based on density, by placing the sample thinly on a coarse screen at or near the surface of a deep vessel of water such as a bucket. As the animals move around they fall through the screen and accumulate at the bottom. Quite high rates of extraction may be achieved by this simple technique, which modification of the method used to obtain Tubificidae for aqua Suppliers of these worms for use as fish food place the usually poll mud from which they are to be extracted on sheets of hessian or bu spread over the water surface of large tubs full of water, and the worms pass through into the clean water below, taking little of the mud with them.

The second method of sorting also depends upon density differences and is based upon differential flotation in solutions of high specific gravity. Such methods have long been used for sorting certain organisms from terrestrial soils, and Beak (1938*b*) first adapted the idea to stream samples. He used a strong solution of calcium chloride, and since then various other salts and sugar have been used. The merits of the different materials are discussed by Anderson (1959), who advocates the use of a solution of sucrose of specific gravity 1·12. Almost all animals float on this, while stones, sand, silt, and much vegetable matter sink. Animals do not, however, float indefinitely, as they lose water by exosmosis, and they do so rather more rapidly in salts than they do in sugar. Flotation has therefore to be carried out fairly rapidly, and it requires some judgement and skill as well as several repetitions. It is also more effective, and easier, with live samples than with preserved ones, since dead animals, especially worms, become shrivelled and sink more quickly. In skilled hands, however, it is a remarkably rapid and effective technique. It requires only simple apparatus, such as rectangular pans from which surface pouring is easy, and appropriate sieves, and the skills are easily acquired. It is also always possible to reconstitute the solution and to use it again.

Recently Whitehouse and Lewis (1966) have advocated the use of carbon tetrachloride as an aid to the sorting of samples. This is analogous to the method sometimes used for collecting insects from soil samples by stirring them up in water under oil, when the lipid cuticles of the insects become attached at the oil–water interface. They claim, however, that carbon tetrachloride collects oligochaetes and molluscs as well as insects, and that it is over 99 per cent efficient. The formalin-preserved samples are washed and re-suspended in water, which is then poured into a larger volume of carbon tetrachloride. The animals remain at the interface and can be retrieved in a sieve; they are then washed in hot water to remove the carbon tetrachloride. The authors warn against inhalation of the vapour and suggest that the work should be done in a fume-hood. It seems probable that, as this method avoids all the shrinkage troubles caused by fluids of high density, it may be a very valuable one, but I should add that my own graduate students did not agree with the originators about its efficiency and returned happily to calcium chloride which has some advantages even over sugar.

In all these methods the animals are ultimately retrieved in a sieve, and the mesh size of this can be adjusted to suit the circumstances. The youngest stages of most stream-dwelling arthropods are retained by bolting cloth of 200 threads per centimetre, which is about the smallest that can be successfully used. But samples often contain vast numbers of tiny organisms, and it becomes necessary to sub-sample in order to reduce the labour of counting and examination. A randomized method of doing this has been described by Gerking (1962), who spread samples out on a grid containing sixteen squares, five of which he selected at random. He then placed an open-ended cylinder on each of the squares and siphoned out the animals. Allanson and Kerrich (1961) advocate separation of the sample into macro- and micro-portions with a sieve. The whole of the macro-portion is then examined, and the micro-portion is made up to a definite volume of fluid and sub-sampled with a pipette. Another method by which samples may be divided is to use a screw-topped jar divided at the bottom by thin card embedded in paraffin wax into four equal quadrants each about 2 cm. deep. The sample is placed in the jar, and fluid, preferably alcohol as all animals sink readily in it, is added to a depth of 10–12 cm. The closed jar is inverted several times to mix the sample thoroughly and is then left until all the animals have settled. It is then found that the contents of any two opposite quadrants, which are easily pipetted out, represent half the sample, and it is, of course, possible to repeat the process. This simple device has been used effectively on samples obtained from moss after vigorous washing and flotation (Hynes, 1961), and it is equally applicable to plankton or other samples.

Although so many different sampling methods are in general use there seem to have been few attempts to compare them, but those that have been made indicate, somewhat disturbingly, that different samplers tend to select different animals. Macan (1958*b*) compared the stone-lifting method with a shovel technique in a small English stream, and he found that large species of *Baetis* tended to flee before the shovel and be missed, while other organisms, particularly small specimens of *Rhithrogena* and *Baetis pumilus*, were caught more effectively by the shovel as they live down in the gravel. Hynes (1961) had rather similar results when comparing a method involving kicking up the substratum in front of a net with a shovel device. He also found that large specimens of some species evaded the shovel and that others, apparently because they tended to live on larger stones which the shovel samples could not include, were relatively under-represented. Also, creatures with very effective hold-fast mechanisms, such as *Rhithrogena*, tended not to be dislodged by the kicking method and so were captured in smaller proportions.

Albrecht (1961) made a careful study of three methods, the Surber sampler, the shovel, and Schräder's stone-lifting method, in the Polenz and Kirnitzsch Rivers, Germany. For all three she found similar large variances, reflecting the well-known heterogeneity of the stream habitat; but, in general, the *weights* of organisms recorded per unit area by the three methods were similar. Nevertheless, many organisms appeared to be more or less abundant in samples taken by different methods. Oligochaetes tended to be missed by the stone-lifting method which, however, collected more Chironomidae. Indeed only relatively few of the more abundant animals, *Hydropsyche*, *Rhyacophila*, and *Ancylus*, appeared to

Sampling method	Schräder's 10-stone method			Surber sampler			Shovel sampler		
Sample no.	1	2	3	1	2	3	1	2	3
Dugesia	32	15	17	11	5	16	16	32	14
Oligochaeta	23	105	98	200	278	178	598	148	706
Erpobdella	4	1	8	—	7	1	4	—	—
Acarina	—	—	1	3	1	1	2	2	—
Nemoura	—	3	2	2	3	—	—	6	—
Perlodidae	2	—	2	5	4	2	4	—	2
Baetis	87	118	132	117	51	62	122	74	74
Chitonophora	98	208	100	465	405	329	318	340	158
Habroleptoides	—	—	—	—	—	2	—	2	—
Torleya	—	—	—	—	—	2	—	2	—
Ephemera	—	—	—	—	—	—	1	—	—
Hydropsyche	62	80	69	79	46	36	106	84	42
Tinodes	6	6	5	3	2	2	4	2	4
Rhyacophila	2	—	—	—	—	—	—	—	—
Sericostomatinae	—	1	—	8	8	4	10	2	4
Silo	—	—	—	—	1	—	—	—	—
Elmis	11	13	8	8	5	13	11	24	8
Limnius	—	—	2	1	—	2	5	—	—
Chironomidae	490	1,839	1,076	23	13	7	104	52	8
Atherix	—	6	8	22	13	7	24	20	30
Pericoma, etc.	—	6	—	1	—	—	—	—	2
Ancylus	21	31	33	22	20	35	12	38	18
Total mean number	1,604 ± 808			844 ± 139			1,096 ± 252		
Without Chironomidae	469			—			—		
Without Oligochaeta	—			625			612		
Total mean weight (g.)	0·8145 ± 0·243			1·1040 ± 0·182			1·5711 ± 0·459		
Without Chironomidae	0·7624			1·1035			1·5662		
Without Oligochaeta	0·7442			0·9924			1·2510		

TABLE XII, 2. The numbers and weights of animals per $\frac{1}{10}$ sq. m. collected by three methods from the River Polenz, Germany, on 31 October 1958. Standard errors are shown after some of the means. From data incorporated in pre-print of Albrecht (1961).

be equally collected by all three techniques. Some of these points are illustrated in Table XII, 2, which shows details of one of her series of observations and is derived from figures presented by Dr. Albrecht during the verbal exposition of her paper. From this table it can be seen that, while the total mean numbers and total mean weights of animals in the three series are not significantly different because of the large variances, there were consistent differences between the numbers and weights of Chironomidae and of Oligochaeta obtained by the different methods. We must conclude, therefore, that all methods tend to be selective and that numerical results can be accepted only with reserve.

THE VALIDITY OF QUANTITATIVE SAMPLING

As we saw at the start of this chapter, many investigators have had doubts about the validity of quantitative sampling because not only are the techniques known to be variously selective, but the variances of samples that they do obtain are usually so large that only very wide limits of confidence are possible.

Some of the early workers (e.g. Leonard, 1939) found, as we have seen also from Albrecht's results (Table XII, 2), that quite a small number of samples gives a fair idea of the total biomass of the organisms, but the variances of numbers and of weights of individual taxa are always proportionately higher than those of the total population. It has been suggested that satisfactory results would be obtained if sufficient samples could be taken to reduce the standard error to within 10 per cent of the mean of total number or weight per unit area (Hess and Swartz, 1941). However, study of large series of samples taken at one place and time have shown that this is hardly possible. Mottley *et al.* (1939) took 37 samples and reported a great variation in results. They showed, moreover, that the distribution of the numbers and volumes collected in each sample was far from statistically normal. Needham and Usinger (1956) made a more detailed study of the problem. They collected, from a 30-m. stretch of a 9-m. wide fairly uniform riffle in Prosser Creek, California, 100 Surber samples, 10 from each of 10 longitudinal lines. The numbers of organisms collected varied from 2 to 198 (average 75·7) and their weights from 0·015 to 2·31 g. Statistical study showed that at least 194 samples would have been needed to obtain a reasonable 95 per cent confidence value for weight, and 73 for numbers per unit area. This is clearly an impossibly large number, especially when it is considered that this was a very uniform habitat, which differed from place to place only to some extent in depth of water (9·5–39·6 cm.) and current speed (36–138 cm./sec. at the surface). Such variations occur everywhere in all streams and they exert, as is shown by Needham and Usinger's results, some influence on the faunal density and composition. Moreover, this study showed that significant differences can occur between sets of samples collected by different people.

On the other hand, the 100 samples demonstrated that any 2 or 3 of them contained at least one specimen of each of the more abundant species, so that not very extensive collecting is needed to list the commoner animals, and this has been confirmed by a more recent study in South Africa (Chutter and Noble, 1966).

From these considerations it is clear that all quantitative estimates of the numbers or biomass of animals on stream beds are, at best, only very rough estimates. It is not really valid to collect a few samples and then to assume that one has a good measure of biomass.

THE 'FOOD GRADE' OF THE FISHERY BIOLOGISTS

It has, nevertheless, been a widespread practice, especially in North America, to classify streams on the basis of the biomass of invertebrates as determined from benthic samples. Results are often quoted in that extraordinary, but fortunately geographically restricted unit, the gram per square foot. Let us hope that it will die out fairly rapidly as it is a cause of much hilarity among European limnologists and non-biological scientists, and it adds an air of quaint out-of-touchness to works in which it occurs.

In the early days of fishery work in the United States it was thought that study of the density of invertebrates, together with consideration of stream size and the distribution and nature of pools, could be used to determine the number of fish that could be added to streams for fishing purposes (Embody, 1928). This became codified in a pamphlet issued by the U.S. Bureau of Fisheries (Hazzard, 1935) and then by a Fisheries Circular (Davis, 1938). The so-called food grade of streams was based on the number, or grams or volume, per square foot of invertebrates collected with a Surber sampler with netting of 30 meshes to the inch. The three grades were >2, 1–2, and <1 g. or ml./sq. ft (approximately >22, 11–22, and <11 g./sq.m.); these were still in use among fishery biologists in the forties (Smith and Moyle, 1944; Shoup, 1948) and may indeed still be so. Albrecht (1953*a*) attempted to apply the same criteria to German streams and she produced rather different figures for the three grades (30–70, 6–30, and <6 g./sq. m.), but in the course of her study she convinced herself that this sort of comparison has no absolute value, and that it is of practical comparative value only when all samples are taken at the same season. Indeed, as we shall see, the biomass of invertebrates varies greatly from place to place and from time to time, and some American workers showed quite early on that the whole concept was dubious, as a single spate could drop the grade from 2 to 3 (Mottley *et al.*, 1939). Nevertheless, the statement is still sometimes made that biomass can be regarded as some measure of productivity (e.g. Grimås, 1963). As we shall see in Chapter XXII this would seem to be quite unjustified.

VARIATIONS IN BIOMASS

We have seen earlier (p. 198) that swift and varying currents are often correlated with low numbers of animals. This is reflected in the fact that

Stream width ft. (m.)	Year of investi-gation	Average weight in g./sq. ft. (g./sq. m.) at sides	Average weight in g./sq. ft. (g./sq. m.) in centre	Percentage increase or decrease from sides to centre
Below 18 (6·0)	1932	1·4 (12·8)	3·18 (29·2)	+127
	1933	1·7 (15·6)	3·6 (33·1)	+114
20–50 (6–16·5)	1932	1·89 (17·4)	1·22 (11·2)	−54
	1933	3·12 (28·7)	2·15 (19·7)	−45
50–100 (16·5–33)	1932	1·96 (18·0)	1·29 (11·8)	−51
	1933	3·2 (29·4)	2·2 (20·2)	−45
Over 100 (33)	1932	1·85 (17·0)	1·39 (12·8)	−33
	1933	2·25 (20·7)	1·72 (15·8)	−30

TABLE XII, 3. The weights of invertebrates collected with a Surber sampler at the sides and centres of various streams in New York State. Data from Pate (1932, 1933, and 1934) quoted from Needham (1934) and Albrecht (1959).

in larger streams the fauna tends to be less plentiful in the middle than at the sides (Needham, 1934). Early studies of streams in New York State (Table XII, 3) showed that this applies only to wide streams, more than about 20 ft. (7 m.) wide. In smaller streams the centre is more densely populated than the sides, and the critical width seems to be about 8–10 m. in stony trout streams (Behney, 1937). In swift sandy streams, however, the effect is more intense and the critical width is smaller. Albrecht (1953*a*, 1959) found no animals at all in the centre of the River Plane, where it was only 7 m. wide but flowing swiftly over sand, although there were large numbers nearer the banks (Table XII, 4). In sluggish water the effect is not apparent; Jónasson (in Berg *et al.*, 1948) found virtually no difference in density or weight of organisms between the areas near the bank and in the centre of the Danish River Susaa, except that the large, and heavy, clams of the genus *Unio* tended

Distance from bank, m.	Surface current speed, cm./sec.	Approximate numbers (and g.) of invertebrates/sq. m.
2·5–3	45	nil
2	40	210 (12)
1·5	37	680 (18)
1	30	1,200 (65)
0·25	18	780 (27)

TABLE XII, 4. The numbers and weights of organisms per square metre of sandy substratum in the River Plane, in September 1950, at different distances from the bank of the 7-m.-wide stream. Data from Albrecht (1953*b*).

to be more frequent near the bank. One must conclude therefore that this 'width effect' on biomass is the result of differences in current speed and the availability of shelter.

Table XII, 4 also shows that the extreme edge of the River Plane supported fewer animals than areas further towards the centre. This is a feature which has been noted also in other European streams in summertime (Albrecht, 1959), and it is probably caused by the low current and the tendency of the water to warm up in dry periods and so to become less suitable for rheophile animals. It would be instructive to know if the same applies in tropical streams at all seasons.

We have seen earlier (p. 206) that the type of substratum controls the types of invertebrate which occur there, and that the effect is numerical as well as specific (Tables XI, 1 and XI, 2). This has been confirmed statistically by Gaufin *et al.* (1956), who showed that even where a common organism occurs on various substrata it tends to cluster in particular micro-habitats. Many studies subsequent to the classical paper of Percival and Whitehead (1929) have shown that in the same stream the numbers and weights per unit area vary with the type of substratum. For example, Pennak and Van Gerpen (1947) found that for a mountain stream in Colorado the average numbers (and wet weight in grams) per square metre in summertime were: on rubble, 610 (2·5); on bedrock, 551 (1·7); on coarse gravel, 575 (1·3); and on coarse sand, 202 (0·6). Similarly Denham (1938) found the following average numbers per square metre in the sluggish, meandering White River in Indiana: on gravel, 2,652·5; on sandy gravel, 1,033·3; on sand, 967·2; and on mud 1,146·0; and Murray (1938) obtained similar results from other streams in northern Indiana.

Even quite small differences in substratum result in different concentrations of animals. Pentelow *et al.* (1938) record populations of 995 and 2,325 per square metre from each of two stations in the Bristol Avon in England, which differed primarily in that one contained some 'pebbles' mixed with the sand and gravel. Other workers have shown experimentally that the addition of rubble to sandy areas can increase the population density by a factor of between 3 and 5 (Wene and Wickliff, 1940).

In large rivers also there are big differences between different substrata. Behning (1924) in his early study of the Volga found 20–40 animals per square metre on sand, 50–100 on clay, and up to 800 on muddy sand. Details from many more Russian rivers are given by Shadin (1956) and from rivers and streams all over the world by Albrecht (1959).

From these it is possible to conclude that, in general, sand is the poorest habitat, that bedrock, gravel, and rubble on the one hand and clay and mud on the other, especially when mixed with sand, support increasing biomasses. It is also clear that islands of solid material, such as rock or rubble, and doubtless also trees and stick debris, on sandy areas are concentration points for the fauna (Mikulski, 1961).

Some eastern European workers have, however, stressed that sand is not so poor a habitat as is commonly supposed, as it supports a large micro-fauna and, in that part of the world, also often large numbers of *Propappus* and other small oligochaetes (Neiswestnowa-Shadina, 1937; Shadin, 1956). These are usually missed by ordinary sieving methods, but their biomass is probably relatively insignificant.

The fact that rubble supports more animals than does sand is almost certainly correlated with the amount of available living space (Scott and Rushforth, 1959) and with the greater probability that organic matter will lodge among stones and provide food. These are points we have already discussed (p. 211). Similarly, as was stressed some years ago by Wene (1940), the addition of silt to sand increases its food content, and this applies even more to the addition of organic mud, although there the factor of liability to de-oxygenation tends to come into play. The system is therefore variable and complex, and rules hold in only a general way.

Another cause of complexity is that some types of substratum are temporary or are subject to seasonal alteration. Thus, for instance, the stony areas in some rivers are alternately covered with sand and cleared, and these support a very small biomass (Shadin, 1956); in the River Vistula masses of silt accumulate in times of low water and are rapidly colonized by large numbers of Chironomidae—up to 1,500 per square metre. But these accumulations are temporary and are swept away by spring floods, which therefore greatly and rapidly alter the faunal density (Mikulski, 1961).

Many studies have shown, and we have already seen (p. 213), that the presence of plants or moss often greatly increases the density of the fauna and that this varies according to the species of plant. Uniform stands of plants are exceptional in running water, so any type of vegetation adds greatly to the complexity of obtaining quantitative estimates of the fauna.

We have also already seen that, in general, one result of the complex interaction of local factors on faunal density is that in streams with pool and riffle structure, the fauna is considerably denser on the latter. Some data of this kind have already been cited (Table XI, 3), and the general fact was established early in the history of quantitative stream-sampling, when Surber (1939) showed that it applied to the Shenandoah and Potomac Rivers, and Lyman and Dendy (1943) found that in the Holston River the fauna in the pools averaged less than 20 animals per square foot as opposed to over 100 at all their stations on riffles.

When it comes to comparing the biomass present in different streams so many possible factors are involved that it is exceedingly difficult to generalize. Albrecht (1959), after tabulating results of many studies in North America and Europe, concludes that there is some indication that hard waters are more heavily populated than soft waters, but the evidence is far from clear, and it is often conflicting. Moreover, there is some indication that different groups of organisms may be affected in opposite ways. Armitage (1958) found that, in the Firehole River, Wyoming, there was some correlation between the standing biomass and the alkalinity of the water, which is itself, of course, usually directly correlated with the hardness. But while the correlation between the numbers of Trichoptera and the alkalinity was positive, that of the Ephemeroptera appeared to be negative. In terms of total biomass and its relation to water hardness therefore, these two important insect orders tend to cancel one another out.

On the other hand, quantitative studies of 49 streams in the Scottish Highlands showed that, disregarding *Simulium* which feeds rather differently from most stream-invertebrates, the fauna was consistently small, in both spring and summer, in streams in which the water contained less that 400 micro-equivalents of cations per litre. Streams with between 401 and 800 μ-e./l. of cations had, on average, significantly richer faunas, although some poor streams were found in this group, and there was no significant difference between these and even harder waters (Egglishaw and Morgan, 1965). Indeed, the effect of water hardness on total biomass, if it exists at all, may be due, at least in part, to quite other factors. Müller (1954*d*) made observations in an acid forest stream in North Sweden in which the stream-bed over a length of 20 m. had been replaced by calcareous rock, and he was able to demonstrate a definite increase in biomass as compared with above and below the experimental reach, although there was, of course, little change in the water. Much of the increase was caused by *Simulium*, and why this particle-feeding organism should have been affected is difficult to understand except that perhaps the calcareous rock may have been better for its hold-fast mechanism than the native stone. But Plecoptera also increased in numbers, and Müller suggests that this may have been because the new substratum was rapidly colonized by algae which could serve as food.

Rock type	Granite	Schist	Basalt	Limestone	Sandstone
Number of streams	5	11	6[1]	7	5
SPRING FAUNA					
Platyhelminthes	4	2	3	4	3
Oligochaeta	9	3	3	4	16[2]
Plecoptera					
Brachyptera risi	4	9	25[3]	6	2
Amphinemura sulcicollis	24	25	10	25	32
Protonemura meyeri	10[4]	1	1	2	1
Leuctra inermis	69[5]	33	18	22	25
L. hippopus	8	7	3	3	+
Chloroperla torrentium	9	5	4	5	11
Isoperla grammatica	16[6]	6	3	6	3
Ephemeroptera					
Baetis rhodani	45	23	66	49	31
B. pumilus	6	7	12	6	5
Baetis spp. total	53	41	85	59	52
Rhithrogena sp.	16	20	25	20	46
Trichoptera	27	10	7	18	20
Hydropsychidae	3	3	4	5	5
Polycentropidae	9	1	2	5	7
Coleoptera	15	3	2	12	26[7]
Diptera					
Tanypodinae	7	+	1	2	2
Tanytarsini	5	2	1	2	13
Other Chironomidae	21	7	15	19	14
Mollusca	1	+	+	1	5
Mean total catch less *Simulium* spp.	309	198	222	222	290
Mean weight (mg.) of catches	601	525	577	569	814
SUMMER FAUNA					
Leuctra fusca	25	27	31	24	34
Ephemerella ignita	51[8]	12	14	22	22
Baetus scambus/bioculatus	20	13	5	8	11
Mollusca, Gastropoda	3	0	1	5	29
Mean total catch less *Simulium* spp.	910	447	565	749	1,191
Mean weight (mg.) of catches	1,091	578	508	826	987

1. Five streams in summer. 2. 55 from one stream. 3. 129 from one stream.
4. 38 from one stream. 5. 185 from one stream. 6. 45 from one stream.
7. 87 from one stream. 8. 200 from one stream.

TABLE XII, 5. The average number of animals collected by a standard netting technique in spring and summer from streams in the Scottish Highlands, which were similar in type, which all had water with more than 400 μ-e./l. of cations, but which differed in the rock from which their drainage basins were formed. Data from Egglishaw and Morgan (1965).

This result is, however, rather at variance with the observations of Egglishaw and Morgan (1965) from the Scottish Highlands. Their results, summarized in Table XII, 5, show that, when they disregarded both *Simulium*, because of its peculiar feeding mechanism, and streams with less than 400 μ-e./l of cations, which they had already shown to possess poor faunas, there was little difference between the faunas of streams on, and originating from, five contrasting types of rocks. This applied not only to the total numbers and weights, but also, as can be seen from the table, to the individual common taxa. In fact no statistically significant differences occurred except that Gastropoda were significantly commoner in the sandstone streams. Clearly many experiments similar to Müller's need to be performed, but the one cited above indicates that, in respect of calcium, the type of rock may be as important as the ionic content of the water.

Other factors that have been considered to account broadly for differences of invertebrate biomass in streams or reaches of the same stream are differences in the uniformity of gradient and the consequent liability to spates (Jones, 1941, 1948), differences in the proportions of various types of substratum (Tarzwell, 1937), and differences in vegetation on the banks, which, of course, supplies food to the biota.

It has already been pointed out (p. 224) that individual spates may greatly reduce the biomass of invertebrates and that this effect may be selective. As an additional example the effect of the typhoon of September 1959 on the Yoshimo River in Japan may be cited (Tsuda and Komatsu, 1964*b*). In August 1953 this river had a standing biomass of 128 g./sq. m., the highest ever recorded in Japan, and very high by any standard. The floods associated with the typhoon must have done enormous damage, because several years later in September 1963 the biomass was only 2 g./sq. m. The effect was, however, not only long-lasting, but it was biologically quite subtle since it involved a change of dominance as well as of biomass. In 1953 the fauna was dominated by, and most of the biomass was contributed by, the net-spinning caddis-worm *Parastenopsyche*; in 1963 this insect was scarce and *Hydropsyche* had to a large extent replaced it. Single spates can therefore lead to long-standing changes, and it is easy to appreciate why rivers and streams which differ in their liability to them should have different faunal densities and compositions.

The effect of varying proportions of the different types of substratum will be clear from what has been said earlier, but it is worth recording that in Michigan and Connecticut it has been experimentally proved that it is possible to alter the density of the invertebrate fauna by the erection of small dams, baffles, etc., which increase the proportion of certain substrata (Tarzwell, 1937; Hunter *et al.*, 1941).

Some mention has already been made of the effect of riparian vegetation on the faunal density of streams in Europe (p. 231). In New Zealand it has been suggested that reaches with bush-covered banks support more benthic biomass than those in cleared areas (Phillips, 1931), and

Jewell (1927) as a result of her early study of streams on the open prairies in Kansas suggested that one reason for the paucity of their faunas is that they receive little in the way of leaves, sticks, etc., which are important in streams elsewhere.

Under some circumstances other biotic factors outside the stream may also influence faunal density. Ruggles (1959) concludes that this may have occurred in the Wenatchee River in the State of Washington. In 1940 the average number of invertebrates in this river was 233 per square foot, but in 1955 and 1956 it had dropped to 50; all the samples were collected in the fall. During this period of years the run of adult sockeye salmon had increased sevenfold, and the resulting young fish had all passed through the river on their way to the sea from the lakes in which they had grown. Ruggles suggests that increased predation by the young migrating fish could account for the reduction in the benthic fauna, especially as mayflies, which are not much eaten by young sockeye, were less affected than other groups.

Another extrinsic factor which may have far-reaching effects is pollution. We are not concerned here with the gross effects which lead to readily detectable qualitative changes in the flora and fauna, but normal

Approximate altitude, m.	Distance from sea, km.	Range and mean number of invertebrates per sq. m.	Range and mean weight of invertebrates in g. per sq. m.
50	13·5	207–347/285	0·9–7·2/3·0
20	8	148–620/383	3·0–8·1/5·6
10	5	173–4,187/1773	5·5–65·9/21·5
2	1	243–1,200/694	7·5–26·2/15·9

TABLE XII, 6. The numbers and weights per square metre at various points on the River Håelva, Norway, based on summertime studies extending over two years. Slight domestic pollution occurred above the 5-km. station. Data from Økland (1963).

waters usually respond to a mild degree of enrichment by an increase in density of the normal fauna. Several examples are given by Hynes (1960), and there are others in the literature. For instance, Lyman and Dendy (1943) found especially large numbers of insects at one of their riffle stations on the Holston River, and they were able to correlate this with slight pollution, and Table XII, 6 gives some more recent data from Norway. Similarly, because of the production of plankton, the fauna below lakes is often much denser than it is above them. We shall consider this point in the next chapter.

Finally the density of the fauna varies with the seasons, and this applies even in the tropics where, as we have seen (Table XI, 9), there are differences between wet and dry seasons. In temperate latitudes all investigators who have sampled at various times during the year have found definite seasonal trends in faunal density. Such studies are those of Behney (1937), Surber (1951), Armitage (1958), Gaufin (1959), Nelson and Scott (1962), and Logan (1963) in various parts of North America and of Hynes (1961) in Britain. All are agreed that under normal conditions the numbers fall in spring and early summer, primarily because of the emergence of insects, that they rise again in late summer and autumn as new specimens hatch from eggs, and that they decline during the winter period of little or no recruitment. We shall discuss this point more fully in Chapter XIV.

A further seasonal complication is that some of the larger invertebrates apparently change their location with the seasons. Nelson and Scott (1962) record that the prosobranch *Goniobasis* wandered off a rocky outcrop in the Ocanee River, Georgia, and Wickliff (1940) noted that crayfish moved off the riffles in Blacklick Creek, Ohio, during the winter. Both these are large and heavy organisms, and their presence or absence, even in small numbers, would have a great effect on estimates of biomass. Finally, it should be stressed here that all the available seasonal information has been obtained from fairly small streams and rivers. There is a great deal of information on the standing biomass in the large rivers of the U.S.S.R. (Shadin, 1956) but this has nearly all been obtained during the summer. Possibly, because of the preponderance of molluscs and crustaceans in the fauna of large rivers, there is little change in invertebrate biomass, but this, very largely, remains to be studied.

From all the above considerations it will be clear that, as stated early in this chapter, it is extremely difficult to obtain quantitative data on the benthic fauna of running water, and such data as can be obtained are bound to be very approximate. It is therefore at best a dubious procedure to multiply up from, say grams per one-tenth of a square metre to kilograms per hectare. Even though this may have to be done if we are to begin to understand the biological productivity of running water, we should not forget its very shaky foundations.

CHAPTER XIII

Effects of downstream movements of organisms on the benthos

We have seen in Chapter VI that considerable quantities of plankton are carried out of lakes by their outflows, and it is also clear that any benthic animal which moves up into the water is liable to be swept away. Both these downstream movements of living material can exert a great influence on the benthic fauna.

LAKE OUTFLOWS

It was discovered in early studies of streams that lakes have marked effects on the faunas of their outflows since they supply food in the form of plankton (Brehm and Ruttner, 1926). Krawany (1930) noted that *Hydropsyche angustipennis* was particularly abundant below lakes in Austria, although he apparently failed to appreciate that this was because they were feeding on plankton. Since then it has been repeatedly shown in many parts of the world that filter-feeding organisms, particularly Simuliidae, Polycentropidae, and Hydropsychidae, are especially abundant below lakes (e.g. Müller, 1955*b*), and other organisms, e.g. *Spongilla* (Badcock, 1949), *Ephydatia* (Müller, 1954*c*), and the filter-feeding mayfly *Tricorythus* (Corbet, 1958), have also been reported to be particularly encouraged. Müller (1956) found that *Hydropsyche* larvae occurred at a density of over 50 million per hectare below a lake in Jämtland, Sweden, and he calculated that this represented 2,620 sq. m. of net, or over a quarter of the occupied area. Also, both he and Badcock found that other, non-filter-feeding animals, e.g. stoneflies, were more abundant below lakes, presumably because of an increased food supply.

Similarly, as we have already seen (p. 227), Chutter (1963) found that a barrage on the Vaal River in South Africa greatly increased the number of Hydropsychidae, mostly *Cheumatopsyche* and *Amphipsyche*, as compared with other South African rivers. Their numbers decreased as he sampled further downstream from the dam, indicating clearly that their great abundance was dependent upon it. Dense populations of Hydropsychidae are not, however, always associated with lakes. For instance, Tsuda and Komatsu (1964*a*), found very large numbers in the Torii River, Japan, as compared with few in the nearby Kusu River, and their map shows a lake on neither. It seems, however, that net-spinning caddis-worms are particularly common in some Japanese streams (Uéno, 1952; Tanaka, 1966); the reasons for this are unknown, but presumably they are connected with amounts of drifting material.

Even quite small ponds on a stream may alter the fauna markedly. The outflow of a small pond near Darmstadt, Germany, was found to be dominated by *Hydra*, *Ephydatia*, *Plumatella*, *Sphaerium*, *Simulium*, and *Hydropsyche*, all of which feed on small organisms or detritus in the water, together with Tubificidae, *Gammarus*, and *Erpobdella*. In contrast the fauna of the inflow was dominated by Ephemeroptera. Moreover, the numbers of some of the organisms fluctuated in ways which could be correlated with changing conditions in the pond. Thus *Hydra* was most common when Cladocera and Copepoda were abundant, *Plumatella* seemed to be controlled by fluctuations in detritus, *Simulium* was influenced by temperature and became scarce in warm water, and the abundance of *Ephydatia* was correlated with the amount of dissolved oxygen (Knöpp, 1952).

Among the Simuliidae it seems that at least some species are confined to lake outlets, and this represents a high degree of specialization to a peculiar set of conditions. These species are *Cnephia eremites* in Alaska (Sommerman *et al.*, 1955), *Simulium decorum* in Wisconsin (Anderson and Dicke, 1960), the closely related *S. argyreatum* in the French Alps (Dorier, 1961), and *S. sublacustre* in Britain (Davies, 1966).

The gross effect of the plankton from a lake is to increase the biomass of the benthos in the stream. Briggs (1948) found that a dam on Stevens Creek, California, more than doubled the average density of the fauna, as determined at weekly intervals from October to June with a Surber sampler, and that most of the increase was caused by Trichoptera. Müller (1955*b*) obtained similar results in Sweden, and Økland (1963) gives some figures from Norway where he found that the numbers and biomasses of invertebrates from above and below the Tisleifjord were 272–642 (mean 392) animals per square metre (1·5–4·9, mean 3·1 g.) above the lake and 259–2,025 (mean 665) (3·0–7·9, mean 5·2 g.) below; the figures are based on samples collected during two summers.

More detailed figures are given by Cushing (1963), who studied two very similar rapids in the Montreal River, Saskatchewan, above and below a series of four lakes. Some of his data are shown in Table XIII, 1, from which it can be seen that the presence of the lakes greatly increased

Upper rapids							
Date of sampling	17 June	27 June	11 July	25 July	8 Aug.	22 Aug.	3 Sept.
Campodeiform Trichoptera	8	11	2	45	26	126	84
Eruciform Trichoptera	4	3	4	9	8	46	28
Ephemeroptera	25	37	50	20	26	75	86
Plecoptera	2	0	0	2	1	7	10
Total	39	51	56	76	61	254	208
Lower rapids							
Date of sampling	16 June	30 June	10 July	23 July	11 Aug.	27 Aug.	5 Sept.
Campodeiform Trichoptera	33	90	28	868	520	3,340	1,060
Eruciform Trichoptera	20	43	20	11	26	23	57
Ephemeroptera	23	47	92	144	72	119	67
Plectoptera	0	0	0	0	0	2	4
Total	76	180	140	1,023	618	3,484	1,188

Table xiii, 1. The numbers of animals per square foot collected with a Surber sampler, on various dates in 1960, from rapids above and below a series of four lakes on the Montreal River, Saskatchewan. Data selected from Cushing (1963).

the numbers of campodeiform Trichoptera, mostly *Hydropsyche* and *Cheumatopsyche*, and also to some extent those of the Ephemeroptera, mostly *Baetis* and *Tricorythodes*. It can also be seen that the effect became more marked after about mid July. This was presumably caused by the life cycles of the insects, as it is not apparently correlated with his findings, cited earlier in Table vi, 1, on the plankton in the water.

The enriching effect on the fauna does not, however, usually persist very far. Müller (1956) showed that 6 km. below the lake on the stream which he studied, the density of the benthos had fallen to only one-twelfth of that at the outflow, and the drifting planktonic Cladocera and Copepoda to one-twenty-fourth. Illies (1956), moreover, showed that there can be a large change in the benthos over a very short distance, and that it involves not only the biomass but also the specific composition. In the Norvijokk, just north of the Arctic circle in Lapland, he found that where the water left the lake the benthic fauna was dominated by *Polycentropus flavomaculatus*. Ten metres further downstream there were large numbers of *Neureclipsis bimaculata* and *Hydropsyche*, as well as great masses of *Simulium* (up to 38,000/1,000 sq. cm.), and 10 m. further down *Spongilla lacustris* appeared, together with many *Athripsodes nigronervosus*. There can therefore be a very marked succession of species over quite short distances. Some of his data, reproduced in Table XIII, 2, show how rapidly the total biomass, and the proportion of passive filter-feeders to active feeders, can change.

	15 m. below lake		215 m. below lake	
	no.	mg.	no.	mg.
Hydropsyche sp.	27	1,265·0	16·7	751·5
Polycentropus flavomaculatus	2	70·0	—	—
Simulium venustum	490	1,862·0	4·2	16·0
Total of passive feeders	519	3,197·0	20·9	767·5
Rhyacophila septentrionis	5·5	165·0	4·2	126·0
Athripsodes nigronervosus	2	11·4	4·2	23·9
Stenophylax sp.	0·5	30·0	—	—
Heptagenia delecarlica	1	12·3	29·1	357·9
Chitonophora aronii	0·5	1·7	—	—
Baetis sp.	—	—	45·0	153·0
Chironomidae	20·5	30·8	16·7	25·1
Total of active feeders	30	251·2	99·2	685·9
Total	549	3,448·2	120·1	1,453·4

TABLE XIII, 2. The numbers and weights of benthic organisms per 1,000 sq. cm. at two stations below a lake on Norvijokk, Lapland. Data from Illies (1956).

We can conclude, therefore, that although a lake or pond on the course of a river or stream produces a great increase of passively feeding organisms, and to some extent of others as well, possibly indirectly, the effect is manifest over only a fairly short distance. This is doubtless because, as we have seen in Chapter VI, the still-water plankton is rapidly eliminated, and presumably much of this elimination is caused by the benthic fauna itself.

DRIFT

Several early workers noted that a plankton net suspended in a stream catches considerable numbers of benthic invertebrates, and it is now well established that this phenomenon occurs in all types of running water. Müller (1954*d*, 1956) recorded it in small trout streams in Sweden, Denham (1938) observed it in the White River, Indiana, and it has also been reported from large rivers such as the Nile at Khartoum (Lewis, 1957), the Missouri (Berner, 1951), and the Amur (Klyuchareva, 1963). All kinds of benthic invertebrate may be involved and various species of Oligochaeta, Amphipoda, Isopoda, Ephemeroptera, Plecoptera, Odonata, Hemiptera, Diptera, Coleoptera, Hydracarina, and Mollusca have been collected in this way. Indeed, drift has been suggested as offering a convenient debris-free source of specimens on which to base comparative studies of reaches in polluted waters (Besch, 1966). It has, however, been noted that the drift, or syrton as Berner called it, although this name does not seem to have found favour, differs in composition from the benthic fauna, in that some types of animal, notably Oligochaeta, Coleoptera, and Hydracarina are relatively uncommon and Ephemeroptera, Chironomidae, and sometimes Plecoptera are relatively abundant. Heavy creatures such as snails and caddis-worms with stony cases tend to be rare. Some of these points are illustrated by figures given by Berner (1951) which, although they are based on rather small numbers, are given in Table XIII, 3.

	Benthos	Drift
Diptera		
Chironomidae larvae	31·3	45·9
Chironomidae pupae	0·2	2·6
Chaoborus larvae	4·8	2·9
Chaoborus pupae	0	1·6
Bezzia larvae	2·4	0·3
Other immature Diptera	4·5	0·3
(Total immature Diptera)	(43·2)	(53·6)
Plecoptera nymphs	2·5	9·4
Ephemeroptera nymphs	3·0	11·5
Odonata nymphs	1·9	0·5
Trichoptera larvae	19·1	12·1
(Total aquatic insects)	(69·7)	(87·1)
Oligochaeta	21·1	0·5
Other benthic organisms	0.8	1·1
(Total benthic organisms)	(91·6)	(88·7)
Terrestrial organisms	8·4	11·3
Total number of animals	629	381

TABLE XIII, 3. The percentage composition of the benthic fauna and the drift in the Missouri River at Boonville, as determined by several samples collected by Petersen grab and plankton net from May to October 1945. Data from Berner (1951).

Very large numbers of drifting animals have been observed at times. Berner calculated that 64 million, weighing 200 kg., passed under Boonville Bridge during the twenty-four hours of 18 April 1946, and Horton (1961), as a result of a detailed study of a swift trout stream on Dartmoor, England, concluded that during a year 19·6 g. of organisms drifted over each square metre which, on the bottom, carried a biomass of between

1·5 and 3·6 g. at any one time. Rather lower figures were obtained from a less turbulent stream in the same area (Bailey, 1966), and it seems that riffles produce more drift than do pools. Horton also found that high water increases the amount of drift, a fact that was also noted by Denham (1938) in the White River and by Logan (1963) in Bridger Creek, a trout stream in Montana. Denham, in fact, observed that there was very little drift during periods of low discharge, and Logan found the highest numbers in May, at the time of the spring melt, and the lowest in August, during low water. In August Elliott (1965*b*) found that only a very small percentage of the fauna in a Norwegian stream was in the water column at any given time; his figures range from 0·02–0·03 per cent in the daytime to 0·11 per cent at night.

It will, of course, be clear from our previous consideration of the effect of spates on the benthos that periods of high water inevitably carry many animals downstream. We have seen that a sudden flood in Doe Run carried great numbers of *Gammarus* and *Asellus* away from the source (p. 226), and Van Someren (1952) reports that spates on the Sagana River bring down great numbers of *Simulium* and Baetidae. Heavy rainstorms on Mount Kenya sometimes result in bores up to 60 cm. high which sweep down the valley and carry much debris with them. These wash-out phenomena are, however, distinct from the steady drift which goes on at times of more normal flow. Also distinct are what appear to be downstream movements of some species of Simuliidae, Ephemeroptera, and Plecoptera shortly before pupation or the emergence of the adults. Such migrations have been inferred to occur in a small stony stream in England, where the species involved were *Baetis rhodani*, *B. pumilus*, *Rhithrogena semicolorata*, and *Leuctra hippopus* (Macan, 1957*b*, 1964), and in the Mississippi where the species was the mud-dwelling *Hexagenia rigida* (Dorris and Copeland, 1962), and they have also been demonstrated as occurring in *Simulium costatum*, *Baetis rhodani*, and *B. vernus* in Germany (Müller, 1966*a*). Possibly this is a widespread phenomenon, and it may be correlated with some change of behaviour in the short pre-emergence period.

Most investigators studying drift have used fairly simple plankton nets, and these were in use as early as the 1920s for investigation of the aerial insects, on and in the water, which might serve as fish food (Needham, 1928). Müller (1958) installed a paddle with a counter in the mouth of his net so that he could measure the amount of water passing through it, and other more elaborate devices have also been used. The so-called 'wolf-trap' was originally designed to catch fish moving downstream, and it consisted of a nearly horizontal sheet of screening through which the whole flow of the stream was passed below a weir and from which the animals were swept by the current into a screen box (Wolf, 1950); and Mundie (1964) designed a sampler into which insects could emerge as adults after having been caught. This consists essentially of a length of pipe directed upstream which leads into a screen box, part of which is above the water surface. Such apparatus can be left for relatively long periods except during leaf-fall when it tends to get clogged. Müller (1965, 1966*a*) has described an automatic sampler in which the stream flow over a small fall is led through a pipe on to a sieve which is changed automatically several times during a 24-hour period.

Nevertheless, much good work has been done with fairly simple apparatus. With plain rectangular nets Waters (1965) has been able to show that at least some drifting animals are as common up in the water as they are near to the substratum, and Elliott (1965*a*) using plankton samplers of the Hardy type, which have a small mouth and a large area of net and were originally designed to be towed fast through the open sea, has demonstrated that more drifts down the centre of a stream than near to the banks.

Detailed study has revealed that the amount of drifting varies considerably. We have seen that it varies with the water level; another factor is ice. Maciolek and Needham (1951), in their winter studies on Convict Creek at 2,200 m. in the Sierra Nevada of California, observed that anchor ice, formed on the riffles at night, dammed the pools and reduced the rate of flow. Each day it melted and broke up, greatly increasing the rate of flow, and the churning action of the ice detached benthic animals and increased the amount of drift. Waters (1962*a*), working at various seasons in Valley Creek, Minnesota, found that *Dixa*, an insect associated with stream edges, became much more common in the drift after a rainstorm and that the drift rate of *Simulium* was greatly increased by children playing in the stream. An interesting case of large spatial variation on a single stream is that of Berry Creek, Oregon (Warren *et al.*, 1964). This is an experimental stream with an artificially controlled constant level, and part of it had been enriched by the continual addition of sucrose. The enrichment resulted in a greatly increased biomass of benthic invertebrates, but in a decrease in the amount of drift. In the two untreated reaches, shaded and cleared of bush respectively, the ratios of total annual drift to annual mean biomass (in kcal./sq. m.) were 11 and 6 respectively, whereas in the two corresponding sugared reaches they were only 0·8 and 1·2 (Table XXII, 5). Possibly one reason for this difference lies in the type of invertebrate which was encouraged by the sugar and the consequent growth of *Sphaerotilus*. Two important organisms were the snail *Oxytrema* and the crayfish *Pacifastacus*, neither of which normally drift.

Much more generally important, however, than these rather special cases of variation of drift are diurnal variations. These have been reported from Japan (Tanaka, 1960) and they have been intensively studied in England (Elliott, 1965*a*), Germany (Müller, 1963, 1965, 1966*a*, *b*), Oregon (Anderson, 1966), and Minnesota (Waters, 1962*a*, 1965). All these studies have shown that the amount of drift increases at night, and particularly during the period soon after sunset (Figs. XIII, 1 and XIII, 2). Most of the results quoted refer to *Gammarus* (*pulex* in

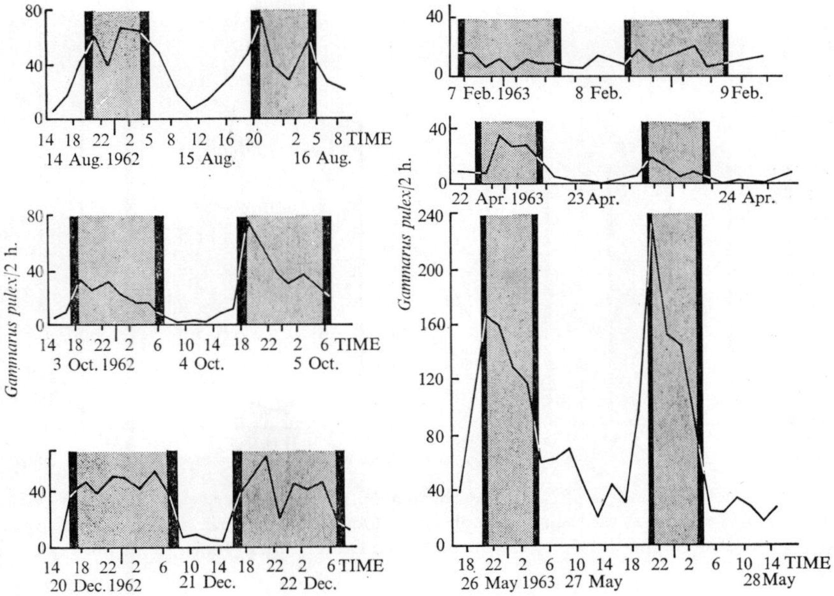

FIG. XIII, 1. The numbers of *Gammarus pulex* caught in a drift sampler every two hours in a small stream in Germany at different times of day and at different times of the year. The darkened areas represent dusk, night-time, and dawn. Redrawn from Müller (1966*a*) with permission.

Germany and *pseudolimnaeus* in America) and *Baetis* (*rhodani* and *vernus* in Europe and *vagans* in America), although other genera are reported as showing the same phenomenon, e.g. *Leptophlebia*, *Leuctra*, *Elmis*, and Polycentropidae in England, and *Glossosoma* in North America. Results with *Simulium* are conflicting, probably because of specific differences, and the Chironomidae, although important in the drift, seem to show no clear temporal variation (Anderson, 1966). This diurnal variation in drift is found at all times of the year and it is always related to the light and dark periods, usually with a peak shortly after the dark begins. Müller (1965, 1966*a*, *b*), however, found that in *G. pulex* there was also a lower one later in the night in spring and fall, and two later peaks in winter (Fig. XIII, 1) and that in continued darkness, caused by artificial shading of a stream, the drift was continuous. When he darkened the stream after long, artificial periods of light there was always a peak of drift similar to that following a natural sunset. In *Baetis* the main peak is, rather exceptionally, late in the night, and it is single during short nights, but it has one or two earlier small peaks during medium-length and long nights respectively (Fig. XIII, 2).

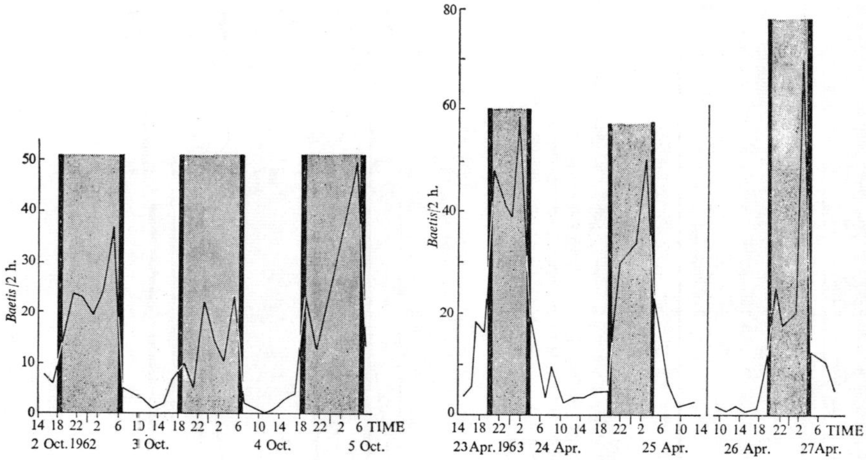

FIG. XIII, 2. The numbers of *Baetis* caught in a drift sampler every two hours in a small stream in Germany at different times of day in October and April. The darkened areas represent dusk, night-time, and dawn. Redrawn from Müller (1966*a*) with permission.

Clearly at least part of the diurnal variation is because light exerts an inhibiting effect. Elliott showed that illumination of an artificial stream at night reduced the amount of drift to the daytime level, and that advancement of the onset of darkness by artificial shading caused the peak to appear earlier; and Anderson found that the drift did not increase at all on the night of a full moon. It would seem therefore that light causes some behaviour pattern which lessens the chances that an animal will be entrained by the current. This could be because negative phototaxis keeps it down among shelter on the bottom, as has been observed of *Ephemerella ignita* (Madsen, 1966), or because, as has been long known of some swimming invertebrates, e.g. *Notonecta*, it maintains its position in running water by reference to visible fixed objects (Schulz, 1931). The situation is, however, obviously more complex than this. Anderson showed that the increase of drift at night is very largely caused by specimens longer than 3 mm. Some of his data are given in Table XIII, 4, from which it would seem that the behaviour pattern which causes night-time drifting is characteristic of only the older stages; indeed Müller had already found that *Baetis* and *Simulium* drift more readily when they are fully grown.

Temp. °C.	4–5 Feb. dark night 8·3	26–27 Feb. moonlit night 6·7
8–11 a.m.	17·9	18·6
11 a.m.–2 p.m.	12·4	17·4
2–5 p.m.	16·2	11·5
5–8 p.m.	42·3	18·3
8–11 p.m.	41·5	28·1
11 p.m.–2 a.m.	30·3	15·7
2–5 a.m.	23·0	19·3
5–8 a.m.	11·5	12·4

TABLE XIII, 4. The percentages of total numbers of Ephemeroptera, Plecoptera, and Simuliidae, which were longer than 3 mm., caught in drift samples in each three-hour period during two days in Berry Creek, Corvallis, Oregon. Sunset and sunrise were at approximately 6 p.m. and 6.30 a.m. Data from Anderson (1966).

Drifting is also controlled by temperature, and different species react differently at different seasons. Müller (1966*a*) in a lengthy study of a small spring-stream in Germany showed that *Gammarus pulex* always drifted mostly at night, but with a maximum in summer, its time of maximum breeding (Fig. XIII, 1), and that a rise in temperature always increased the amount of drifting. *Baetis* also drifted most in midsummer, when *B. vernus* was the common species, and at night, but during the period November to March, when only *B. rhodani* was present, the maximum drift was during the day and this changed to night-time maxima in April, the month in which *B. rhodani* emerges (Fig. XIII, 2). There is thus a complicated relationship between species, temperature, and life history, and at least some species, e.g. the flatworm *Polycelis felina*, appear to drift more in the winter than in the summer.

The relationship to light is also far from simple. When Müller shaded 60 m. of stream above his sampling point with plastic sheeting he obtained a great increase in the drift of *Baetis vernus*, but still at night, but *Simulium costatum* almost ceased to drift under these conditions. In contrast *B. vernus* showed no cycle and very little drifting in the 24-hour day of the Arctic in North Sweden. It seems then that the rhythm is set by light, and that in some species it persists and is enhanced when the inhibitory effect of light is removed, but that in others the recurring stimulus of the change from light to dark is needed for the pattern to persist.

Waters (1965), in a very careful study of *Gammarus pseudolimnaeus* and *Baetis vagans* in Valley Creek, has demonstrated that, despite the considerable number of specimens collected in his drift nets, there was normally no change in the density of the population immediately above them. He found that, below a pool where the amphipod was common and the mayfly was scarce, he caught, in a single 24-hour period, the equivalent of the population in strips as wide as the net and 1 m. and 77 m. long respectively. In a riffle where it was commoner the catch of *Baetis* was equivalent to the population of a strip of stream as wide as the net and 16 m. long. These figures indicated that a large proportion of the population was being carried downstream every day, and Waters then set out to discover how far the animals move and if a significant number of them also moves upstream. When he cleared the animals from small areas, which he then surrounded with screens in such a way that recolonization could occur either only from upstream or only from downstream, very few animals moved in from downstream although many came in from above. When he blocked the stream completely with a row of nets so that no drift could pass, he found that the drift further downstream was markedly reduced. Even 38 m. downstream, after the water had flowed over two riffles and through a pool, the drift was only about 80 per cent of that at the barrier. From these observations he concluded that the normal distance drifted each night by any one individual is probably of the order of 50–60 m.

One consequence of drift is that artificially cleared areas are rapidly recolonized by animals from upstream. Two reports indicate that newly constructed channels are rapidly populated. Leonard (1942) observed that a new channel cut for Hunt Creek, Michigan, in mid October, contained *Simulium* within a week, over 2,000 animals per square metre in December and over 17,000 in twenty-one species by early February. Patrick (1959) watched colonization of a similar channel near Philadelphia, and she also noted the early appearance of *Simulium*, followed by *Baetis* and *Stenonema*. The channel was opened in August, and by early November there were large numbers of young stoneflies of the genera *Taeniopteryx* and *Brachyptera*, which must have been, as we shall see, in the egg stage when the channel was cut and so must have moved down very soon after hatching. It is also interesting to note that *Physa heterostropha* became established within six months, because snails are rarely reported as drifting.

Similarly, the larvae of *Deuterophlebia* returned rapidly to a reach of

Convict Creek, California, from which they had been eliminated by drying resulting from a diversion connected with fisheries work (Kennedy, 1958); and Coutant (1962) observed that areas in the Delaware River, which in summertime were heated to too high a temperature for many stream animals by a power station effluent, were recolonized during the winter.

Müller (1954*f*) and Waters (1964) have made detailed studies of recolonization by drift. Müller studied a stretch of 150 m. of the Swedish Skravellbacken which had been cleared with a caterpillar tractor, thus removing the fauna completely and increasing the rate of flow. Within eleven days the population was up to 9,240 animals per square metre, indicating that over 4 million invertebrates, weighing over $4\frac{1}{2}$ kg., had moved in. Most of them were Chironomidae, *Simulium*, Ephemeroptera, Trichoptera, and Plecoptera; mites, beetles, and molluscs were relatively rare. Waters cleared only small areas and studied only *Gammarus* and *Baetis*. The first returned to its former abundance in one day in the fall and four in the winter. *Baetis* was slower, requiring four days in autumn and ten in winter. The rate of recolonization of the cleared area was directly related to the rate of drift until full recolonization had occurred, but after that point had been reached the carrying capacity appeared to have been attained and no further increase occurred. An interesting observation connected with this study was that, when the population of *Gammarus* was below full capacity, *Baetis* settled in larger numbers than usual and overshot its normal density. This seems to indicate some sort of mutual interference between the two animals.

Some species of fish, most notably the brown trout, *Salmo trutta*, have been shown to feed very largely on drifting organisms, and individual fish can often be observed occupying a sheltered station from which they dart at intervals to catch drifting animals (Nilsson, 1957; McCormack, 1962). This habit is indeed the basis of the technique of angling for trout with a wet artificial fly. The influence of drift on the food of trout is also seen in the observation made in Wales by Thomas (1964), that they eat more on the first day of a spate than at other times. This is a reflection of the fact that high water increases drift. It would seem, however, that as with plankton (Chapter VI) very high rates are not maintained if the spate is prolonged, because after more than two days of high water Thomas found that the trout were eating less than normal. This is an aspect of invertebrate drift which merits further study.

Presumably, under normal circumstances, drifting invertebrates either find an empty niche in which to settle, or they are eaten by such predators as fish or net-spinning caddis-worms. Great numbers must, however, be swept out of areas which are habitually suitable for them and must ultimately perish. Morduchai-Boltovskoi (1961*b*) reports that the Amphipoda and Cumacea of the rivers in the Ponto-Caspian area tend to accumulate in the deltas, presumably as the result of river drift, but Dendy (1944) showed that stream animals swept into a lake did not survive for long. He studied the inflow of Carp Creek into Burt Lake, Michigan, and found that stream animals occurred in the lake only to a distance of about 30 m. from the mouth, where they tended to become concentrated in a zone of plant debris deposited by the stream. They could apparently survive there for some while in calm weather, but at other times wave action dislodged them and stranded them on the lake shore.

Müller (1954*d, f*) and Waters (1961*a*) have argued that the drift represents, as it were, excess production from the habitat. As the animals reproduce and grow the area becomes overcrowded and some specimens are displaced and drift away. Some of these will replace specimens further down that have died or been eaten, and the rest are swept on to be eaten or lost. They point to the established fact that stream insects tend to fly upstream to oviposit (p. 155), and they suggest that this leads, initially, to overstocking in the upstream areas which is then eased by drift. This argument cannot, of course, apply to animals such as *Gammarus* which have no flying stage, but presumably the tendency of each individual to move upstream whenever possible, coupled with a high rate of reproduction, would serve to keep the headwaters populated. Indeed, it is nearly always true that populations of *Gammarus* are denser in headwater streams than further down the same watercourse, but this may be for other reasons (p. 232).

Waters (1962*b*) has attempted to exploit this idea by measuring the rates of drift into and from two riffles and two pools in Valley Creek during two 24-hour periods. He found that the riffles were producing 0·28 g. of *Baetis vagans* per square metre per day and that the pools were losing 0·45 g. He interpreted the latter as consumption in the pools and the former as minimal production—minimal as it does not include death from all causes on the riffles nor emergence of adults there. Perhaps excess production would be a better term. I understand from Dr. Waters (personal communication) that he is pursuing similar work with *Gammarus* and that he has had some preliminary success in equating known rates of drift with calculated production-rates based on life-history studies and biomass; we shall consider his more recent work on *Baetis* in Chapter XXII.

Undoubtedly, therefore, the drift of invertebrates is of great importance in stream biology, particularly because of its considerable bearing on secondary production. If it is indeed true that excess production does not accumulate on the stream bed, but is drifted off to become food for fish or is otherwise lost, the relationship between benthic biomass and fish production would appear to be a loose one. Moreover, attempts such as those of Hynes (1961), which we will discuss later (p. 424), to determine biomass production from changes in population structure on the stream bed would seem to be assured of only limited success.

There remain, however, some still unanswered questions. We know from many observations (Chapter VIII) that stream invertebrates do move upstream, and although Waters's work indicates that this is insignificant in terms of biomass this, I feel, remains to be proved by detailed study over a period and the entire width of a stream. It is possible also that much upstream movement occurs deep down in the gravel. Proof is especially necessary in view of the established upstream spread of radio-phosphorus (Ball *et al.*, 1963). When P_{32} incorporated into cells of *Escherichia coli* was introduced into the Sturgeon River, a Michigan trout stream, it appeared in significant amounts 100 yd. (91 m.) upstream in *Simulium*, *Isoperla*, and *Physa* after one week. After two weeks it had spread 200 yd., and *Nigronia*, *Pteronarcys*, and Baetidae also showed significant amounts of radioactivity. After seven weeks many of these animals were radioactive at 300 yd., and *Nigronia* and *Isoperla* were found to be affected after five weeks as much as 500 yd. above the point of introduction. This is clear evidence of significant upstream movement, and as it was slow it seems to have been waterborne, especially as it was slower over a long riffle than over easier stretches.

Another problem is the well-established phenomenon of quite rapid faunal changes over short distances of stream. We have, a few pages ago, seen that the faunal composition changes greatly in short stretches below lakes, and it is a well-confirmed tenet of the biologists who study water pollution that mild pollution produces subtle changes in the fauna which shows differences over quite short distances (Hynes, 1960, 1965; Brinkhurst, 1965). They may even persist for some while after the pollution has ceased (Hynes, 1965), and it is difficult to reconcile these observations with the suggestion that great numbers of animals regularly drift downstream for distances measured in tens of metres a day. This would be possible if they were all, or mostly, rapidly destroyed as soon as they reach areas where they cannot persist, but this seems unlikely in view of the numbers involved and of Dendy's observation that stream animals survive for an appreciable while even in the presumably quite inhospitable environment of a lake. Clearly this whole, and very important, matter is in need of much more study. Possibly some of the anomalies arise because the excellent and convincing studies of Waters and Müller have been almost confined to *Baetis* and *Gammarus*. Baetidae are unique among stream invertebrates in that they often hold their streamlined bodies up above the still boundary layer, and they also swim about a great deal (Chapter VIII). *Gammarus* also is a clumsy creature which swims very readily and seems ill-adapted to life in streams. Possibly these two animals are particularly liable to drift and not to regain lost ground; and what applies to them may not be true of stream invertebrates generally.

CHAPTER XIV

Life histories and seasonal cycles of benthic invertebrates

In running water in temperate climates, and particularly markedly in small streams, there is a very clear sequence of events in the invertebrate benthos. Many species appear and disappear from collections made at intervals throughout the year as one species after another completes its development.

Much of what we know of life cycles is based on such serial sets of field observations, and there has been comparatively little in the way of laboratory study. Such as there has been, however, has led to some surprises and this indicates a rich field for further research. It cannot be emphasized too strongly that in this area of study a combination of field and laboratory work yields much more information than does either alone. Field-work leaves many gaps in knowledge which are difficult to fill, such as the fate of small eggs and the lengths of life of terrestrial adults, and laboratory results always suffer from the uncertainty as to how they can be applied to the natural situation. This is a point which has been made several times in this book and it is an important one.

TYPES OF LIFE HISTORY

Many species in several orders of the insects have a single generation each year and are to be found in the water only at a certain, often quite limited, season. The assumption is that only the eggs are present when the young stages are not to be found; when such species first appear in collections they are small and they increase in size during their period of occurrence to become mature at the end of it (Fig. XIV, 1).

Such univoltine life cycles have been found in many Ephemeroptera, e.g. species of *Habroleptoides*, *Habrophlebia*, *Ephemerella*, *Baetis*, *Rhithrogena*, *Heptagenia*, and *Oligoneurilla* in Europe (Frost, 1942; Harker, 1952; Pleskot, 1953, 1958, 1960; Macan, 1957*b*; Hynes, 1961; Pinet, 1962), of *Ephemerella*, *Epeorus*, and *Hexagenia* in North America

(Nelson and Scott, 1962; Hartland-Rowe, 1964; Lyman, 1943), and of *Epeorus*, *Cinygma*, and *Ephemerella* in Japan (Tanaka, 1966), and the same doubtless applies in other genera and in other temperate continents. Among Plecoptera this type of life-cycle is known in many genera on both sides of the North Atlantic and in Japan, Australia, and New Zealand (e.g. Frison, 1929; Hynes, 1941, 1962, 1964*b*; Nelson and Scott,

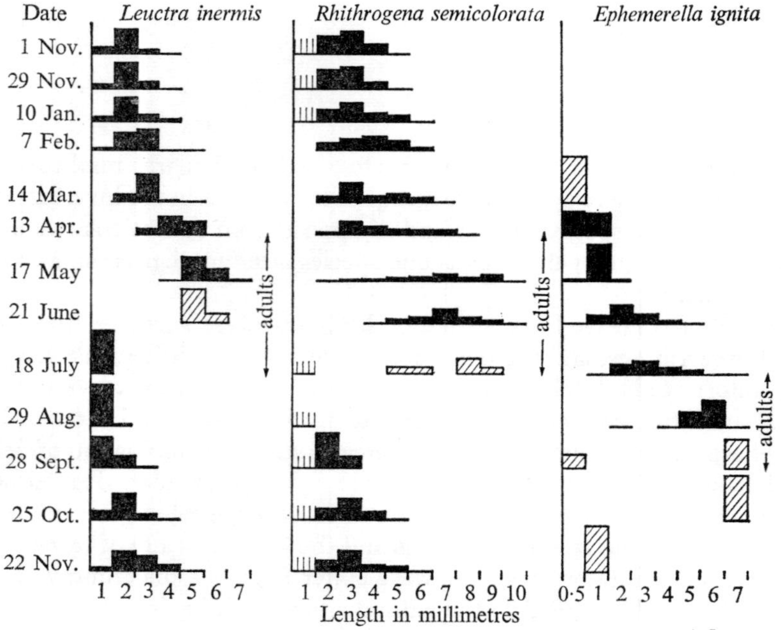

FIG. XIV, 1. The percentage size-distributions in approximately monthly collections of three common species from a stream in Wales. Histograms based on less than five specimens are cross-hatched. The flight period of the adults is shown by a vertical line between arrowheads. Specimens of *Rhithrogena* in the 1-mm. size group were not quantitatively collected, as they could not all be certainly distinguished from allied species, but their presence is shown by vertical lines.

1962; Minshall and Minshall, 1966; Tanaka, 1966; Winterbourn, 1966). Rather fewer genera of Trichoptera are actually missing for long periods, as young larvae in this order tend to appear while older larvae or pupae are still present even where there is a clear seasonal change in size distribution indicating an annual cycle (e.g. Nielsen, 1942; Hynes, 1961; Cummins, 1964). This applies also to some species of mayfly and stonefly, for instance the common European species *Torleya major* (Pleskot, 1958) and *Isoperla grammatica* (Hynes, 1962); and in the former the young nymphs remain for several months as small individuals before embarking upon their growth period.

Many Diptera, most notably species of Simuliidae but also at least some Blepharoceridae and Deuterophlebiidae, have a single generation a year. *Liponeura* is absent for only about a month during the late summer in European mountain streams (Hubault, 1927), but *Deuterophlebia inyoensis* occurs for only a few weeks in high-level streams in California, and each individual goes through its development from egg to adult in about twenty-seven days (Kennedy, 1960). Similarly, the larvae of *Simulium arcticum*, a well-known pest-species on cattle in Canada, are absent from the water during the winter, but they appear soon after the melting of the ice to grow and produce one generation of viciously blood-hungry adults in the summer (Fredeen, 1958; Fredeen *et al.*, 1951; Peterson and Wolfe, 1958). In contrast the larvae of most species of the genus *Prosimulium* are absent in the summer, but they appear in the fall and grow at low winter temperatures to produce adults in the spring (Zahar, 1951; Anderson and Dicke, 1960).

We know very few of the life histories of the ecologically very important Chironomidae. This is mainly because of our inability to identify the larvae. Many species seem to produce several generations in a year, but some running-water species have been reported as having only one, e.g. *Clinotanypus nervosus* (Hall, 1960) and *Chironomus plumosus* (Kajak, 1958), from a small stream in England and a large river in Poland respectively.

Members of this family, and to a lesser extent of the Ceratopogonidae, are almost universally present in running water, and we know very little about their ecology there although they usually constitute a substantial proportion of the fauna. Almost all the information that is available is from Europe (Thienemann, 1954*a*), and, as can be seen in Table XIV, 1, the numbers of species involved are considerable. Ecological study of these is badly needed for a complete understanding of stream biology. Rivers doubtless contain even more species than do the types of stream shown in the table.

The Hydracarina also contain annual species which are found only at certain seasons. Such are *Sperchon clupeifer* and *Sperchonopsis verrucosa* which, in European mountain streams, appear as larvae in the spring, grow into nymphs in June and into adults in August, and these lay, presumably, overwintering eggs (Angelier *et al.*, 1963).

In other groups also there are species which have clearly defined annual generations, but, like some of the insects mentioned above, the generations overlap, and some specimens are always present. These include many North American amphipods of the genus *Crangonyx*

	Glacier streams	High mountain streams			Streams in the forest zone			Low mountain streams		
Type of substratum	Stones	Stones	Plants	Mud	Stones	Plants	Mud	Stones	Plants	Mud
Ceratopogonidae									5(3)	2(2)
Tanypodinae						1(1)	2(2)		7(1)	5(3)
Orthocladiinae s.l.	8(5)	12(7)	11(7)		24(8)	18(6)	3(3)	40(13)	41(17)	5(5)
Tanytarsini			1(1)	1(1)	1(1)	1(1)	1(1)	5(4)	3(2)	7(4)
Chironomini									8(4)	5(3)
Total	8(5)	12(7)	12(8)	1(1)	25(9)	20(8)	6(6)	45(17)	64(27)	24(17)
Grand total	8(5)		21(12)			51(22)			133(48)	

TABLE XIV, 1. The numbers of species (and genera) of midges occurring in different types of stream in Europe. From Thienemann (1954*a*).

(Bousfield, 1958), the common European limpet *Ancylus fluviatilis* (Geldiay, 1956; Hunter, 1953*b*), the leech *Erpobdella testacea* (Mann, 1961*b*), and doubtless many others.

Animals with more than one generation a year are, as far as we know, relatively scarce in the running-water habitat. Doubtless some small crustaceans such as copepods produce a series of generations; for instance, *Canthocamptus zschokkei* can be found breeding at all times of year in hill streams in Wales, but little is known of its life history. Some small malacostracans definitely produce a series of overlapping generations and this can result in one and a half or two or more, generations within a twelve-month period, as occurs, for example in the common European *Gammarus pulex* (Hynes, 1955*b*). But straightforward multivoltine cycles, which are so common in the terrestrial habitat and are by no means scarce in still water, appear to be confined to rather few groups in streams. They occur in some mites, e.g. *Hygrobates calliger* which seems to have two generations, and *Atractides nodipalpis* which seems to have three, in French mountain streams (Angelier *et al.*, 1963). Also some, but by no means all, species of the mayfly *Baetis* (Macan, 1957*b*; Pleskot, 1958, 1960), some species of *Deuterophlebia* (Kennedy, 1960) and some species of Chironomidae (Kajak, 1958; Hall, 1960) have multivoltine cycles. Probably a considerable number of Chironomidae are at least bivoltine, but this has still to be determined. The only important group in which it is well established that there is a substantial number of species with two or more generations is the Simuliidae (Grenier, 1949; Davies, 1950; Fredeen, 1958; Peterson and Wolfe, 1958; Anderson and Dicke, 1960). In this family many species have two generations, and some of the commoner ones, e.g. *Simulium ornatum* and *S. vittatum*, have three or even four during the summer. No Plecoptera or Trichoptera seem to produce more than one generation in the year in temperate latitudes, and, although there are reports of two generations of species of the latter order from Europe, Nielsen (1942), after his careful study of stream-dwelling species in Denmark, concluded that they are erroneous. There are also suggestions that some *Hydropsyche* and *Cheumatopsyche* species may produce two generations in Canada (Sprules, 1947), but these are unconfirmed and are based only on data obtained by the trapping of adults. At present it therefore seems that only a small proportion of the benthos is multivoltine, although some of the animals which are—*Baetis*, *Simulium* and probably many Chironomidae—are important members of the fauna.

In contrast, several species in many groups are definitely known to have life cycles exceeding a year in length. This applies to some leeches, e.g. *Erpobdella octoculata* (Mann, 1953), *Glossiphonia complanata* (Mann,

1957), and *Trocheta subviridis* (Hartley, 1962), to some snails, e.g. *Pleurocera* and *Goniobasis* (Dazo, 1965), and to unionid clams (Hendelberg, 1960; Björk, 1962; Negus, 1966). *G. complanata* and most specimens of *E. octoculata* lay eggs in the year of their birth, and they survive to do so again in the following year. *T. subviridis* leaves the stream after one year to return and breed twice in later years. The two prosobranchs do not breed in the year of their birth, but they do so in the following two years, and the clam *Margaritifera margaritifera* does not mature until it is aged about twenty and it may live for over a century.

Among crustaceans at least some crayfishes, e.g. *Cambarus longulus* (Smart, 1962), live for several years, although others, e.g. *Orconectes clypeatus*, breed within a year of their birth and live only about one and a half years (Smith, 1953). The mitten-crab *Eriocheir sinensis* spends, probably, four or five years in fresh water before returning to the sea to breed (Hoestlandt, 1948). It also seems that African crabs of the family Potamonidae live for many years, as they survive for long periods in captivity without any evidence of growth, and all sizes occur at any one time.

Several insect orders contain species which are known to take more than a year to develop. This applies to *Sialis* in Europe (Kaiser, 1950, 1961), *Archichaulioides* in New Zealand (Hamilton, 1940) and doubtless to other Megaloptera elsewhere, and also to *Agrion virgo* and *Cordulegaster annulatus*, two of the most typical stream-dwelling dragonflies of western Europe (Carré, 1957; Nicolau-Guillamet, 1959). Many stoneflies, including even some large species in the family Perlodidae, have a one-year cycle, but others take longer. *Perla* and *Dinocras* take three years to grow to full size (Hynes, 1941); at least one Australian eustheniid, *Eustheniopsis venosa*, takes two (Hynes, 1964*b*), and *Pteronarcys* takes three, possibly four, years (Holdsworth, 1941; Tebo and Hassler, 1961).

Information on Ephemeroptera and Trichoptera is less conclusive. At least one, lake-dwelling, species of *Hexagenia* is reported to take two years to mature, but there is conflicting evidence on this point, and there are indications that some river-dwelling species of *Ephemera* take two years. This, however, also remains unsubstantiated. Similarly, there have been suggestions that some net-spinning caddis-worms, e.g. *Wormaldia* (Mackereth, 1960) and *Polycentropus* (Frost, 1942), take two years to grow, but this also is uncertain, and in both groups a univoltine cycle seems to be the most usual. Larvae of several species of campodeiform trichopterans and of elminthid beetles are often to be found at all seasons of the year in all or almost all size groups, so the simple serial-histogram technique, illustrated by Fig. XIV, 1, and on which most studies have been based, does not give a clear indication of the life history

(Beier, 1948; Hynes, 1961). Similarly, adults flying in from other habitats where the life cycle is out of phase with the locality being studied can confuse the evidence on the numbers of generations of multivoltine species (Davies, 1950).

RESTING STAGES

In many univoltine arthropods there is a fairly long period during the year, usually the winter or the summer, when no specimens can be found, and the assumption is that such species are then in a resting stage. Often, as in many of the instances cited above, there are no data to back the assumption that it is actually the egg stage which is involved, although this seems reasonable on the available evidence. Sometimes, however, this approach could lead to error; for example during my early work on British stoneflies (Hynes, 1941) I could not find nymphs of *Capnia bifrons* during several months after the flight period of the adults even though I knew from laboratory studies that the eggs hatched very soon after laying. We now know the reason for this (p. 280) but there may be many similar undiscovered examples.

Some direct observations have been made on eggs and they have proved to be very illuminating. To date, however, they have been concerned with running water species in only three groups of insects—the Plecoptera, Ephemeroptera, and Simuliidae—although there is some information about Trichoptera (Nielsen, 1942) and it is known that the eggs of Megaloptera, which are laid above the water, hatch fairly rapidly.

In the Simuliidae the eggs of many species also hatch within a short period, but in others a definite egg-diapause has been demonstrated, mainly by Fredeen (1959*a*, *b*; Fredeen *et al.*, 1951). The eggs of *Cnephia dacotensis* remain unhatched for many months, and it is clear that this species aestivates in this stage. Other species, e.g. *Simulium arcticum*, *S. aureum*, *S. decorum* and *S. venustum*, overwinter as eggs, and these can be dredged up and floated out of the substratum with brine. Such eggs continue development even at low temperatures and they hatch when they are exposed to warm water. They can also be stored for long periods at low temperatures and remain viable. This is in contrast to species like *S. vittatum* whose eggs normally hatch soon after laying and are able to survive for only a few weeks at low temperatures. It seems then that there are at least three types of simuliid eggs: those which must hatch fairly soon, those which develop slowly and are held back by low winter temperatures but hatch as soon as the water begins to warm up, and those which aestivate and have a built-in true diapause lasting for several months independently of temperature. These, therefore, together with the seasonal changes of temperature, control the various types of life history found in this important family.

In the mayflies it has been demonstrated that the eggs of *Ephemerella ignita*, a well-known summer species of western Europe, hatch only after about six months (Percival and Whitehead, 1928), and this is in agreement with field observations on the times of appearance of the nymphs. Field observations also indicate that the eggs of some species of *Baetis*, most notably *B. rhodani*, remain unhatched for long periods (Macan, 1957*b*; Pleskot, 1961), and they have even been found overgrown by the sponge *Ephydatia*, demonstrating that they had been in place for a long while (Berg *et al.*, 1948). This genus lays flat plates of serried rows of eggs which are easily recognized on the undersides of stones. Illies (1959) brought a stone bearing 36 egg-patches into the laboratory in the autumn and watched the hatching process. One patch began to hatch soon, and half the eggs had emerged after 12 days, but the last egg hatched after 161 days. Twenty-six patches began hatching after about 50 days; 50 per cent. of the eggs had hatched by day 200 and the rest hatched soon thereafter, and 9 patches did not begin to hatch for 150 days, but once started they all hatched fairly rapidly. This then is clear evidence of a long egg-diapause in this genus. Unfortunately the eggs were of unknown species although probably most, if not all, were *B. rhodani*. It also shows a point to which we shall return, and which had also been inferred from field collections of nymphs, namely that the hatching period of any given batch of eggs can be very extended, far longer than the flight period of the adults.

Information on the stoneflies is more extensive, although some of it remains to be published. Some species, e.g. *Capnia bifrons*, *Allocapnia vivipara*, and sometimes *A. pygmaea*, are oviviviparus (Frison, 1929; Hynes, 1941; Khoo, 1964*b*; Brinck, 1949; Clifford, 1966*a*) and many species lay eggs which hatch in 2–4 weeks. Such are several species of the genera *Taeniopteryx*, *Nemurella*, *Nemoura*, *Protonemura*, *Capnia*, *Leuctra*, and *Chloroperla* (Hynes, 1941; Brinck, 1949; Khoo, 1964*a*). In other, mostly rather larger, species the developmental period is about two (*Isoperla grammatica*) to three or even four months (*Perlodes microcephala*, *Isogenus nebecula*, *Perla bipunctata*, and *Dinocras cephalotes*) (Hynes, 1941; Brinck, 1949; Khoo, 1964*a*), and the eggs of *Pteronarcys* remain for 10–11 months before hatching (Holdsworth, 1941). The last is clearly an example of egg-diapause which in this instance has no immediately obvious function, because, as we have already seen, *Pteronarcys* nymphs live for several years. The egg-diapause thus does not enable the nymphs to avoid high or low temperatures. Other examples of long diapause found by Khoo are: *Nemoura cinerea* 3–4 months at room temperature, but less at low temperatures and so clearly a case of temperature-controlled diapause, *Amphinemura standfussi* 4–5 months at stream temperatures and with little or no development at higher temperatures, and *Brachyptera risi* 3–6 months with high temperatures inhibiting hatching. He also found that in *Diura bicaudata* the eggs were of two types, both of which can occur in a single batch. Non-diapausing eggs hatch after about 3 months, but diapausing eggs remain 9–11 months before hatching. The estimates of Hynes (1941) and Brinck (1949) from field observations of nymphs, of an incubation period of 2 or 1 months respectively are therefore wrong, and both must have found young nymphs hatched from eggs laid by the previous generation a year earlier. This, like the example of *Capnia bifrons* cited above, demonstrates a weakness in the purely observational approach.

Khoo's study also revealed that in many species, e.g. *Leuctra fusca*, *L. moselyi*, *Nemoura cinerea*, *Brachyptera risi*, *Amphinemura standfussi*, and *Isoperla grammatica*, hatching of a single batch of eggs goes on for a long period. This is either because of irregular rates of development of the individual embryos or because of irregular breaking of the diapause, which in the truly diapausing eggs occurs at a definite stage of development, which varies according to the species. In this respect therefore the behaviour of the eggs of some stoneflies resembles that of some mayflies.

It has also been inferred from field studies that many species living in temporary streams or surviving a particularly severe drought do so in the egg stage (Hynes, 1958; Burks, 1953; Clifford, 1966*a*). Khoo was able to demonstrate that the gelatinous coat which occurs on the eggs of many stoneflies protects them from drying, and that they can survive for many days, and probably for long periods, while the humidity remains high. This, of course, it would always be down among the stones of even a dry stream bed.

The eggs of many species of insect serve therefore as resting stages in the life cycle and allow them to evade unfavourable periods in a quiescent condition. This applies also to other stages in some groups. The most spectacular example from freshwater biology is the larva of *Polypedilum vanderplankei* which lives in rock pools in northern Nigeria and can stand complete dessication for twenty months and temperatures up to 65 °C. (Hinton, 1952). No such examples have been reported from running water, but it seems not improbable that they occur in hot dry climates in temporary streams which are often as ephemeral as rain-filled rock pools.

Some species of mayfly, e.g. *Habroleptoides modesta* and *Torleya major*, spend the warm summer months in early instars and do not grow until temperatures fall (Pleskot, 1958), and in some *Baetis* species, e.g. *B. subalpinus* and *B. rhodani*, high temperatures seem to inhibit growth. In warm streams, therefore, spring- and fall-flying generations are clearly separated, whereas in cooler streams they overlap (Pleskot, 1961). Other species overwinter as very small larvae and grow only when the temperature rises, e.g. *Chitonophora krieghoffi*, (Pleskot, 1960, 1961), but in no mayfly has a really quiescent nymphal stage been reported. Similarly in at least some stream-dwelling dragonflies the last winter is spent in the final instar, but the nymphs remain active, e.g. *Cordulegaster annulatus* (Nicolau-Guillamet, 1959).

Many animals, of course, cease to grow, or grow only very slowly, at low temperatures, and this is a point we shall discuss later. At least one species of stonefly, the common European *Nemoura cinerea*, ceases moulting even at room temperatures in the winter (Khoo, 1964*a*). It therefore displays a true winter-diapause which may be light-controlled, but such animals are not quiescent.

In contrast, many species of Trichoptera spend long quiescent periods in their last larval instar often already sealed up in their pupal houses, and they may hibernate or aestivate in this condition. An example of an overwintering species is *Apatidea muliebris* in a Danish spring (p. 288) and of an oversummering one, the North American *Pycnopsyche guttifer* (Cummins, 1964). Sometimes, however, as in the common European *Rhyacophila septentrionis*, overwintering, resting, last-stage larvae occur together with active larvae of the same or earlier instars (Nielsen, 1942). In such species, as in those with diapausing eggs combined with a nymphal life of more than a year, the quiescent stage would seem at first sight to be of little ecological consequence. It may, however, serve as an insurance (p. 289).

The limnephilid *Pycnopsyche lepida* is exceptional; or perhaps that statement merely reflects our ignorance of life histories. Unlike many other caddis-worms which go into diapause where they will finally pupate, its full-grown larvae burrow down into the gravel where they remain inactive from February to September (in Michigan); they then come up again to pupate at the surface of the substratum and emerge in the fall (Cummins, 1964).

At least one species of stonefly, *Capnia bifrons* of western Europe, has a true nymphal diapause which carries it through the warm period of the year (Khoo, 1964*b*). There are indications that the same phenomenon occurs also in some other Capniidae, but this remains to be more thoroughly investigated. As already stated *C. bifrons* is oviviviparous but the nymphs tend to disappear from collections for several months. Khoo showed that in the 4th or 5th instar, when they are still very small, the nymphs become sluggish and opaquely white because of a great accumulation of fat globules. They also lose the whorls of bristles on most of their distal cercal segments, and they finally become immobilized in a very characteristic position (Fig. XIV, 2). This occurs at some time in the early summer and they remain in this condition until August or September, although the period is shortened by low temperature. Diapausing nymphs have been found very rarely in field collections, and it seems likely that they are hidden deep in the substratum where they would be relatively safe from predators. Nymphs of *Allocapnia pygmaea* in a condition closely resembling that of diapausing *C. bifrons* have recently been found in considerable numbers deep in the gravel of a stream bed in Ontario, by Mrs. Mary Coleman.

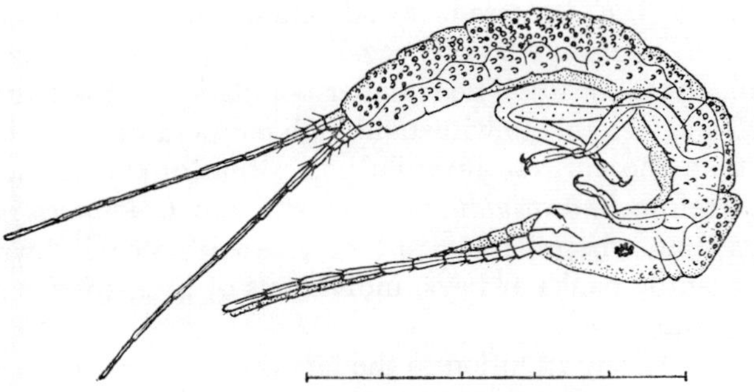

FIG. XIV, 2. A diapausing nymph of *Capnia bifrons*. From Khoo (1964*b*) with permission.

Finally quite a number of animals pass over unfavourable seasons by simply remaining quiescent. The North American prosobranchs *Pleurocera* and *Goniobasis* become inactive at temperatures below 5 °C. (Dazo, 1965), and Unionidae cease growing in winter (Negus, 1966). The crayfish *Cambarus longulus* burrows down into the substratum of Virginian streams and spends the winter there (Smart, 1962), and its relative *Orconectes clypeatus* evades dry periods in the same way in streams in Louisiana (Smith, 1953).

Thus many varied devices have been evolved among the benthic invertebrates which enable them to avoid unfavourable conditions. This is, of course, a commonplace in biology, but what is unusual in running water is that it is the *warm* period of the year which is avoided by a great number of the animals, especially in small streams. This is in considerable contrast to the land habitat where winter-active species are the exception, and it is particularly odd when one recalls that extremes of summer heat are much less marked in water than they are on land.

COEXISTENCE OF CLOSELY ALLIED SPECIES

The competitive exclusion principle is an article of faith among animal ecologists, who hold, with good reason, that no two species which occupy the same ecological niche can occur together indefinitely in the same habitat. This situation is avoided by the fact that species in the same genus usually have different distributions in space or in time. There are many examples in running water of two or more closely allied species which occur together but which are separated by differences of life history.

The genus *Erpobdella* has already been mentioned as having two species which often occur together in Europe but differ in their life histories (p. 217). *E. testacea* is strictly annual and produces cocoons in June, while *E. octoculata* lays its eggs later at 1, 2, or sometimes 3 years old (Mann, 1953, 1961*b*). The summer population of the latter is therefore mixed and less able to withstand high temperatures and low oxygen contents than the entirely juvenile population of the former. In cool streams therefore *E. octoculata* is successful and *E. testacea* fares better in swamps; but, where habitats are diverse with a cool flowing central area and swampy banks or bays, individuals of both species may occur under the same stone.

Illies (1952*a*) drew attention to the fact that in many genera of Plecoptera, most notably *Protonemura*, *Nemoura*, *Amphinemura*, and *Leuctra*, and also in some caddis-flies, e.g. *Rhyacophila*, there are pairs of species, or sometimes three or four, which occur together but emerge at different times. Thus, as they are at different stages of their life cycle at any one time, they do not come into direct competition. Illies stressed that, as male stoneflies tend to emerge before females, there is complete reproductive isolation even in species whose emergence periods overlap. He also made the point that the temperature of the water at any given time appears to have very little effect on the rate of growth of the nymphs, so that emergence dates tend to be constant, helping to keep the species isolated. Actually the prior emergence of the males does not seem to be as definite as Illies supposed, as was shown by a long emergence-trap study on an English stream, but the constancy of the emergence dates in this order is well established. In a six-year study of Ford Wood Beck in north-west England, Gledhill (1960) found only fourteen days between the first emergence dates of *Leuctra inermis* and nineteen between those of *Chloroperla torrentium*.

Hynes (1961, 1962) has commented upon the existence of a number of species pairs in small British streams in the genera *Protonemura*, *Leuctra* (in which two serial pairs may be involved in a single stream), *Chloroperla*, and *Baetis*, and he has shown that, even where the emergence periods overlap a great deal, the young are, throughout their life, at different stages of growth, so that one is always smaller than the other. A similar species pair has been demonstrated in the caddis-fly *Pycnopsyche* in North America (Cummins, 1964), and here the larvae also become physically separated. Both hatch on and live on silty substrata in their early stages, but *P. lepida* moves out on to gravel at about the time that *P. guttifer* hatches. The latter remains on the silt, so the two species come to occupy different areas as well as being of different sizes. It seems likely that other closely allied species which are separated by size at most times of the year, but which do not become spatially separated, are able to co-exist because of differences in requirements for shelter. Smaller specimens can inhabit smaller crevices and so do not compete with larger ones for a feature of the environment which appears to be of prime importance in the stream habitat.

Probably many such examples of successions of species caused by differing life histories remain to be uncovered. Several workers, e.g. Ide (1940), Sprules (1947), and Robert (1960) in Canada, and Hynes (1961) and Gledhill (1960) in Britain, have reported well-defined serial emergences of species from streams. As we have seen, as many as four species of *Leuctra* may succeed one another in this way in western European streams. There are similar examples from North America, e.g. five species of *Alloperla* in Gorge Creek in the Albertan Rocky Mountains (Hartland Rowe, 1964), and there are others from the Ephemeroptera. In Wales *Baetis rhodani* (first generation), *B. scambus*, *B. tenax*, and *B. rhodani* (second generation) follow one another through the summer (Hynes, 1961), and in some Austrian and Czechoslovakian streams three species of Ephemerellidae share out a single habitat in time (Pleskot, 1961; Petr, 1961). During the winter partially grown nymphs of *Torleya belgica* are present together with diapausing larvulae of *Chitonophora krieghoffi*. In early spring *Chitonophora* grows fast and emerges, followed in May by *Torleya*. By this time the eggs of *Ephemerella ignita* have hatched and the nymphs grow fast to produce adults in the late summer, by which time larvulae of *Torleya* are again present. Thus ephemerellid nymphs are present all the year round, and most sizes are present at all times. This, presumably makes full use of that part of the habitat which can be exploited by this particular life-form.

FACTORS AFFECTING LIFE CYCLES

An obvious factor affecting life histories is temperature, and it is well established that low temperatures decrease the growth rates of many animals. Illies (1952*a*) suggested that a major difference between the

Ephemeroptera and the Plecoptera is that growth in the former is stopped by cold weather, while the developmental temperature-threshold (Entwicklungsnullpunkt) of the latter is very low indeed. The threshold temperature of many stoneflies does indeed appear to be very low. Raušer (1962) has suggested that it is near 0 °C., and in most species which have been studied growth is only slowed, not stopped, during cold weather, even in the Canadian winter (Hartland-Rowe, 1964). Some species, which emerge very early in the year and so spend their entire active life in cold weather, are very little affected by cold, e.g. *Capnia bifrons* and *Brachyptera risi*, *Leuctra hippopus*, and *Protonemura praecox* (Hynes, 1962; Khoo, 1964*a*, *b*). On the other hand, there are species, e.g. the European *Nemurella picteti* and the already-mentioned *Nemoura cinerea*, which cease to moult at winter temperatures.

Similarly, while it is true that many Ephemeroptera do cease to grow during cold weather, e.g. *Cinygmula remaleyi* and *Ephemerella inermis* in the Canadian Rockies (Hartland-Rowe, 1964), *Heptagenia lateralis* and *Siphlonurus lacustris* in England (Macan, 1957*b*; Hynes, 1961), and *Hexagenia* in the Tennessee River (Lyman, 1943); others, e.g. *Habroleptoides* and *Torleya* in Austria (Pleskot, 1960), *Rhithrogena semicolorata* in England (Macan, 1960*b*; Hynes, 1961), and *Epeorus* in West Virginia (Maxwell and Benson, 1963), grow right through the winter. Indeed the development of the wing-pads of *Epeorus* is at its fastest in the coldest weather, and in Africa there are species which appear to be confined to high altitudes where the water is very cold at all times (Hynes and Williams, 1962). The developmental threshold is therefore low for a considerable number of mayflies also.

The same applies to other groups. Some species of Simuliidae, particularly of the genus *Prosimulium*, continue to grow in very cold weather (Zahar, 1951) as do some Trichoptera, although Mackereth (1960) found that in *Agapetus fuscipes* this can vary from year to year for no obvious reason. There are also indications that the same variation may occur in some mayflies; Harker (1952) reported no growth of *Rhithrogena semicolorata* below 6 °C., but the workers cited above did not find this to be so in other years in streams only a few tens of kilometres away from the site where she made her study. Many other animals, e.g. the bug *Aphelocheirus* (Larsén, 1932) and crayfishes, do cease to grow in cold water, and, in the former, cold inhibits the hatching of the eggs, so that the young nymphs appear either in the fall or in the spring from eggs laid during the previous summer.

We can conclude then that the emergence dates of stream insects are only partly determined by the rates of development of young stages. Some seem fairly definitely to be controlled by the temperature at the time of emergence; e.g. the critical temperature for overwintering Simuliidae in Scotland seems to be about 10 °C. (Zahar, 1951). Such species, and this seems to include many mayflies, therefore have variable beginnings to their emergence periods. This cannot, however, always apply, as definite seasonal cycles of species in many groups (Plecoptera, Ephemeroptera, Trichoptera, and Diptera) occur in springs with year-round even temperatures (Thorup, 1963), and, as we have seen, some Plecoptera emerge at about the same time year after year. It seems likely therefore, that day-length or its rate of change is an important time-setter, and that the relative importance of temperature and change of day-length as a signal for emergence may be more important in establishing differences in constancy of emergence times than the developmental threshold suggested by Illies.

Khoo (1964*a*) was able to show that while a rise in temperature does stimulate stoneflies to emerge they also emerge from water at winter temperatures if they are exposed to long days, and this, coupled with the established fact that many stoneflies and mayflies emerge earlier at lower altitudes and lower latitudes, indicates that both factors interact. Their relative importance probably varies with the group and the species.

Both factors almost certainly also control the length of the emergence periods. These tend to be shorter at lower altitudes, and there are indications that, at any rate for early species, rising temperatures may eliminate laggard specimens (Pleskot, 1953; Macan 1960*b*; Hynes, 1961). Khoo (1964*b*) has shown that in *Capnia bifrons* there is a steady decrease in size of the emerging specimens as the season progresses, as if the increasing day-length were hurrying on the later specimens to emerge before

	February		March		April		May	
Half of month	1	2	1	2	1	2	1	
Males	6·0	6·0	6·5	6·1	5·5	5·3	4·7	Size in mm.
	4	19	60	48	30	19	4	No. of specimens
Females	9·0	9·0	8·8	8·3	7·8	7·3	7·1	Size in mm.
	3	8	39	50	24	23	15	No. of specimens

TABLE XIV, 2. The mean length in millimetres of specimens of *Capnia bifrons* emerging in successive half-monthly periods in 1962 and 1963 in a cold room maintained at stream temperature and ambient day-length. From Khoo (1964*b*).

their growth is completed (Table XIV, 2). He found that the small specimens had relatively shorter wings, and he had some evidence that they had been through fewer instars before becoming adult (Khoo, 1964*a*). This sort of relationship between the size attained by specimens and the time of arrival of the signal for emergence may contain the explanation for the brachyptery of many species of Plecoptera at high altitudes (p. 145). Strength is added to this suggestion by the fact that all the species in which the phenomenon has been recorded emerge early in the year (Kühtreiber, 1934; Hynes, 1941; Brinck, 1949). It may also account for the well-known fact that in bivoltine species of *Baetis*, specimens of the early generation, which have had many months in which to grow, are always larger than those of the rapidly growing summer-generation (Harker, 1952; Pleskot, 1958; Macan, 1957*b*). The same phenomenon is also known to occur in some species of Simuliidae (Grenier, 1949) and it may be caused in the same way.

At the other end of the year falling temperatures and/or decreasing day-length may account for the observation that some small specimens of late-flying species apparently miss the flight season, and persist for months after their fellows have gone. This has been observed in *Ephemerella ignita* in England, Wales, and Denmark (Percival and Whitehead, 1930; Hynes, 1961), and it is to be seen in Fig. XIV, 1 where one specimen in the 1-mm. size group was found in late November long after it should have flown.

Much of the above is, however, largely speculation and, although the explanations seem reasonable, there would seem here to be much scope for further study. We have at present no hint of an explanation for the remarkable phenomenon observed on the Mississippi where, between mid June and mid August, *Hexagenia bilineata* emerges in waves at 6- to 11-day intervals, the waves being synchronized along at least 1,000 km. of river (Fremling, 1960*a*, 1964). Neither do we have any good information on the effect of nutrition on life cycles. Zahar (1951) noted that pupae of Simuliidae tended to be small during a dry summer in Scotland, when food was presumably in short supply, but not that emergence was delayed. This family, however, seems to be very varied in its food requirements as is shown by the two species *Prosimulium fuscum* and *P. mixtum* which occur together in streams near Ottawa. Both emerge in the spring, but the former is able to lay its first batch of eggs without a blood meal, which is essential for the latter (Davies, 1961). Thus the two species living in the same places clearly differ greatly in their ability to obtain foodstuffs in the larval stage, and this sort of difference must have some influence on rates of growth. *P. ursinum* of the far north of Europe and North America does not feed at all as an adult and, unlike most members of its genus, it emerges very late in the year from the cold infertile streams. Possibly this is a change from the normal pattern in the genus which is imposed upon it by the necessity to procure enough food during the larval stages to form a sufficiency of eggs (Davies, 1954).

FLEXIBILITY OF LIFE HISTORIES

We have already seen that temperature and day-length may alter life cycles and that there are certain species, e.g. *Aphelocheirus aestivalis* and *Erpobdella octoculata*, which merely cease to develop during cold periods and take up where they left off when temperatures are again suitable. Such animals have life cycles which are rather loosely tied to the annual cycle, although their reproductive period is inevitably restricted to a particular, but usually extended, time of year. Most stages are, however, present at most seasons, and this situation is characteristic of many amphipods of the genus *Gammarus* and some species of *Crangonyx* (Hynes, 1955*b*) and of many isopods, e.g. *Asellus intermedius* of North America (Ellis, 1961) and *A. aquaticus* and *A. meridianus* of Europe. It also occurs in many crayfishes, e.g. *Cambarus propinquus* of North America (Van Deventer, 1937) of which two or three overlapping generations occur together. Pulmonate snails show similar plasticity, and they may or may not breed during the summer of their birth, according to the conditions and whether they arise from eggs laid early or late in the season. Such a situation is found, for instance, in *Physa fontinalis*, a common European snail (Duncan, 1959), and, as in the Peracarida, it permits a very flexible response to the vagaries of climate.

Another plastic response to different climates is displayed by species which go through, or are able to go through, more than one generation a year. Many species of Simuliidae can do this and they respond to different latitudes or altitudes by alterations in the number of generations passed through during the year (Grenier, 1949; Zahar 1951; Sommerman *et al.*, 1955). *Simulium ornatum* is well known for its ability to adjust in this way and it is a widespread and very opportunist species because of this flexibility. At least some mayflies of the genus *Baetis*, e.g. *B. bioculatus*, *B. subalpinus*, and *B. vernus*, can respond to warm water by producing more than one summer-generation, and *B. rhodani* can dispense with its summer-generation in high cool streams (Pleskot, 1958, 1960).

Univoltine species have also been found to display different life cycles in different localities and this may represent flexibility of response to different conditions. For instance, Petr (1961) records nymphs of *Rhithrogena semicolorata* at all seasons in streams flowing into Sedlice Reservoir in Czechoslovakia, whereas in Britain there is always a period in late summer when they do not occur. We do not know the significance of such differences, and it may be trivial, but one well-documented example which appears to be an evolutionary response to changed conditions is reported from Himmerland, North Jutland (Nielsen, 1950*a*, *c*,

1951*b*). One spring, the Rold Kilde, is inhabited by the caddis-fly *Apatidea muliebris* which there has what appears to be a quite inappropriate life cycle for an environment with an almost constant temperature. The flies emerge in late April and early May, the larvae grow fast during the summer and then immure themselves in their pupal houses, where they spend the whole winter as inactive last-instar larvae. Only 10 km. away in another spring, the Lille Blaakilde, there occur two other species of *Apatidea*, presumably derived by mutation in this parthenogenetic genus, which have a life cycle much more suited to the even conditions of spring life. They, and populations of *A. muliebris* in England, fly from early spring to late fall, all stages occur together at the same time and there is no winter diapause. Nielsen suggests that *A. muliebris* was originally an Arctic species brought south by the Pleistocene glaciation which has for some reason retained its Arctic life cycle in Rold Kilde, but elsewhere has given rise to forms, recognizable as distinct species in Lille Blaakilde but not in England, with a life cycle more appropriate to the localities they now inhabit.

Another difference in the life histories of a species in two localities in a single English county, Lancashire, is not so easily accounted for. In Wayoh Stream Harker (1952) found that *Ecdyonurus torrentis* went through three overlapping generations in two years. Adults emerging in February–March gave rise to adults in June–July, which gave rise to adults in April–May, the parents of the February–March adults. A similar cycle had earlier been reported as occurring in the North American *Baetis posticus* and had been confirmed by the rearing of specimens in cages in a creek (Murphy, 1922). However, only a few tens of kilometres away from Wayoh Stream, in Ford Wood Beck, Macan (1957*b*) found that *E. torrentis* has a simple univoltine cycle, but with a very extended flight period, from May to September. One wonders if, in fact, Harker did not misinterpret her data, especially in view of the knowledge that another member of the Ecdyonuridae, the North American *Stenonema femoratum*, which has a univoltine cycle, has three peaks of emergence giving the impression of three almost separate populations (Lyman, 1955).

There remains clearly much work to be done on ephemeropteran life cycles, and workers in the same area seem quite often not to agree. For instance, Hunt (1953) and Lyman (1955), who both worked on lakes in Michigan, disagree as to whether *Hexagenia limbata* takes one year or two to develop, and as stated earlier there seems to be much confusion about the life cycles of *Ephemera*, which is possibly caused by the lack of very small specimens in collections made by the usual techniques.

EXTENDED HATCHING PERIODS OF INSECT EGGS

Many field studies have shown that very small specimens of certain species of Plecoptera and Ephemeroptera occur in collections over periods greatly exceeding the flight period of the adults, and there is good reason to believe that this is caused by irregular hatching of the eggs (Fig. XIV, 1). This phenomenon has been recorded in the genera *Baetis*, *Torleya*, *Habroleptoides*, *Rhithrogena*, and *Heptagenia* among the mayflies (Macan, 1957*b*, 1958*a*; Pleskot, 1960, 1961; Hynes, 1961) and in *Brachyptera*, *Amphinemura*, *Protonemura*, *Leuctra*, *Isoperla*, and *Chloroperla* among stoneflies (Hynes, 1961, 1962). It is also known in the blackfly genera *Prosimulium* and *Simulium*, of which young larvae may occur during several months while adults are present for only a few weeks (Grenier, 1949; Fredeen *et al.*, 1951; Davies, 1961).

In other species, e.g. the mayfly *Ameletus* (Gledhill, 1959), and, as we have already seen, the bug *Aphelocheirus*, young specimens may appear in autumn and again in early spring without any intervening generation of adults. This indicates cessation of hatching during the cold period, and in *Ameletus*, of which the nymphs continue to grow in the cold weather, it results in two sibling growth-streams in a univoltine animal, where, as in bivoltine species of *Baetis*, there are two peaks of emergence with larger individuals in the first one.

We have also seen (p. 278) that there is good *direct* evidence that the incubation period of many species of Ephemeroptera and Plecoptera is indeed very irregular so that hatching continues for a long period. To this can be added the observation of Grenier (1949) that one batch of eggs of *Simulium nolleri* continued to hatch for fifty days, which is long enough for the first-hatched specimen to have pupated before the last had hatched.

We can take it then that extended hatching periods, both with and without initial or intervening diapause, are common in the insects of streams. It would seem that they are a form of insurance in this very unreliable habitat, since eggs are less affected by drought and spates than are young insects. Some are thus likely to survive to replace an eliminated or decimated population of nymphs or larvae. Perhaps here lies the significance of the very long diapause of the eggs of *Pteronarcys* and *Diura*, which, as their nymphs are present for all or most of the year, seems to have no obvious function in the avoidance of climatic pessima, except possibly the rather doubtful one of preventing exposure of very small nymphs to very cold weather. This, however, would seem to be of little advantage to these cold stenothermic animals, whereas insurance against periodic disaster would be of great importance.

LIFE CYCLES IN THE TROPICS

We have very little information on life cycles in the tropics, and in view of the unvarying temperature and light conditions it may be that reproduction and growth are continuous, with all stages present at all times. A corollary of this would be, if our earlier speculations about co-occurrence of closely allied species are correct, that in any one section of a tropical stream each genus or life-form would have a greater tendency to be represented by only one species. For most groups we do not know if this is so as too little work has been done, but it is certainly not true of *Simulium* in streams on Mount Elgon, as can be seen in Table XIV, 3.

Altitude in m.	1,830	1,590	1,400	1,340	1,310	1,280	1,265	1,205
Simulium alcocki		0·5						
S. aureosimile	3	4						
S. dentulosum	97	89·5	11					
S. vorax	P*	P	24	2·5	0·5			
S. hirsutum		2·5	2	1·5	3	6	3*	8
S. medusaeforme			47·5	85*	96·5*	89·5*	69*	21
S. cervicornutum				1·5	P	4·5	28	69·5*
S. alcocki group								P
S. copleyi	P	3·5	15·5	1·5	P	P		
S. lumbwanus				8				1
S. neavei			P	P	P	P		0·5
Total number of larvae in collection	2,176	1,939	290	353	2,330	2,100	654	328

TABLE XIV, 3. The percentage composition, by species, of 200 larvae of *Simulium* selected at random from each of eight samples collected during three days (31 December–2 January 1960–1) at various points on the River Manafwa, Mount Elgon, Uganda. P = present but not among the 200 selected. * = pupae or pupal skins present. From Hynes and Williams (1962).

There the last three species listed are perhaps not involved as the first two are phoretic on mayfly nymphs and the last on crabs, but, even allowing for this, several ordinary species occurred at most stations. It may, however, be significant that the extensive net-collections from which the larvae were selected contained significant numbers of pupae or empty pupal cases of only *S. medusaeforme* (sixty-eight specimens) with the other species represented only by single individuals. It seems therefore that only one species was actively emerging at the time when the collections were made and this does perhaps indicate a succession of species despite the even climatic conditions.

Van Someren (1952) after his extensive study of the Sagana River on Mount Kenya stated that insects there emerge at all seasons with no mass emergencies such as occur in temperate latitudes. He admitted, however, that his data were incomplete, and Gillies (1954), who studied *Prosopistoma* on the Usumbara Mountains in Tanzania, wrote of an emergence season for this mayfly although he was not very definite about this. Similarly, Cridland (1958) found that the snail *Limnaea exserta* ceases to breed in the Kibibi River, Uganda, during the dry season. That definite flight periods do occur in at least some groups of insects is indicated by collections made by the de Witte Mission to the Upemba National Park in the Congo. They sought insects throughout the year 1947–8 in the valley of the River Lufira, and they collected 127 specimens of *Neoperla spio*, only one of which was taken outside the period August–December. In that part of Africa the rainy season begins with gales and storms in August followed by rain in September or October through May. It would seem then that the main flight period of this stonefly is the first half of the rainy season, and that it emerges rarely at other times (Hynes, 1952*c*).

We similarly know little about rates of growth in the tropics, but some information was derived from study of reaches of the River Manafwa which had been poisoned by D.D.T. (Hynes and Williams, 1962). One month after the poisoning many animals were fully grown, e.g. *Caenis*, *Acentrella*, *Baetis*, *Centroptilum*, and *Simulium*. Presumably these had hatched from surviving eggs and had gone through their development within the month. On the other hand, other insects had reached only small sizes as shown below:

	Maximum length attained in one month: mm.	Maximum length in collections from nearly unaffected reaches: mm.
Oligoneuriella	2	19½
Tricorythus	1¼	4½–5½[1]
Ephemerythus	1¾	2½–6[1]
Prosopistoma	1¼	2½
Euthraulus	3¾	5½
Afronurus	2	8½
Neoperla	4½	14–16
Hydropsyche	3½	13
Cheumatopsyche	4½	8½

1. Probably at least two species.

It can be seen, therefore, that at least some tropical stream insects take more than a month, and possibly several, to grow to full size. Perhaps the judicious use of single doses of D.D.T. would be a useful technique for determining the growth rates of insects in tropical streams although results might be invalidated by drift. In the example cited above, however, the first affected station studied was only 1·7 km. below the dosing point and there was, at any rate in respect of the insects tabulated above, no evidence of recolonization by drifting. It is possible though that all or most of the specimens of species which appear to have grown up within a month had drifted in. In that event the estimate of less than a month would be wrong, but in view of what is known of at any rate *Baetis* and *Simulium* under summer conditions in temperate climates this seems improbable.

SEASONAL CHANGES IN THE ENTIRE FAUNA

In view of the various types of life history discussed above it is possible to sketch an outline of the series of events which recur annually in streams in temperate latitudes. Several such streams have been sampled at intervals throughout the year with varying degrees of thoroughness and by varying methods. Although the results do not agree in detail, it is possible to observe a general pattern and to explain differences by reference to specific local conditions, e.g. flood seasons, exceptionally important species, etc., and particularly to the technique used to obtain the samples.

Many studies have been made with fairly coarse-meshed nets, and this has resulted in failure to collect young stages. One consequence of this is that sometimes minimum numbers have been recorded in the autumn, when in fact they were probably at or approaching a maximum, and another is that increases have been recorded during periods when numbers were probably falling, but an increasing proportion of the specimens which were surviving were attaining catchable size.

The following is a list of territories and references to year-round studies involving the whole fauna:

North America	Alberta	Hartland-Rowe, 1964
	Michigan	Ellis and Gowing, 1957
	Utah	Gaufin, 1959
	Wyoming	Armitage, 1958
	Missouri	Clifford, 1966*b*
	North Carolina	Tebo and Hassler, 1961
	Virginia	Surber, 1951
	Georgia	Nelson and Scott, 1962
Europe	England	Whitehead, 1935; Pentelow *et al.*, 1938
	Wales	Badcock, 1949; Hynes, 1961
	Ireland	Frost, 1942; Humphries and Frost, 1937

Europe	Sweden	Badcock, 1953*a*
	Poland	Sowa, 1965
South Africa	Cape Province	Harrison and Ellsworth, 1958
	Natal	Oliff, 1960

The list could probably be expanded somewhat as it is a matter of judgement as to what constitutes a year-round study of the entire fauna, but all the authors included above have attempted to take identifications as far as possible, to sample at frequent and regular intervals and to present the data in such a way as to illustrate over-all change. One thing the list does show very clearly is how much this kind of study has been confined to the English-speaking world. Even continental Europe which has for so long been prominent in freshwater biology seems to have produced few studies of this kind; the Swedish study was done by an English woman!

Considering then a typical stream in temperate latitudes which is not unduly influenced by extensive floods, drought, or other factor which will seriously alter the fauna, it is possible to describe the general pattern of seasonal change. A speculative diagram of this pattern is given in Fig. XIV,5; it follows a synoptic discussion of life-histories but it may also prove helpful at this point.

The autumn is a period of hatching of the eggs of species which will grow during the winter, and of growth of permanent species such as malacostracans born during the summer. It is thus a period of increasing, often maximum, numbers, including many small individuals, and of increasing biomass, despite losses caused by late-flying species of insect. Losses caused by predation and other factors in the environment tend to be made up by the continued hatching of eggs.

In winter the growth of many animals slows down and the rate of recruitment from newly hatched eggs also declines. Numbers therefore tend to decline as individuals die or are eaten, and they may decline steeply if severe climatic conditions such as floods or underwater ice intervene. Biomass, however, may not decline as the growth of surviving individuals offsets losses. It may indeed rise quite steeply if many winter-growing species, chiefly Plecoptera, but some Ephemeroptera also, are present.

In early spring there is a loss of biomass and numbers caused by the emergence of early stoneflies, e.g. *Capnia*, *Allocapnia*, and *Taeniopteryx*, but the loss in numbers may be offset to some extent by the hatching of the eggs of species which grow during spring and early summer, e.g. some North American species of *Ephemerella*. In the absence, or without the dominance, of very early species, the biomass rises in early spring as renewed growth occurs in individuals which have overwintered, although the numbers continue to decline as losses are not replaced. This is a season, however, during which several workers (e.g. Needham, 1934; Tebo and Hassler, 1961) have recorded increases in numbers. This was caused, probably, partly by growth to catchable size and partly by more satisfactory collecting conditions.

During late spring many species of insect emerge, and the biomass falls steeply. Numbers also fall, but towards the end of the period they begin to rise again as the eggs which give rise to summer species, or to summer generations of multivoltine species, hatch. Where amphipods or isopods are present their breeding period also augments the numbers.

Through the summer the growth of the summer species adds to the biomass, and the numbers rise or fall depending upon the fauna. In streams dominated by *Gammarus* they may continue to rise as the amphipods produce brood after brood (Beak, 1938*a*), but where insects are dominant the hatching of the eggs of summer- and fall-flying species declines as summer progresses, and there is a tendency for the numbers to decrease after a while, first because of loss of individuals by death and then because of emergence. Streams in the latter category also seem not to reach as high a peak of biomass and numbers in early summer as in autumn and early winter (Hynes, 1961). This may, however, not be true of places where Simuliidae or Chironomidae, of which there are many summer species, are abundant.

Late summer is a second period of low biomass in many streams because of the emergence of insects, but this is probably not true of streams dominated by malacostracans. It is also a period of recruitment from the hatching of the eggs of species which will overwinter, so the decrease caused by emergence is offset by, and may be eliminated by, recruitment of small insects, newly hatched or, like *Capnia bifrons*, newly emerged from diapause.

A MODEL OF THE ANNUAL CYCLE IN STREAMS

In the present state of our knowledge of the life histories of benthic animals we are almost overwhelmed by facts, and it is difficult to discern any clear over-all pattern. The final section of this chapter is therefore devoted to setting up a model of the sequence of events in temperate streams, which, it is hoped, may help to marshal the facts into some sort of order and indicate points which are still unresolved.

Hynes (1961) has drawn up a semi-quantitative diagram of the sequence of events in a mountain stream in Wales, the main features of which are reproduced here in modified form in Fig. XIV, 3. It seems that this concept has general validity and that in temperate streams it is possible to distinguish three main types of life cycle. These are:

1. Non-seasonal, so called because individuals of all stages are present at all times. This may be because the life history is long and spans more

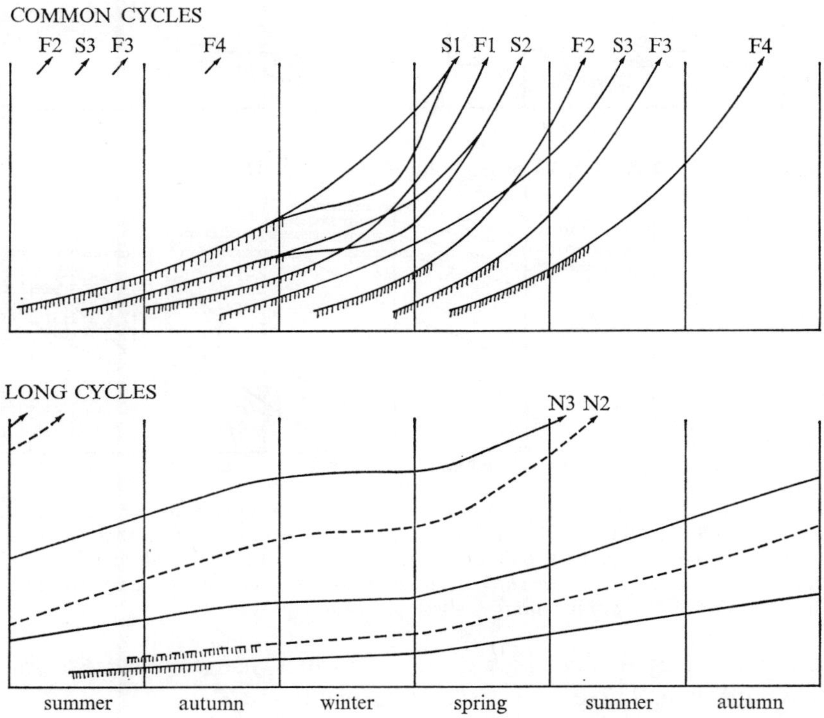

FIG. XIV, 3. A diagrammatic presentation of the various types of life cycle which are commonly found in stream-dwelling insects. Growth to full size is upwards and arrows indicate emergence. Recruitment of hatching young is shown by vertical hatching. S and F are slow and fast cycles respectively and only one of each type is shown in full. For further explanation see text.

than a year, e.g. some large stoneflies and the Megaloptera (Fig. XIV, 3), because a series of overlapping generations occurs, as in many Peracarida and Mollusca (Fig. XIV, 4), or for other unexplained reasons. For example, specimens of all, or almost all, stages of certain Rhyacophilidae and Polycentropidae occur at all seasons and the life cycles remain to be worked out; and in some groups, most notably the Chironomidae, our ignorance of the young stages prevents our separating the individual species.

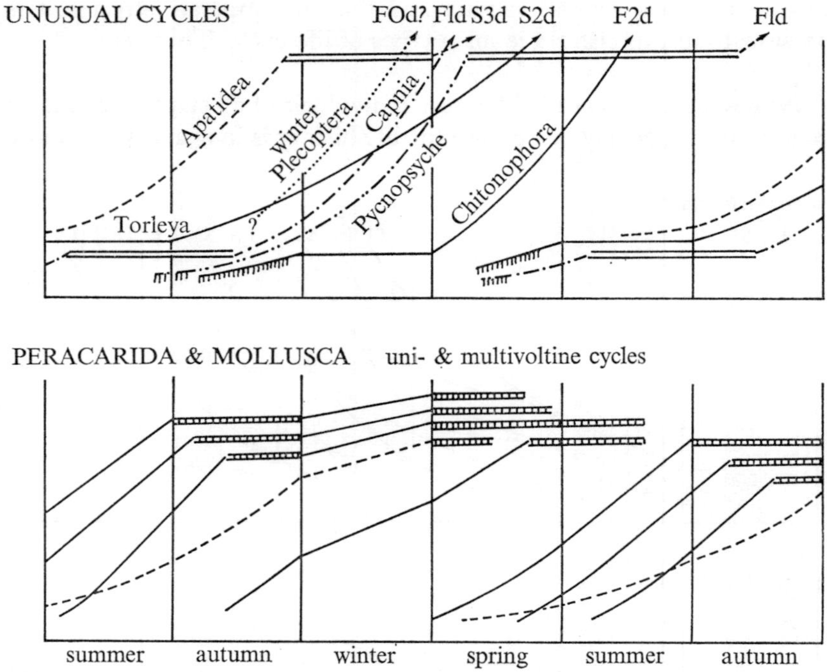

FIG. XIV, 4. A diagrammatic presentation of: *Above*. Some unusual life cycles found in stream-dwelling insects. Parallel horizontal lines indicate larval or nymphal diapause; other symbols as in Fig XIV, 3. *Below*. Typical life cycles of univoltine (dashed line) and multivoltine (solid lines) Peracarida and Mollusca. Parallel horizontal crossed lines indicate periods of reproduction. For further explanation see text.

2. **Seasonal cycles** in which there is a distinct change of size distribution with time. These are divisible into two types:

S. *Slow seasonal cycles* in which the eggs begin to hatch soon after laying and where growth occurs over a long period to maturity nearly a year later. In many insect species with slow cycles there is a very long hatching-period with steady recruitment over a lengthy period, and these cycles span the winter. In its original form this concept applied only to insects, and in the Welsh stream on which it was worked out all the species were found to grow during the winter. Clearly, however, univoltine viviparous crustaceans with a single breeding period, e.g. many North American species of *Crangonyx* (dashed line on lower Fig. XIV, 4), can properly be included here, and Hartland-Rowe (1964) has pointed out that the slow cycles also include species which cease to grow in winter. This point has therefore been included in Fig. XIV, 3 where slow cycles are shown as taking two possible routes through the winter.

Three, or perhaps more, slow cycles follow one another in time, differing in whether maturity is attained in early, or late spring or in early summer: cycles S1, S2, and S3 on the diagram. These are the normal slow cycles, but a few unusual ones also belong in this category. Such are the mayflies which spend a long period as small larvulae, e.g. *Torleya* and *Chitonophora*, (Fig. xiv, 4) and the stonefly *Capnia bifrons* with its summer nymphal diapause. Possibly the winter stoneflies, e.g. *Allocapnia* and *Taeniopteryx*, should also be included here but we know little of their early stages. It is, however, established that the eggs of the early emerging European *Taeniopteryx nebulosa*, which English fly-fishermen call the 'February Red', hatch fairly rapidly, and *Allocapnia vivipara* is oviviviparous, so at least some of them do, but the group as a whole is indicated with a query in Fig. xiv, 4.

Note that species with slow cycles mature early in the year, not in late summer or fall.

F. *Fast seasonal cycles* in which growth is rapid after a long egg-diapause or one, or more, intermediate generations. These are the cycles labelled F1 to F4 in Fig. xiv, 3. The earliest, F1, cycle is slower than the others and it has a longer hatching period; its slowness is doubtless because the early stages span the winter. This does not, however, make it intermediate between the fast and slow cycles, as all the species included in it have either an intermediate generation or a long egg-diapause.

Like the slow cycles, several fast cycles succeed one another in time. Four are shown in Fig. xiv, 3, but this number is arbitrary, although it is based on the number actually identified in the stream on which the model was worked out. A point which is, however, important is that fast cycles reach full term not only in spring and early summer but also in late summer and in fall. The later fast cycles are therefore those of the summer-growing species, and they appear to contain few carnivorous animals. Two or more fast cycles may be exhibited by the same species when rapid generations succeed one another, as in some species of *Baetis* and *Simulium*. Possibly there are earlier cycles than the one shown as F1, occupied by some autumn and winter stoneflies in the genus *Allocapnia*, but as already stated, we have insufficient information on their early stages and they have tentatively been labelled FO in Fig. xiv, 4.

There are also some very early-flying species of Chironomidae of which we know little. The peculiar cycles of some Trichoptera with a long larval diapause shown in Fig. xiv, 4 are also fast cycles which differ from normal ones only in that the diapause is prepupal instead of being in the egg-stage.

It would seem then that nearly all the life cycles displayed by running-water invertebrates can be fitted into one of the three categories, non-seasonal, slow, and fast seasonal. It is clear also that the non-seasonal category includes three groups; these are cycles longer than a year, cycles with multiple, overlapping, flexible generations, and cycles which result in wide size-distributions at all seasons but are as yet unexplained.

Having earlier adopted the letters N, S, and F for the three types of

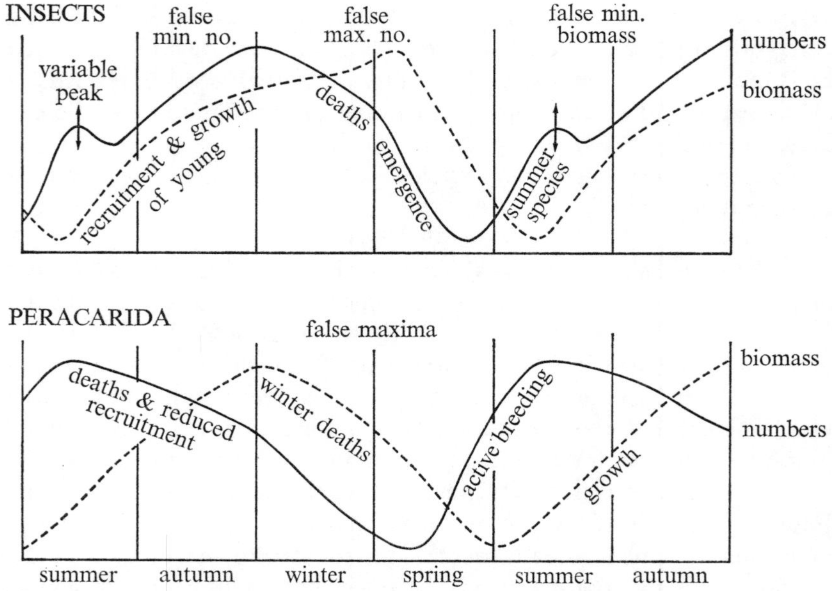

FIG. XIV, 5. A diagrammatic presentation of the supposed seasonal fluctuations in numbers and biomass of invertebrates in streams dominated by insects and peracaridans (or multivoltine snails) respectively.

cycle (Hynes, 1961) with numerical suffixes for the last two it is perhaps appropriate here to suggest N2, N3, N4 for two, three-, and four-year cycles, Nm for the multiple overlapping cycles of Peracarida and some snails and Nu for unexplained non-seasonal cycles. It would also seem appropriate to add the letter 'd' to the symbol for cycles which contain a diapause in a stage other than the egg. Thus the cycle of *Capnia bifrons* would become F1d and that of *Torleya* S2d (Fig. XIV, 4). Similarly, that of *Pycnopsyche lepida* would be F1d—not F4d, despite the period of emergence of the fly—in recognition of the fact that it is the period of growth which is significant in the stream biota. Some of these symbols have been incorporated into Figs. XIV, 3 and XIV, 4.

Finally Fig. XIV, 5 shows in diagrammatic form the supposed normal fluctuations in biomass and numbers of insects and Peracarida resulting from these cycles in an undisturbed stream. It also shows points where false findings of numbers may result from investigations made with coarse-meshed nets. These curves are speculative, but they are probably something like the true state of affairs. It will also be clear that they can occur in combination, but that however this combination is made the late spring and early summer seem likely to be periods of minimum biomass in all streams. Perhaps at this point it would be well to emphasize that this model, and indeed much of this chapter, applies to streams rather than to rivers. We have hardly any information on invertebrate cycles in large rivers, although it seems probable from consideration of faunal lists that non-seasonal cycles are much more important in them than are seasonal cycles.

CHAPTER XV

The fishes of running water

All groups and nearly all genera and species of the fishes which inhabit fresh water are to be found at times in rivers and streams. Indeed one can say with fair certainty that fishes moved into fresh water from the sea by way of the rivers, and that many still do so as a normal part of their life history.

Many of the ancient groups have survived only in fresh water, and most of these are inhabitants of rivers and streams. They include most of the cyclostomes (the world-wide lampreys), the Dipnoi (the lung fishes of Africa, Australia, and South America), the Chondrostei (the sturgeons and paddlefish of Europe, North Asia, and North America and the bichir of Africa), and the Holostei (the garpikes and bowfin of North America); and several whole orders of bony fishes are almost confined to fresh water. Such are the Salmopercae (the pirate perches and trout perch of North America), the Haplomi (the pikes and mudfish of the northern temperate regions), the Microcyprini (the top minnows and various viviparous fishes of Africa, South Asia, South America, and southern North America), and most important of all the Ostariophysi.

This large order is the dominant freshwater group and it contains over 5,000 species in three sub-orders: the Characoidei (the characins of Africa, South and Central America), the Siluroidei (the true catfishes of Asia, Africa, and South America, with one family in each of Europe and North America), and the Cyprinoidei (the minnows and carps of most of the world except South America and Australia, the loaches of Eurasia, and the suckers of North America).

In other orders of bony fishes whole families, or substantial sections of them, have moved into fresh water, or are associated with it at some stage in their life history. These include the shads (Clupeidae), the salmon, trout, grayling, and whitefish (Salmonidae and Coregonidae), and the smelts (Osmeridae) of the Northern Hemisphere, the southern trout, smelts, and grayling (Galaxiidae, Retropinnidae, and Haplochitonidae) of temperate South America and Australia, the eels (Anguillidae) of almost world wide distribution, the sticklebacks (Gasterosteidae) of the Northern Hemisphere, the sunfish and black basses (Centrarchidae) of North America, the rather similar Cichlidae of South and Central America, Africa, and India, the perches and darters (Percidae) of the Northern Hemisphere, the climbing perches (Anabantidae) of southern Asia and Africa, and the snout fishes (Mormyridae) and spiny eels (Mastacembelidae) of Africa and southern Asia. In addition some representatives of many other families have entered fresh water in many parts of the world, e.g. sand smelts (Atherinidae), burbots (Gadidae), bass (Serranidae), gobies (Gobiidae and Eleotridae), blennies (Blenniidae), sculpins (Cottidae), flounders (Pleuronectidae), drums (Sciaenidae), and mullets (Mugilidae). For a full discussion of the origin and distribution of the freshwater fish faunas of the world the reader is referred to Lagler *et al.* (1962). The object of listing the most important of them briefly here is to illustrate the great variety of types that may occur in rivers and the great differences between the various continents.

Despite the fact that many families must have inhabited fresh water for very long periods there are indications that evolution is continuing at a fairly rapid rate. In Italy various populations of trout seem to belong to a number of distinct forms, suggesting incipient speciation (D'Ancona and Merlo, 1959); in North America many species have local or regional races, and there are indications that the Atlantic salmon has quite recently evolved land-locked races because of climatic changes (Power, 1958). The perch of Eurasia and North America, which must have been in contact during the Pleistocene, are specifically distinct although still very similar (Weatherley, 1963), and there are indications that evolution at the generic level occurred in the Coregonidae during the Pleistocene (Smith, 1957). There are also racial differences in the physiology of populations of *Cyprinodon* in cool and warm springs respectively in California and Nevada (Sumner and Sargent, 1940). We can take it then that at least many species of fish are plastic in their responses to changes in the environment, and that this accounts for their, on the whole, much wider distribution than most species of running-water invertebrate. It also makes identification very much easier and is at least part of the reason why much more is known about fishes than about invertebrates.

In addition to the presence of true freshwater fishes and those that move in to breed, the lower parts of rivers are often invaded by marine species which remain for some time in fresh water. This applies, for example, in Europe to the flounder *Pleuronectes flesus*, and long lists of marine fishes have been recorded in such situations elsewhere: e.g. menhaden, anchovy, needle fish, and atherines in Virginia (Raney and

Massman, 1953), and sharks, gobies, and pipefishes in the Zambesi (Jackson, 1961). Even sting-rays have been recorded far upstream in Nigerian and South American rivers (e.g. Welman, 1948), and these occurrences indicate how easily fishes of recent marine origin can come to colonize freshwater. In Hawaii, for example, before the introduction of rainbow trout, the only stream fish was the climbing goby *Awaous guamensis*, a catadromous species which has clearly moved in from rocky sea-shores to the lower parts of the streams (Needham and Welsh, 1953), and in New Zealand the torrent fish *Cheimarrichthys fosteri* is certainly of independent marine origin (Allen, 1956).

There is also a number of species which regularly enter the lowest reaches of rivers to breed or to grow. These include, from the sea, the chum and pink salmons of the Pacific and some Galaxiidae in southern latitudes. Similarly many lake-dwelling species move into the lower parts of inflowing streams to breed. Such are lacustrine populations of the brown trout (Stuart, 1953*a*) and many species of Cyprinidae in Central Africa. We shall be considering these as well as longer distance migrants in Chapter XVII.

ANATOMICAL ADAPTATIONS OF RUNNING WATER SPECIES

Many species of fish, and of anuran tadpoles which in many respects resemble fishes, have anatomical adaptations which clearly fit them for life in rapidly flowing water.

It is well known that fast-swimming fishes in all aquatic habitats are streamlined and round in cross-section, and that laterally flattened fishes of the same size swim less rapidly. It has been shown experimentally that some common European species, the bream (*Abramis brama*), rudd (*Scardinius eryophthalmus*), perch (*Perca fluviatilis*), pikeperch (*Lucioperca lucioperca*), and pike (*Esox lucius*), are able to swim increasingly fast in that order (Ohlmer and Schwartzkopff, 1959); and each is slightly more rounded in cross-section than the one before. Moreover, it was found that the speed of flight from danger of the first four of these is just a little faster than the current speed in which they can hold station, ranging from 80 to 110 cm./sec. There is thus a definite correlation between the shape of the cross-section of the body and the ability of fishes to resist current. It has indeed often been noted, e.g. by Hubbs (1941), that the fishes which regularly *swim* in fast water are round in cross-section and streamlined. Such are, for example, the salmon and trout of the Northern Hemisphere, the Australian 'trout' (*Galaxias*), the piranha (*Serrasalmus*) of South America, and the large barbels, particularly *Barbus altianus* of the African rivers and *B. tor* of India. Nikolsky

(1933) showed long ago that in the rivers of Central Asia the swimming cyprinids of fast water are streamlined forms such as *Shizothorax* and *Leuciscus*, and that these are replaced in slower water by flatter forms such as *Rutilus* and *Abramis*. Many other studies elsewhere indicate the same correlation, which is illustrated in Figs. xv, 1 and xv, 2.

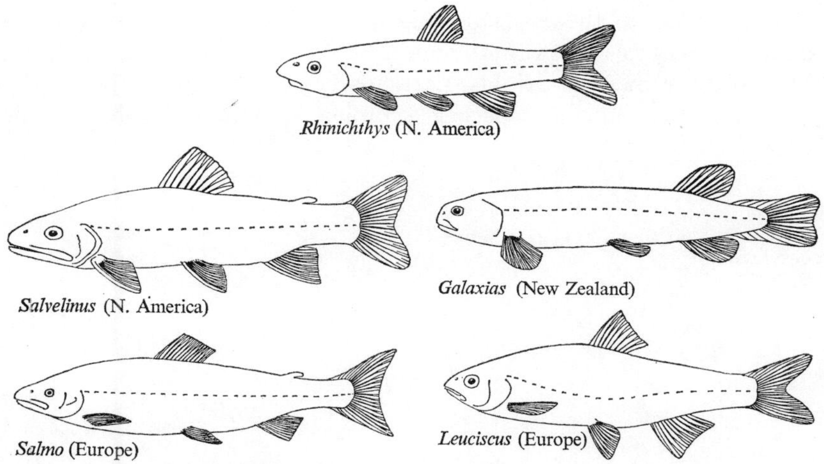

FIG. xv, 1. Outlines of fishes which live in fast-flowing water. Redrawn from various sources.

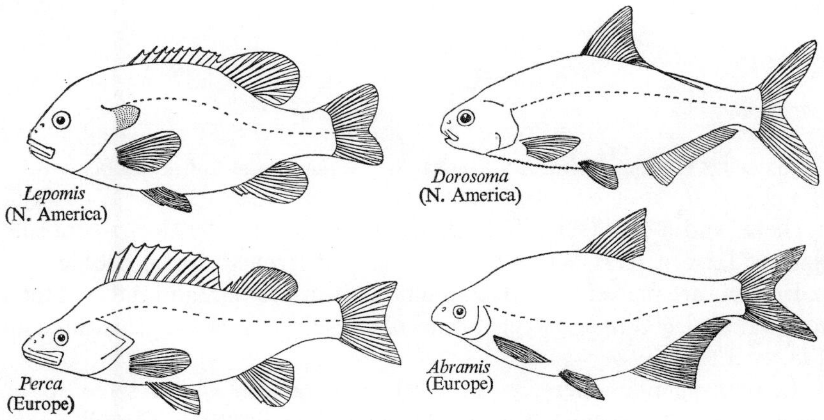

FIG. xv, 2. Outlines of fishes which live in slow-flowing water. Redrawn from various sources.

An interesting complication in the general picture seems, however, worthy of further study. Müller (1952) reports that two forms of *Salmo trutta* can be recognized in many streams in Germany. These are the relatively short and dark-coloured 'Waldforelle' and the longer and paler

'Steinforelle'. They often occur together and nothing is known of their genetics, but as they are differently streamlined they are, presumably, differently able to resist rapid current. Perhaps this is a case of polymorphism, such as is known in other animals, which enables the species to exploit a greater range of habitat. If so it is a further indication of the plasticity of freshwater fishes.

However, not all the fishes of rapid water are swimmers; many maintain themselves close to the bottom or in shelter under and behind stones. These include many cyprinids, particularly in Asia and Africa, and a great variety of catfishes, loaches, sculpins, darters and gobies, some

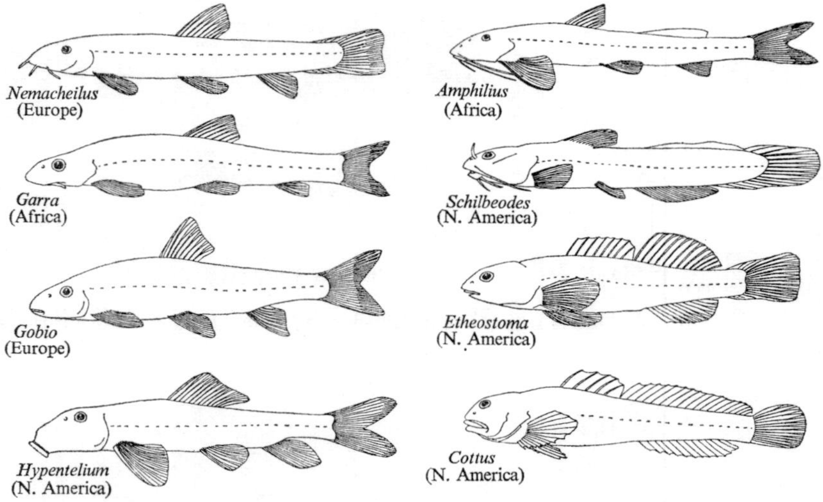

FIG. XV, 3. Outlines of small fishes which live on the bottom in fast-flowing water.

suckers, and a few Galaxiidae. Such fishes are usually dorso-ventrally flattened, or at least have flat bellies and an arched dorsal profile; the paired fins are placed laterally and tend to be muscular, and the eyes tend to be dorsal (Hubbs, 1941). Some of these features are illustrated in Fig. XV, 3.

In many genera there are also other characters which seem to be associated with the benthic way of life (Hora, 1922, 1930). The gill openings tend to be laterally placed rather than latero-ventral; this is inefficient for respiratory function, but these fish are always in well-aerated water. There is often a reduction in the scale covering, and perhaps correlated with this is the presence of poisonous spines such as occur on the opercula of *Cottus* and on the pectoral and dorsal fins of some catfishes; the little madtom *Schilbeodes* of North American rapids is well known to be able to inflict a painful sting (Clugston and Cooper, 1960). There are also changes in the position of the mouth which has moved more or less to the ventral side, so that when it is opened it does not disrupt the arched dorsal profile.

Hora (1930) also noted that in some Asian fishes of this type the lower lobe of the caudal fins is enlarged, and he suggested that this serves to raise the head when the fish swims forward. This is, of course, incorrect as has been established by study of the action of the heterocercal fins of sharks; such a fin raises the tail and drives the head down. In sharks this diving motion is corrected by the laterally placed pectoral fins which function as ailerons; presumably the same applies in stream fish. In any event for a small fish beset with all the above modifications, which inevitably make it an inefficient swimmer, to raise the head inadvertently out of the boundary layer into swift water a centimetre or so above the substratum would loop it head over tail and it would be swept away. The enlarged ventral lobe keeps the head down and prevents such disasters.

A further point observed by Hora is that in loaches of the families Cobitidae and Homalopteridae in Indian hill-streams the inner rays of the laterally spread pectoral fins are in constant motion. This serves to pump out water from below the fish and so to keep it down on the bottom. The same behaviour can be observed in sculpins and darters.

Many of the above characteristics of fishes which live on stony substrata are also shared by benthic fishes from sandy rivers. Such are a number of species of darter and some minnows of the genera *Notropis* and *Ericymba* which occur in sandy streams in south-eastern North America (Hubbs, 1941; Hubbs and Walker, 1942; Winn and Picciolo, 1960). They also have flat almost naked bellies, laterally placed fins, and dorsally located eyes; in addition they are very pale and translucent. It is also true that many active streamlined fish use their outspread pectoral fins to hold their station when they rest on the bottom in fast-flowing water. Kalleberg (1958) and Keenleyside (1962) observed that small salmon and trout do this, and that their fins are then so placed that the leading edge faces the current and is held downwards. The fin thus functions as a hydrofoil and presses the fish down, and more of the fin is brought into contact with the substratum the faster the current.

Another, internal, modification found in bottom-dwelling fishes is a reduction of the swim-bladder. All normal bony fish have swim-bladders and young trout fill them with air at the surface when they begin to feed and so require nearly neutral buoyancy (Tait, 1960). In bottom-dwelling fish, however, the swim-bladders are reduced and they are often divided and encased in bone (Hora, 1922); their contribution to lift is therefore eliminated. This reduction is found in many groups, e.g. cyprinids, loaches, catfishes, suckers, and darters (Hora, 1922, 1923; Burton and Odum, 1945; Nelson, 1961; Winn, 1958*b*). It is obviously a response to benthic life in fast water which has evolved many times, because in all the groups mentioned above there are other, unadapted, species which retain normal bladders.

It is even sometimes possible to show that closely allied species differ in the size of their swim-bladders, and that this is correlated with their way of life. In the lower Fraser River of British Columbia the leopard dace and the longnose dace (*Rhinichthys falcatus* and *R. cataractae*) occur together; but the former lives in shelter along the banks and the latter occurs in the current. The swim-bladder of *R. cataractae* is definitely smaller than that of *R. falcatus* (Gee and Northcote, 1963). Somewhat

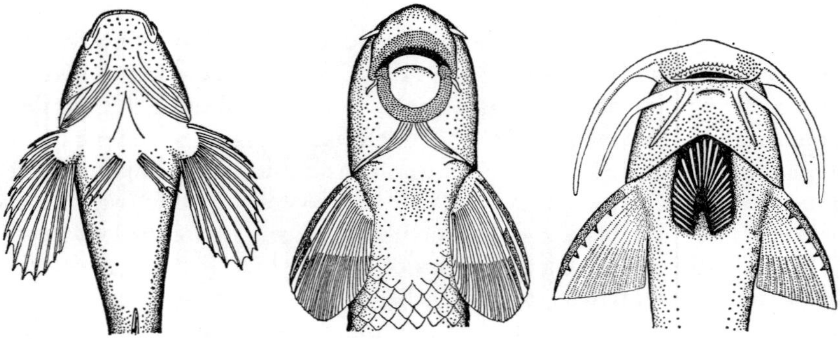

FIG. XV, 4. Ventral views of fishes with friction devices and suckers. *Glyptothorax* redrawn from Saxena and Chandy (1966).

analogously, frog tadpoles which live in torrential waters have very small lungs; this applies, for example, to *Rana afghana* in India (Hora, 1923).

Many fishes from torrential waters also have ventral suckers or friction pads with which they adhere to rocks. They have been reported from catfishes, loaches, gobies and cyprinids, and tadpoles from South America, Africa, and South-east Asia (Kleerekoper, 1955; Marlier, 1953; Jackson, 1961; Johnson, 1957; Annandale and Hora, 1922). These structures have been the subject of extensive study by Indian biologists, and there are several types of device (Hora, 1922, 1923, 1930; Saxena, 1961, 1962; Saxena and Chandy, 1966).

The simplest involves only the thickening of the anterior rays of the pectoral fins, and the bringing together and forward of the pelvic fins, so that all four come to form a friction plate below the body. This can be seen, for example, in *Cottus* (Fig. xv, 4) and in many species of darter and loach. Another, which is well displayed by various tadpoles and by the cyprinid genus *Garra*, involves the thickening and papillation of the front lip and the formation of a button-like disc with a papillated edge behind the downwardly directed mouth. The centre of the disc forms a hydraulic sucker, and with this and the front lip the fish (Fig. xv, 4) or tadpole (Fig. xv, 5, *Rana*) can crawl like a leech. The suction produced by

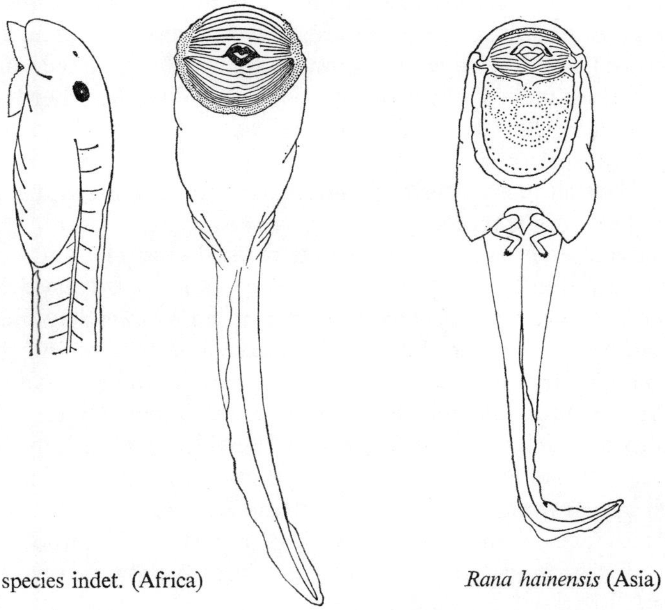

FIG. xv, 5. Tadpoles from rapid streams which have series of ridges on the lips, and, in *Rana*, a sucker behind the mouth. *Rana* redrawn from Hora (1930).

the disc is such that stones five times the weight of a tadpole can be lifted out of the water by using the animal as a handle.

In some fish, e.g. some species of loach, some catfishes, and also some tadpoles, the enlarged lips of the mouth themselves form suckers or friction plates (Fig. xv, 5), and in other catfishes, particularly the Asian Sisoridae, e.g. *Glyptothorax* (Fig. xv, 4), there is a series of ridges behind the isthmus. These are variously arranged in the different genera; they are covered with small spines, and the grooves between them can function as suckers. These small fish can therefore attach themselves very firmly to rocks.

Mouth suckers are also characteristic of lampreys, and these combined with their eel-like form enable the animals to occur in quite rapid water. It would seem that the sinuous motion of eel-like forms permits them to take the maximum advantage of any available cover on stony substrata, because, although they occur in many types of water, they are often abundant in swift stony streams. Fully grown European eels can in fact swim upstream against a current of 24 cm./sec. but they tire rapidly in currents faster than 50 cm./sec. (Lowe, 1952). Smaller specimens can swim against faster currents, up to 150 cm./sec. (Sörensen, 1951), and New Zealand eels are considered to be pests in trout waters because of their predatory habits. Lampreys are, however, poor swimmers (Bainbridge, 1960), and a strange fact about them, which may have something to do with their powers of movement, is that, unlike the other fish, they become shorter when they are starved as well as thinner (Larsen, 1962). They thus preserve their length/weight ratio, and this may be connected with the fact that many species do not feed at all as adults and are therefore always slowly starved.

We have already seen (p. 305) that fishes living on sand are pale-coloured and translucent. Those living on stony substrata are either dark-coloured, or, more usually, boldly mottled so that they blend into the background. This mottling is particularly conspicuous on sculpins, darters, and suckers such as *Hypentelium*, which live on riffles, and in the Pacific region there are mottled species of eel which match the stony substrata on which they normally live (Bertin, 1956). Many of these fishes can exert some control over their colour, becoming paler on pale-coloured backgrounds, and, while no stream fish can match the spectacular performance of the plaice, they can adjust fairly rapidly. *Cottus cognatus*, for instance, can change from dark to light in two or three minutes (Miller and Kennedy, 1946).

It is well known that fish avoid obstacles in the dark by echo-location with their lateral-line system, and that this is also their primary distance-receptor in muddy water. Many rivers such as the Missouri are always turbid, and that this is not a recently man-made phenomenon is certain because the early explorers of North America reported upon the muddiness of the rivers in the Great Plains. One can thus infer that fishes in these rivers have had to move about for millenia in at least very dim light. It is therefore perhaps not surprising that Moore (1950) was able to show that, when he compared species of *Erimystax*, *Platygobio*, and *Extrarius* from clear and from silty water, he found that in each genus the species from the silty habitat had larger or more numerous cutaneous sense organs. This is therefore yet another adaptation shown by fishes to a peculiar condition associated with running water.

BEHAVIOUR

There are several aspects of the behaviour of fishes which are important in their ecology in running water. These include such obvious things as their reaction to current and the seeking of shelter in the dead water behind obstacles, which even good swimmers do for much of the time. Every skilled angler knows that the best place to try for a trout is downstream of a large obstacle, and that the larger the obstacle the larger the fish he may catch.

Fishes accumulate lactic acid in their tissues very rapidly so they tire easily, which is why anglers 'play' a hooked fish which is too large to land directly. They actually accumulate lactic acid three to eight times as fast as mammals and take six to twelve times as long to get rid of it (Vibert, 1962); they thus need shelter fairly often, even though they can swim well in short bursts. Bainbridge (1960) discovered to his surprise that goldfish could keep swimming for longer than either the dace *Leuciscus* or trout, but perhaps this merely reflects the differences of the way of life of these species. The stream-dwelling trout and dace normally swim rapidly in short bursts and then rest; the goldfish in its pond is normally on the move much of the time, but it can stop when and where it pleases.

Another facet of behaviour is that swimming fish spend most of their lives facing into the current. This has the two-fold advantage of helping them to maintain their position and of making respiration easier. They have only to open their mouths to obtain a boost to their respiratory current. Indeed, so ingrained is this habit of facing upstream that salmon smolts and kelts moving downstream in swift water usually do so tail first (Hoar, 1951; Jones, 1959).

The reference above to large fish behind large obstacles brings us to the important subject of territoriality and dominance. Many species defend territories while they are breeding, but in still water it is rare for them to do so at other times. In rivers and streams on the other hand territoriality unassociated with breeding is quite common. It occurs among such riffle-dwelling fishes as sculpins and darters (Smyly, 1957; Seifert, 1963), and it is particularly obvious among the Salmonidae. Vibert (1962) suggests that it results from the need for shelter from the current, and it is significant that it ceased in Kalleberg's (1958) experimental streams when he reduced the current to slow speeds. Keenleyside (1962) also observed, while skin-diving among brook trout in natural habitats in New Brunswick, that although young trout are territorial in shallow rapids they tend to form shoals in pools and back waters. It also seems that no territories are defended by trout which have taken to living in lakes.

The territorial behaviour consists of snapping at and chasing off intruders, and the size of the area defended increases with the size of the fish. In a contest the largest fish usually wins, which is why the best shelter is usually held by the largest fish. It is the stream bed which is defended not the overlying water, and the size of the defended area depends upon the line of sight. This is shown by the fact that where there are many large stones or artificial barriers more fish amicably occupy the same area (Kalleberg, 1958; Stuart, 1953*a*). The behaviour begins very soon after the young fish emerge from the gravel and start to feed (Stuart, 1953*a*), and when there are too many fish for the available space some specimens fail to obtain territories and are continually chased around if they are unable to leave the area (Kalleberg, 1958). It also persists through the winter in very cold water (Needham and Jones, 1959), so it is a very constant feature of the life of those fishes which display it.

The opposite to territorial behaviour is shoaling, and this is a characteristic social pattern of many species, particularly in the Cyprinidae. It occurs also not infrequently in territorial fishes under some conditions, as we have already seen. Nikolsky (1963) suggests that the function of shoals is protection of the individual; the presence of a large number of specimens tends to confuse predators, and when one fish in a shoal is damaged the rest disperse to safety. He states that experimental studies show that single fish are eaten more readily by predators than members of a shoal. Clearly the instinct to shoal is very strong and it results in some alteration in behaviour. In their experiments on the swimming speeds of fish, Ohlmer and Schwartzkopff (1959) found that bream, dace, perch, and pikeperch swim more slowly when in shoals than they do when they are alone, and these authors suggest that this permits the smaller specimens to keep up, thus preserving the integrity of the group. In many species, however, e.g. the European minnow *Phoxinus* (Frost, 1943), there is a distinct tendency for shoals to be composed of specimens of fairly uniform size.

The fact that fishes such as young trout can switch from territorial behaviour to shoaling to suit conditions, indicates considerable flexibility in social pattern, and an interesting study of this type of flexibility has been made by Kawanabe (1958). *Plecoglossus altivalis* is a common fish of Japanese rivers which breeds in the lower reaches; the young spend some months in the sea and then return to the river to graze on periphyton. There the fish may shoal or be solitary, wander or be sedentary, and in the latter event they are usually territorial. They thus show the full gamut of social patterns and can adjust them to suit the circumstances. The numbers present in the River Ukawa vary from year to year, and Kawanabe was able to observe that when wandering fish, shoaled or solitary, are relatively scarce, the sedentary fish, which each defend about a square metre, keep them off the best feeding areas. They are then forced to feed in marginal areas and they grow more slowly than the sedentary residents. When, however, the wandering fish are abundant they tend to overwhelm the resistance of the sedentaries which then join the shoals, and all the fish grow equally well. They also, of course, grow equally well when both types are scarce and there is room for all. The existence of the two main patterns of behaviour therefore, in theory at least, permits the maximum exploitation of the available grazing. Surprisingly though, Kawanabe found that the growth in a particularly crowded year (over 5 fish/sq. m.) was as good as that in a less crowded one (less than 1 fish/sq. m.). Possibly in none of his years of observation did the number of fish exceed the carrying capacity of the environment. In such a year one would expect that only those few fish which successfully defended their territory would show satisfactory growth.

Territoriality is primarily intraspecific, and, while it has the negative effect of eliminating, driving away, or reducing the growth rate of, fishes that fail to maintain a territory, it also seems that association with other specimens is advantageous to those individuals that do survive and grow. Brown (1946*a*, *b*), in her studies of the growth of brown trout, *Salmo trutta*, under carefully controlled conditions, found that, although there were various social reactions which resulted in larger fishes growing faster than small ones, there was an optimum degree of crowding for the maximum growth rate. When crowding was less than this optimum the fish ate and grew erratically. Undoubtedly, therefore, some sort of social intercourse, in territoriality or in shoals, is needed by at any rate this one species, and it seems likely that this is general in all socially interacting species.

Territoriality is, however, also sometimes interspecific. The sculpin *Cottus gobio* will not share the same stone with the loach *Nemacheilus barbatula* (Smyly, 1957), and many darters interact with one another as if they were the same species (Winn, 1958*b*). For example, *Poecilichthys zonata*, *P. variatus*, and *Etheostoma blennioides*, the banded, variegated, and greenside darters, seem to share out the riffle areas between them in creeks in western Pennsylvania (Lachner *et al.*, 1950), and such dissimilar fishes as trout and sculpins have been seen to interact territorially in Sagehen Creek, California (Needham and Jones, 1959). It is, however, particularly within the Salmonidae that interspecific reactions have been noted (Kalleberg, 1958; Gerking, 1959; Vibert, 1962; Le Cren, 1965). *Salmo salar*, *S. trutta*, *S. gairdneri*, and *Salvelinus fontinalis*, the Atlantic salmon, the brown, the rainbow, and the brook trouts, seem to regard one another as the same species. In this complex the larger fish dominate the smaller and the healthy dominate the unhealthy; but, other things being equal, brown trout dominate the other species, particularly the salmon, and this must have important ecological consequences. Probably as a result of this interaction young salmon are usually found in deeper water than are young trout (Le Cren, 1965), but success in territorial behaviour does not necessarily result in success as a species. The brown trout dominates the rainbow trout territorially and in South Island and the southern part of North Island, New Zealand, it is the most abundant species; but in the northern part of North Island the rainbow trout is predominant (Allen, 1956). Both species were introduced during the second half of the nineteenth century and so have had a long time to establish themselves, and they illustrate very well how some other factor, in this instance probably climatic, can over-ride the effects of a purely social relationship.

In experiments it is also found that resident fish dominate newcomers (Vibert, 1962), and this has important implications in the stocking of streams. Miller (1952, 1954*a*, 1958) made a particular study of this problem in Gorge Creek, Alberta. He found that hatchery-reared 2- and 3-year-old cutthroat trout, *S. clarkii*, drifted downstream for about two weeks after their introduction into the stream, and that great numbers died quite soon. Only the older fish began to feed and very few of them survived the first winter. In a second experiment, fishes which had been reared in streams fared rather better than pond-reared fish, but not as well as wild fish which had been merely transplanted to, and confined in, another section of the stream. Even the last lost weight for the first thirty to forty days after transplantation, and Miller concluded that this, and the poor performance of the introduced fish, was attributable to harrying by the original residents. He noted that the lactic acid content of the hatchery-reared fish rose to high levels, and he suggested that they died of simple exhaustion resulting from continually being chased from occupied territories. Lack of success in finding territories may also account for the ease with which newly introduced trout are caught by anglers. In Belgium for example, Timmermans (1961*a*) recorded up to 60 per cent of stocked legal-sized trout as being caught during their first season, many of them in the first few days after introduction, as compared with only 24–8 per cent of stocked roach, *Rutilus*, a non-territorial fish.

A final aspect of fish behaviour which is important in running water, is the existence of diurnal rhythms. Many species of stream and river fish are more active at night or in twilight than they are in the daytime. These include trout (Needham and Jones, 1959), mountain whitefish *Prosopium williamsoni* (Sigler, 1951), the common European loach and sculpin *Nemacheilus barbatula* and *Cottus gobio* (Smyly, 1955, 1957), and some minnows. Similarly, the spotfin shiner, *Notropis spilopterus*, the sand shiner, *N. stramineus*, the bigmouth shiner *N. dorsalis*, and the sucker-mouth minnow, *Phenacobius mirabilis*, have been observed to move into shallow water in the Des Moines River, Iowa, to feed at night (Starret, 1950*a*); and many migrating fishes, e.g. ascending larval eels and descending fully grown adults (Bertin, 1956; Lowe, 1952) and descending salmon smolts (Jones, 1959; Hoar, 1951), move only at night. Movement of adult silver eels is inhibited by moonlight or artificial light so it is clearly a direct response to darkness. On the other hand many species such as the common shiner of North America, *Notropis cornutus*, are very active in the day-time, and swimming activity and feeding may not occur at the same time. Young salmon and trout, for instance, which swim actively at night, do most of their feeding early and late in the day, with little activity of any kind round about midday (Hoar, 1942; Kalleberg, 1958), and it has been shown that yearling rainbow trout very definitely seek and remain under cover when their habitat is well lighted (McCrimmon and Kwain, 1966). As it is also known that the removal of natural cover decreases populations of trout (Boussu, 1954) it can be seen that this diurnal pattern of behaviour is important to them.

THE COLLECTION OF FISHES FOR STUDY

The problems of collecting representative samples of fish for research purposes are always great because all methods of catching fish are to some extent selective of species or size, and fishes of particular size- or age-classes tend to shoal together. This difficulty is bad enough in lakes, but it is worse in rivers and streams for various reasons. Not only are there large local differences between swift and slow reaches, or riffles and pools, which tend to segregate species and size groups, but many species migrate along rivers at different stages of their life-histories. Thus the situation is rarely static, and it is difficult, if not impossible, to discover in detail the population structure of many species (Cleary and Greenback, 1954).

The chief methods which have been used for research purposes are described by Lagler (1952) whose book is a mine of information on fishery techniques. The most commonly used techniques in running water are netting, electro-fishing, and poison, and more rarely trapping or draining sections of stream. We will consider each of these briefly.

Allen (1951) based his well-known study of the trout in a New Zealand stream on a combination of seine-netting and stop-nets. The sink-line of the seine was made of chain as this lies more closely to the bottom than does a rope with weights at intervals. Shallow reaches were seined down to the stop-net, and deeper water was netted upstream; special areas of shelter were netted individually and netting was continued until no more fish were caught. This technique probably lands a large proportion of the fish catchable with the mesh-size used, but studies made in Michigan suggest that seining for trout is only 80–90 per cent effective as indicated by subsequent poisoning (Shetter and Hazzard, 1939; Shetter and Leonard, 1943). It has also been shown that seining downstream catches more fish than seining upstream, that the catch varies inversely with the water-level and that it is greater at night than by day (Paloumpis, 1958*a*). There are, therefore, quite a number of variables and uncertainties in netting even a small stream.

In recent years a popular method of catching fish in running water has been electro-fishing. A generator is used to produce 200–300 volts between electrodes, one of which, appropriately insulated, is carried by the operator; the other lies in the stream. The current is either alternating or direct, but A.C. has not found great favour as it merely stuns the fish, which are then easily overlooked (Cleary and Greenbank, 1954; Timmermans, 1954). Direct current has the remarkable effect of drawing the fish towards the positive pole, which is therefore the one carried by the investigator.

Vibert (1963; Vibert *et al.*, 1960) has explained in neurophysiological terms how the attraction to the positive pole occurs. At increasing distances from the anode there are decreasing voltages, so at a greater distance from the electrode the voltage drop over the length of the fish is smaller. At some distance away therefore, a fish facing the electrode experiences only a small voltage drop (+ve to −ve) along its body. This inhibits the long motor-fibres running backwards from the brain to the tail muscles, and swimming ceases or becomes weak. A little nearer the electrode the shorter sensory nerve-fibres running in series forward to the brain become facilitated by the voltage change along their length. They thus force the brain to override the inhibition of the motor-fibres, and the fish swims forwards vigorously. As it approaches the electrode the stronger inhibition of the increasing voltage-drop along the motor-fibres overrides the brain and the fish becomes limp. But it is usually still moving forward as the result of its previous swimming, although by now on its side or back, so it goes on increasing the voltage drop along its body as it nears the electrode. At this stage the electrical potential gradient becomes steep enough to facilitate the short oblique sensory-fibres of the spinal nerves, and unco-ordinated spinal reflexes cause jerky swimming to resume and to carry the fish even further. Finally tetany sets in very near the electrode when the short muscle-fibres are directly excited.

Should the fish in the first position be facing away from the positive electrode the motor nerves are directly stimulated, but the short sensory-fibres are partly inhibited and slow swimming away from the electrode results. Should it be nearer the electrode when the current is switched on complete tetany results, and the chances are that the drifting body of the fish will be turned so that it is no longer pointing directly away from the anode. Then it, and any fish lying across the electric field, is turned towards the anode because the body is bent by stimulation of the spinal nerves facing the positive pole. The probability is therefore that a large proportion of the fishes are induced to move towards the anode, where they are easily caught with a hand net.

Fish species, however, vary in their reaction to an electric field. Strong swimmers such as trout move as described above, but many bottom-dwelling fish hide before swimming, and so may be overcome by tetany, as the electrode is brought close, before they move. Small fish, because they are shorter and so have a smaller voltage-drop along their bodies, are less affected than large ones (Timmermans, 1954). The critical length at which efficiency falls steeply seems to be about 17 cm. for trout (McFadden, 1961). Efficiency also varies with the colour and hence the visibility of the stunned fish, and with the conductivity and the temperature of the water, which both affect ion-flow and hence the extension of the field about the electrode. Alabaster and Hartley (1962), who based their estimates on mark and re-capture experiments (see below), calculated that the efficiency rose to about a 68 per cent catch at a conductivity of 279 μ-mho, although it was variable. They found that in trout waters of lower conductivity in Cornwall, England, an increase in the voltage improved the efficiency.

It is necessary therefore to match the machine to the water and to the species to be caught (Vibert *et al.*, 1960; Burnet, 1952*b*), and the process of fishing is itself something of an art. Merely to move the electrode about scares fish away, as they are stimulated to flight by the irritation of the edge of the field. The current has therefore to be switched on and off, or better, the electrode repeatedly raised from the water and replaced elsewhere; and to fish effectively one must have a good knowledge of where to expect the fish one is seeking and how to draw them out of cover without just stunning them where they are hidden.

The technique though has many advantages; it is flexible and can be used where no net can be fished, and the fish soon recover and can be freed unharmed. There have been reports of spinal damage caused by excessive tetany, but these are mostly caused by A.C., and a careful study of Ontario streams has shown that spinal abnormalities of the type which have been blamed on electro-fishing are common in rainbow trout and that there is no convincing evidence of damage by electro-fishing (McCrimmon and Bidgood, 1965). I can indeed state that during many hours of using D.C. apparatus I have never seen a fish damaged except by being trodden upon while it was helpless.

One important restriction is, however, that it is really only possible to electro-fish in water in which one can wade, and as the operator must be protected from the electric current by rubber waders there must be no great risk of falling. In an attempt to overcome this limitation Funk (1949, 1957*b*) has designed an 'electric seine', which is in effect a series of alternating positive and negative electrodes hanging like a widely spaced bamboo-curtain from floats on the water surface. This is dragged through deep water by the boat carrying the generator and by operators on the bank, or by two boats; another boat containing the netsman cruises up and down behind it. This device is probably worthy of more study and use than it has received.

Another, rather drastic, technique for collecting fish is to use poisons. The most common of these is rotenone, which is relatively non-poisonous to mammals and can be removed from the water by the addition of appropriate amounts of potassium permanganate. It is thus possible to add rotenone to a river or stream, allow it to pass with the current through a given reach and then prevent its doing further damage downstream by destroying it with permanganate. The affected fish die fairly soon and can be collected, and it seems often to be assumed that it is very efficient. Krumholz (1950), however, in a thoughtful discussion, expresses doubts about its supposed efficiency even for work in ponds, and Boccardy and Cooper (1963) found that it was only 64–94 per cent efficient in collecting brook trout from several streams in the Allegheny Mountains. They based these estimates on mark and recapture methods using fish caught with an A.C. electro-fisher, and they concluded that rotenone was the more efficient method.

In spite, however, of its ease of operation, it is, as has been said, a drastic method and is one which should be used only rarely. It is difficult in practice to avoid some loss of the poison downstream, and it kills great numbers of fish, often far in excess of the requirements for a reasonable sample. Moreover, rotenone is also very poisonous to arthropods; it was used as an insecticide before the advent of synthetic compounds, so the unrecorded and unnoticed damage to the whole biotic community must be very great. It is to be hoped that some of the, now many, fishery workers, particularly in North America, who use this lethal material far too readily, will take note of these strictures and use it less often and with more respect for our diminishing natural resources. Its use can only be morally justified in very small amounts in remote areas, with great care and a sense of personal responsibility.

Traps, although often used in Africa, Asia, and South America for the collection of river fish for food, have rarely been used for research work in rivers. This is because they are so obviously selective of species and size-groups, and they depend on fish movement which varies from place to place and from time to time (Cleary and Greenbank, 1954). They are nevertheless sometimes useful for catching large specimens of sluggish fish from deep or difficult sites, and they have been effectively used in studying the reactions of fish to pollution (Allan *et al.*, 1958).

Just occasionally it is possible to drain a section of stream by diversion, and this, of course, really does permit one to assess the whole fauna. Usually, however, one has to rely on the relatively inefficient techniques outlined above. Several studies have compared the efficiencies of various methods. Some have already been mentioned, and others are those of Funk (1958; Funk and Campbell, 1953) comparing seine-nets, electric seine, and hoop-nets, Larimore (1961) comparing electro-fishing with subsequent draining and poisoning of remaining pools, and Scott (1951) comparing hoop-nets, wire baskets, and rotenone. In the last study, carried out in the Coosa River, Alabama, each of the three methods failed completely to reveal the presence of one important species. There is indeed little doubt that differences in behaviour, colour, and habitat of the various species inevitably result in different catches by different methods.

How then is one to attempt to find out how many specimens are present? Two methods have been suggested and have been used in many field studies (Timmermans, 1957). One is to apply a particular technique over and over again in the same reach and to plot the catch per unit effort for each species, or size class, separately against the number of trials. This will give a curve which tends towards zero, and from this one can estimate, by extrapolation, the total number that would have been caught if the collections had gone on until no fish were left. Another is to catch as many fish as possible, mark them in some way—fin-clipping is often used—return them to the water, and, after an interval, fish again. Then from the proportion of marked fish in the second catch, or third, fourth, etc., after marking each unmarked fish each time, it is possible to calculate the total population.

Both these methods rely, however, on the assumption that any given fish has an equal opportunity of being caught by the method being used, and the second assumes that after an appropriate interval the marked fish become randomly distributed among the population at large. Neither is justified. The fish which is cryptic or difficult to catch by any one of the techniques remains so no matter how often that technique is applied, and when an electro-fisher is being used it is known that any fish shocked but not collected is less susceptible to being caught by this method than it was before (Libosvársky, 1966). Moreover, as we shall see, many, probably most, fish in rivers have homes to which they return. So a fish caught the first time because its home is in a susceptible place is more likely to be caught again than one which lives in an inaccessible spot. There is always therefore a tendency to underestimate the population. Thus the problem of really adequate sampling of fishes, like most other problems of biological sampling in running water, remains to be solved.

CHAPTER XVI

Ecological factors affecting fishes

Many ecological and, to a lesser extent, physiological studies have provided some insight into the factors which control the occurrence of running-water fishes, their interspecific relationships, their numbers, and their growth rates. Many of the climatic, or abiotic, factors such as oxygen content of the water, temperature, and substratum interact with one another, but, for convenience, we will consider them separately here, followed by a consideration of biotic factors, the numbers of fishes, and the ecological value of study of their rate of growth. The important subjects of migration, breeding, and food are dealt with separately in Chapters XVII and XVIII.

OXYGEN

The major work on the use of oxygen by fishes is that of Winberg (1956), who makes the point that there is no reason to *expect* fishes from different environments to show differences in their basal rates of oxygen consumption, but who suggests that one would anticipate that the amounts used in active life would vary from environment to environment. This is in fact found to be so; Schlieper (1952), for example, showed that the eurythermic chub, *Squalius cephalus*, from rivers is less active and uses less oxygen than the lively stream-dwelling rainbow trout.

Winberg also showed that fishes—he cites the carp in particular—become acclimated to the habitat or temperature in which they live, and that specimens from running water are more prodigal in their use of oxygen than those from still water, where it may at times be in short supply. As we have seen, running water is usually well oxygenated and only rarely are low levels of saturation encountered. We have already mentioned in Chapter III the exceptional example of the River Ob in Siberia where oxygen falls to low levels in the winter because of drainage from the marshes of the Vasiugan basin (Mosevich, 1947; Mossewitch,

1961). There the oxygen begins to decline even before the water is quite frozen over; and, when it reaches 50–5 per cent saturation, salmonids and the sterlet, *Acipenser ruthenus*, move out into the brackish Ob Bay. As ice-formation increases the oxygen continues to fall, and between 30 and 20 per cent saturation, bream, perch, and dace, *Abramis*, *Perca*, and *Leuciscus*, move to areas where the water is still open. At 21–14 per cent saturation roach and pike, *Rutilus* and *Esox*, follow them, and finally carp, *Cyprinus*, move out at saturations below about 14 per cent. Thus all the fishes leave the affected areas during a period, which extends over about a month, and long reaches of iced-over water are deserted by the time really low saturations are attained. The fish by then have congregated into a few areas where they remain fairly inactive throughout the cold winter. So here, in this one exceptional situation, although some fishes die, most of the population evades the consequences of lowered oxygen in a way that is not open to the fishes of frozen-over still water.

Severe lack of oxygen occurs in only one other type of environment in running water—the upper reaches of streams fed by de-oxygenated ground water. This lack does not, of course, persist far downstream because of rapid solution from the air, but it can often be seen that the upper parts of spring streams are devoid of fish, probably for this reason. In Florida, however, it has been observed that the top-minnows *Gambusia* and *Mollienisia* run far up de-oxygenated spring streams and obtain oxygen by gulping air at the surface in the same way as do pond fishes in hot weather (Odum and Caldwell, 1955).

Normally therefore oxygen is no problem, except for species in which the rate of metabolism rises very rapidly with temperature. Trout are such fishes, and among them the brook trout, *Salvelinus fontinalis*, has been the subject of much study. Graham (1949) measured the oxygen consumption of this species at various temperatures, both when the fish were resting and when they were kept actively swimming in a rotating chamber. He found that the minimum resting-rate of consumption rises from 5 to 25·3 °C., the lethal temperature, but that the active rate rises only up to 19 °C. and then declines. Thus the difference between the two, representing as it were the extra metabolism which the fish can mobilize for activity, is maximal at about 19 °C. This would therefore appear to be the optimum temperature in respect of oxygen relations. In fact Graham found that the maximum swimming speeds which his experimental fish could maintain occurred in the temperature range 16–20 °C., which was where resting and active oxygen-consumptions were most different. This was therefore in good agreement with the experimental results.

Brook trout normally live in well-aerated water and so are prodigal of oxygen, and Graham found that their rate of oxygen uptake becomes 'dependent' at air saturations only slightly below 100 per cent. That is to say it falls as the oxygen concentration falls, and, because the gas is less soluble in warmer water and the oxygen consumption of the fish is higher, this dependence becomes more marked the higher the temperature. So, while less than 75 per cent air saturation reduces activity even at low temperatures, full saturation is necessary for full activity at 20 °C. and above. Thus any conditions which even slightly decrease oxygen tensions in warm weather are deleterious for this species, quite apart from its intolerance of temperatures above 25 °C. Many other cold-water headstream species are probably limited in this way. It is, for instance, known that the sustainable swimming speed of young coho and chinook salmon, *Onchorhynchus kisutch* and *O. tshawytscha*, varies inversely with the oxygen concentration of the water (Davis *et al.*, 1963).

However, as is usual with laboratory studies, one should be cautious about applying numerical results in the field. Power (1959) tried the experiment of confining freshly caught young salmon in respiratory chambers fed with water from, and submerged in, the river from which they came. He found that their basal respiratory rate was high for the first few hours after handling, but, as they were in a near-natural environment, he could hold them for longer periods than had been possible for most experimenters. He then found that the basal rate fell to about three-quarters of that reported from laboratory trials. Thus while there is no reason to doubt the *principles* of the findings from physiological studies it is possible that *figures* should be accepted with reserve.

While no work appears to have been done with adult fish on the effect of current on oxygen consumption, such as has been reported on invertebrates (Chapter IX, p. 163), there would seem to be little doubt that the speed of water flow and the limiting concentration of oxygen are inversely related for fishes as they are for other organisms. This has been shown to be true for the eggs of steelhead trout, anadromous *Salmo gairdneri*, and coho and chinook salmon, of which the embryos become larger and hatch earlier the higher the oxygen content of the water and the faster its flow (Silver *et al.*, 1963; Shumway *et al.*, 1964). Even at high oxygen concentrations the current still continues to exert its effect, and this is, of course, in agreement with the findings on many invertebrates. Current therefore, like temperature, is intimately linked with oxygen.

TEMPERATURE

Temperature is always an obvious ecological factor, and for stream fishes it is an important one which limits both broad geographical distributions and local occurrences within a single watercourse.

Some species can tolerate remarkably warm water. *Cyprinodon* and *Crenichthys* are found in warm springs in the south-western United States with water temperatures ranging up to 33 and 37 °C. respectively (Sumner and Sargent, 1940), and *Barbus collensis* has been collected from water between 37 and 38 °C. in streams fed by the hot springs of Hamman in Algeria (Mason 1939). Temperatures in the low 30s must be commonly endured by tropical species but some fishes from the running waters of middle latitudes can also support temperatures approaching these levels. Thompson and Hunt (1930) report that several species of minnow, the creek chub *Semotilus atromaculatus*, the white sucker *Catastomus commersonii*, and the bullhead *Ameiurus melas* survive temperatures up to 32 °C. in pools in intermittent streams in Illinois, and carp will tolerate 36–7 °C. (Huet, 1962).

Many species are, however, more limited in their tolerance of warm water. The American and European perches, *Perca flavescens* and *P. fluviatilis*, seem to have an upper limit of tolerance of around 30 °C. (Weatherley, 1963), many European cyprinids die at 29–31 °C., the pike *Esox lucius* survives only to about 29 °C. (Huet, 1962), and trout are very definitely cold-water fish. The brook trout cannot for long sustain temperatures above 25·3 °C. (Fry *et al.*, 1946) and for the rainbow trout this figure is 24·5 °C. (Huet, 1962).

In temperate climates most fishes of running water are annually exposed to water at 0 °C., and some at least remain active at these low temperatures. Needham and Jones (1959) observed that Californian trout continue to feed at near freezing temperatures, but the swimming speeds of young salmon are reduced by very cold water (Davis *et al.*, 1963). However, despite the fact that the fishes of temperate rivers and streams can endure a fairly wide range of temperature they need to adjust themselves and they cannot withstand sudden changes. Fry *et al.* (1946) showed this clearly in their careful study of the temperature tolerance of the brook trout.

This species can withstand temperatures from 0 to 25·3 °C., but it needs acclimation. For instance the upper tolerable limit is raised by about 1° for every 7° rise of acclimation temperature up to 18 °C. where it levels off at the absolute limit of 25·3 °C., and fishes kept at 24 °C. or above cannot tolerate 0 °C. (Fig. XVI, 1). The response is therefore adjusted to, and by, the seasons, and in some other species the adjustment is greater than it is in the brook trout.

Fishes from warm spring-streams also clearly adjust in this way, as the upper tolerable limit for specimens of *Barbus collensis* depends upon the temperature of their locality of origin (Mason, 1939). Such adjustments are probably connected with basic alterations of metabolic rate (Sumner and Sargent, 1940). Even fish eggs are known to adjust in this way;

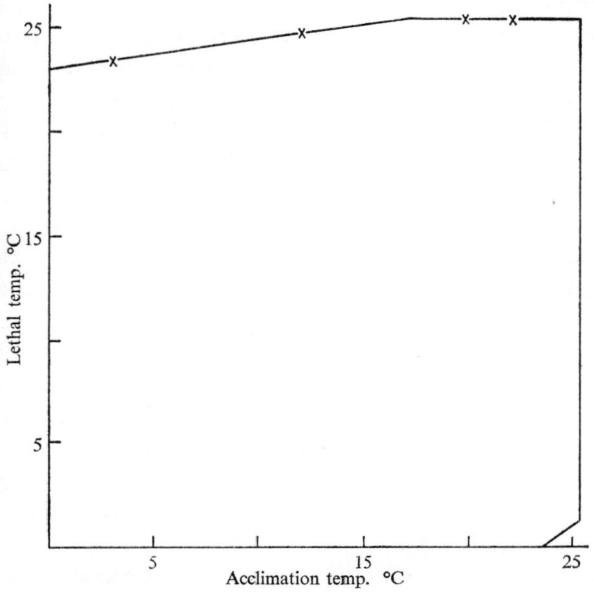

FIG. XVI, 1. The relationship between the lethal temperature and the temperature to which specimens of *Salvelinus fontinalis* have been acclimated. Redrawn from Fry *et al.* (1946).

Hubbs (1964) found that, when he compared the development of eggs of the greenthroat darter, *Etheostoma lepidum*, from a spring with a constant temperature with those from a normal stream, the latter did much better in fluctuating temperatures than the former. He attributed this to greater resistance to thermal shock of the eggs from fishes which are normally exposed to variations in temperature.

Limits of temperature tolerance clearly control distribution in the broad sense, for example the gudgeon, *Gobio gobio*, seems to be limited in Europe to the area between the July isotherms of 15 and 27 °C. (Bernet, 1960). They also control it in the narrow sense; trout rivers in Ontario are distinguishable from bass, or centrarchid, rivers by a maximum temperature of about 24 °C. (Ricker, 1934). They similarly seem to determine, at any rate in part, the relative distributions of closely allied species; *Cottus bairdi* lives in warmer waters in streams along the north shore of Lake Superior than does *C. cognatus* (Smith and Moyle, 1944). Probably, in this last and similar instances, interspecific competition is directly affected by the temperature regime rather than that each species lives up to, or down to, its physiological limit. This as we have seen probably applies to invertebrates also (p. 233).

It will also be noted that the lower limits for distribution, e.g. 15 °C. for gudgeon and 24 °C. for bass, bear no relation to the lower limits of *tolerance*, since winter temperatures are much lower. They are the limits for success as a species not as an individual, and they probably represent the lowest temperature at which some activity, such as successful breeding or adequate feeding, can occur. It was long ago noted that trout introduced into the very cold streams of the Glacier National Park, Montana, fared badly, probably because of inadequate feeding activity (Hazzard, 1933). Similarly, Tarzwell (1939) found that the Norris Dam of the Tennessee Valley Authority had eliminated many warm-water species from the Clinch River, because during the summer the water was discharged from below the thermocline, at a temperature of about 10 °C. In both these examples the fishes were not exposed to lethal temperatures but it was nevertheless too cold for successful life. An example of the inverse phenomenon is the upstream displacement of trout populations in streams in the Smoky Mountains, Virginia, following clearance of the forests and the resulting summertime warming of the water (Burton and Odum, 1945).

The timing and extent of temperature changes are also important, as they often control breeding. Although the so-called warm-water fishes can tolerate winter temperatures, the water must warm up early enough and far enough to allow them to breed at the proper time. Similarly temperatures must fall below 14·4 °C. at some time of year, or in some accessible area, to allow trout to breed (Huet, 1962). In their natural habitat in temperate climates this is ensured by the seasons, but where they have been introduced into the tropics, as on Mount Kenya, they must have unimpeded access to the cold waters at high altitudes (Van Someren, 1952). Under these conditions 14·4 °C. still remains the upper limit for successful reproduction. Similarly, trout do not breed at altitudes below about 1,800 m. in India, and this is undoubtedly related to temperature (Chacko and Ganapati, 1952).

Influxes of groundwater, which keep up winter temperatures and so make favourable areas in cold streams, may also have important local effects. Benson (1953) found that brook and brown trout are commoner in some parts of the Pigeon River, Michigan, simply because they congregate and breed in areas kept warm by groundwater, and being rather sedentary fish they tend to remain there all the year round. The chum salmon, *Onchorhynchus keta*, is similarly attracted to such areas in Siberia to spawn, and this has the advantage of ensuring that the eggs do not freeze in that severe climate (Nikolsky, 1963). Groundwater inflow probably, therefore, enables this salmon to breed in waters that would otherwise be closed to it.

Growth rate is closely related to temperature, and many, probably

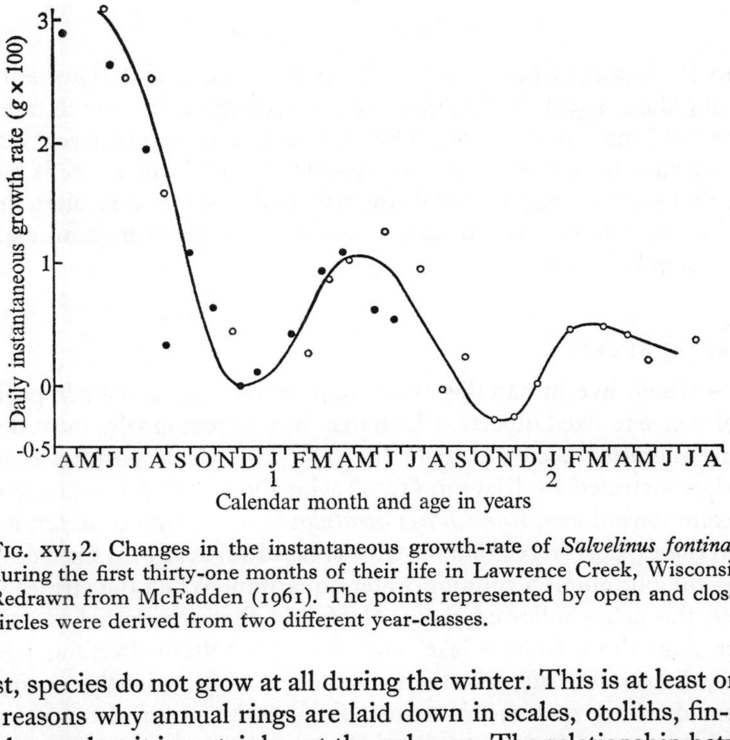

FIG. XVI, 2. Changes in the instantaneous growth-rate of *Salvelinus fontinalis* during the first thirty-one months of their life in Lawrence Creek, Wisconsin. Redrawn from McFadden (1961). The points represented by open and closed circles were derived from two different year-classes.

most, species do not grow at all during the winter. This is at least one of the reasons why annual rings are laid down in scales, otoliths, fin-rays, and bones, but it is certainly not the only one. The relationship between temperature and growth is, however, by no means simple, as has been shown by several studies on trout. Brown (1946*b*) found that when she kept brown trout at a uniform temperature of 11·5 °C. they still maintained an annual growth cycle, with the growth-rate falling in the autumn, rising to a maximum in the spring and falling a little during the summer. Even in the uniform temperatures of the high-level equatorial streams on Mount Kenya, both brown and rainbow trout produce annual rings on their scales (Van Someren, 1952), and so perhaps they do not grow as fast during some part of the year as at others. Careful study of the growth of brook trout in Michigan streams shows that it is faster from April to June then it is thereafter, even though temperatures continue to rise through the summer (Cooper, 1953). Many similar studies could be cited, and an excellent example is that of McFadden (1961), who made a detailed investigation of the brook trout in Lawrence Creek, Wisconsin. Fig. XVI, 2, from his paper, shows very clearly the seasonal change in growth rate of this fish and the way in which it rises in the spring and falls through June to December. The instantaneous growth rate (g) is calculated from the formula.

$$g = \log_e Wt - \log_e Wo$$

where Wt is weight (in grams) at the end of time t, and Wo the initial weight (Allen, 1951). In this instance the instantaneous growth rate was calculated from samples taken at monthly intervals, and two-year classes of trout have been used to plot the points on which the curve is based. Note that g can be negative in young fish, and that it is very often designated by the letter i, as on p. 425, to avoid confusion with grams and the gravitational constant.

CURRENT SPEED

Fishes which live in running water apparently maintain their position by reference to fixed objects, which may be one reason why most downstream migration occurs at night (p. 313). This visual position-holding was demonstrated by Clausen (1931) who showed that blacknose dace and common shiners, *Rhinichthys atratulus* and *Notropis cornutus*, maintained their places in relation to stripes placed vertically up the sides and across the bottom of a running-water aquarium. When the stripes were moved the fishes followed them. He found that stream fishes reacted better than those from a lake, and that the bottom-dwelling johnny darter, *Etheostoma nigrum*, did not react at all. Obviously, therefore, bottom-dwelling species maintain their place by physical contact with the substratum, and we have seen that swimming species go down to the bottom at times (p. 305). The mere removal of this simple aspect of behaviour would therefore probably result in downstream displacement, and this seems to be what happens to migrants. As has already been mentioned (p. 308), it is probable that fish species living in permanently muddy rivers make greater use of their lateral-line systems than their eyes, and this must be important in position-holding. In any event all the species living in running water must hold their position in the face of the current and this often involves swimming against it.

We have seen that fishes are easily fatigued (p. 309) so even streamlined species which are well adapted to fast-flowing water cannot swim rapidly for long periods. The maximum speed of swimming of Kamloops trout in fact falls from 440 to 89 cm./sec. after only three minutes of continuous swimming (Vibert, 1962), and all fishes spend most of their time resting in shelter. But there are differences between even closely allied species. Several observers have reported that brook trout and brown trout spend more time swimming than do young Atlantic salmon (Stuart, 1953*a*; Kalleberg, 1958; Keenleyside, 1962), and this must be of significance in the ecology of these two types of fish which so frequently occur together.

Studies in central Germany have indicated that brown trout grow better in slower flowing streams than in very swift ones, and better in natural streams with varied depths and shelter than in straightened artificial channels. Their growth rate is also decreased by spates (Tesch and Albrecht, 1961). So even this very streamlined fish expends much energy in fighting the current. It is not surprising, therefore, to find that different species are characteristic of waters of different velocities, and that flattened species are absent from fast water. Huet (1962) gives the following list of the maximum swimming-speeds of a number of species of European fishes and these are closely related to the relative speeds of flow of places where they are to be found.

Salmon, *Salmo salar* (adult)	800 cm./sec.
Trout, *S. trutta*	440
Nase, *Chondrostoma nasus*	350
Chub, *Squalius cephalus*	270
Barbel, *Barbus barbus*	240
Souffe, *Telestes soufia*	180
Bream, *Abramis brama*	55–65
Bleak, *Alburnus alburnus*	60
Tench, *Tinca tinca*	45–50
Pike, *Esox lucius*	45
Carp, *Cyprinus carpio*	40

As we shall see in Chapter XX (p. 383), a zonal classification of European running waters is based very largely on the relationship between the current and the common species of fish. In other parts of the world it is also well established that the various species are distributed according to the current. For example, after his survey of large rivers in Alabama, Swingle (1954) reported a change in important species from swifter to more sluggish waters, from drum, *Aplinodotus grunniens*, to gizzard shad, *Dorosoma cepedianum*, to blue and channel catfish, and smallmouth buffalo, *Ictalurus furcatus*, *I. punctatus*, and *Ictiobus bubalus*, to carp, bluegill, and crappies, *Cyprinus carpio*, *Lepomis macrochirus*, and *Pomoxis* spp. in fairly still places. Similar lists of fishes characteristic of waters of decreasing rates of flow in North America can be gleaned from the papers of Gersbacher (1937), Gerking (1945), Martin and Campbell (1953), and Lachner (1956), and from such general works on freshwater fishes as Hubbs and Lagler (1958). Minckley (1963) has produced a table showing this kind of distribution in the Ohio River drainage in Kentucky (Table XVI, 1).

FLUCTUATIONS IN DISCHARGE

The instability of flow regime in many rivers makes them difficult habitats for fish. We have just seen that spates are deleterious to trout. Eggs and small specimens can be destroyed by unusually high water in the winter (McFadden and Cooper, 1962), and even large trout and salmon may be killed by exceptional floods such as those caused in the spring of 1956 in rivers on the Kamchatka Peninsula by a volcanic eruption (Kurenkov, 1957). On the other hand, many species inhabiting large rivers depend upon floods for spawning. In the Qu'Appelle River, Saskatchewan, the bigmouth buffalo (*Ictiobus cyprinellus*) spawns successfully only at times of high discharge (Johnson, 1963), so it is definitely controlled by the lack of high water in some years.

A very nice study by Starret (1951) of the Des Moines River, Iowa, revealed the great importance of spates in the biology of small cyprinids. This medium-sized river rises rapidly after rain and floods, usually in May and June, but it may also do so at other times. Many species of minnow occur there and they show great variations in relative and absolute abundance. Starret showed that this depends to a great extent upon the irregularities of discharge. The species fall into three groups, early spawners breeding in late spring and early summer, intermediate spawners which may breed all through the summer, and late spawners breeding in late July and August. High water favours spawning and breeding fails in its absence, but if it is too high much oviposition occurs on flooded land away from the river bed, and the young fish become stranded. The species, however, vary in their ability to avoid stranding; for example spotfin and sand shiners, *Notropis spilopterus* and *N. stramineus*, become stranded more readily than do the speckled dace, *Extrarius aestivalis*. High water after spawning also tends to remove eggs and young fish. Thus the complex interactions between when spates occur, the times at which the species spawn, which of them can avoid the consequences of being encouraged to spawn off the river bed, and the deleterious effects on breeding success caused by overcrowding, which is to some extent offset by stranding, very largely determine the year to year specific composition and total population of these small short-lived fishes. It seems very likely that similar interspecific relationships occur in all unstable rivers and streams where numerous species occur.

Another important aspect of fluctuating discharge is the occasional or annual, drying out of streams, leaving only isolated pools or no water at all. This phenomenon has been studied extensively in the American Middle West where it occurs in many small streams (Stehr and Branson, 1938; Starret, 1950*a*; Schneller, 1955; Paloumpis, 1958*b*; Larimore *et al.*, 1959; Metcalf, 1959), and it is clearly important in many parts of the world. Complete dryness seems to eliminate most fish species, although there are indications that at least one African species may be able to persist as eggs (Jackson, 1961), and lung fishes, and some Galaxiidae in New Zealand (Woods, 1963), burrow down into mud to aestivate. Reduction to stagnant pools exposes fish to high temperatures and land-based predators, and often to de-oxygenation, especially if autumn leaf-fall occurs before flow is restored. Most species cannot survive these conditions, although the creek chub, *Semotilus atromaculatus*, white crappie, *Pomoxis annularis*, bluegill and longear sunfish, *Lepomis macrochirus* and *L. megalotis*, bigmouth shiner, *Notropis dorsalis*, and the bullhead, *Ameiurus melas*, may do so. But usually populations of these and other species survive somewhere in favoured locations or further downstream, and as soon as flow is restored there is rapid recolonization from such havens.

Larimore *et al.*, observed that the species returned in seven waves, often through very shallow water, after an exceptional drought in Smiths Branch, Illinois, and that recolonization was mainly by adult fish. Twenty-one of the normal complement of 29 species were back within two weeks of full restoration of flow in February, and 25 were back by the end of the following summer. The specific compositions of the waves were:

1. Bluntnose minnow, *Pimephales notatus*; white sucker, *Catastomus commersonii*; creek chub, *Semotilus atromaculatus*; and silver jaw minnow *Ericymba buccata.*

2. Redfin shiner, *Notropis umbratilis*; orangethroat darter, *Etheostoma spectabile*; hog sucker, *Hypentelium nigricans*; and blackside darter, *Hadropterus maculatus.*

3. Common shiner, *Notropis cornutus*; spotfin shiner, *N. spilopterus*; and creek chubsucker, *Erimyzon oblongus.*

TABLE XVI, 1. The ecological distribution of various species of fish in the Ohio River, and small streams flowing into it, in Kentucky. Fishes are grouped according to the apparent relations to current, depth, and substratum, beginning with those of relatively swift, deep (in the Ohio River), or shallow (in streams), hard-bottomed areas to the left, and terminating with those inhabiting relatively deep, quiet, soft-bottomed areas to the right in each section. From Minckley (1963).

RIVER FISHES

River channel ——————→ Backwater

Alosa chrysochloris				
Hybopsis storeriana				
Moxostoma breviceps				
Stizostedion canadense	*Hiodon tergisus*			
	Notropis atherinoides			
	N. blennius			
	N. stramineus	*Lepisosteus osseus*		
	Ictalurus punctatus	*Hybognathus nuchalis*		
		Notropis buchanani		
		N. volucellus		
		N. whipplei		
		Pylodictis olivaris	*Dorosoma cepedianum*	
		Aplodinotus grunniens	*Cyprinus carpio*	
			Carpiodes carpio	
			Ictiobus bubalus	
			Moxostoma anisurum	*Notemigonus crysoleucas*
			Micropterus punctulatus	*Carpiodes cyprinus*
				C. velifer
				Ictiobus cyprinellus
				Minytrema melanops
				Lepomis macrochirus
				Micropterus salmoides
				Pomoxis annularis
				P. nigromaculatus

STREAM FISHES

Riffles ⟶ Pools

Phenacobius mirabilis				
Noturus flavus				
Etheostoma blennioides	*Salmo gairdneri*			
Cottus carolinae	*Campostoma anomalum*			
	Notropis photogenis			
	Rhinichthys atratulus			
	Micropterus dolomieui			
	Etheostoma caeruleum	*Semotilus atromaculatus*		
		Hypentelium nigricans	*Hybopsis amblops*	
		Etheostoma flabellare	*Notropis ardens*	
			N. cornutus	
			N. spilopterus	
			Catastomus commersonii	
			Moxostoma duquesnei	
			Ameiurus natalis	
			Anguilla rostrata	
			Ambloplites rupestris	
			Lepomis cyanellus	
			L. megalotis	
			Hadropterus maculatus	*Esox americanus*
			Percina caprodes	*Chrosomus erythrogaster*
				Ericymba buccata
				Pimephales notatus
				P. promelas
				Moxostoma erythrurum
				Ameiurus melas
				Chaenobryttus gulosus
				Lepomis humilis

4. Stoneroller, *Campostoma anomalum*; golden redhorse, *Moxostoma erythrurum*; and green sunfish, *Lepomis cyanellus*.

5. Rainbow darter, *Etheostoma caeruleum*; sand shiner, *Notropis stramineus*; and rock bass, *Ambloplites rupestris*.

6. Smallmouth bass, *Micropterus dolomieui*, and suckermouth minnow, *Phenacobius mirabilis*.

7. Greenside darter, *Etheostoma blennioides*, and fantail darter, *E. flabellare*.

Great numbers of species are therefore able to exploit these difficult habitats, and they do so primarily by opportunistic and rapid recolonization from refuges.

DISSOLVED SALTS

Most species of fish have a wide tolerance of pH, and hence, presumably, of water hardness and specific conductivity. Brook trout, for example, occur throughout the range of 4·1–9·5 pH units (Creaser, 1930), and wide ranges are also reported for other species. A few of the smaller species from acid waters in South America and south-east Asia require acidity, or some related factor, for breeding, as is well known to tropical fish fanciers, so this may be a factor in their, virtually unstudied, ecology. Occasionally it has been suggested that water softness, or low pH, may account for the absence of certain species from areas where they were expected. For instance, Jones (1949*a*) thought that the absence of the common European minnow, *Phoxinus phoxinus*, from the Welsh River Rheidol might be because of the soft acid water; but there is little evidence to support such ideas. It seems just as likely that the river is too swift and too liable to spate for this rather fragile fish.

It has, however, frequently been observed that brown trout grow faster and to a larger size in hard than in soft water. This seems to apply on both sides of the Atlantic (Frost, 1939, 1945*a*; Went and Frost, 1942; Tesch and Albrecht, 1961; McFadden and Cooper, 1962) although Bush (1933) did not find this to be so among rainbow trout in Natal. The range of pH available to him for study was, however, rather narrower than that studied by other workers.

Why hardness, or pH or specific conductivity, should affect the growth rate of trout has remained a puzzle, although it has been much studied because of its importance for sport fishing. It is sometimes suggested that the benthic food supply is better in harder waters, and in particular that intervertebrates with non-seasonal cycles, e.g. crustaceans and molluscs, are more common and so provide a more consistent food supply than the mostly seasonal insects of soft water. There may be something in this, but Frost's (1939, 1945*a*) careful study of the growth, food, and food supply of trout at two stations on the Irish River Liffey, which differ considerably in hardness, did not reveal any very clear differences in food supply. Another suggestion is that it is a result of the relatively greater breeding-success in softer waters. Campbell (1961) found no correlation between pH or alkalinity and the growth of trout in several Scottish lochs; but he did find that fast-growing fish occurred where the spawning facilities were poor, and that where they were good the fish were crowded and slow-growing. He points out that acid waters are usually on hard rocks, which provide good spawning gravel, and that alkaline waters are usually on softer rocks and have much more limited areas for breeding. There is some support for this in the observation that larger streams in Germany contain larger and fewer trout per unit area than do smaller ones (Tesch and Albrecht, 1961). These fish tend to be sedentary and to run upstream to breed; they would thus be more crowded from birth in the smaller waters. On the other hand, enrichment of a soft-water stream with sewage has been found to compensate for softness, and to bring brown trout production up to the level of hard-water streams (McFadden *et al.*, 1965), which suggests that nutrition rather than crowding is the factor. If, however, the slow growth in soft water is caused by lack of food, a reduction of the population should increase the rate of growth of the remaining fish; but this does not apparently happen. Cooper *et al.* (1962) removed a large number of brook trout from a soft-water stream in Pennsylvania, and they found two years later that the growth of the survivors had been very little increased. Similarly, if the fishes are exploiting acid waters to such an extent that they are reducing their growth rate, one might expect that their total biomass in the maximally exploited habitat would be higher than that in the less fully exploited alkaline water. This, of course, assumes that total food production per unit area is the same in the two habitats and that there is no great loss of weight caused by intraspecific competition. The latter assumption is probably not justified, but it would seem that this approach to the problem is worth consideration. McFadden and Cooper (1962) compared the populations of fish in three soft-water and three hard-water streams in the Susquehanna watershed and found, in fact, that there were indications that the biomass was larger in the harder streams. This applied, however, to the sum of all the species present; it may be that the indications would have been different if only the one species had been considered.

To date all the reports of a negative correlation between water hardness and growth have been concerned with trout, mostly the brown trout. It would seem that the situation might be clarified if some attention were given to other types of fish which similarly live in a wide range of hardnesses. Obvious candidates for such study would be the North American creek chub, *Semotilus atromaculatus*, and some species of *Cottus*, e.g. *C. gobio* of western Europe or *C. bairdi* of North America. One further point that may have a bearing on this problem is the experimental observation of Schlieper *et al.* (1952) that calcium and magnesium raise the respiratory rate of the rainbow trout and the chub *Squalius* at low temperatures and decrease it at high ones. This may mean that fishes in hard water feed more actively in winter and use up less energy in summer, thus having more available for growth.

As we have seen running water is rarely saline, but occasionally human activity makes it so, and this gives us some insight into the salinity tolerance of freshwater fishes. Many can, of course, migrate down into brackish or even sea water, and various normally freshwater species occur in the peculiarly unvarying brackish waters of the Baltic Sea. The water of the German River Werra has become polluted by a mixture of salts resembling sea water, and occasionally the chloride concentration rises to over 5,000 mg./l. (Schmitz, 1956). Some fish die when this occurs—mostly pike, barbel, and chub, but not perch or eels. Schmitz experimented with several species and established the following upper limits of endurance in 'artificial Werra water'. The figures are in grams of total salts per kilogram (per mille) comparable to sea water of various concentrations.

39·8	*Anguilla anguilla*
20	*Acerina cernua* and *Lucioperca lucioperca*
17–18·5	*Esox lucius*, *Tinca tinca*, and *Cyprinus carpio*
15·5–17	*Rutilus rutilus*, *Abramis brama*, *Alburnus alburnus*, *Chondrostoma nasus*, *Scardinius erythrophthalmus*, and *Gobio gobio*—plus a few specimens of pike, carp, and tench from the 17–18·5 category.
13	*Nemacheilus barbatula* and *Carassius vulgaris*—plus a few specimens of species in the 15·5–18·5 categories.

It is clear, therefore, that at least many species of fish are very resistant to raised salinity and that it must be an insignificant factor in their ecology in most running waters.

THE SUBSTRATUM AND TURBIDITY

A few fishes, particularly small benthic species, are more or less confined to rocky or stony substrata. These include, for example, all those with ventral suckers and friction plates, and cryptic fishes such as sculpins and loaches. Many others are also fairly definitely associated with a definite type of substratum, e.g. the gudgeon *Gobio gobio*, with gravel in Europe (Bernet, 1960), the sand-darter *Ammocrypta beanii*, and a number of small cyprinids in North America with sand (Hubbs and Walker, 1942; Metcalf, 1959), and the mudfish, *Umbra limi*, with thick marginal vegetation (Peckham and Dineen, 1957). But for the great majority of species the nature of the substratum is apparently of little consequence except at times of breeding, and the current velocity and the depth of the water seem to be more important (Cleary and Greenbank, 1954). In rivers and streams the greatest numbers of specimens and of species, are usually associated with beds of gravel or sand, as was shown by Allen and Clark (1943) in their analysis of a large number of collections from Kentucky. This is, however, hardly surprising, as these two types of substratum are by far the commonest in running water.

Shelter is also important, and even swift-water fishes such as trout occur more abundantly where there are more hiding places (Saunders and Smith, 1962; Hartman, 1963). In larger rivers where the bed is relatively smooth this becomes increasingly clear, and fish of many species tend to congregate near obstructions, in bays and along the banks, or anywhere else that offers some shelter. Reynolds (1965) noted, for example, that several species in sandy turbid reaches of the Des Moines River, Iowa, were usually associated with the few small rubble areas; and the flathead catfish, *Pylodictis olivaris*, moves from riffles to larger and larger shelter in increasingly deeper water as it grows, and large specimens are particularly associated with fallen trees (Minckley and Deacon, 1959).

Such observations are fairly general, and superimposed upon this complex of stony or rocky substratum, current speed, depth, and the need for local shelter is the inability of some species to tolerate continuously turbid water or heavy siltation. This factor has attracted particular attention in North America because increased siltation is one of the more obvious results of human alteration of the landscape. Quite a variety of cyprinids, catostomids, and cottids and the brook stickleback, *Eucalia inconstans*, have been shown to be affected (Gerking, 1945; Lachner, 1956), but others, e.g. the fathead, bluntnose, and suckermouth minnows, *Pimephales promelas*, *P. notatus*, and *Phenacobius mirabilis*, have extended their ranges into formerly clear waters. Doubtless the same applies in many other parts of the world, but it is particularly obvious in North America because so much of the alteration is comparatively recent.

An interesting point associated with the substratum is the ability of many small species to penetrate far down into gravel. A few such as *Erimystax harperi*, a minnow of caves and springs in Florida, move in and out of underground water quite readily (Marshall, 1947*b*), as do some members of the cave-fish family, Amblyopsidae. But it is less generally known that many stream fish can penetrate deeply into the gravel. Stegman and Minckley (1959) report finding several species 10 cm. or more down in the gravel of a dry riffle between pools in Illinois, and darters may sometimes be collected at 20 cm. or more down, even when there is no shortage of overlying water. Phillips and Claire (1966) observed that small *Cottus perplexus* burrowed down to the maximum possible extent (36 cm.) in coarse gravel, in aquaria and that even specimens 5–7·5 cm. long went down at least 18 cm. This is an aspect of behaviour which has been little studied, and it is one which may be important in the ecology of at least some benthic species.

BIOTIC FACTORS

We have already discussed interspecific territorial relationships between trout species and their possible effects in determining which species becomes dominant. The inability of brook trout to compete with brown trout is sometimes suggested as being the reason for failure of the former to become firmly established in Europe (Macan, 1961*a*), although this clearly cannot be the only factor determining the relationship of these two species. If it were, the widespread introductions of brown trout into North America would have eliminated the brook trout and they have not done so.

Some complex of factors must therefore determine the way in which species coexist, and, while we know little about this, there are many examples of closely allied species sharing out the available territory between them. We have elsewhere mentioned the eels of Australia, New Zealand, and Tahiti which succeed one another from large river to head-stream (p. 345), the *Cottus* species in streams flowing into Lake Superior which replace one another according to the summer temperature of the water (p. 324) and the two species of *Rhinichthys* which occupy swift and slower water in the Fraser River (p. 306). Many other instances of more or less completely vicarious occurrences are known, but we have no explanation for them and it is probable that each is different. For example:

1. In the genus *Notropis*, *N. galacturus* and *N. rubricroceus* inhabit deep and shallow pools respectively in the upper Tennessee River system (Outten, 1958).

2. Among the darters, *Etheostoma flabellare* of the riffles in small streams is replaced by *Percina phoxocephala* in riffles of large streams in Iowa (Karr, 1964).

3. In the redhorses, *Moxostoma macrolepidotum*, *M. erythrurum*, and *M. anisurum* succeed one another in water of decreasing swiftness (Meyer, 1962).

4. In the genus *Galaxias* in New Zealand, *G. attenuatus* occurs near the sea, probably because it is katadromous, *G. alepidotus* inhabits quiet inland streams, *G. brevipennis* swift water, and *G. fasciatus* shaded pools (Allen, 1960*b*).

5. Also in New Zealand among the Eleotridae, *Gobiomorphus gobioides* is a species of lower reaches, and it is replaced upstream by *G. radiata*, which is itself replaced in small streams by *Philypnodon hubbsi* in rapid water and by *P. breviceps* in pools (Allen, 1960*b*).

It seems also that some species are intolerant of competition from any others and tend always to occur more or less alone. Starret (1950*a*) states that the fathead minnow, *Pimephales promelas*, appears to be such a species in Iowa. The whole subject of the mechanisms of competition between animal species is still remarkably unexplored, which is odd when one considers its importance in ecology and its central position in evolutionary theory.

Relationships between fish species and other organisms are perhaps even less well understood. Many mammals such as otters and bears eat fish, as do many birds, but no thorough studies have been made of their effects on fish populations in running water. Fishes are also heavily parasitized by a great variety of protozoans and helminths and by the glochidia of clams, many of which are highly specific (Howard, 1951). Here also there is little information on which to base ecological conclusions. It would seem, however, that the effect of parasites is not very great as they are almost always present, and there are indications that they may be more abundant in good and varied environments where fishes grow well than they are in poorer more limited ones (Thomas, 1964).

NUMBERS

The number of species of fish varies greatly from one part of the world to another, although generally the main continents have rich faunas of freshwater species, as well as recently acquired species of marine origin. Islands, such as New Zealand and Australia, have only the latter.

Western Europe is exceptional in that it has a very limited freshwater fish fauna, probably because of the Pleistocene glaciations. There the Alps prevented both retreat and recolonization, and many species must have been eliminated. In contrast, most of the species in North America could move down the Mississippi or the coastal areas and back north again as the ice retreated. The result is that quite small rivers there contain large numbers of species, e.g. 92 from the Illinois River (Moore and Paden, 1950), 89 from the Kaskaskia River, Illinois (Luce, 1933), and 68 from the Black River, Missouri (Martin and Campbell, 1953), whereas even the Danube has only 83 species (Bănărescu, 1961). Müller (1955*a*) lists the following numbers of species recorded in various European rivers:

Rhine	30+7 migrants	Susaa	32
Weser	33+11 migrants	Volga	39+30 migrants
Fulda	31	Sula	33

Even small streams on most continents contain almost as many species as this; Clear Creek in southern Illinois has 23 species (Lewis and Elder, 1953), which is very nearly as many as could be collected from all the rivers and streams in the British Isles. It will be appreciated, therefore, that the ecology of fishes in European rivers is possibly somewhat simpler than it is in most other continents and that one cannot with impunity apply results obtained there to other areas.

It is everywhere a fairly general rule that the number of species increases from the source to the mouth of a river, as has been established in western and eastern Europe (Huet, 1954, 1959; Backiel, 1964), South America (Kleerekoper, 1955), and North America (Hallam, 1959; Larimore and Smith, 1963). In an early study in Illinois, Thompson and Hunt (1930) noted that there is a more or less linear relationship between the logarithm of the drainage area upstream of a collecting station and the number of fish species occurring there. They also found, as has been confirmed by others (e.g. Müller, 1955*a*; Hallam, 1959; Tesch and Albrecht, 1961; Larimore and Smith, 1963), that the total number of individuals of all species per unit area declines in a downstream direction. There are thus fewer fishes per square metre, and per cubic metre, of water as one proceeds downstream, although the individuals are usually larger and the total weight per unit area remains more or less constant.

The numbers of specimens per square metre vary, however, considerably from place to place and from river to river. Backiel (1964) found only one fish per 3 sq. m. in the Polish River Drweca, whereas Shetter and Leonard (1943) and Gerking (1949) record nearly 1·5 and nearly 3·0/sq. m. in Hunt Creek, Michigan, and a small stream in

Indiana, and Williams (1965) calculates that there are about 10/sq. m. in the Thames at Reading. These four studies therefore cover a 30-fold difference. There are similarly very great differences in the weight of fish. The following figures have been obtained from Shetter and Leonard (1943), Gerking (1949), Smith *et al.* (1949), Allen (1951), Swingle (1954), Larsen (1961), Timmermans (1961*b*), McFadden and Cooper (1962), and Williams (1965), and recalculated where necessary to bring them to kilograms/hectare.

Horokiwi River, New Zealand	225
River Thames, England	476
Belgian trout stream	300
Brandstrup Brook, Denmark	176 and 315 (see below)
Minnesota trout streams	56–280
Susquehanna trout streams	51–392
Michigan trout streams	4·6–200
Small streams in Indiana	51–1,050
Various rivers in Alabama	51–1,730

The last figure, for the Tensaw River, may be too high, but 1,192 was recorded for the Black Warrior River by Swingle so we can take the range as 4·6–1,192, or at least a 200-fold difference between a poor northern trout stream and a rich southern river. Even allowing for sampling inaccuracies and differences of method it is clear that there are large differences in fish biomass. There are also considerable differences from time to time. The two figures given above for Brandstrup Brook are for June, when Larsen caught 176 kg./ha., and September, when he caught 315 kg./ha. But in June there were only 101 trout/ha. and this number had risen to 240 in September. It would seem fairly certain, therefore, that the later collection included specimens which had moved in to spawn, and this perhaps illustrates the difficulties of measuring total biomass.

GROWTH

The annual rings which fishes so conveniently lay down in scales, otoliths, fin-rays, and bones have proved to be of enormous value in fishery work, as they allow accurate assessments of growth rate, population structure, comparison of different waters, etc. Discussion of them is outside the scope of this book and the reader is referred to such standard works on fishery biology as Lagler (1952), Rounsefell and Everhart (1953), and Carlander (1950, 1953) for details, and to Ricker (1958) and references in Williams (1965) for methods of handling numerical data obtained from fishes.

From a purely ecological standpoint, however, it is perhaps a pity that this one aspect of the biology of fishes has attracted so much attention, that so much of the now enormous literature on freshwater fishes is devoted to studies of age and growth. This is particularly so in that many such studies are regrettably uncritical, and are based on the assumption that observable 'rings' are always annual and are always laid down in the winter. This is not true, as has been shown for many species, and unless it is established when a particular species lays down its ring, and how often, one can only accept the results of such work with scepticism.

These strictures are, however, not intended to decry work on fishery biology, their purpose is merely to point out that some of it is not very good, and that, despite the wealth of literature on the subject, much of it is remarkably uninformative about the ecology of the species with which it deals. Growth rates are certainly important as they enable one to compare waters and perhaps to decide that a particular species does better in one than in another. But what do we mean by does better? Is it of *ecological* significance that in one place each kilogram of fish swims around in two bits while in another it is divided into four? What is the significance of the fact that many small fishes, e.g. sculpins, darters, and small minnows, live for only a few years (Lachner *et al.*, 1950; Bailey, 1952; Smyly, 1955, 1957; Gunning and Lewis, 1956) while trout in the same water live and grow for longer periods?

It seems that it is perhaps time that we made a conscious effort to separate ecology from fishery biology, and that the ecologists should forget that fishes can be eaten and angled for and start to try to find out more about them as animals. It is, for instance, of considerable theoretical interest that trout appear to grow faster in harder water, but the ecological significance of this would be much reduced if it were, for example, shown that this is peculiar to trout and does not apply to other fishes.

This then is a plea for a more general approach by many students of fishes to the problems of fish ecology. To date we are almost overwhelmed by studies of one rather limited type.

CONCLUSIONS

From this brief survey of the factors which affect fishes in running water we can conclude that the most important abiotic factors are temperature, both directly and, for cold-water fish, indirectly through its influence on oxygen consumption, rate of flow and fluctuation in discharge, and the availability of suitable shelter. The chemical content of the water seems to be of relatively minor importance.

It is thus possible, in areas which have been well studied, to describe rivers and streams in fairly general terms and to list with some degree of confidence the fish species which are likely to be found there.

We are much less well informed on the effect of biotic factors, although it is clear that closely related species interact with one another in such a way as to tend to divide up the habitat between them.

CHAPTER XVII

Movements and breeding of fishes

Many species of fish move some distance to breed and they must be able to do so if they are to survive in certain bodies of water. Similarly, most species have fairly definite requirements for successful reproduction, and clearly these also have to be met. These two aspects of fish ecology are closely related, but we will consider them separately as both have features which do not involve the other.

MOVEMENT

It is well known that many fish migrate for great distances through rivers, but there is always some uncertainty as to how much less-obvious movement occurs. This means that the boundaries of populations are difficult to define, and it adds to the problems of sampling. It appears, however, that many fishes are remarkably sedentary, at least for long periods, and migrations may be either for breeding or for other purposes, or they may result merely from population pressure, and lead to the occurrence of specimens in waters far outside their normal ecological range. Many of these aspects of movement, or the lack of it, are of considerable ecological importance, and we will consider them under a number of headings.

HOME TERRITORY

We have seen that many species are territorial in running water, although they may cease to be so when the flow is reduced, and that, in at any rate the salmonids, the acquisition of a territory begins very early in life, as soon as the newly hatched fish can swim. Such fishes clearly have a home territory which they guard and remain in for long periods.

Several studies, involving the capture, marking, release, and recapture of fishes, have shown that many species move very little. Nearly all the work has been done in North America where it has been established that trout of various species and the mountain whitefish, *Prosopium williamsoni*, are very sedentary (Sigler, 1951; Brown, 1952; Newell, 1957), and there is confirmation that the same applies to brown trout in Europe and in New Zealand (Timmermans, 1960; Allen, 1951). This is not surprising as these fish are well known to be territorial throughout their lives, but it is also found that the catostomids, *Catostomus*, *Moxostoma*, and *Hypentelium*, the centrarchids, *Micropterus*, *Ambloplites*, and *Lepomis*, and the ictalurids, *Ictalurus* and *Ameiurus*, are similarly stationary for long periods (Bangham and Bennington, 1938; Scott, 1949; Stefanich, 1952; Larimore, 1954; Funk, 1957*a*; Muncy, 1958; Gerking, 1950), as is the European cyprinid *Gobio* (Stott *et al.*, 1963). Even the eel, *Anguilla rostrata*, which migrates enormous distances to breed, remains in one place for long periods. Gunning and Shoop (1962) found that individuals remained in reaches 100–50 m. long in two streams in Louisiana for 10–13 months. Populations of fishes may, therefore, be so static for long periods that it is possible to observe different rates of growth in different parts of the same river (Went and Frost, 1942; Purkett, 1958).

Gerking (1950, 1953, 1959) made a particularly detailed study of the lack of movement in Richland Creek, Indiana, and has added considerably to our knowledge. He established that many species have very definite home ranges in which they may remain for several years, and that even very high spates do not remove them. He estimated the home range of the rock bass, *Ambloplites*, and the longear and green sunfish, *Lepomis* spp., to be about 30–60 m. of stream, and that of the smallmouth and spotted bass, *Micropterus* spp., and hog sucker, *Hypentelium*, to be 60–150 m.; he also found that, in general, riffles were respected as boundaries to home ranges. A riffle and pool stream therefore forms a series of fairly discrete units for these pool-inhabiting fishes.

Quite often a few of the fishes which have been marked in one place have been caught far upstream or downstream, even though the majority have remained where they were originally caught. This has been observed, for example, in trout (McFadden, 1961; Logan, 1963), rock bass (Scott, 1949), smallmouth bass (Reynolds, 1965), and channel catfish, *Ictalurus punctatus* (Behmer, 1964). Funk (1957*a*) has suggested that in any population at any given time there are two groups of individuals—the majority which are sedentary and a few which are actively mobile—and that the proportion of the two may vary from place to place and from time to time. The same suggestion has been made in England quite independently, and apparently without knowledge of the American work (Stott *et al.*, 1963). The significance of the two groups is still somewhat in dispute, but, whether or not one refers to the mobile specimens as strays, it seems undoubted that they occur, at any rate in small numbers, and they must be the means of spread of the species even if they have little effect upon the population as a whole.

Many of the species with a definite home-range move very little at any time, even to spawn, but some, e.g. the green sunfish and largemouth bass, are known to move into deeper water in lakes during the winter and to return to their home area in the spring (Hasler and Wisby, 1958); the same applies to many species in rivers and streams. Trout, for example, move downstream into deeper water during cold weather, and they also move some distance upstream to spawn. But they return to their home territories, and individuals have been recorded as doing so in several successive years (Shetter, 1937; Watts *et al.*, 1942; Schuck, 1945; Timmermans, 1960).

Fishes are, therefore, able to find their homes when they have been away from them, and this has been confirmed by experimental displacement of individuals of a variety of species in both upstream and downstream directions (Larimore, 1952; Gerking, 1953, 1959; Gunning, 1959; Stott *et al.*, 1963). Miller (1954*b*) carried out a large-scale experiment on the cutthroat trout, *Salmo clarkii*, in Gorge Creek, Alberta, during which he confined individually marked fish for a period in a fenced-off reach about 0·8 km. long. Then, two or three weeks after releasing them, he used poison along a great length of the stream to find out how they had moved. The fish had been collected over a period and placed in the enclosure, so some had been confined there for a long while and others for only a short time, and over a hundred fish had been gathered from each of the upstream and downstream areas. The results of this study were that individuals which had been held for more than thirty days had mostly remained in the new area, while those which had been confined for less than thirty days had moved. If they originated from upstream they had returned to, or moved towards, their homes, but if they had been brought from downstream they had moved erratically, and some had even overshot their original homes and gone on further downstream.

We can conclude from this experiment that attachment to a site lasts only about thirty days, although clearly it must last longer than this in winter in species which move down into deeper water and return in spring, and that the location of home is connected with something brought down by the water. We shall be returning to these points, but it can be noted here that homing upstream is more efficient than homing downstream, and this fits well with the possibly not uncommon downstream displacement by spates. It is difficult to think of any natural phenomenon that would forcibly displace a live fish any considerable distance upstream.

MIGRATIONS FOR BREEDING

Many species migrate for breeding purposes, and such migrations may be long or short, within fresh water or to and from sea water, or they may involve movement into running water from still water or from rivers on to flooded land.

Migrations to or from the sea for breeding occur in many species and all over the world. They fall into two classes: *katadromous*, going down to the sea, and *anadromous*, going up into fresh water. Katadromous behaviour is the less common in fishes that are found at all far inland, although it involves, of course, all those species already mentioned (p. 301) which wander into the most seaward parts of rivers. It is best known in the eels, *Anguilla* spp., which breed in ocean deeps and of which species occur in rivers and streams almost everywhere where drainage to the sea is adjacent to deep water. Various species occur on both sides of the Atlantic, in China and Japan, in Australia and New Zealand, and in Africa, but none is found in Malaya which is far from deep oceanic water (Bertin, 1956). The young eels, after a marine planktonic stage, migrate to the coasts and swim upstream, often in great numbers, and they ascend barriers such as falls to move inland. In Europe, *A. anguilla* arrives all through the winter and early spring, and the young elvers arrive later the further are the mouths of their destination streams from the edge of the continental shelf. Where only one species occurs, as in Europe and North America, it penetrates far inland, except where it is prevented from doing so by such obstacles as the Niagara Falls, and it has often been stated that there is a tendency for females to occur further inland than males (Bertin, 1956). This seems, however, not to be so. Eels grow slowly (Frost, 1945*b*), and the sex of even quite large specimens is difficult to determine. Sinha and Jones (1966) have shown that when the sex of all specimens is correctly determined it is found that both occur together far inland.

In regions where two or more species occur they seem to share out the area between them. Thus, in Australia and New Zealand, the short-finned *A. australis* lives in the larger rivers and lakes, and the long-finned *A. dieffenbachi* penetrates to smaller waters (Cairns, 1941, 1942*a*, *b*), and in Tahiti *A. obscura* remains near the coast, *A. marmorata* occurs in rivers on the plains, and *A. megastoma* lives in mountain streams (Bertin, 1956).

The other important group of katadromous fishes comprises the southern trouts or jollytails, *Galaxias*, several species of which migrate down to breed in tidal marshes. The larvae then go to sea for a while and return to grow in the rivers. *G. attenuatus* of Australia, New Zealand, and South America is such a species (Burnet, 1965), and it is noticeable that it occurs on all these land-masses despite the hundreds of kilometres of open sea between them. Some species of *Galaxias* have, however, severed their connection with the sea, and remain always in fresh water. Such are *G. vulgaris* of New Zealand and *G. findlayi* of the Australian Alps (Burnet, 1965; Lake, 1959). Other katadromous species are *Plecoglossus altivalis* of Japan (Kawanabe, 1958), and the smelts and graylings of southern continents, the Retropinnidae and Haplochitonidae, which have life histories resembling that of *G. attenuatus* (Allen, 1956).

Many katadromous fishes, therefore, spend much of their lives in rivers and streams, and each year the adults pass down to breed and the young ascend, often in great numbers. So numerous are the little elvers in Europe that in western England they used to be caught for human food, and the catching of descending adults is quite an industry, especially in Ireland. Returning *Galaxias* young support a whitebait fishery in New Zealand.

Anadromous migrants are better known because of the commercial importance of salmon and smelts. Some of these, e.g. the chum and pink salmons, *Onchorhynchus keta* and *O. gorbuscha*, may enter fresh water for only short distances and their young leave very soon after hatching (Hoar, 1951), so they are not very important in river biology; but the coho salmon, *O. kisutch*, the Atlantic salmon, *Salmo salar*, several migratory races of trout, e.g. *S. trutta*, *S. gairdneri*, and *S. clarkii*, and the arctic char, *Salvelinus*, move far inland, and their young remain and grow there for a long time (Jones, 1959; Shapovalov and Taft, 1954; Lowry, 1965). Other important fishes which move in from the sea or brackish water to breed include the famous sturgeons of eastern Europe, *Acipenser* spp. and *Huso huso*, specimens of which may weigh up to a ton, the shads, e.g. *Alosa* and similar genera of many parts of the world, various herring-like fish, e.g. *Clupeonella* in the Danube and *Harengula* in the Volga, some coregonids, e.g. *Stenodus* and *Coregonus* in the U.S.S.R., and many species of lamprey (Behning, 1928, 1929*b*; Beling, 1929, 1949; Raney and Massman, 1953; Bailey *et al.*, 1954; Shadin, 1956; Hubbs and Lagler, 1958; Buşnită, 1961). In the Volga there is also an anadromous race of roach, *Rutilus rutilus caspicus*, which lives in the Caspian Sea and spawns in the delta of the river; and in northern Europe the anadromous race of the three-spined stickleback, *Gasterosteus* *aculeatus*, migrates in such great numbers that it used to be caught in quantities sufficient to be rendered down for lamp oil before the exploitation of petroleum.

Many of these fishes still support important fisheries, and it is noticeable that there are many more of them in the northern temperate and Arctic regions than there are elsewhere. Although sea-going lampreys occur in southern regions and there are some anadromous freshwater herrings there is no corresponding mass of abundant species to equal the salmonids and their close relatives. Perhaps this is a result of some effect of the Pleistocene glaciations.

Many migrants do not feed at all while on their reproductive journeys, although they often remain for long periods in fresh water. Post-spawning mortality is therefore high, and in many species it is normal. The masses of dead fish remaining in running water, but originating in the sea, must have quite an influence on the availability of organic matter, especially in small streams.

The distances covered by migrants are often enormous. Sturgeons swim 2,000 km. up the Danube (Buşnită, 1961), and the white salmon, *Stenodus leucicthys*, goes 2,800 km. up the Volga and spends a year on the journey (Behning, 1929*b*). Even some lampreys move great distances; *Caspiomyzon* goes 2,000–2,400 km. up the Volga (Behning, 1928) Distances travelled, however, vary with the river; thus, in the U.S.S.R., salmon go 2,000 km. up the River Lena but only 650 km. up the Lana, and the omul, *Coregonus autumnalis*, goes 1,600 km. up the Lena and only 500 km. up the Jana (Shadin, 1956). They also vary with the time of year of arrival at the river mouth. Autumn-run chum salmon, *O. keta*, are large and migrate more than 1,500 km. up the Siberian River Amur, but the spring fish are smaller and go only 400–500 km. Nikolsky (1963) has shown that this is connected with their initial fat-content. The autumn fish contain 10–11 per cent while the spring fish have only 8–9 per cent. Most of the fat is used up on the journey—that of *Stenodus leucichthys* drops from 21 to 2 per cent during its long journey up the Volga—and there is a good correlation between the initial fat contents of species of *Caspialosa* and the maximum distances they go up that river from the Caspian Sea. Nikolsky gives the following figures.

	% fat	Maximum distance travelled in km.
Caspialosa kessleri	16·00	1,000
C. volgensis	8·71	*c.* 500
C. caspia	7·48	*c.* 100
C. soposhnikovi	5·61	0

Although in many of the anadromous species the migratory habit is well established and the fishes do all or much of their post-larval feeding and growth in the sea, there are others, particularly in the genus *Salmo*, in which the habit is by no means fixed, and there are two races, one sea-going and one freshwater, which are not clearly separated. Such are the sea and brown trout, *S. trutta*, the steelhead and rainbow trout, *S. gairdneri*, and the coastal and inland races of the cutthroat trout, *S. clarkii*. Sometimes populations of these races are fairly distinct and discrete (Le Cren, 1958), and they may differ a little morphologically and breed true (Neave, 1943*a*), but it is often unclear how much they inter-breed and how much their morphological differences are caused by their habit. The migratory race may not even go to sea, as, for example, the steelhead trout which now live in the Laurentian Great Lakes, and, as we have seen, the Atlantic salmon appears quite readily to evolve land-locked races. *Salmo* then is a good example of a genus which is still in a very ambivalent position, and is in process of evolving entirely freshwater species.

Migrations within fresh water. Just as a large number of species move from the sea into rivers and streams to breed, so do many species whose normal habitat is lakes. Many of them, such as the brown trout and the mountain whitefish, *Prosopium williamsoni*, which are regular inhabitants of streams, also live in lakes, but such lake-dwelling populations migrate out into streams to spawn, the stimulus to do so being apparently a change of temperature (Brown, 1952; Sigler, 1951; Stuart, 1953*a*). The young fish remain in the streams after hatching, and brown trout spend the first one or two years of their lives there before descending to the lake. Some such species also spawn in the outflow streams, up which the young fish then ascend to their final habitat (Stanković, 1960); and a number of species which are normally completely confined to lakes have local populations which spawn regularly in streams. Such are some lake trout, *Salvelinus namaycush*, in Lake Superior (Loftus, 1958), one race of the char, *S. willughbii*, in Windermere (Frost, 1952), and the whitefish, *Coregonus lavaretus*, in the fresher parts of the Baltic Sea (Fabricius and Lindroth, 1954).

In Africa many varied fishes move out of lakes to breed in rivers; these include the cyprinids *Barbus*, *Varicorhinus*, *Barilius*, and *Labeo*, the cat-fish *Clarias*, and the characid tiger-fish *Hydrocyon* (Jackson, 1961; Marlier, 1952, 1953). Doubtless the same applies to many other genera there and elsewhere, because, although many species have taken to living in lakes, there seems to be little doubt that their original habitat was running water, and that they tend to return to it to breed. At the appropriate time and place it is therefore possible to find most species of freshwater fishes in rivers and streams.

Equally, however, quite a number of riverine species, and some from very sheltered habitats in streams, e.g. the mudfish *Umbra* (Peckham and Dineen, 1957), move on to submerged areas of flood plain and into recently filled backwaters to spawn. This applies to many species from large rivers, e.g. the Mississippi, to species in some of the African genera mentioned above, e.g. *Hydrocyon*, *Labeo*, and *Clarias*, and sometimes to the sterlet in eastern Europe. Indeed, in many species the impetus to migrate and breed is provided by rising water, as has been proved by artificial freshets (Huntsman, 1945). It is thus not surprising that species which need shallows, plants, or some shelter for breeding should actually come to deposit their eggs outside the river bed. We shall return to this point later in this chapter.

We have seen that trout normally move upstream to breed, and they often go up into very small streams. In most species the upstream movement begins in the fall, and the spent fish return during the winter and early spring. The males tend to move up first, and as with other fish the stimulus seems to be a rise in water level; the migration occurs at all times of day when the water is high and turbid, but as it clears the movement becomes primarily nocturnal (Munro and Balmain, 1956). Mortality is high, even though dead specimens are not often seen, and small streams contain mainly only very young fry at times other than the breeding season (Le Cren, 1958). Huet (1961) trapped all the brown trout running up and down a small tributary of the River Lesse in the Belgian Ardennes, and he found that only about 50 per cent of the 500 fish which had ascended returned to the river. Five hundred fish, incidentally, represents the population of adult fish from about 600 m. of the river, so the small streams attract a great proportion of the breeding fish.

A great many other types of river-fish also move upstream to spawn. These include sturgeons, paddlefish, garpikes, shads, suckers, barbels and other cyprinids, including even the small European minnow, catfish, and some darters (Hora, 1923; Coker, 1929, 1930; Tack, 1940; Frost, 1943; Beling, 1949; Carter, 1950; Carlander, 1954; Winn, 1958*b*; Jackson, 1961). Even the sluggish carp has been observed to move up the Des Moines River in Iowa, and to jump over lowhead dams (Rehder, 1959). It seems undoubted that the strong urge to go upstream to breed, especially as it is often initiated by rising water, is an important factor in the recolonization of intermittent streams or streams depopulated by unusual droughts.

NON-BREEDING MIGRATIONS

We have already seen that some species make regular diurnal movements (p. 312) and it is probable that this occurs much more widely than has been observed and recorded.

There would seem also to be a general readiness to move among certain individuals of even sedentary species. This accounts for the fact that freshwater fish are often found in weakly brackish water which they cannot inhabit indefinitely. Bailey *et al.* (1954) report that they frequently found garpikes, *Lepisosteus*, minnows, *Notropis*, top-minnows, *Gambusia*, black bass, *Micropterus*, and sunfish, *Lepomis*, near the mouth of the Escambia River, Florida, in water with salinities ranging at least up to 4·5 per mille, and so well outside their normal habitat. Rapid recolonization also probably accounts for the lack of depletion of small species when large numbers are collected at frequent intervals for bait (Larimore, 1955).

Constant wandering of specimens is also undoubtedly responsible for the finding that irrigation canals from the Gallatin River, Montana, cause considerable mortality of fishes which wander down them and become stranded (Clothier, 1953), and this must apply generally to irrigation systems unless provision is made to exclude fishes. Similarly, fishes will wander into streams in some numbers even if they are not suitable for them. This resembles the occurrence 'out of place', of invertebrates in habitats near to a suitable one, which we discussed in Chapter XI.

The stray fish which are ready to move, under, perhaps, population pressure or because they have unsuitable home territories, are therefore at least numerous enough to be observable as a general phenomenon, and they can, in fact, result in very rapid recolonization. Wickliff (1940) removed, as far as possible by netting, all the fish from a riffle in a stream in Ohio, and he found that the darter population was back to 45·3 per cent of its previous level two days later. Similarly, he collected darters at a rate of 13,087/ha. from a headwater riffle which had been dry in the summer, and had presumably been repopulated by reinvasion when the water returned. Thus the wandering specimens, even if they are a small proportion of the population or represent only over-production of individuals, play a significant role in ensuring that all the available areas are explored and colonized.

There is also some movement associated with increasing age and size. We have seen that outside the breeding season very small trout-streams contain mainly small fish and relatively few larger ones. This is because trout, and also salmon which according to species and latitude may spend one to seven or more years in fresh water, drift downwards into ever larger stream reaches and deeper water as they grow, and they are limited at the downstream end only by the suitability of the available territories. Anadromous species such as salmon are, therefore, already some way on their journey down to the sea when they change to the migratory smolt stage, cease to be territorial and move out, usually in response to rising discharge. This movement into deeper water, and hence often also downstream, or into the deeper quieter pools, seems to be general among most species of fish that grow to any considerable size. It is well known to anglers that the large specimens are always in the deep pools except when they are migrating to breed, a time which is outside the legal fishing-season of most countries.

Many species also move into deeper water for the winter, darters move downstream off riffles (Winn, 1958*b*), the European and many other minnows disappear from the shallows (e.g. Tack, 1940), brook trout move downstream (Watts *et al.*, 1942), young brown trout may move into lakes (Stuart, 1957), and the sterlet moves into the deeps in the Volga (Behning, 1928). This is probably a general pattern of behaviour connected with reduced activity in the cold water; in the River Amur in Siberia, however, a number of species, most notably the grass carp (*Ctenopharyngodon idella*) and the bighead, *Hypophthalmichthys molitrix*, have a more or less complete hibernation. They move into deep pools and aggregate on the river bed, where they become enveloped in slime (Nikolsky, 1963). Similarly, in dry climates, there is often a marked downstream migration during the dry season when headwaters and swamps begin to dry out. Welman (1948) records that, near Lake Chad in Nigeria, he observed definite tracks in the sandy beds of shallow streams worn by the passage of multitudes of descending fishes of diverse species. These were so densely packed in the channels, which were about 30 cm. wide, that they could be easily caught with a fisherman's landing net. He also observed that the catfish *Clarias* leaves the water at night to migrate long distances overland to avoid becoming stranded.

NAVIGATION

Displaced fishes return to their homes, as do individuals which have dropped downstream for the winter or gone upstream to spawn. Salmon return to the river of their birth to spawn, with only a very small proportion running up other rivers (Jones, 1959; Hasler, 1960), and trout of several species return to the same stream or tributary year after year to spawn (Stuart, 1957; Taft and Shapovalov, 1938), as do lake trout and the walleye, *Stizostedion vitreum*, from lakes (Loftus, 1958; Gerking, 1959).

We have seen from Miller's experiment with cutthroat trout that homing is more successful, or is at any rate more rapidly accomplished, when it involves upstream movement (p. 344). This indicates that it is connected with some property of the water, and Hasler (1954) and his co-workers have shown that the bluntnose minnow, *Pimephales notatus*, is able to distinguish odours at low dilutions and can learn to distinguish between the water from two streams. Similarly Gunning (1959) showed that longear sunfish, *Lepomis megalotis*, could find their way home in a stream if they were blinded, but not if their olfactory organs were damaged. Hasler suggests that the odour is carried by volatile nitrogenous organic matter in the water, and, in an informative review paper (1960), he has discussed the whole subject of navigation by fishes. It seems undoubted that river fishes navigate by smell, at any rate when they are in rivers. Adult coho salmon caught in one branch of a river and returned downstream below the junction invariably returned to the branch they had first ascended, unless their olfactory orifices were plugged, when they swam up the various branches at random. The return of salmon to the river and even to the tributary of their birth is, therefore, directed by the smell of the water. This, however, can begin to apply only when they have arrived in the general geographical area of the river mouth. There is evidence that navigation to this point is celestial, since several lake-dwelling fishes are known to orientate themselves to the sun (Hasler *et al.*, 1958). Clearly, however, navigation by the sun up the winding course of a river is not practicable, and it is at this point that the sense of smell takes over. It also seems clear that the smell is imprinted on the fish during their early life as it is possible to establish populations of salmon in places where they did not previously occur, by the introduction of eggs or fry.

The remarkable features of this method of homing, especially in fishes which, like the salmon and other species return after long intervals to breed, are the ability of fishes to 'remember' a smell for so long, the fact that apparently very similar streams should differ detectably in smell, and that smells should be sufficiently constant from year to year for the fishes rarely to err.

SURMOUNTING OF OBSTACLES

Movement either up- or downstream inevitably often involves the overcoming of obstacles such as rapids and waterfalls, yet many species pass up and down even tumultuous rivers. The little benthic species can move in swift water by keeping in shelter, as do loaches; and species with suckers or friction discs can often ascend wet vertical faces beside falls, as does *Gobiomorphus coxii* in Australia (Lake, 1959).

Most fish are stimulated to move by rising water, and when the movement is to be upstream this enables them to pass over riffles with greater safety, because the increased width at such points spreads out the discharge, and provides zones of slower water which are nevertheless deep enough to swim through. In rapids even the most powerful swimmers such as trout or the marinka, *Schizothorax*, remain close to the bed where the current is slowest (Shadin, 1956); but falls necessitate jumping, as no fish can swim against falling water. Many fish jump very effectively; Atlantic salmon can clear 11 ft. (3·3 m.) (Jones, 1959), small falls and beaver dams are no obstacles to trout (Gard, 1961), and, as we have seen, even carp can jump over low dams.

Stuart (1962) has made a study of the leaping of fish at falls, and he has shown that they always leap from the crest of the standing wave where there is an upward component of the turbulence (Fig. 1, 7). They swim up through the surface with their tails vibrating, so that as soon as they land on or above the crest of the fall, at which their aim is visual, they are swimming. They thus move rapidly upstream away from the lip of the fall. The stimulus to leap is the impact of the water on the pool, and larger fish respond to greater impacts. Thus, as the discharge increases during a spate, larger and larger fishes are stimulated to leap. But if the water continues to rise to the point where entrained bubbles significantly reduce the density of the water in the wave, or if it is cold, leaping is inhibited.

In contrast to the actively swimming fishes, young eels ascend relentlessly and are much less affected by discharge. More, however, ascend successfully in wet years, and their movement is inhibited by low temperatures (Lowe, 1951; Sörensen, 1951). The elvers keep close to cover, often along the very edge of the water, and they clamber up falls and chutes like little wet worms beside the water. In the Irish River Bann, where there is an important eel fishery, special passes have been constructed for them on barrages. These consist of a sloping trench down each side of the face of the dam, and they contain a rough rope made of straw through which water trickles continuously. When the elvers are running it is an astonishing sight to see many tens of metres of these ropes literally writhing with little blackish elvers on their way up to Lough Neagh.

Descending fish, such as smolts, adult eels, and lampreys, are also stimulated to move by rising water (Allen, 1944, 1951; Jones, 1959; Bertin, 1956; Applegate and Brynildson, 1952), and, as has already been stated, most descend tail first as long as the current remains swift. They thus drift rather than swim down, and the migration is really more a loss of position than an active process. Eels, however, seem to swim actively downstream, and salmon smolts will also swim if the current is slack or the water is unusually warm (Hoar, 1951).

Under normal circumstances descending fish readily overcome obstacles, and the cushioning of the water prevents damage at falls, or at any rate at falls which are small enough for them or their parents to have ascended. But descending fishes follow the bottom contour not the surface (Stuart, 1962). This presents no problems in a natural stream, but where man has erected dams the habit leads them not over the fall, but to the bottom of the upper side of the dam, where they tend to become held up. This has been shown seriously to interfere with the production of smolts in salmon rivers in Scotland (Mills, 1964), and it may be an important factor in the reduction of migratory fishes in many parts of the world which has been attributed to barrages, even when they incorporate fish-passes allowing the adults to ascend.

CONTROL OF MIGRATION

We have seen that both upstream and downstream migrations are usually initiated by rising water, and for large fishes these increases in discharge have to be considerable for successful movement. In Scottish rivers for example, salmon need 30–50 per cent of the mean annual flow to ascend the lower and middle reaches, and up to 70 per cent to ascend head-waters (Baxter, 1961). As we have seen in Chapter I, there are long periods during which discharges of these magnitudes do not occur, so ascending fish are held back and then go through in large numbers when suitable conditions obtain. An exception, however, is the West African barbel, *Barbus occidentalis*, which begins its migration at the end of the dry season, before there is any increase of discharge (Welman, 1948).

Water temperature is also important, but the response varies with the species. Young eels move up only when the temperature is above 15 °C., young rainbow trout move down only when the water is colder than 14 °C. (Sörensen, 1951; Northcote, 1962), but brown trout may be stimulated to descend by rising temperature (Needham and Cramer, 1943). The last conclusion is, however, possibly spurious, as the observation was made in Convict Creek, California, in May, when both day-length and temperature were increasing, and it has more recently been shown experimentally that young rainbow trout respond to long days by moving downstream, and to short days by moving up (Northcote, 1958). This, coupled with an inhibitory upper temperature, say the 14 °C. cited above, would explain the well-known fact that the smolts of salmonids usually descend in the spring or very early summer.

Generally speaking light is inhibitory, so most movement occurs at night or when the water is turbid, and migrating fish are thus not readily seen; but there are exceptions to this general rule. Small elvers move at all times irrespective of light, and leaping fish depend upon light to aim their jumps. Salmon therefore leap at falls during the day-time and little if at all at night (Neave, 1943*b*). In general, however, when fish are migrating they are quiescent by day and active by night, and as they all start moving as darkness falls the migration tends to be maximal in the early hours of the night. This point was nicely demonstrated by Elliott (1966) from his catches of brown trout fry in drift nets fished in an English trout-stream.

The physiological changes which cause fishes to respond to stimuli by migration are outside the scope of this book, but we can note here that it is well established that they are connected with increased thyroid activity, followed by altered responses to the environment (e.g. Fontaine, 1961, 1964). They also often involve morphological changes so that the migrating 'silver eel', for example, is clearly distinguishable from the growing 'brown eel' stage which precedes it, and the descending 'smolt' of the migrating salmonids is very different in appearance from the earlier 'parr' stage. Similarly, fishes migrating to breeding grounds are often distinguishable from non-migrating specimens by the development of breeding coloration, head tubercles, etc., according to the group to which they belong.

Finally, as a footnote to this section on migration, we may draw attention to the odd discovery of Bainbridge (1960) that, in the European dace, the rainbow trout, and the goldfish, the percentage of the body weight which is muscle increases up to a total length of fish of about 17·5 cm., and it then becomes more or less constant in all three of these rather dissimilar fishes. It is perhaps a significant fact that a total length of about 17·5 cm. is the size of a number of fish species at the time that they make long journeys, and it may be that fishes which remain on the breeding grounds for some while after hatching must attain this maximum of muscular development before they are fit to face a long journey. It is, for instance, known that the age of salmon smolts from any one river can vary considerably, but they are all more or less the same size when they migrate (Frost and Went, 1940). Similarly female eels, *A. anguilla*, are mostly 54–60 cm. in length when they migrate although they may be from 9 to 19 years old and males are mostly about 40 cm. long but from 7 to 12 years old (Frost, 1945*b*). It would be interesting to know if in this unfish-shaped fish the percentage of muscle increases in a different way, and if it is different in the two sexes.

BREEDING HABITS

Nearly all species of fish have fairly well-defined breeding habits and requirements, and, as the latter are usually much more restricted than those needed for ordinary life, they determine to a large extent the suitability of particular rivers and streams for the various species. Those such as the common shiner, *Notropis cornutus* of North America, which will breed almost anywhere on gravel or sand, in shallow or in deep water (Raney, 1940), or the blacknose dace, *Rhinichthys atratulus*, which has three races with different breeding requirements (Schwartz, 1958), are exceptional. More usually suitable breeding sites for each species are narrowly definable, as they are, for example, among the darters (Winn, 1958*a*, *b*).

The great majority of freshwater fishes spawn on a solid surface which they select fairly carefully. In still waters rooted vegetation is the site favoured by most species, but this is of much less importance than the bare substratum among running-water species. Unexpectedly, there are also some riverine fish which lay pelagic, or perhaps we should call them drifting, eggs. This is quite out of line with the usual tenet that freshwater animals lack planktonic stages because they would be carried to sea before having time to settle in the parental environment. We have seen that at least one very common eastern European bivalve, *Dreissena*, has pelagic larvae and yet lives in rivers (p. 99); so this, coupled with river fishes which lay drifting eggs, emphasizes the complexity of biological phenomena and the absence of any clear rules which bear close scrutiny. However, after this digression we will consider the various breeding sites separately.

STONES AND GRAVEL

Many species select or prepare definite nest-sites on or in stony or gravel substrata, which are then guarded by the male. A common site among stream fish is a flat area under a large stone, to which the eggs are stuck in a cluster after a more or less elaborate courtship. This habit is common to the sculpins (Smyly, 1957; Zarbock, 1951), several species of *Etheostoma*, such as the johnny, the spotted and the fantail darters (Raney and Lachner, 1939; Atz, 1940; Lake, 1936; Winn, 1958*a*, *b*), the minnow *Pimephales* (Hankinson, 1919; Markus, 1934), and *Gobiomorphus*, the red-finned bully of the family Eleotridae in New Zealand (McDowall, 1965*a*). It requires the presence of suitably elevated stones, and thus restricts these fishes to habitats where such stones are available. Other fishes which guard nests are the longear sunfish, *Lepomis megalotis*, which excavates a saucer-like depression in gravel, which must be both shaded and in shallow water (Hankinson, 1919; Witt and Marzolf, 1954), and the sticklebacks *Eucalia*, *Gasterosteus*, and *Pygosteus*, which require algal filaments for making their nests and a suitable substratum, of sand or plants according to genus (references in Hynes, 1950; Winn, 1960).

Other species dig pits in gravel in which the eggs are laid but are not guarded. The simplest are made by the stoneroller, *Campostoma anomalum*, of which the males carry stones away from the breeding area, thus forming a pit in which the eggs are laid (Hankinson, 1919; Lennon and Parker, 1960; Miller, 1962). Some lampreys make pits by carrying stones in a similar manner, although others also waft small stones out of the area by holding on to a fixed stone and fanning with the tail (Raney, 1939; Dendy and Scott, 1953; Hagelin and Steffner, 1958; Seversmith, 1953). For these fish the gravel must be of a suitable size and reasonably free of sand and silt, and the fanning out of a pit clearly depends upon the current to carry the displaced stones away.

Similar to fanning is the vigorous 'cutting' of pits performed by salmonids. This has been described many times and for most species of salmon and trout which live in running water (Hazzard, 1932; Smith, 1941; Jones and King, 1949; Jones and Ball, 1954; Shetter, 1961; Needham, 1961; Needham and Taft, 1934). The female turns on her side, and by vigorous flapping of the tail she lifts stones by negative pressure, and they are then carried downstream by the current. After each spawning act, during which the male joins the female in the so-called 'redd', the female moves forward and digs again, thus burying the newly laid eggs. This procedure is well described by Jones (1959), to whom the reader should turn for details.

An odd fact about breeding in the Atlantic salmon is that young male fish, which are still in the pre-migratory stage, take part in the spawning and can fertilize eggs, although their presence, unlike that of adult males, does not stimulate the females to dig (Jones and Orton, 1940; Jones and King, 1952). The reason for the participation of the small males is obscure, but Jones (1959) suggests that it may ensure that the bottom-most eggs are fertilized. The deep hole forms a pocket of dead water from which eggs and sperm are not swept out by the current. This must apply to all stream fishes which dig a hole, but the salmon hole is 15–20 cm. deep, and the adults are longer than the hole and so cannot get their vents right down to the bottom. The eggs, however, fall there, and the little male can swim right down to ejaculate very close to them and this probably ensures more complete fertilization.

The quantities of gravel disturbed by these large fish are so great that the spawning site is clearly marked, and it is possible to count the redds from the banks of small rivers, or even to estimate their numbers from aerial photographs, as has been done for *Onchorhynchus nerka* in Alaska (Eicher, 1953).

Some North American cyprinids, most notably the chubs of the genera *Nocomis* and *Semotilus* and minnows of the genera *Parexoglossum* and *Exoglossum*, make piles of pebbles which are carried by the male to form spawning sites (Hankinson, 1919, 1932; Van Duzer, 1939; Raney, 1947; Lachner, 1946, 1952). The males carry the stones in their mouths and can make very large structures. *Nocomis* piles may be over 30 cm. high and a metre or more in diameter, and the fish begins first by excavating a pit, removing stones and sand, then replacing the stones and completing the pile with other stones. Reighard (1943) calculated that this involves swimming over 25 km. and carrying stones weighing under water a total of 40 kg.—a truly herculean task for a small fish.

An important point about these constructions, which perhaps explains the newly assembled piles of stones and the sand-free excavation beneath them made by *Nocomis*, is that loose gravel allows the water to pass through and hence to bring oxygen to the buried eggs. The pile in the current provides this, but fishes which merely dig in gravel have to find places where movement of water through the gravel occurs before they start to dig. This has been particularly studied with reference to spawning by salmonids and to the oxygen content of the interstitial water, which has been shown to be related to the percentage hatch of pink and chum salmon eggs (Pollard, 1955; McNeil, 1962). In years with low discharges the survival may be quite low because of death caused by oxygen shortage.

Brown trout, steelhead trout, and Atlantic salmon select places for spawning where there is a down-flow of water into the gravel (Needham and Taft, 1934; Stuart, 1953*a*, *b*, 1954). Such places occur at the downstream ends of pools, where the water flows into riffles; but there are reports that the brook trout selects places where there is an upward flow out of the gravel, as at the tail end of riffles. In either event the buried eggs are constantly bathed in new water, and any considerable introduction of silt which blocks its passage is usually fatal to them.

Species which construct nests or redds are, therefore, restricted not only in respect of the size of material of the substratum, which they must be able to move, but by the need to be free of silt; and salmonids, and probably some other fishes, are also restricted to places where there is na intra-gravel flow of water. Suitable sites are therefore often limited, and when great numbers of breeding adults are present they may destroy eggs already laid by digging them up. This so-called 'overcutting' must act as a density-dependent control mechanism which restricts the numbers of salmonid fishes that can breed successfully in many streams (Hobbs, 1940).

A great many species breed on gravel or stones but construct no nests. The eggs are partially buried by the tumult of shoals of spawning fishes or by the activity of pairs, or they are merely swept under cover by the turbulence made by the excitedly swimming fish. This method of oviposition is found in the sterlet, *Acipenser ruthenus* (Janković, 1958), the paddlefish, *Polyodon spathula* (Purkett, 1961), the whitefish, *Coregonus* and *Prosopium* (Fabricius and Lindroth, 1954; Brown, 1952), the grayling, *Thymallus* (Janković, 1960; Müller, 1961), several species of darter in which the female shuffles herself down into the gravel and so buries the eggs (Reeves, 1907; Hankinson, 1932; Petravicz, 1938; Winn, 1958*a*) the loach, *Nemacheilus* (Smyly, 1955), many small cyprinids (Smith, 1908; Tack, 1940; Frost, 1943; Weisel and Newman, 1951), and several genera of catostomids, *Catostomus*, *Moxostoma*, and *Hypentelium* (Hankinson, 1919; Reighard, 1920; Stewart, 1926; Raney and Webster, 1942; Raney and Lachner, 1946). Doubtless the list could be greatly extended, since this is probably the most common pattern of breeding among running-water species.

Nearly all fishes which spawn in this way move on to clean gravel to do so, often upstream for some distance and into shallower and swifter water than is their normal adult habitat. They thus definitely select a rather specialized breeding-site which differs from their normal requirement. The eggs of some species are slightly sticky and adhere to the stones until they hatch, but those of many species merely roll into interstices and small pockets of dead water, and remain undisturbed by the current.

An interesting feature of fish-spawning on gravel, which has been observed in North America but which probably also obtains elsewhere, is that the elaborate pebble-nests of the chubs, *Nocomis* and *Hybopsis*, are often also used as spawning sites by other fishes. These include several species of *Notropis*, the stoneroller, *Campostoma*, and some darters, *Hadropterus* (Hankinson, 1932; Raney, 1940, 1947; Reighard, 1943; Lachner, 1952, 1956; Outten, 1957, 1958). Thus definite associations of species tend to be built up, with different chubs and minnows spawning together in various geographical areas. This is perhaps a form of commensalism, as many of these species are territorial while they are breeding and so may help to keep the nest clear of predators, but it also, because of the external fertilization, leads to hybrid offspring.

OTHER SUBSTRATA

Some species which live over sandy substrata spawn on sand, either making a small fanned-out depression, as does the log-perch, *Percina caprodes* (Reighard, 1913), or scattering the eggs over the bottom as do several species of *Notropis* (Hankinson, 1930; Marshall, 1947*a*; Hubbs and Walker, 1942) and the burbot, *Lota* (Fabricius, 1954). The eggs of these fishes are markedly sticky, and they soon become coated with sand and invisible on the bottom. But sand is not a very usual oviposition site, and even some species which are adapted to sandy substrata move to faster water or to rocks to spawn; such for example is the glassy darter, *Etheostoma vitreum* (Winn and Picciolo, 1960).

Mud appears to be an unpopular site used by few species although the Murray cod, *Maccullochella*, is said to make shallow nests on muddy banks of rivers in Australia (Lake, 1959). Most fishes which live in muddy reaches either move upstream to gravel, or, like still-water species, spawn on plants which are either aquatic or, very often, terrestrial but covered with flood water. This applies to many African and South American species (Jackson, 1961; Kleerekoper, 1955), to the gizzard shad, *Dorosoma cepedianum* (Miller, 1960), the bridled shiner, *Notropis bifrenatus* (Harrington, 1947) and to some other fishes in North America. It also applies to the carp, perch, bream, pike, and various other fishes of sluggish waters in Europe (e.g. Svärdson, 1949; Fabricius, 1951). The availability of inundated areas of flood plain is, therefore, important to many non-migratory species of base-level rivers.

Among the darters, however, there are a few species that spawn on vegetation even though they do not inhabit base-level rivers. Possibly this reflects the enormous proliferation of species in this sub-family, and, because some are very restricted in their choice of breeding site, it serves as a mechanism ensuring specific isolation. The Iowa darter, *Etheostoma exile*, spawns in deep pools on rotting vegetation (Jaffa, 1917), rather a strange place for eggs. The greenside darter, *E. blennioides*, spawns only on *Cladophora* (Fahy, 1954), the greenthroat darter, *E. lepidum*, is perhaps unique among fishes breeding on riffles in that it lays sticky eggs on vegetation (Hubbs and Strawn, 1957), and the least darter, *E. punctulata*, usually spawns on plants (Petravicz, 1936).

DRIFTING EGGS AND LARVAE

Quite a number of riverine species lay buoyant or semi-buoyant eggs which float in the water and are carried downstream by the current while they develop. These include the callop, *Plectroplites*, the silver perch,

Bidyanus, and the macquarie perch, *Macquaria*, of Australia (Lake, 1959), the goldeye, *Hiodon alosoides*, the shad, *Alosa sapidissima*, and the white perch, *Roccus* of North America (Battle and Sprules, 1960), the herrings of the Volga, *Caspialosa* (Behning, 1928), and a number of species in the rivers of eastern Asia—the grass carp, *Ctenopharyngodon*, the bigheads, *Hypophthalmichthys* and *Aristichthys*, some gudgeons, *Saurogobio* and *Rostrogobio*, and the razor fish, *Pelecus* (Nikolsky, 1963). Probably this list could be extended, as other freshwater fishes, e.g. the drum, *Aplodinotus*, are known to lay pelagic eggs in lakes; they probably do so also in rivers.

The eggs of most of these species are not unusually large at laying, but they soon swell to a diameter of several millimetres and often become very transparent and difficult to see. They also hatch fairly soon, and, as there is a tendency for spawning to occur when the discharge is increasing, the eggs are often carried out of the channel on to flooded land. This presumably prevents some downstream loss and enables hatching to occur in suitable nursery areas. It should also be noted that most, possibly all, the species which lay drifting eggs migrate upstream to do so, thus there is no net downstream movement of the species as a whole. Clearly also, as long as they do not get lost into unsuitable habitats, transparent drifting eggs are less liable to predation than are demersal ones. It seems probable that this type of egg is more common than is at present realized, especially as their invisibility and short development-periods tend to militate against their discovery. *Hiodon* eggs, for instance, are laid shortly after the ice goes out of Canadian rivers, and this is not a time of year when much work has been done on river plankton. It seems likely, therefore, that similar habits in other fishes await detection.

Even some species with demersal eggs have a drifting stage soon after hatching, as the larvae swim up into the water and are carried downstream. Such are the spoonbill, *Polyodon*, of which the larvae leave the shallow bars used for spawning as soon as they hatch (Purkett, 1961), the whitefish, *Coregonus lavaretus*, which moves into some Swedish rivers from the Baltic Sea to spawn (Fabricius and Lindroth, 1954), and the bully, *Gobiomorphus*, of New Zealand, of which the fry become positively phototropic after leaving the nest and drift down probably to the sea (McDowall, 1965*a*).

FACTORS CONTROLLING BREEDING

Nearly all fish species breed at a definite season, and while the factors controlling this undoubtedly vary from species to species the most important seem to be day-length and temperature.

It has long been known that decreasing day-length causes trout to come into spawning condition, and that they can be caused to spawn out of season by manipulation of the light regime (Hoover and Hubbard, 1937; Hazard and Eddy, 1951). Day-length also controls the breeding cycle of at least some cyprinids; *Notropis bifrenatus* can be brought into spawning condition by exposure to long days (Harrington, 1957). On the other hand some species, e.g. the greenthroat darter, *Etheostoma lepidum*, seem to be uninfluenced by day-length (Hubbs and Strawn, 1957), and even the trout may possibly be influenced by other factors in addition to day-length and suitable temperature. Van Someren (1952) noted that on Mount Kenya, right on the equator, brown trout appear to breed in January and April and rainbow trout in August. This, if it is true, does not correlate with any change in temperature or day-length, nor with alterations in discharge. The long rains in the mountain occur in April to June and the short rains in November and December. As has been stated elsewhere in this book, high-level tropical streams are clearly ideal situations for study of the effects of ecological factors, and they deserve much more attention than they have received. For instance, the cut-throat trout spawns in May in the Logan River, Utah (Fleener, 1951) which is late for a trout, although it may be caused by the absence of winter increases in discharge. It would probably be very instructive to compare its timing with other species in tropical streams.

Many species spawn only above or below certain temperatures, and this is the main distinction between the so-called warm-water and cold-water fishes. In the latter breeding may occur at very low temperatures —only a few degrees above freezing-point in salmon and trout and the burbot (Fabricius, 1954), 6–10 °C. in the grayling (Janković, 1960), 3–12 °C. in the American smelt, *Osmerus mordax* (Hoover, 1936), 8·7 °C. in the river lamprey, *Lampetra fluviatilis*, and 10–11 °C. in *L. planeri* (Hardisty, 1961*a*). Many warm-water species start to breed only at much higher temperatures, and so they are successful only in places where high temperatures are available for long enough for breeding and early development.

Another factor which is known to influence breeding is discharge, a rise in which is often the signal for movement and reproductive activity, as we have seen above. These three factors, alone or in various combinations, are primarily responsible for the timing of breeding in most species.

LARVAL LIFE OF LAMPREYS

The great majority of fishes behave in very much the same way in both their young and adult stages, although there are, as we have seen, some differences connected with rapid migration away from spawning sites. Larval fish do not, however, react normally to current and visual stimuli until the yolk sac is almost fully absorbed (Stuart, 1953*a*; Bishai, 1960). They remain hidden until this occurs, and then begin their normal activities.

Lampreys are exceptional in that they have a very long period as filter-feeding ammocoetes in silty areas along the borders of streams. There seems to be some doubt as to how long this period is. Seversmith (1953) reports three years for *Lampetra aepyptera*, with the last two years spent in increasingly coarse bottom materials. Hardisty (1961*b*), from his own work and a review of the literature, reports five and a half years for *L. planeri*, *L. fluviatilis*, *Eudontomyzon danfordi*, and *Petromyzon marinus*, and Leach (1940) reports six and a quarter years for *Ichthyomyzon fossor*, with the last year as an inactive period. On the other hand, Stauffer (1962) records at least seven years for *P. marinus*. Probably different species grow more or less slowly, and it seems likely that the growth rate varies from place to place and from individual to individual, but it is worth stressing that an enormous proportion of the life of a lamprey is spent as a filter-feeding larva, and that the role of even parasitic species in this capacity is probably of greater ecological significance than that of the conspicuous and much more frequently observed adult stage.

CHAPTER XVIII

Feeding habits of fishes

Although there is no very definite evidence that any running-water species of fish is limited by food supply it is clear that the availability of suitable food must be a factor of considerable ecological importance, and the diets of fishes have received much attention. It is also, of course, through their feeding habits that fish, which when present form a considerable proportion of the total biomass, exercise their major influence on the biotic communities in rivers and streams.

Most studies have been based on analysis of stomach contents, and few direct observations of feeding habits have been made in the field. This is because of the obvious difficulty of watching fishes under water in their natural habitats; but two notable exceptions are the studies of Needham and Jones (1959) and Keenleyside (1962). The former watched trout through a glass plate so placed that part of a stream could be observed as through the wall of an aquarium, and the latter studied them while scuba-diving in streams. It would seem that more work of the latter type might be profitable, as fishes permit divers to approach them closely, and direct observation might produce information which can only be guessed at from analysis of stomach contents.

Examination of stomach contents presents many problems, not the least of which is the recognition of the organisms which have been eaten. Only too often this work has been done by people whose knowledge of the local flora and fauna was clearly inadequate to obtain any but a very general impression of the diet. This is especially true when, as apparently often happens, no attempt is made to look at organisms other than those recovered from the fish. Even a skilled phycologist or entomologist is at a disadvantage if he has nothing but mangled remains to study without the supplement of well-preserved general collections from the habitat where the fishes were feeding. With such collections to hand identifications of eaten organisms are greatly facilitated as no habitat has the full range of species which a damaged specimen taken from a fish may resemble. It cannot therefore be too strongly stressed that stomach analysis should *follow* a thorough faunistic and floristic study of the habitat.

Once organisms are identified there remains the problem of how the data should be handled. Is it meaningful to record numbers of specimens eaten even if this can be done? Often it can not because of breakage and digestion, and accompanying this there is always the doubt that perhaps digestion very rapidly renders some things, e.g. flagellates, protozoans, and worms, unrecognizable. And even if one can count numbers how does one equate say diatoms and snails? Ideally one should work with some absolute unit such as the calorie, or some approach to this such as the volume of each type of organism eaten. It is, however, very tedious, and none too accurate, to attempt to measure volumes, simply because of the mechanical difficulty of sorting crushed and often chewed and mucus-encased materials. The whole subject of the presentation of data has been discussed by Hynes (1950), who advocates the use of points, assigned to each fish on the basis of the fullness of its stomach, which are then divided between the various items in proportion to their estimated volumes. Thus, if 20 points are allotted to a full stomach, a half-full stomach, containing what are judged to be equal volumes of organisms A and B, is recorded as A, 5 points and B, 5 points. This appears to be a quick and fairly accurate measure of the food found in a stomach. It enables a large number of specimens to be analysed in a reasonable time, and it gives results for the percentage composition of the diet similar to those of more laborious methods. The method used must, however, be adapted to the inquiry in hand. If one is interested in the effect of a fish species on one particular invertebrate, say *Simulium*, the rest of the food may be of little consequence, but it is clearly necessary to count the *Simulium* in the stomachs.

The way in which fishes feed varies greatly, and this affects their diet. Most seem to hunt by sight and swim around seeking food, but many, such as sturgeons, loaches, and many cyprinids, feel for their prey with barbels trailed on or through the substratum. The little amblyopsid, *Chologaster*, which comes out of underground water at night into spring streams, also feeds by touch (Weise, 1957), as do presumably its quite eyeless relatives. Some fishes, particularly catostomids and some cyprinids, scrape stones and roll them over and scrape them again, and the North American cyprinid, *Campostoma*, is called the stoneroller because of this habit. The hog sucker, *Hypentelium*, similarly rolls stones over and sucks up ooze from beneath them (Raney and Lachner, 1946). Trout feed primarily on drift, and do not often seek food down on stream beds, in contrast to the grayling, *Thymallus*, which searches the substratum and feeds there. Thus, although these two species often occur together and eat the same prey, the differences in their methods of feeding result in very different proportions of the various food organisms in their stomachs (Müller, 1954*e*).

Feeding habits, of which we know far too little, therefore greatly influence the composition of the diets of fishes. Active and fugitive prey are less likely to be detected by a trailed barbel than are sluggish cryptic species. They are, however, more likely to be seen and caught by a fish which hunts by sight, and so are more often eaten by such fishes, even though as a source of energy they may be satisfactory for either type of fish.

DIETS

Generally speaking most species of fish have fairly well-defined diets; they eat plants or mostly plants, or invertebrates or mostly invertebrates, or fish or detritus. But within these generalities the diets are often very varied; they also change with locality, age, and season, so the influence of any one species is spread widely in the biotic community.

Some species feed largely or entirely upon algae. These include the mochokid *Chiloglanis* of Africa (Jackson, 1961), *Plecoglossus* of Japan (Kawanabe, 1959), and several cyprinids in North America, e.g. *Notropis* spp. and *Campostoma* (Kraatz, 1923; Marshall, 1947*a*; Griffith and Vorhees, 1960). Some catostomids, e.g. *Carpiodes* (Buchholz, 1957), and the gizzard-shad, *Dorosoma* (Miller, 1960), are also reported to feed very largely on algae. Similarly the filter-feeding ammocoete larvae eat large quantities of diatoms at some seasons, and they appear to be selective in that each age-group tends to ingest a particular species at any given time (Schroll, 1959). They also ingest a great deal of detritus, particularly in winter when the silt banks in which they live are relatively free of diatoms. The same probably applies to Catostomidae.

In fact though, species which are mainly algal feeding are comparatively rare in running water, but many cyprinids among those which have been studied in Europe and Japan supplement a diet of invertebrates with some algae (Radforth, 1940; Hartley, 1947; Kawanabe, 1959). This can be quite an important component, e.g. over 20 per cent by volume in the European minnow (Maitland, 1965*a*; Frost, 1943), but it is usually much less. Other species combine invertebrates with higher plants. The roach, *Rutilus*, which is very common in Europe, eats substantial amounts of higher plant material of both aquatic and terrestrial origin (Hartley 1947), as does the carp (Rehder, 1959). The latter, and the smallmouth buffalo, *Ictiobus bubalis*, also eat seeds which have fallen into the water. Berner (1951) records that 29 per cent of the food in carp and 59 per cent in the buffalo was seeds in his samples from the Missouri River; and the channel catfish, *Ictalurus punctatus*, also eats considerable quantities of seeds (Bailey and Harrison, 1945). It is clear, therefore, that some species which are typical of fairly slow-flowing waters subsist to a great extent directly upon allochthonous organic matter.

The most widespread and important foodstuff of running-water fishes is undoubtedly invertebrates which form most of the diet of a great range of types of fish. These include spoonbills (Hoopes, 1961), sturgeons (Behning, 1928; Petropavlovskaya, 1951; Greze, 1953; Janković, 1958; Hoopes, 1961); salmonids (e.g. Southern, 1935; Carpenter, 1940; Allen, 1941*a*, 1951; Idyll, 1942; Van Someren, 1952; Müller, 1954*e*, 1961; Nilsson, 1957; Ellis and Gowing, 1957; Kawai, 1959; Tebo and Hassler, 1963; Reed and Bear, 1966), coregonids (Laakso, 1951; Sigler, 1951), galaxiids (Butcher, A. D., 1946), eels (Frost, 1946; Burnet, 1952*a*), the great majority of cyprinids, many of which supplement their invertebrate diet with some plant material (e.g. Radforth, 1940; Frost, 1943; Lachner, 1946; Hartley, 1947; Schwartz and Norvell, 1958; Kawanabe, 1959; Müller, 1954*b*; Rehder, 1959; Maitland, 1966), cobitids (Smyly, 1955; Kawanabe, 1959; Maitland, 1965*a*), cottids (Zarbock, 1951; Smyly, 1957), perches and darters (e.g. Karr, 1963, 1964), catfishes (e.g. Bailey and Harrison, 1945; Hoopes, 1961), and most members of groups which have rather few representatives in fresh water. Such are the true basses, drums, macquarie perch, and gobies (e.g. Butcher, A. D., 1946; Kawanabe, 1959; Hoopes, 1961; McDowall, 1965*b*). It is indeed far easier to list groups of fishes which do not feed on invertebrates than those that do. Such a list would include only lampreys, some purely piscivorous families such as the Esocidae, the algal eaters we have discussed above, and probably some of the Catostomidae.

The range of invertebrates eaten is very wide; most species are very catholic in their tastes, and in general it can be said that they eat what is available. There are, however, some fairly well-defined influences of feeding habits on the diets. Trout of all kinds, for instance, which feed on drift, take great quantities of food at the water surface, so that in summer time 40–50 per cent or even more of their food is terrestrial insects (Allen, 1941*a*; Kawai, 1959; Tebo and Hassler, 1963; Reed and Bear, 1966). Similarly, many swift-swimming cyprinids feed very often at the surface and so live to a great extent on aerial insects. Such are the dace of North America and Europe, *Clinostomus* and *Leuciscus* (Hartley, 1947; Schwartz and Norwell, 1958), and *Thymallus* also feeds at the surface in the summer (Müller, 1961). We can take it then that many invertebrate-eating fish obtain a great deal of their food from sources outside the water, and that in temperate climates this is particularly important during the summer. It is also known that even such strictly bottom-feeding fishes as carp and eels are often found to have eaten terrestrial worms and caterpillars brought in by high water (Burnet, 1952*a*; Rehder, 1959); so direct feeding on allochthonous material is by no means confined to surface-feeding species.

A few species are piscivorous from an early age, e.g. the pikes, and many become so after a period of growth on invertebrates; such are the black basses of North America (McLane, 1948; Tate, 1949) and the tiger fish, *Hydrocyon*, of Africa. These are definite piscivores, but a great many other carnivorous fishes turn to smaller fish as food to a greater or lesser extent as they grow to a large size. This applies not only to active fishes such as the perch, the drum, and sometimes the carp, but also to fairly inactive types such as catfishes and, at any rate in New Zealand, to eels. It is thus impossible to assign a definite position in the food nexe to a large proportion of the species. Some eat invertebrates and plants and so are both primary and secondary consumers, and some eat both invertebrates and fishes and so are both secondary and tertiary consumers. Darnell (1961, 1964) has indeed suggested that the only meaningful way to present data on the food of many species is in the form of a trophic spectrum, in which each major item is shown as a band of which the width is proportional to its importance in the diet.

Some species occasionally eat terrestrial vertebrates. The piranha, *Serrasalmus*, of South America is legendary because of its attacks on large mammals, including man; but large pike and the European catfish, *Siluris*, have been known to catch and eat swimming mammals and birds, and Nikolsky (1963) states that grayling feed on shrews migrating across the River Pechora in northern Russia.

The last important source of food is detritus. We have seen that ammocoetes take in a great deal of detritus despite the fact that they appear able to select different diatoms (p. 366). In wintertime when diatoms are relatively scarce detritus forms the bulk of the gut contents (Schroll, 1959). Similarly detritus often forms the bulk of the stomach contents of quite a range of species. These include the gizzard-shad, *Dorosoma*, the cyprinids *Campostoma*, *Pimephales*, and several catostomids, e.g. *Erimyzon*, *Catostomus*, *Hypentelium*, and *Carpiodes*, and some top minnows, e.g. *Mollienisia* and *Xiphophorus* (Hankinson, 1910; Kraatz, 1923; Stewart, 1926; Thompson and Hunt, 1930; Raney and Lachner, 1946; Starret, 1950*b*; Buchholz, 1957; Minckley, 1963; Darnell, 1964; Keast, 1966). Stomachs of these fishes usually also contain small organisms such as protozoans, rotifers, and algae, but the preponderance, at any rate at times, of detritus indicates that it is eaten for its own sake and that nourishment is derived from it.

CHANGES WITH SIZE AND SEASON

Some fishes, e.g. sturgeons and brown trout, begin to feed at an early age on the type of food, here benthic invertebrates, which they will eat for some time (Petropavlovskaya, 1951; McCormack, 1962; Yukhimenko, 1963); but even in invertebrate-feeders there is a clear tendency to eat larger prey as the fish grows. This is shown by some numerical data given by Tebo and Hassler (1963) for rainbow trout (Table XVIII, 1).

Length of fish in cm.	2·6–7·0	7·1–12·0	>12·1
No. of specimens	67	168	6
Average volume of stomach contents in ml.	0·56 ± 0·12	1·15 ± 0·10	1·68 ± 0·77
Average number of organisms per stomach	16·5 ± 2·04	16·0 ± 0·43	15·1 ± 7·82

TABLE XVIII, 1. The average total volumes and numbers (± standard error) of invertebrates in stomachs of rainbow trout of different sizes from North Carolina. Data from Tebo and Hassler (1963).

There the differences between the volumes of the stomach contents of the smallest fishes and those of each of the larger groups are significantly different, but the numbers of organisms are not significantly different. We can see, therefore, that there is a statistically significant increase in the size of prey taken as the fish grows. This is a quite general phenomenon and it has been observed by many workers to apply to such varied fishes as salmon and trout, black basses, minnows, dace, darters, catfishes, and gobies (Tack, 1940; Surber, 1941; McLane, 1948; Kuehn, 1949; Lachner, 1950; Daiber, 1956; McDowall, 1965*b*; Maitland, 1965*a*; Keast, 1966). Indeed the change of size can lead to quite a large proportional change in the diet, as was found by Maitland for salmon and trout in a Scottish river. Table XVIII, 2 shows that during the first year of their lives there is a considerable shift in the importance of various major items.

In fishes which ultimately become carnivorous this change continues to increasingly large invertebrates and thence, often through crayfish, to fishes, e.g. in the black basses, *Micropterus* (McLane, 1948; Lachner, 1950) and the flathead catfish, *Pylodictis* (Minckley and Deacon, 1959). Table XVIII, 3 illustrates this change in another fish with a varied diet, the burbot, which however, begins to eat some fishes while it is still

	S. salar Fry	*S. salar* Parr	*S. trutta* Fry	*S. trutta* Parr
AQUATIC INVERTEBRATES				
Oligochaeta	4	—	1	2
Crustacea	1	1	19	6
Plecoptera	5	8	11	7
Ephemeroptera	66	46	20	13
Trichoptera	6	15	8	12
Diptera	9	3	7	1
Coleoptera	2	2	4	1
Mollusca	5	10	—	1
others	—	1	—	—
TERRESTRIAL INVERTEBRATES				
Hemiptera	+	2	2	9
Diptera	+	6	13	26
Coleoptera	—	1	5	10
others	1	4	9	12
No. of fish examined	78	32	60	60

TABLE XVIII, 2. The percentage composition by volume of the stomach contents of young salmon and trout from the River Endrick, Scotland, in 1961. 'Fry' are younger than, and 'parr' older than, six months. + = <0·5 per cent. Data from Maitland (1965*a*).

Length of fish in cm.	2–3·9	4–6·9	7·9–9·0	10–12·9
No. of stomachs with food	35	68	43	5
Oligochaeta	3·1	2·2	23·7	—
Gammarus	30·6	16·9	2·2	—
Cambarus	1·6	10·3	29·7	22·5
Ephemeroptera	16·3	12·4	3·2	+
Hydropsyche	2·3	21·2	4·9	—
Other Trichoptera	2·7	3·7	1·3	+
Diptera	4·6	0·8	3·3	0·3
Other insects	1·7	6·4	3·1	0·5
Other invertebrates	—	0·2	0·1	—
Fishes	3·9	1·4	12·9	72·1
Plant and animal debris	33·2	24·3	15·5	4·6

TABLE XVIII, 3. The percentage composition by volume of the stomach contents of burbot, *Lota lota*, of different sizes from the White River, Michigan. + = <0·05 per cent. Data from Beeton (1956).

quite small. The pike, *Esox*, also begins to eat fishes at a very early age, and unlike the species mentioned above it is then exclusively piscivorous (Spanovskaya, 1963; Keast, 1966).

At least some of the species of which the adults eat algae or detritus also begin life as carnivores and subsist on small animals, and it is quite likely that this applies to most such fishes. The gizzard-shad, *Dorosoma cepedianum*, feeds for a few weeks on zooplankton before taking to its diet of algae and detritus (Miller, 1960), and the white sucker, *Catastomus commersonii*, feeds at the surface on small animals and diatoms before taking to its diet of benthic ooze (Stewart, 1926). Similarly, the European minnow, *Phoxinus*, of which the food of the adult is about 20 per cent algae, begins life as a pure carnivore (Tack, 1940). It would seem, therefore, that at least many of the species which feed on other things have evolved from invertebrate feeders.

In the white sucker the change of diet involves morphological changes in the intestine. In specimens less than 17 mm. long this organ is fairly straight as it is in most fishes; but later it becomes complexly coiled and three times as long as the fish. This increase in length is clearly concerned with the diet, as very long intestines occur in other detritus-eating species, e.g. *Campostoma anomalum*. It seems probable that a metamorphosis takes place in them similar to that occurring in the other direction when detritus-eating tadpoles transform into frogs. That the length of the gut of fishes is related to diet was long ago noted by Nikolsky (1937), who found that in Asian rivers the intestines of piscivorous fishes are shorter, in terms of body length, than those of species feeding on benthic fauna, and that species with long guts—he cites carp and barbel, the first of which certainly eats a lot of plant material—spend much of their time away from main river channels, feeding on plants. It would seem that this subject might be worth experimental study in a species with a varied diet.

Feeding also varies to some extent with the season and the availability of food. For instance, terrestrial insects are scarce in winter and hence not normally present to be eaten, and the changing population of the stream bed caused by invertebrate life-histories is reflected in the diet.

A few species such as the European eel (Thomas, 1962) cease to feed altogether in the winter, and this probably applies to all species which become inactive in cold weather. But the majority of river and stream fishes continue to feed all the year round; darters, sculpins, and trout have been proved to continue feeding in very cold water (Slack, 1934; Maciolek and Needham, 1951; Kajewski, 1955; Daiber, 1956). It would seem, therefore, that earlier impressions that salmon parr feed hardly at all at low temperatures are erroneous (Allen, 1940, 1941*b*), but there is definite evidence that salmonids feed less in the wintertime than during the summer. This is based both on the amount of food found in the stomachs (Lake, 1957) and on experimental studies of feeding and growth of *Salmo trutta* and *Salvelinus fontinalis*. Brown (1946*c*) found that the amounts of food eaten by *S. trutta* were maximal between the temperatures of 10 and 19 °C., but that because their activity was at its peak from 10 to 12 °C., their maximum growth occurred at 7–9 °C. and 16–19 °C. Between 9 and 16 °C. their increased activity used up available energy sufficiently to depress the growth rate. *S. fontinalis*, on the other hand, has been found to eat most and grow fastest at 13 °C., with only one growth-rate maximum (Baldwin, 1956), and this fits well with field data, which indicate that maximum volumes of food are to be found in specimens of this species collected at temperatures between 11 and 18 °C (Benson, 1954). Temperature therefore influences food intake and through it, although not always simply, growth rate.

The changing seasons also indicate very clearly that fishes are very opportunist and eat what is available at the time. Some, such as the creek chub, *Semotilus*, eat almost anything that is available (Dinsmore, 1962), but even species which feed largely on one general type of food switch their attention to particular items as they become seasonally available. Thus young salmon in the Welsh Dee eat *Baetis* in the spring and substitute *Ephemerella* for it in the fall, in conformity with the life histories of the principal species of these genera in that river (Carpenter, 1940); and the mountain whitefish, *Prosopium*, eats mainly Trichoptera in the fall and winter, and moves to Plecoptera in the late winter and to Ephemeroptera in the spring, in conformity with the availability of specimens of suitable sizes at the various seasons (Laakso, 1951). Many other similar studies could be cited, e.g. Neill (1938) on brown trout in Scotland, Thomas (1964) on the same species in Wales, and Janković (1960) on the European grayling.

AVAILABILITY AND SELECTION

When a prey species is abundant in a habitat it does not necessarily form an important part of the diet of a fish which eats it. Fishes are to some extent selective in what they eat, and once a particular specimen has started feeding on one type of organism it tends to continue to do so. This fact is well known to anglers who exploit it by attempting to find out what is being eaten by their quarry on the day that they are fishing and then using that organism, or a facsimile of it, as bait. It is also well illustrated by the fact that the stomachs of individual fishes are often quite filled with only one type of organism. This has led to the concept of the 'availability factor' of individual prey species, which is the ratio of the percentage of that species in the food of the fish to its percentage in the fauna (Allen, 1941*a*, 1942; Hess and Swartz, 1941). When this ratio is 1·0 there is no selection, but if it is more than or less than 1·0 the prey is being preferentially selected or rejected. Several studies have shown that, for prey organisms eaten by trout, high ratios are associated with species which are active, live exposed lives on stones, or are large, and that low ratios are found for cryptic or burrowing species (Allen, 1941*a*; Hess and Swartz, 1941; Tebo and Hassler, 1963). This correlates well with what is known of the feeding habits of trout with its tendency to feed on drift. The ratio may, however, change with the size of the fish, and it may be high for reasons other than availability. Maitland (1965*a*), in a thoughtful discussion, points out that high ratios may result from the feeding of fish in areas away from those where the faunal samples are taken, and that very abundant species may be important in the diet even if their ratio is low. It is, however, a useful way of demonstrating differences in feeding habits between species of fish and of stimulating thought on this aspect of fish behaviour. For instance, when loach, which seek out food with their barbels, were compared with trout, salmon, and minnows in the same habitat, the ratio for chironomid larvae was much higher for the loach than it was for the other species (Maitland, 1965*a*).

The concept of this ratio has not been applied to piscivorous fish, but, as it is known that they also are selective and sometimes feed much more on some species than on others, its use might be found to be thought-provoking.

INTERSPECIFIC COMPETITION

It has frequently been noticed that fish of different species living in the same habitat often eat the same kinds of food, so that one has pairs or trios of species apparently making identical, or very similar, demands on the habitat. It is tempting to conclude from such findings that the species are in competition with one another, and this has often been suggested.

Such groupings have been observed between *Abramis* and *Blicca*, *Perca* and *Acerina*, *Gobio* and *Gasterosteus* (Hartley, 1947, 1948), *Salmo trutta* and young *S. salar* (Frost, 1950; Thomas, 1962), *Gasterosteus* and *Pygosteus* (Hynes 1950), *S. trutta* and *Thymallus* (Müller, 1954*e*), and the young of three species of *Acipenser* (Petropavlovskaya, 1951) in Europe and of three species of trout in Canada (Idyll, 1942). Closer examination of the data, however, always shows that there are differences in the proportions of the various foodstuffs in the diet. And even where such differences are small or, as with the Canadian trout and the three species of sturgeon, become apparent only in the larger specimens, there is no reason to suppose that the similarity of diet indicates direct competition. Other factors may be of greater importance and food in no way limiting. It seems probable, for example, that the two sticklebacks are more regulated by suitable breeding sites than they are by food, and that only when food is even scarcer than breeding sites does the ability of *Gasterosteus* to eat plant material, which *Pygosteus* does not share, give the former an advantage (Hynes, 1950). Similarly, Keast (1966) showed that twelve species shared out the food resources in a smal warm-water stream in Ontario partly by specialization of feeding habit and partly by occupation of different habitats within the stream. Only detritus, algae and Chironomidae appeared to be eaten in large amounts by species living together, and it seems probable that none of these three items was in short supply.

Two recent studies, from which some data are reproduced in Tables XVIII,4 and XVIII,5, illustrate very well that, although lists of food

	Salmo salar	*Salmo trutta*	*Phoxinus phoxinus*	*Nemacheilus barbatula*
Oligochaeta	1	4	6	3
Gammarus	4	13	7	8
Plecoptera	16	6	4	18
Ephemeroptera	46	14	14	7
Trichoptera	6	10	9	25
Simulium	4	1	2	2
Chironomidae	10	8	14	29
Other Diptera	2	3	4	3
Elminthidae	1	+	1	+
Hydroporus	+	+	1	+
Ancylus	2	+	+	—
other aquatics	4	3	1	1
Algae	—	—	21	—
Surface food	1	37	11	+
No. of fish examined	222	458	370	305

TABLE XVIII,4. The percentage composition by volume in the stomach contents of important items in the diet of four species of fishes from the River Endrick, Scotland. + = <0·5 per cent. The specimens were collected throughout the year. Greatly simplified from Maitland (1965*a*).

organisms may be similar between species in one locality, they may be very different within species between localities, e.g. *S. trutta* and *P. phoxinus*. They also show that even when the same animals are eaten by two or more species of fish their relative importances are mostly very

	Salmo trutta	*Phoxinus phoxinus*	*Cottus poecilopus*
Trichoptera			
Hydropsyche	—	+	10
Polycentropus	—	+	5
Sericostoma	17·2	—	—
Drusus	16·3	—	—
Stenophylax	0·3	—	—
(Total)	(33·8)	(0·1)	(16)
Ephemeroptera			
Baetis	+	6·2	7
Ecdyonurus	+	+	3
(Total)	(10·4)	(7·0)	(10)
Plecoptera (total)	2·0	—	19
Chironomidae			
Eukiefferiella	+	40·9	—
Euorthocladius	—	2·8	—
Rheorthocladius	—	4·5	—
Thienemaniella	—	35·4	—
Ablabesmyia	+	—	+
Trichocladius	—	—	+
Glyptotendipes	—	—	+
Pentapedilum	+	—	—
(Total)	(5·5)	(85·2)	(50)
Simuliidae			
Larvae	11·6	3·1	5
Pupae	—	0·5	—
Imagines	10·5	0·8	—
(Total)	(22·1)	(4·4)	(5)
Coleoptera	2·5	2·0	—
Other insects	5·7	—	—
Gammarus	10·6	—	—
Ancylus	5·3	—	—
Fishes	0·3	—	—
Other items	1·8	1·3	—
No. of fish examined	168	83	93

TABLE XVIII, 5. The percentage composition, based on numbers of specimens in the guts, of the food of three species of fish caught in July 1962 from the Morávka River, Czechoslovakia. Modified from Straškraba *et. al.* (1966).

different. In the Scottish study (Table xviii, 4) Ephemeroptera were far more important to the salmon than to the trout, which fed to a greater extent on the surface; and the loach had presumably been seeking its three major food items from beneath stones. In the Czechoslovakian study (Table xviii, 5) the minnows and sculpins, although both feeding heavily on Chironomidae, were taking different genera. Often, therefore, more detailed analysis reveals differences which are not at first apparent.

A most informative study and discussion of competition for food among stream fishes is that of Kawanabe (1959) who worked on several rivers in Kyoto Prefecture, Japan. These rivers contain a number of species of which the most important are *Plecoglossus*, several cyprinids (*Pseudogobio*, *Tribolodon*, *Zacco*, and *Carassius*) loaches (*Cobitis*), and gobies (*Rhinogobius* and *Chaenogobius*). *Plecoglossus* eats algae, and the others invertebrates and algae. Kawanabe found that the free-swimming cyprinids ate more surface food (terrestrial insects) when *Plecoglossus* was abundant and more algae when it was rare, that the chub *Zacco* was displaced from the centre of the river bed when the katadromous *Plecoglossus* migrated in; when the latter were very abundant it was pushed into areas where there are no algae. Similarly the diet of the gobies, loaches, and crucian carp varied according to the presence or absence of the bottom-dwelling cyprinids. When these cyprinids were present in a river the other species tended to eat more algae and fewer insects than they did when they were absent. Thus, Kawanabe argues, surely correctly, that *change* of diet caused by the presence of another species indicated competition for food, not *similarity* of diet. Indeed, one can probably go further and argue that similarity, or near identity, of diet between two species in the same habitat indicates that, however else they may be competing, they are not competing for food.

Returning now to Tables xviii, 4 and xviii, 5, we can perhaps conclude on this basis that the trout and minnows were perhaps competing to some extent for food in the Morávka River but less so, if at all, in the River Endrick. This, incidentally, is directly opposite to the conclusions drawn by the authors of the two papers from which the tables were compiled. That it is possible to reach opposing conclusions from the same data illustrates the point made earlier that we know far too little about how interspecific competition occurs. This important and fundamental aspect of ecology is tending to be forgotten in the present trend towards regarding productivity studies as synonymous with ecology.

CHAPTER XIX

Other vertebrates

Many types of higher vertebrate are associated in some way with running water and some are clearly ecologically important. It can, however, be truly said of most of them that we know very little about their place in the running-water community and that much of our knowledge remains in the natural-history stage of development. There is a wide field open for quantitative ecological work on some of these animals in relation to populations of the fish and invertebrates on which so many of them feed.

THE AMPHIBIANS

The running waters of the world contain some very primitive members of the Amphibia; these are the giant salamander, *Megalobatrachus*, of mountain streams in the Far East, and the hellbender, *Cryptobranchus*, and the mudpuppy, *Necturus*, of eastern North America. These creatures never, or rarely, leave the water, and they are general carnivores which include invertebrates, other amphibians and fishes in their diets (Hamilton, 1932). As *Megalobatrachus* grows to a length of 80 cm. or more and lives in very small streams it seems probable that it is an important predator, as must be the others where they are numerous. *Necturus* is so permanently aquatic that it has become the host for the glochidia of the clam *Simpsoniconcha ambigua* (Howard, 1951), thus, as it were, taking on the role of a fish in more senses than one.

Several families of salamanders also contain genera which breed in small streams and lay their eggs under stones. Their larvae are peculiar in that they lack the little lateral balancers when they hatch, they develop only a small area of fin and have short digits with horny claws and short gills; the adults also have reduced lungs (Noble, 1931). We can see here several adaptations parallel to those which we have discussed among invertebrates and fishes. Some of these animals, e.g. the European *Salamandra* (Salamandridae), go to streams only to breed, but others, e.g. species of *Leurognathus*, *Desmognathus*, and *Eurycea* (Plethodontidae) in North America and *Euproctus* in Europe, spend a great deal of time in small streams. They feed there on aquatic invertebrates, and *Eurycea* feeds even during very cold weather and is thus much less tied to summer temperatures than are most amphibians (Hamilton, 1932; Martof and Scott, 1957; Martof, 1962). The larvae of some of the species take many months to develop (Organ, 1961), so many stream salamanders are permanent members of the aquatic community, and like the primitive urodeles they are probably ecologically equivalent to small predatory fishes.

The Anura are probably less important than the Urodela, although, as we have seen, some frog tadpoles are adapted to life in torrents, and others, in both South America and Asia, are very flattened and live at the edges of streams in shallow water (Hora, 1930; Noble, 1931). They seem, however, rarely to be very numerous, and so they are not comparable to the species which breed in still water and occur in swarms at certain seasons. The adults live outside the water, and it is known of at least one species with a highly adapted tadpole, *Amolops larutensis* of Malaya, that, although they remain near the water and often enter it, most of their feeding is done on land (Berry, 1966).

THE REPTILES

Three groups of reptiles are associated with rivers and streams, namely the crocodiles, the terrapins, or turtles as they are called in North America, and some snakes.

Crocodiles, caymans, alligators, and gharials occur in various areas throughout the warmer parts of the world and are often present in great numbers. All are fish-eating, although crocodiles are well known also to attack mammals and birds. We have, however, very little information on their effects on fish populations nor on the numbers which occur. Although they are spectacular animals they are difficult to observe except when they lie out of the water, and then their numbers seem enormous. It is an astonishing sight to travel along some rivers in Africa, and doubtless elsewhere, where they have not been hunted to extinction for their leather, and to see one or more of these huge reptiles every few yards along the bank. It would seem that, in their unexploited state, they must be of great ecological importance, and that their removal, which in Africa is occurring very rapidly, must have a great influence on fish populations. Unfortunately this seems to be going almost unrecorded.

Young crocodiles, and possibly also the young of other genera, feed on insects and other invertebrates, and Corbet (1959, 1960*b*) records that *Crocodrilus niloticus* feeds in this way until it is about 2 m. long. As the insects are mostly large predatory bugs and dragonflies the effect of these young specimens must be considerable. Here again, however, we lack data.

We are rather better informed on the freshwater Chelonia as they have been the subject of some study in North America. Many species occur in slow streams and rivers and most are omnivorous, feeding on invertebrates and some plant matter. But they also eat small fishes on occasions, and the snapping turtles, *Chelydra*, consume quite large amounts of fish as well as other animals. Detailed information on the food of several North American species is given by Lagler (1943) and Shockley (1949).

Several snakes of the family Colubridae are associated with water and feed there. In North America *Natrix sipedon* is known as the water-snake, and it is particularly common in medium-sized streams where it catches fishes and also frogs and toads (Raney and Roecker, 1947; Shockley, 1949). It also eats some insects and other invertebrates, but not apparently crayfish (Lagler and Salyer, 1947). *N. septemvittata*, the queen snake, on the other hand, is recorded as feeding mainly on crayfish (Raney and Roecker, 1947).

It would seem then that the role of the reptiles in rivers and streams, like that of the urodeles, is parallel to that of the fishes and that the crocodrilids are likely to be of major importance.

THE BIRDS

The following groups in several orders of birds are fairly closely associated with rivers and streams.

Gaviiformes	Divers.
Podicipitiformes	Grebes.
Pelecaniformes	Cormorants, shags, darters, and pelicans.
Ciconiiformes	Herons, egrets, bitterns, storks, and ibises.
Anseriformes	Ducks, geese, swans, and mergansers.
Falconiformes	Fish-eagles and ospreys.
Gruiformes	Rails, coots, and waterhens.
Charadriiformes	Waders and gulls.
Strigiformes	Fish-owls.
Piciformes	Kingfishers.
Passeriformes	Wagtails and dippers.

Of these the divers, cormorants, shags, darters, pelicans, herons, mergansers, fish-eagles, ospreys, fish-owls, and kingfishers feed primarily on fish, but mergansers and kingfishers also eat invertebrates. The rest feed primarily on invertebrates, although they often take fish and vegetable matter, and the last is important in the diet of waterhens and coots. Even the small dipper, *Cinclus*, has been observed to eat salmon eggs which it swims down to collect (Jones and King, 1951).

The importance of this great weight of predatory activity on running-water communities is largely unknown, although it is clear that it must be considerable where the birds are at all common. Cormorants and shags in particular are often credited with eating great numbers of fishes, and the same must apply to the fish-eagles which are very common in some parts of the world. An eagle perched on a high tree overlooking the water is part of the normal scene on many African rivers. Similarly, pelicans occur often in great numbers and can be seen fishing in lines, and these large birds must consume great weights of fish.

Here again our data are largely generalities and there is much scope for study of how many birds feed per unit area, how much they eat and how this compares with the total production in the aquatic environment. Birds are clearly significant factors in the life of rivers, possibly almost as significant as fishes, but we are only guessing when we conclude this. At the present time there is a tendency to regard all piscivorous birds as vermin, and in Europe inland cormorants are often shot because they 'ruin the fishing'. Nobody really knows what effect they have, and this is a field in which reliable study is badly needed.

THE MAMMALS

The mammals also are well represented in rivers and streams although the really aquatic species belong to only a few groups.

The monotreme *Ornithorhynchus* lives in sluggish Australian streams and feeds on aquatic insects, and one species of opossum in South America is associated with water, but it, like the beaver, feeds on land vegetation.

The water-shrews, which are members of the Insectivora, are a larger group although they seem nowhere to be abundant. The Rocky Mountain species, *Sorex palustris*, feeds mostly in the water, and Conaway (1952) records 49 per cent of its food as being immature aquatic insects; the Old World water-shrew *Crossopus* has similar habits.

Several carnivores feed fairly regularly in streams. Bears have often been seen to kill migrating salmon, and raccoons, *Procyon*, regularly eat turtles and crayfish. Otters, *Lutra*, eat fishes and large crustaceans and so do mink and some other small species. Like cormorants, otters and mink are often credited with great depredations of fisheries with no very sound data to back the statements.

It is less generally known that there are dolphins which are confined to rivers. The susu, *Platanista*, occurs in the large rivers of India, and in contrast to its marine relatives, but in keeping with its life in almost continuously turbid water, it is blind, and other genera of the Platanistidae range far up the Amazon, La Plata, and Yangtze rivers. In South America and West Africa manatees inhabit sluggish waters and feed on plants, in contrast to the fish- and crustacean-eating dolphins. They consume such large amounts of vegetation that they are being seriously considered as a means of controlling weeds in canals and man-made lakes in the tropics. But they seem to breed very slowly and they are clearly in need of much further study before their usefulness in weed-control and protein production for human consumption is established.

The rodents are probably the most important group of river mammals as there is a whole range of species which are associated with water. Well known among these are the water-vole of Europe, the beaver, and the coypu and muskrat of North America, which are totally or mainly vegetarian and do much of their feeding away from the water. Also vegetarian, but more spectacular, are the giant cavies of South America, and there is also a fish-eating rat, *Icthyomys*, in that continent.

The major effects of rodents on rivers and streams are, however, probably brought about more by their building activities than by their diets. Muskrats, coypus, and voles often undermine the banks by their burrowing activities and thus alter channel patterns, and the beaver entirely changes the nature of streams by the building of dams. The enormous influence of these animals is particularly apparent in North America where their numbers are now increasing rapidly after having been very reduced by trapping. They are reoccupying areas from which they have been absent for long periods, and they are flooding places which have been dry for so long that the forest has taken them over. A very common sight at the present time, particularly on the Canadian Precambrian shield, is a large area of dead trees killed by the drowning of their roots by beavers. Such places are very conspicuous and they demonstrate forcibly how great is the influence of this species on the ecosystem.

We have already discussed the effects of beaver dams on the invertebrate fauna and the way in which they cause a change from typical running water species to Chironomidae. Studies in Maine and Ontario indicate that the biomass per unit area of water is reduced, but the area is so much increased by the dams that the amounts of fish food are more than doubled (Rupp, 1955; Sprules, 1941). The impoundments also encourage the growth of rooted plants as they change the substratum from gravel and rubble to silt, and they provide pools for fish (Rasmussen, 1940; Curry, 1954). Gard (1961) has shown that trout are able to pass over the dams, and there is little doubt that the populations of most types of trout are increased by the activity of the beavers (Neff, 1957). Gard, however, found that in Sagehen Creek, California, *Salvelinus fontinalis* and *Salmo trutta* were more benefited than was *S. gairdneri*, because of their ability to feed on the bottom fauna in the ponds. *S. gairdneri* maintained its drift-feeding habit, and when floods destroyed a dam it replaced *S. trutta* as the dominant species. Beavers have, therefore, the ability to interfere in the interspecific competition among these closely allied species. Doubtless their effects are very far-reaching in other respects also.

CHAPTER XX

Longitudinal zonation

We have seen in earlier chapters that, by their very nature, streams and rivers change along their length in respect of such properties as temperature, depth, current speed, substratum, and turbidity, and that these are all important factors controlling the distribution of the various sections of the biota. Inevitably therefore, running waters display longitudinal biotic zonation. Equally inevitably, biologists, for many of whom categorization is a way of life, have attempted to use this phenomenon as a means of classifying watercourses.

These attempts have met with varying success and acceptance, and they are useful only as long as they do not become articles of faith which, like all firmly held tenets, tend to stifle further inquiry. They provide convenient shorthand descriptions of a complex of conditions, but the zones are rarely, if ever, discrete, and their sequential arrangement is often irregular because of local topography. The 'rejuvenated' river of the geographers which becomes steep and turbulent below a stretch of near base-level flow is biotically as well as physiographically rejuvenated, but not in all respects. For instance, the headwaters often have a temperature regime very different from that of lower rejuvenated reaches.

Many early studies of running water revealed the existence of a zonal distribution of animals, (e.g. Steinmann, 1907; Shelford, 1911; Thienemann, 1912), and Illies and Botosaneanu (1963) have reviewed the early period in some detail. In western Europe the rather limited vertebrate fauna allowed the development of a scheme of classification based very largely upon fishes, and this has been further refined by Huet (1949, 1954, 1959, 1962). Basically four zones are distinguished and are named after characteristic fishes; trout (*Salmo*), grayling (*Thymallus*), barbel (*Barbus*), and bream (*Abramis*) zones usually follow one another in that order down a river. In some ways this use of fish names for the zones is unfortunate because the characteristic species are absent for historical reasons from some areas, and so do not always occur in zones named after them. For example, the grayling occurs in western Britain only where man has introduced it (Carpenter, 1928*b*), the barbel has a limited distribution in western Europe, and the grayling is replaced by *Telestes*

Trout zone	Grayling zone	Barbel zone	Bream zone
Salmonids	Mixed fauna salmonids dominant	Mixed fauna cyprinids dominant	Cyprinid fauna with carnivores
TROUT	TROUT (GRAYLING)	trout (grayling)	
	fast-water cyprinids	FAST-WATER CYPRINIDS	fast-water cyprinids
	accompanying cyprinids	*accompanying cyprinids*	ACCOMPANYING CYPRINIDS
	carnivores	*carnivores*	CARNIVORES
		calm-water cyprinids	CALM-WATER CYPRINIDS

TABLE XX,1. The communities of fishes in the four fish-zones of western Europe (from Huet). Small capitals denotes dominance, and italics sub-dominance. The principal species are as follows:

Salmonids (trout and grayling) *Salmo trutta* and *Thymallus thymallus*.
Fast-water cyprinids *Barbus barbus*, *Squalius cephalus*, and *Chondrostoma nasus*.
Accompanying cyprinids *Rutilus rutilus* and *Scardinius erythrophthalmus*.
Carnivores *Esox lucius*, *Perca fluviatilis*, and *Anguilla anguilla*.
Calm-water cyprinids *Cyprinus carpio*, *Tinca tinca*, and *Abramis brama*.

soufia in southern France and the Pyrenees. Despite this limitation, however, the zones remain recognizable because Huet has carefully defined the fish faunas associated with the indicator species (Table xx,1). He has also shown that there is a fairly clear relationship between the slope of the stream and the fish zone that it harbours, and that this slope decreases for a given zone the wider the stream (Table xx,2).

	Brooklet	Brook	Little river	River	Large river
Width m.	0–1	1–5	5–25	25–100	100–300
Slope in ‰ for breadth of	1 m.	3 m.	15 m.	60 m.	200 m.
Trout zone	50·0–12·5	25·0–7·5	17·5–6·0	12·5–4·5	—
Grayling zone	—	7·5–3·0	6·0–2·0	4·5–1·25	>0·75
Barbel zone	—	3·0–1·0	2·0–0·5	1·25–0·33	0·75–0·25
Bream zone	12·5–0·0	1·0–0·0	0·5–0·0	0·33–0·0	0·25–0·00

TABLE XX,2. Huet's 'slope rule' for western European streams showing the fish-zones which occur in watercourses of different widths and slopes.

This empirical approach has proved to be of great practical value to fishery biologists in the general region in which it was developed, as long as due allowance is made for other factors which may at times be operative. For example, high temperatures may make a 'slope-rule' trout-zone unsuitable for salmonids, and in northern Sweden the presence of a lake on the course of a small river has been shown to lower the temperature of the outflow to such an extent that a trout-zone lies downstream of a more or less typical grayling zone (Müller, 1954*a*). Similarly in northern Italy *Salmo* and *Tinca* occur together, and there are indications that the upstream limit of the latter is determined by temperature rather than by slope; this is a fact which would not be apparent in the cooler climate where the scheme was worked out (Sommani, 1953). There are indeed fairly clear indications that at increasing distances from the area in northern Germany and Belgium, where the concept originated, the fish species involved become different. In the Volga, for example, the upper reaches fit more or less into the scheme, but in the lower parts of the river a host of other species complicates the picture (Lastochkin, 1943), and even in Poland the species lists from the various zones are not quite the same. Backiel (1964), who carried out an extensive study of the River Drweca there, also stresses that the zones change gradually from one to another and that the transitional reaches may be as long as the zones themselves. One can only conclude therefore that fish-zones, while being useful in a descriptive sense, are not ecological entities with anything approaching clear-cut boundaries.

In North America also, many studies, beginning with those in streams around Lake Michigan by Shelford (1911), have stressed that fish species occur in more or less well-defined zones. We have already seen one such scheme from Kentucky (Table XVI, 1), which, however, deals more with the depth of the water than with differences along the lengths of streams. Kuehne (1962) also working in Kentucky and basing his observations on the stream-order system (p. 12) found only *Semotilus* and salamanders in first-order streams. These were joined by darters and then by centrarchids and other cyprinids as he proceeded to streams of higher orders. Similar lists of stream types and the fish species occurring in them have been drawn up for Indiana (Gerking, 1945), Missouri (Funk and Campbell, 1953), the upper Ohio River (Lachner, 1956), Ontario (Ricker, 1934; Hallam, 1959), New Hampshire (Hoover, 1938), and Maryland (Van Deusen, 1954), and Table XX, 3 shows the altitudinal distribution of fish species in two streams in Virginia (Burton and Odum, 1945).

Making allowances for the great range of the fish fauna over the large area bounded by Missouri, Ontario, New Hampshire, and Virginia, one can see some similarities, e.g. trout and *Cottus* and/or *Semotilus* and *Rhinichthys* in headwaters with *Catostomus* and *Notropis* coming in lower down, to be joined at still lower levels by Centrarchidae and Ictaluridae.

	Little Stony Creek														Sinking Creek											
Height of collecting station, m.	1,127	1,070	1,022	990	940	920	915	900	840	760	665	610	565	520	665	655	640	620	610	595	580	552	550	540	525	521
Salvelinus fontinalis	x	x	x	x	x	x	x	x	x	x																
Rhinichthys atratulus					x	x	x		x			x	x		x	x	x	x	x	x	x		x			x
Etheostoma flabellare							x			x			x	x					x				x	x		
Salmo gairdneri								x	x	x	x	x	x	x							x					
Cottus bairdi												x	x	x	x	x	x		x	x	x		x	x		x
Campostoma anomalum													x	x				x		x			x		x	
Notropis albeolus														x											x	
Rhinichthys cataractae														x							x					
Catastomus commersonnii														x				x		x	x	x	x	x	x	
Clinostomus vandoisulus															x	x					x		x			
Hypentelium nigricans																		x	x	x			x	x		
Nocomis micropogon																		x				x			x	
Notropis photogenis																		x	x		x					x
Pimephales notatus																		x				x				x
Semotilus atromaculatus																				x	x					
Etheostoma blennioides																										x
Ambloplites rupestris																						x	x			x
Parexoglossum laurae																							x	x	x	x
Micropterus dolomieui																								x		
Lepomis macrochirus																								x		
Poecilichthys osburni																										x

TABLE XX, 3. The distribution of fish species along the length of two adjacent tributaries of New River, Virginia. Data from Burton and Odum (1945).

But there is no clear general pattern, and, as in Europe, it would seem that fish-zone schemes can apply only over areas of a few thousand square kilometres because of local differences in fauna, altitude, and general climate. It is interesting also to note that, even within the relatively clearly defined trout zone of the mountains, different species of trout seem to share out the zone between them. Holton (1953), who surveyed about 4 km. of a stream in Montana, over which it descended about 1,300 m., found that *Salvelinus fontinalis* was the commoner species at the top (83–96 per cent in different samples) and *Salmo gairdneri* at the bottom (70–92 per cent).

Zonation of fishes, or at least some indication that fish communities change along the lengths of rivers, has been reported from other continents, e.g. South America (Kleerekoper, 1955), Africa (Marlier, 1953, 1954; Oliff, 1960), Asia (Nikolski, 1933), and New Zealand (Allen, 1956, 1960*b*), but it seems probable that further study would show that, as in Europe and North America, the pattern varies from region to region.

Similarly, many algal studies have shown zonal arrangements of species (p. 73), and mixtures of algae, mosses, and higher plants have been shown to be more or less zonally arranged in karst spring streams in Hungary (Entz, 1961). Zonation has also been observed in many groups of invertebrates, and groups on which there is a fair amount of information of this type include:

Tricladida	Thienemann (1912), Carpenter (1928*a*), Beauchamp and Ullyot (1932), Geijskes (1935), Imanishi (1941), Dittmar (1955*a*), An Der Lan (1962), Chandler (1966).
Hirudinea	Pawłowski (1948).
Amphipoda	Thienemann (1950).
Decapoda	Hynes *et al.* (1961).
Ephemeroptera	Ide (1935), Macan (1957*a*), Dodds and Hisaw (1925*b*), Imanishi (1941), Berthélemy (1966).
Plecoptera	Hynes (1941), Brinck (1949), Dodds and Hisaw (1925*b*), Berthélemy (1966).
Odonata	Fittkau (1953), Zahner (1959).
Hemiptera	Hynes (1948, 1955*a*).
Sialidae	Kaiser (1950).
Trichoptera	Dodds and Hisaw (1925*b*), Nielsen (1942).
Dryopidae	Steffan (1960).
Elminthidae	Berthélemy (1966).
Simuliidae	Grenier (1949), Davies and Smith (1958), Dorier (1961).

	Brooks and creeks				Rivers		
	Small brooks	Small creeks	Larger creeks	Small rivers	Medium-sized rivers	Fairly large rivers	Large rivers
Anodonta grandis			r	r	r	r	r
Lampsilis siliquoidea			r	r	r	r	r
Strophitus rugosus		C	C	C	C	r	
Lampsilis ventricosa				A	C	r	r
Elliptio dilatatus			C	A	A	A	r
Lasmigona compressa	C	A	C	C	r		
Anodontoides ferussacianus	r	A	C	C			
Alasmidonta calceolus		A	A	r	r		
Micromya iris		r	C	A	A	C	r
Alasmidonta marginata			r	C	C	r	r
Ptychobranchus fasciolare				A	C	C	r
Lampsilis fasciola				A	C	C	r
Dysnomia triquetra				r	C	r	
Ligumia recta latissima					r	r	r
Lasmigona costata					C	r	r
Cyclonaias tuberculata						A	C
Anodonta imbecillis				r		r	C
Lasmigona complanata							C
Actinonaias carinata							C
Micromya fabalis							C
Fusconaia flava							C
Carunculina parva							C

TABLE XX, 4. The distribution of Unionidae in the main running-water habitats of the Huron River system, Michigan. r = rare, C = common, A = abundant. Data from Van der Schalie (1938).

Chironomidae	Thienemann (1954*a*), Scott, K. F. M. (1958).
Rhagionidae	Pomeisl (1953*b*).
Hydracarina	Angelier (1952), Schwoerbel (1961*b*).
Pelecypoda	Behning (1928), Van der Schalie (1938), Goodrich and Van der Schalie (1944), Clark (1944), Howard (1951), Clarke and Berg (1959).

Without doubt this list of groups and references could be greatly extended by further search in the literature. The point to be emphasized

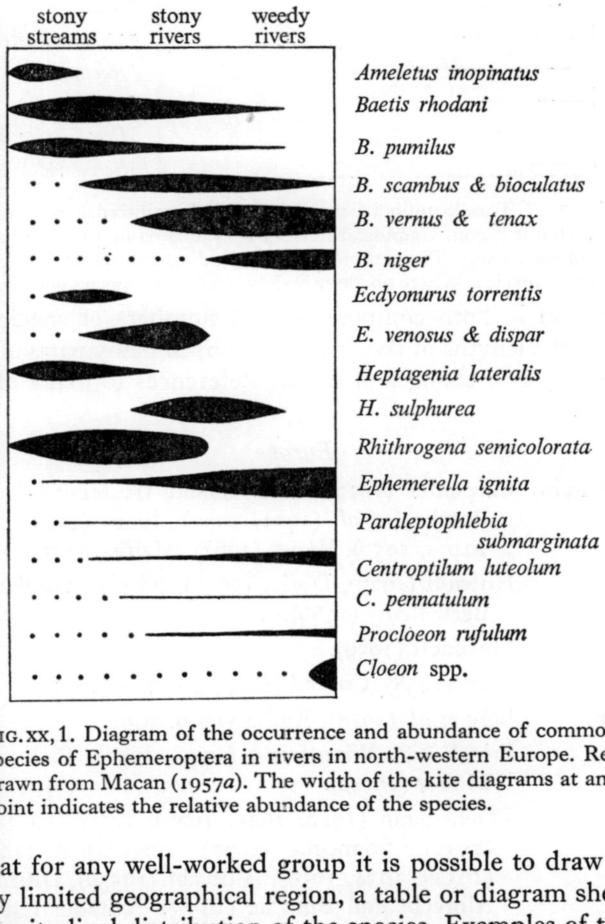

FIG. XX, 1. Diagram of the occurrence and abundance of common species of Ephemeroptera in rivers in north-western Europe. Redrawn from Macan (1957*a*). The width of the kite diagrams at any point indicates the relative abundance of the species.

here is that for any well-worked group it is possible to draw up, for a reasonably limited geographical region, a table or diagram showing the normal longitudinal distribution of the species. Examples of three such exercises are given in Table XX, 4, and Figs. XX, 1 and XX, 2.

This is obviously similar to the situation with the fishes, and when many such groups are combined together it is clear that entire faunas

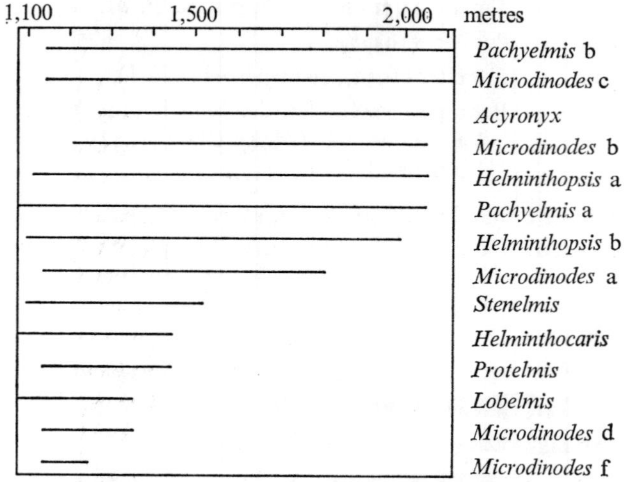

FIG. XX, 2. The altitudinal distribution of the Elminthidae in streams on Mount Elgon, Uganda. From T. R. Williams and Hynes (unpublished data). The letters a, b, etc., indicate different but unnamed species. Where no letter is shown only one species occurred.

show zonation in both composition and numbers of species. Studies made along the lengths of rivers and streams in many parts of the world have revealed such changes of fauna. References to many of them are given below:

Europe

Great Britain	Carpenter (1927), Percival and Whitehead (1929, 1930), Butcher *et al.* (1931, 1937), Jones (1941, 1943, 1948, 1949*a*, 1951), Hynes (1961), Maitland (1966).
France	Hubault (1927), Dorier (1937), Nicolau-Guillamet (1959), Berthélemy (1966).
Holland	Smissaert (1959).
Belgium	Huet (1938), Conrad (1942), Marlier (1951).
Denmark	Berg *et al.* (1948), Kaiser (1949, 1951).
Sweden	Brinck and Wingstrand (1949, 1951), Badcock (1953*a*), Müller (1954*c*).
Germany	Thienemann (1912, 1925, 1926), Eidel (1933), Nietzske (1937), Vonnegut (1937), Engelhardt (1951), Illies (1952*b*, 1954, 1958), Albrecht (1953*b*), Dittmar (1955*a*) Albrecht and Bursche (1957), Schräder (1959), Kothé (1961).
Switzerland	Steinmann (1907), Geijskes (1935).
Austria	Brehm and Ruttner (1926), Von Mitis (1938), Kühn (1940), Pomeisl (1953*a*), Findenegg (1959).

Italy	Moretti (1949, 1953).
Poland	Pawłowski (1959*a*), Sowa (1961, 1965).
U.S.S.R.	Behning (1928, 1929*b*), Shadin (1956, 1962, 1964).
Czechoslovakia	Zelinka (1950), Bilý *et al.* (1952), Winkler (1956, 1963), Dudich (1958), Straškraba (1966).
Rumania	Botosaneanu *et al.* (1959).
Yugoslavia	Matoničhkin and Pavletić (1959).

Asia

India	Hora (1923, 1930), Chacko and Ganapati (1952).
Malaysia	Johnson (1957).
Japan	Uéno (1938), Nakamura *et al.* (1954), Komatsu (1964*a*, *b*), Tanaka (1966).

Africa

Uganda	Hynes and Williams (1962).
Kenya	Van Someren (1952).
Congo	Marlier (1954).
South Africa	Harrison and Elsworth (1958), Harrison (1963, 1965*b*), Oliff (1960, 1963), Oliff and King (1964), Oliff *et al.* (1965).

North America

Ontario	Ricker (1934), Ide (1940), Sprules (1947).
Wyoming	Muttkowski (1929), Armitage (1958, 1961).
Utah	Gaufin (1959).
Illinois	Richardson (1921, 1928), Gersbacher (1937).
Colorado	Dodds and Hisaw (1925*a*, *b*), Pennak and Van Gerpen (1947).
Ohio	Ludwig (1932).
Kentucky	Minckley (1963).
New Mexico	Noel (1954).
Louisiana	Bick *et al.* (1953).
Florida	Beck (1965).

South America

Brazil	Schubart (1946, 1953), Fittkau (1964).
Peru	Illies (1964).

New Zealand

Allen (1951).

Illies (1954, 1955, 1958, 1960, 1961*b*, 1962; Illies and Botosaneanu, 1963) has made an attempt to combine all this information into one universal system which can be applied to all watercourses. In this he has been at least partly successful in the sense that he has produced concepts which permit one to think about types of stream in the same way that one can about types of lake. But hard and fast definitions seem unattainable at present, and it is to be hoped that running-water ecology will escape the decades of semantics and arguments that followed upon the very useful, but equally imprecise, classification of lakes. Already though at least one author has followed up Illies's idea with a long discussion which positively bristles with terms drawn from botanical ecology (Steffan, 1965*b*). This is a branch of biology which at one time almost submerged itself in verbiage to the exclusion of conceptual thought, and it is to be hoped that the same fate does not await the study of running water.

Illies found that when he listed the identifiable species occurring at stations spaced along the length of the Fulda River in North Germany there were points at which more species appeared or disappeared than did at others. In other words, although there were many changes along the length of the stream, there were more at some points, and these seemed to mark faunal divisions. He found that these divisions corresponded very well with the fish zones, and the same data were worked over rather more elaborately to show that the rate of change of faunal composition was greatest at the boundaries of the zones (Schmitz, 1955, 1957). Illies then went on to develop the idea that perhaps these zones are real ecological entities, and to show that the largest faunal change is to be found at the lower end of the grayling-zone. He termed the area above this point the rhithron and that below it the potamon and defined them as follows:

Rhithron is the region extending from the source to the point where the monthly mean temperature rises to 20 °C., where oxygen concentration is always high, flow-rate is fast and turbulent, and the bed is composed of rocks, stones, or gravel with only occasional sandy and silty patches. The fauna is more or less cold stenothermic and is all, or almost all, characteristic of running water, and there is little or no true plankton.

Potamon is the region where the monthly mean temperature rises to over 20 °C., oxygen deficits may occur at times, flow is slower and tends towards becoming laminar, and the bed is mainly sand and mud. The fauna is eurythermic or warm stenothermic, and is composed mostly of types which reach their maximum development in still water, and there is often a rich plankton.

This is a reasonable distinction between streams and rivers, and, although some small streams may be cool and silty and some stretches of rivers rocky and turbulent, it has been tried out in South America (Illies, 1964) and South Africa (Harrison, 1965*b*), as well as elsewhere in Europe and it certainly seems applicable in both southern Canada and Central Africa. Because temperature is an important factor the point of change from rhithron to potamon tends to lie at higher altitudes at lower latitudes, and at very high latitudes it may not occur at all. Similarly at low altitudes in the tropics the water is always warm so a true rhithron is absent (Illies, 1955).

Above the rhithron there is often a spring and spring brook region, and Illies and Botosaneanu (1963) propose that these should also be distinguished. We have, therefore, a series of zones.

Eucrenon The spring region.

Hypocrenon The spring brook.

Rhithron The stony stream to small river, which can often be faunistically divided into *epi-*, *meta-*, and *hyporhithron*. In Europe for example, the meta-hyporhithron boundary is that between the trout and grayling zones.

Potamon Which is also divisible into *epi-* and *metapotamon* both on faunistic grounds, e.g. barbel and bream zones in Europe, and on presence or absence of shallow shoals. The *hypopotamon* is the brackish water region affected by the sea and it is inhibited in Europe by such fishes as the flounder *Pleuronectes* and the sterlet.

Although the scheme has been subjected to some criticism, most of this has been concerned with details of subdivision of the rhithron into parts (e.g. Macan, 1961*b*; Berthélemy, 1966). Thorup (1966) has suggested that the faunal discontinuities which were found by Illies in the Fulda River were artifacts caused by local pollution, and there is indeed some evidence that this may have been so. Similarly, Hynes (1961) has shown that within a short and uniform stretch of the epirhithron of a stream in Wales there were trends in the relative abundances of the commonest species in the benthic fauna (Table xx, 5). These indicate that, although the *list* of species may not change from end to end of a zone, the *ecological structure* of the community undoubtedly does. Thus even as clearly defined a zone as the epirhithron in north-western Europe, from which the Welsh samples were collected, is not an ecological unit which can be treated as a uniform functional entity. Moreover, it seems undoubted that faunal changes from one zone to another are not usually as abrupt as Illies and Schmitz found in the Fulda River, and they may even be complex. For instance, Goodrich and Van der Schalie (1944) stress that, although it has been repeatedly shown that the number of species of molluscs increases as one passes from the headwaters to the

Altitude in m.	235	260	280	325
Brachyptera risi	0·9	2·9	0·3	2·0
Protonemura meyeri	4·7	1·8	2·7	1·5
Amphinemura sulcicollis	24·1	23·1	25·8	7·6
Leuctra inermis	14·3	18·2	16·5	18·6
L. fusca	1·8	2·1	0·7	0·6
Isoperla grammatica	1·0	3·3	5·3	11·6
Baetis 3 spp.	4·6	9·2	10·1	11·7
Ephemerella ignita	3·8	4·1	4·8	5·4
Rhithrogena semicolorata	17·6	11·7	6·9	2·7
Hydropsyche instabilis	7·3	1·8	0·2	+
Elmis maugetti	0·3	1·3	2·8	12·7
Simuliidae	6·3	4·8	5·9	5·0
Orthocladiinae	2·5	3·5	3·6	7·4
Total numbers in samples	7,788	7,351	7,593	7,160

TABLE XX, 5. The percentage composition of the fauna of the Afon Hirnant, Wales, at four altitudes; based on thirteen sets of samples taken around the year. Only taxa which formed at least 2 per cent of the numbers at one station are included. + = < 0·05 per cent. Data from Hynes (1961).

mouth of a river, it is not generally emphasized that in a zone located somewhere between small-river and large-river conditions the total number of species is frequently surprisingly high, and that this has been found to be true of many drainage systems. This point is, presumably, the rhithron/potamon boundary, and they quote as an example the Wabash River, of the American Middle West, with 16 species of mussel in the headwaters, up to 48 in the transition zone, and only 33 in the large river zone, with a further 6 occurring near the outfall into the Ohio. Similarly, Maitland (1966) found increased numbers of species in several groups in the zone of transition from a stony stream to sandy weedy lowland reaches in the Scottish River Endrick. The zone of overlap between rhithron and potamon can be expected to be rich because it contains many of the species from both types of water.

Despite these shortcomings, however, the distinctions between crenon, rhithron, and potamon are valid *in a general sense*, and the terms are a useful descriptive shorthand in the same way as are the terms oligotrophic, dystrophic, and eutrophic as applied to lakes. They only cease to become so when attempts are made to regard them as quite precise—to argue for example whether the critical temperature is 20 or 24 °C., which is where it would be on Ricker's somewhat similar classification of Ontario streams, or to dispute about just how and where subdivisions should be made. Rivers are clines in the ecological sense, and such

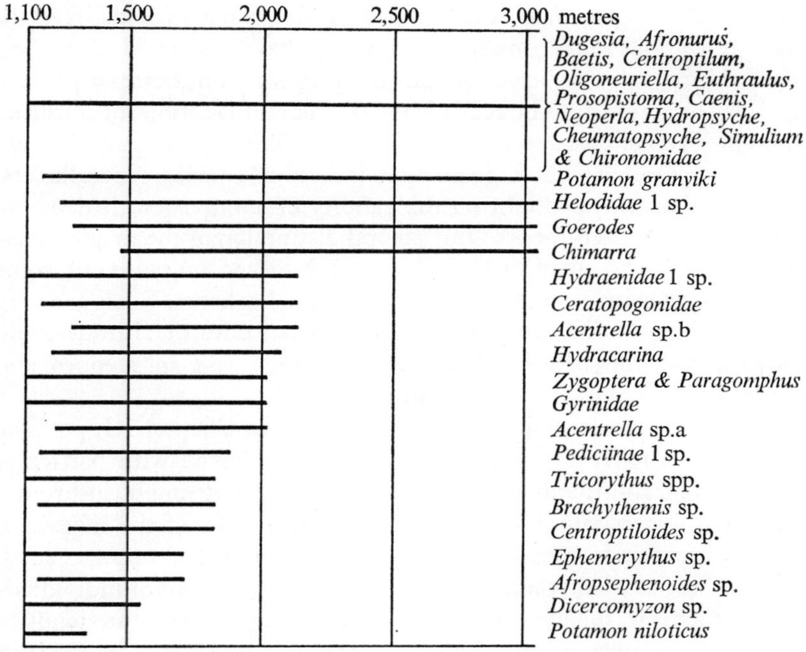

FIG. XX, 3. The altitudinal distribution of the important members of the fauna in streams on Mount Elgon, Uganda. From T. R. Williams and Hynes unpublished data. No collections were made between 2,140 and 3,100 m., so the upper limits of three taxa may have been truncated. The Elminthidae are shown in Fig. XX, 2.

boundaries as do occur in them are ill-defined and are caused by a variety of factors. It is unrealistic to expect that any small group of them will act in the same way all over the globe, and there is no reason to suppose that subdivisions of the major zones such as can be seen in Europe will necessarily occur elsewhere.

For instance, in high streams in the tropics the rhithron zone may extend over thousands of metres of altitude and hence through a great change in temperature but with little local seasonal variation. Temperature is probably the most important factor controlling the occurrence of species in these uniform streams and it affects each one differently. Fig. xx, 3 shows the altitudinal distribution of a large proportion of the invertebrate fauna in the rhithron reaches of streams on Mount Elgon in East Africa. The survey extended from 1,100 m., which is about where potamon begins in most of the streams, to 2,140 m. on many streams round this conical mountain, and there were two stations on one stream-system at 3,100 m. The water was very cold at the top; it was only 6–7 °C. at 2,000 m. and it reached the high 20s at the bottom. It can be seen from the figure that, while much of the fauna extends over the whole altitudinal range, many invertebrates have restricted altitudinal ranges. Moreover, the survey reveals nothing which indicates a point at which the streams could be realistically divided into sections of rhithron, although it is just possible that it would have done so to a limited extent had it been possible to distinguish more individual species. Nevertheless, while this does throw doubt on the validity of subdivision, it does not detract from the usefulness of the general distinction of the major zones.

It should also be noted that, as Illies and Botosaneanu point out, some types of watercourse do not fit into the general scheme. Such are very cold ice- or snow-fed torrents at very high latitudes or altitudes, and probably also the little-studied streams which rise in steppes and prairies and so have no normal rhithron zone.

Within each major zone it is, as we have seen in Chapter XI, possible to distinguish fairly definite benthic faunas associated with particular types of substratum; and as one passes from crenon through rhithron to potamon it is generally true that the individual units of the mozaic of substrata increase in area. Rocks, gravel, stones, and sand occur in small patches in the small stream, and there is much movement of individuals between them. In the hyporhithron the riffle and pool structure tends to produce larger areas of gravel and silt, and this increase in the size of uniform areas becomes more marked in the epipotamon (Illies, 1958). Here then there tend to be two or more sufficiently distinct benthic communities for each to represent something approaching an ecological unit. In the metapotamon also large areas of sand, mud, clay, etc., each have their own fairly characteristic fauna as has been stressed in particular by Danish and Russian workers (i.e. Berg *et al.*, 1948; Behning, 1928, 1929*b*; Lastochkin, 1943; Greze, 1953; Shadin, 1956). It is possible therefore, at least in the lower rhithron and the potamon, to make some meaningful subdivisions within zones, but as with the zones themselves a rigid classificatory approach rapidly loses touch with the complexities of reality. As Maitland (1966) has so clearly put it—'some species are limited by temperature, others by substrate, amount of suspended matter, calcium, etc. Not only are those factors usually transitional in themselves from the extremes normally found at source and mouth, but if by chance there is a sharp change in one of them (e.g. a sudden increase in calcium where a highly calcareous tributary enters the main river) there would rarely be a change in any of the others at the same point. Thus the general theme of the change in biotic associations from source to mouth in a river will tend to be one not of zonation but of transition.'

The conclusion from this chapter is then that the idea of zonation is useful in a general descriptive way but that attempts at precise definition of zones are of doubtful ecological value.

CHAPTER XXI

Special habitats

Most of the biological data dealt with to this point have been concerned with ordinary watercourses such as could be readily included in the concepts of rhithron and potamon. Only scattered mention has been made of less usual types of water body and of special habitats, and the purpose of this chapter is to consider these further, and so to round out our discussion of those habitats which are not, as it were, on the main stream of the normal drainage pattern but which have fairly well-marked biological characteristics.

SPRINGS

Springs and the streams leading from them constitute the crenon in Illies's terminology (p. 393), and they differ from normal streams in a number of ways. They tend towards a very uniform temperature, which is normally the mean annual air temperature of the region, but which may be high in volcanic areas. They are often de-oxygenated and they may then be iron-depositing; and if they are neither they are often in close communication with underground waters which harbour a phreatic fauna.

A spring may discharge directly into the channel of the spring stream so that the water flows directly into the channel, or it may discharge into a small basin or pond which overflows into the channel. These two types have long been distinguished as *rheocrene* and *limnocrene* respectively (Bornhauser, 1913), but the term *helocrene* is used for the latter, when they are marshy.

When no oxygen is present in the water there are usually no visible organisms actually in the spring, although sulphur bacteria and Cyanophyta have been recorded from the 'boil' zone of a de-oxygenated spring in Florida (Odum and Caldwell, 1955). Some insects will, however, enter water which contains very little oxygen. Sloan (1956) found specimens of

Callibaetis and *Caenis* in water containing only 2·5 mg./l. in spring streams in Florida, and normally, in such circumstances, the number of species increases quite rapidly in the downstream direction.

In volcanic areas hot springs may rise at any temperature up to boiling point, and the water as it cools in the channel provides a clear zonation of flora and fauna. Above about 60 °C. there appear to be only bacteria, and *Leptothrix*, or an allied form, has been recorded up to 75·5 °C. in Tibet. Cyanophyta of the genera *Synechocyetis* and *Phormidium* occur up to 60 °C. and are joined by species of *Onconema*, *Spirulina*, and *Oscillatoria* in waters up to 55–8 °C. (Brues, 1927; Round, 1965). Diatoms are not found above 40 °C. Among animals, adults of the bean shrimp *Cypris balnearea* and the mite *Thermacarus*, which are restricted to warm waters, occur up to temperatures of 51·5 °C. (Brues, 1928; Mason, 1939). Other animals occur only at temperatures of 45 °C. and below, as do the young stages of *C. balnearea*. In other species also, e.g. *Bidessus signatellus* of North Africa, the adults penetrate into hotter water than do the larvae (Mason, 1939).

Few other species of animal tolerate temperatures above 40 °C., and those that do are specialists, e.g. species of *Scatella* and *Cricotopus* in Iceland (Tuxen, 1944), but some fishes range up to this point. It is easy, however, to be misled in the field, because the warm water often flows over a cooler layer; in this way *Limnaea pereger* has been recorded from water warmer than it can in fact tolerate (Starmuhlner, 1961). The fauna in water just below 40 °C. is restricted to species in a few groups, which occur, however, in most areas where hot springs are found. They include Nematoda, Oligochaeta, Ostracoda, some Malacostraca including *Potamon* and *Thermasbaena* in Africa, Odonata, Dytiscidae, Hydrophilidae, Ephydridae, Stratiomyidae, Chironomidae, and some snails, e.g. *Limnaea*, *Melanopsis*, and *Paludestrina* (Tuxen, 1944). Thus even before the stream has cooled to normal warm-water temperatures a considerable, although selected, fauna may be present.

In normal springs where the water emerges with some dissolved oxygen many organisms occur right up to the source. Pond species may be present in limnocrenes and there may be aquatic plants. The number of species is, however, restricted as compared with a normal pond, probably because of the uniform temperature, and some species normally found in running water are also usually present. In rheocrenes the algal flora resembles that of small streams (Whitford, 1956), and important constituents are species of *Achnanthes*, *Fragilaria*, and *Melosira* (Round, 1965). There is also a seasonal succession of dominant species (Teal, 1957), indicating that light is an important factor which operates despite the very constant temperature.

The fauna in and near the boil often includes specimens of phreatic animals such as *Niphargus*, *Bathynella*, and white *Phagocata* in Europe, and blind *Asellus* and *Crangonyx* there and in North America (Thienemann, 1912; Carpenter, 1928*a*; Beyer, 1932; Geijskes, 1935; Dahm, 1949; Dittmar, 1955*a*; Matoničkin and Pavletić, 1960; Minckley, 1961, 1963; Motas *et al.*, 1962). These animals may indeed form a fairly important part of the fauna, and they persist in sheltered places for some distance downstream. Another important faunal element is composed of animals whose normal habitat is wet soil. Such are, for example, *Eiseniella*, *Anacaena*, *Osmylus*, and *Pedicia*, which elsewhere can be found in cool shady places at the edges of streams (Thienemann, 1912; Demel, 1923; Kühn, 1940). A third element consists of species which, while being truly aquatic and epigean, are largely confined to springs and cool spring streams. These include some flatworms, e.g. *Crenobia alpina* in Europe and *Phagocata gracilis* in North America, some Trichoptera, e.g. *Apatidea*, *Agapetus*, and *Synagapetus*, some Plecoptera, e.g. *Leuctra nigra* and *Nemurella* in Europe and *Scopura* in Japan, some beetles, e.g. *Hydroporus ferrugineus* and *Helodes*, and some snails, e.g. *Bithynella* in Europe and *Paludestrina* in North America (Thienemann, 1912; Geijskes, 1935; Kühn, 1940; Davidson and Wilding, 1943; Nielson, 1950*a*; Noel, 1954; Dittmar, 1955*a*; Matoničkin and Pavletić, 1960; Komatsu, 1964*a*; Chandler, 1966).

Ordinary stream-dwelling species appear further downstream, and some of these may penetrate up to the spring itself. This applies particularly to amphipods and isopods, which are often the dominant element in the fauna of hardwater spring streams (Demel, 1923; Eidel, 1933; Geijskes, 1935; Kühn, 1940; Noel, 1954; Stella, 1956; Teal, 1957; Motas *et al.*, 1962; Minckley, 1963).

One feature of springs which is of considerable interest is that they provide very uniform conditions in areas which are subject to great seasonal changes. They are thus places where relicts of former times have survived, protected from cold winters or warm summers as the case may be. Many so-called crenobionts (species confined to springs) are therefore species which occur far outside their normal geographical range. We have already discussed *Apatidea muliebris*, which retains an Arctic life-cycle in a Danish spring (p. 288). Other examples are: *Crenobia alpina* left by the ice age in many parts of central Europe, the snail *Melanopsis* left behind by warmer times in springs in southern Europe (Thienemann, 1949, 1950), *Capnia bifrons* left by the ice age in Austria (Findenegg, 1959), a long list of northern species left by the ice age in springs in the central plain and elsewhere in Rumania (Motas *et al.*, 1962; Botosaneanu *et al.*, 1959; Botosaneanu and Negrea, 1961), and *Gammarus fasciatus* far from its present north-eastern range, in a spring in New Mexico (Noel, 1954). Indeed, some animals, e.g. *Gammarus bousfieldi* and *Asellus bivittatus* from Doe Run, Kentucky, are so far known from only one spring stream (Cole and Minckley, 1961; Walker, 1961), which suggests that they are relicts from long ago, which have either died out elsewhere or changed in their isolated habitat. Because so many crenobionts are relicts, the faunas of springs in relatively recently glaciated areas tend to be less rich than those of springs which have been longer in existence. Engelhardt (1957), for example, found fewer species in springs in Lapland than occur in springs in the Alps, and the same relative paucity of fauna is found in springs in the glaciated area of North America.

As would be expected, there are some floral and faunal differences between hard-water and soft-water springs, and between rheocrenes and limnocrenes (Schwoerbel, 1959), and the nature of the substratum is also important (Thorup, 1966). Less expected, however, is that, like the algae, spring-dwelling insects, snails, and fish show distinct periodicity. They breed and emerge at the proper season for the species despite the lack of temperature change, and one can only conclude that their periodicity is controlled by day-length (Demel, 1923; Odum, 1957; Thorup, 1963).

Another interesting problem concerns the possibility of dispersal from one spring to another. Maguire (1963) has reviewed our rather scanty knowledge of the mechanisms of passive dispersal of aquatic organisms, and has shown that micro-organisms travel fairly readily. But for larger invertebrates, especially those with habitats as restricted as springs, the problems would seem to be almost insuperable. Possibly there really is very little movement, and Thienemann (1949) has suggested that, when *Crenobia alpina* appeared in some German springs after being apparently absent for many years, it did so from a retreat in the groundwater. It is also perhaps significant that *Hydroporus ferrugineus*, a beetle of springs and spring streams all over Europe, including the temporary 'winterbournes' of the chalk hills of south-east England, is apparently quite unable to fly (Jackson, 1958). It must, therefore, exist in small isolated populations and retreat into the groundwater near temporary springs. Since most water-beetles fly well there must be some advantage in flightlessness, possibly connected with the smallness and scattered nature of the habitats. It would seem, therefore, that springs may offer almost ideal sites for evolutionary genetic studies, especially in glaciated areas where their ages can be fairly accurately known.

VERY COLD STREAMS

At high altitudes and in high latitudes there are streams which never warm far above freezing point, and which thus present an exceedingly severe aspect of rhithron conditions. We have seen that glacier torrents above 1,500 m. in the Alps are inhabited only by *Diamesa* (p. 201), which Thienemann (1954*a*) identifies as *D. steinbocki*. A similar species, *D. ruwenzoriensis*, is found below glaciers on the Ruwenzori Mountains and Mount Kenya in tropical Africa (Harrison, 1965*b*). The genus *Diamesa* is confined to cold water and it occurs even in meltwater channels on Spitzbergen (Kureck, 1966), but the fact that it occurs alone, or almost alone, in glacier water is probably because of some property of that water other than merely low temperature. In equally cold, but snow-fed, high streams *Euorthocladius* is also present, as are mites, flatworms, and some mayflies and stoneflies (Hubault, 1927; Dorier, 1937; Thienemann, 1954*a*; Berthélemy, 1966). Similarly, in high, very cold, spring streams in Austria, Pleskot (1951) found *Crenobia alpina*, *Ecdyonurus helveticus*, *Rhithrogena alpestris*, and *Dictyogenus alpinus*, all of which are characteristic of very cold places and which are replaced by other allied species in warmer water.

Streams in the far north also harbour a fauna made up of a limited number of cold-adapted species, particularly of stoneflies. Some of these are circumpolar in distribution, e.g. *Diura bicaudata* and *Nemoura arctica*, as are some blackflies, e.g. *Prosimulium ursinum* (Davies, 1954). Brinck and Wingstrand (1949) found that the fauna of cold torrents in Swedish Lapland is dominated by *Deronectes*, *Stenophylax*, *Diura*, *Arcynopteryx*, *Isoperla*, and *Nemoura*, and they noted that there was little difference between the fauna of streams which dried out during the summer and those that did not. This indicates that, despite their ability to withstand low temperatures, most of these animals are able to complete their life cycles in less than a year and to survive the dry period as eggs. We have, however, already noted that *Prosimulium ursinum* grows more slowly than other species of the genus and does not fly until September (p. 286); it is thus presumably confined to permanent streams.

There is, therefore, at any rate in the arctic, a definite, although limited, very-cold-water fauna. Perhaps more surprising is the existence of such a fauna in the tropics even on young mountains. In streams in the permanently cold tundra zone on Mount Kenya, Van Someren (1952) found *Dugesia*, Chironomidae, and *Simulium*, although he noted that Ephemeroptera, Plecoptera, Trichoptera, and Hemiptera were absent. This absence he attributed to the occurrence of frost every night which would possibly eliminate adults, and he noted that the mayflies appeared at 2,900 m. at the upper edge of the forest which presumably provides some shelter from the frost. This explanation seems very probable, and it is perhaps to some extent confirmed by the occurrence of a variety of species of three of these groups at 3,100 m. on nearby Mount Elgon (Fig. xx, 3), where the forest extends to higher altitudes. Nevertheless, many of these species live through all their developmental stages in water that is never more than a few degrees above freezing, and some of them, e.g. *Chimarra*, one species of *Acentrella*, *Goerodes*, and a member of the Helodidae, seem to eschew warmer water (Fig. xx, 3) (Hynes and Williams, 1962). We must conclude therefore that even in the tropics there are species which can, and some which must, complete their life histories in very cold water, and that these occur even on quite young and isolated mountains. It seems likely that they spread to their present habitats during cold periods of the Pleistocene era.

INTERMITTENT STREAMS

We have seen in Chapters XI and XVI that periodic interruption of water flow is an important factor in the ecology of animals. Truly intermittent streams are nevertheless always inhabited by aquatic organisms unless they flow for only very short periods, and they may provide habitats for such pests as *Simulium damnosum*, an African vector of human onchocerciasis (Crisp, 1956). Often, however, intermittent streams retain pools somewhere along their length, and, although conditions in these soon cease to resemble the stream habitat, they do serve as refuges for some species and temporary habitats for others (Schneller, 1955).

Generally speaking, normal stream animals which would be in an active aquatic stage during the dry period are eliminated from the fauna. Such are late-emerging insects in temperate zones, which would be growing while the stream is dry (Hynes, 1941; Sprules, 1947). This can make striking alterations to the fauna of large areas, as for example in south-western Spain where, because most of the streams are temporary, the stonefly fauna is dominated by *Tyrrhenoleuctra* and *Hemimeleana*, which emerge very early, instead of the more usual *Leuctra*, *Protonemura*, and *Perla*, which form most of that fauna elsewhere in the country (Aubert, 1963).

The faunas of temporary streams are therefore comprised of six main groups.

1. Species which survive in pools despite high temperatures and low oxygen concentrations. These include a few species of fish, e.g. some

Cyprinidae and Centrarchidae in North America (Shelford, 1937; Gersbacher, 1937; Schneller, 1955), as well as some invertebrates, particularly flatworms, crustaceans, some Trichoptera, and snails (Stehr and Branson, 1938; Kamler and Riedel, 1960*a*). It seems that tolerance of poor conditions may be augmented by special physiological provisions in some species. Jackson (1956*a*) noted that *Deronectes elegans* survives in drying pools, and that this flightless beetle stores fat in the space normally occupied by the wing muscles. Perhaps some similar provision of emergency energy-supply is available to the larvae of *Chaetopteryx villosa*, which has been observed to survive pool conditions particularly well in streams in the Tatra Mountains (Kamler and Riedel, 1960*a*).

A particularly striking instance of an animal which is adapted to the pool habitat in temporary streams is the atyid *Paratya australiensis*, for information on which I am indebted to Dr. W. D. Williams. This prawn occurs widely in south-eastern Australia, particularly in lowland creeks and rivers which become reduced to isolated or barely connected pools during the summer. The adults are able to maintain their position in a current, but the larvae are planktonic and so would be swept away by any considerable discharge, and the life history is geared to the regime of flow of the streams. The females become berried, that is with eggs attached to the pleopods, in November, when the stream flow has already become greatly reduced, and the larvae are released in December (mid-summer) when the streams have become reduced to still pools. They then develop as plankton in the pools, and by March, when the water again begins to flow, the young have become benthic and, like the adults, are able to maintain their position in the running water.

2. Species which survive by burrowing down into the substratum. These include flatworms, nematodes, oligochaetes, some amphipods and isopods, crayfish, elminthid beetles, some Chironomidae, some caseless Trichoptera, snails, mites, and small crustaceans (Stehr and Branson, 1938; Engelhardt, 1951; Smith, 1953; Cridland, 1957; Hynes, 1958; Clifford, 1966*a*). Clifford showed that in a small stream in Indiana, in which a species of each of *Lirceus* and *Crangonyx* were the dominant invertebrates, only small specimens survived the summer and autumnal dry period, and they burrowed as much as 60 cm. down into the damp substratum.

3. Species with eggs which survive long periods of drought. In cool climates this tends to favour stoneflies as against mayflies. This is probably because the former appear as a group to be more cold-adapted and so many of them are able to complete their growth before summer droughts occur; they have early fast life-cycles (p. 297). Burks (1953)

states that mayflies are absent from temporary streams in Illinois, although stoneflies are common, whereas Berner (1950) records species of *Callibaetis* and *Caenis* from such streams in the warmer climate of Florida. It should be noted that the survival of stoneflies, particularly of *Allocapnia* which is abundant in intermittent streams in eastern North America, is possibly as very small diapausing nymphs rather than as eggs. *A. pygmaea* is known to be ovoviparous in Indiana, and yet its nymphs disappear very soon from the water and do not reappear until fall or winter rain restores some flow to the streams (Clifford, 1966*a*). Possibly a nymph which can actively seek out a place to aestivate has a better chance of survival than an egg which must stay where it falls.

4. Species which reinvade from elsewhere, usually downstream, as soon as the water returns. These include fishes, which we have already discussed at some length in Chapter XVI (p. 329), and some insects particularly Ephemeroptera (Neave, 1930; Verrier, 1953). Other insects fly in to oviposit as soon as the water reappears. This applies especially to some species of *Baetis* and *Simulium* (Harrison, 1958; Fredeen, 1958), and Harrison (1966) suggests that in Africa most, if not all, the insect fauna of temporary streams originates in this way. He studied a temporary stream in Rhodesia, an area with the long dry season which is typical of much of Africa, and he found little difference between the fauna during the wet season and that of nearby permanent streams. Soon after the flow was resumed *Chironomus satchelli* and *Simulium ruficorne* appeared; but they were rapidly joined by species of *Baetis*, *Pseudocloeon*, and *Cheumatopsyche* as well as by other Chironomidae and *Simulium*, and there was no evidence that any had survived as eggs. This appears to be in contrast to the fairly clear evidence of aestivation of eggs or very young stages in cooler climates, and it perhaps indicates an important difference. In cool climates streams refill late in the autumn or in early winter, when adult insects are not available to oviposit, whereas in the warmer climate of dry Africa the early part of the rainy season is probably particularly favourable for ovipositing adults and their dispersal. Doubtless also the dry season in Africa is far more severe than it is in many cooler climates.

5. Species which occupy pools or the damp parts of the river bed only during the dry period or its early stages. Such are mosquitoes, several species of which may succeed one another (Abell, 1959) and which in the tropics include such vectors of malaria as *Anopheles gambiae*, Chironomidae, and other flies which live in muddy silt, e.g. *Tubifora*, and a variety of beetles and bugs which fly in to live in the pools. Some of these beetles, e.g. *Limnebius* and *Hydroporus*, may burrow down into the

substratum when the water has gone (Clifford, 1966*a*). When the bed becomes quite dry it may form a habitat for a variety of beetles, fly larvae, and worms which belong normally to the soil fauna (Moon, 1956).

6. A few species which appear to be highly specialized inhabitants of temporary waters. In streams these include a few snails, such as the African *Bulinus* and *Limnaea*, which can close over their shell openings with dry mucus (Cridland, 1957; Harrison, 1966) and some caddisworms. Harrison found that *Setodes* seals off its case in Rhodesia, and Clifford (1966*a*) believes that three species of Trichoptera in Indiana, *Ironoquia punctatissima*, *Lepidostoma* sp., and *Neophylax concinnus*, aestivate as pre-pupae, and that they pupate and emerge before the water returns. Apparently *Ironoquia* moves to the banks to aestivate and the adult oviposits on dry stones. This then is an animal which is highly adapted to these peculiar, but in some areas common, streams.

It will be appreciated then that temporary streams, although unsuitable for many stream organisms, support a varied fauna, and they may be the only habitat of some animals such as *Tyrrhenoleuctra* and *Hemimeleana*, presumably because they cannot compete with related forms in normal habitats. They are also quite suitable for any species which has a resting stage and a life cycle which fits in with the regime of flow, and in warm climates they may contain a quite normal fauna. Many species with long life-cycles or which are multivoltine, are eliminated by periodic dryness, and univoltine species of which the growing season coincides in part with the dry period are inevitably absent. The extent to which the occurrence of species with aestivating eggs is correlated with the severity of the dry season would seem worthy of investigation, in view of the different findings in Rhodesia, Indiana, and Wales.

THE HYPORHEIC ZONE

Water flows through the interstices of any loosely packed coarse material lying beneath a river, and we have seen that this interstitial flow is important for the well-being of the buried eggs of fishes lying at some depth in the substratum (p. 358). Even where the particle size is small there is still some movement in the upper few centimetres, and we have seen that this allows the development of a psammophile microfauna peculiar to river beds which has been studied mainly in Russia (Chapters VII and IX, pp. 114 and 175). The oxygen concentration in fine sediments, however, falls steeply with the depth, so the habitable layer for small animals is a thin one (Eriksen, 1966).

Beneath much of the rhithron the coarser sediments allow movement to a much greater depth, the extent of which has hardly been explored. There is thus a link between the rhithron and the groundwater or phreatic zone which was designated the hyporheic zone by Orghidan (1959). Here the water moves slowly, at rates measurable in only centimetres per hour, and although its oxygen content is lower and its carbon dioxide content higher than that of the overlying stream water, because of local decay and respiration, it provides a suitable habitat for a variety of animals. It also, as has already been mentioned, provides a refuge for stream invertebrates from extremes of current and temperature.

The indigenous fauna, for which Schwoerbel (1964) proposes the name hyporheal, consists of *Hydra*, Turbellaria, Nematoda, Oligochaeta, Tardigrada, Copepoda particularly Harpacticoidea, Ostracoda, Chydoridae, Syncarida (*Bathynella*), small Amphipoda, Hydracarina, and Porohalacarida. Large numbers of very small specimens of Chironomidae and Plecoptera and considerable numbers of small Trichoptera and Ephemeroptera are also always present.

The total numbers may be enormous. Schwoerbel (1961*a*, 1964) found over 34,000/sq. m. in the upper 10 cm., including as many as 8,000 young stages of truly stream animals. The community lies mainly below the bed of the stream, but in suitable situations it also spreads out laterally under the banks, and although the normal stream animals become less common there this extends their habitat beyond the margin of the stream.

Some of the organisms in this community, e.g. tardigrades, porohalacarids, and *Bathynella* occur nowhere else in the running-water environment, and many of the ostracods and other mites are groundwater forms. There is thus a mixing of faunas, and Schwoerbel suggests that this is a major site of evolution of the groundwater animals. Light penetrates for only a short distance, and the deeper down the more uniform is the temperature; the habitat therefore presents a cline of conditions.

Schwoerbel found that this fauna is most abundant in areas of hard crystalline rock, where the pore spaces are large, and less abundant on softer rocks, such as limestones, where fine rock-debris tends to fill the interstices. He also points out that most of the inhabitants feed on detritus, so there is a fairly direct relationship between the density of the population and the amount of organic detritus in the deposits. Angelier (1962), whose study was concerned mainly with mites, found that the coarser the sediments and the swifter the current the further under the banks the species extended. He also recorded downward movements in winter and at times of flood, and he noted that the altitude affects the depth at which some species live. *Torrenticola remiger* and *Paraxonopsis vietsi* live deep in the sediments in the Alps and the Pyrenees, but near to the surface in Corsica. He concludes, therefore, that both a rapid current and a low temperature, acting probably through the dissolved oxygen content of the water, cause deeper penetration.

Although it has received so little attention it is clear that the hyporheal is an important feature in the ecology of streams, and that it contains a biomass equal to about the same area of moss. Moreover, many of the ordinary organisms of the benthos clearly spend much of their early lives there. This community lives in a habitat which should receive much further study.

THE PSAMMON

It has been discovered relatively recently that wet and damp sand at the margins of lakes and marine habitats contains a fauna of specialized species, and also often a flora of algae which may make the sand green. A similar habitat occurs along the margins of running water where there are sandy beaches, and it has been studied by Angelier (1952, 1953) and Ruttner-Kolisko (1961, 1962) in Europe.

Natural sand is usually packed so that it encloses about 40 per cent of its volume as pore space, which is rather less than the space which would be enclosed by cubically packed spherical balls. This percentage falls, however, if the grain sizes are very mixed, and it is primarily in sands of grain diameters between 0·2 and 2·0 mm. that the pores are large enough to shelter a significant population of organisms. The pore sizes are too small to permit much interstitial water movement when the sand is quite submerged, and then the presence of organic matter readily reduces the oxygen content to zero. Thus, even where the water flows rapidly, the oxygenated layer is thin, and is inhabited as we have seen, by a specialized microfauna well adapted to survive turbulent entrainment.

At the water's edge, however, the water is drawn upwards by capillary attraction and it evaporates from the wet surface, and this causes a slow current through the pore spaces. The slow seepage coupled with the presence of air-filled spaces, maintains a supply of oxygen and creates a narrow zone along the beach which is habitable by small animals. This zone is the psammon, and it moves up and down with the water level.

It contains a specialized fauna of ciliates, Turbellaria, Nematoda, Rotifera, Gastrotricha, Oligochaeta, Tardigrada, Harpacticoidea, and mites, many species of which occur only in this type of habitat. These do not concern us here, except in so far as specimens of them may occur in samples taken from the river bed, but it also contains specimens of the truly aquatic fauna, primarily very small specimens of Chironomidae, Ephemeroptera, and Plecoptera, but also some Turbellaria. It therefore, like the hyporheic zone in the rhithron, shelters many young stages of true aquatics in the potamon, and it should not be overlooked in studies of life histories and productivity in areas where it occurs. Even though it is a narrow zone it appears to be densely populated, and in many places it is continuous for long distances. Its fauna, therefore, like the hyporheal, deserves more attention in its relationship to the ordinary benthos than it has yet received.

THE MADICOLOUS HABITAT

The madicolous, or hygropetric, habitat consists of thin sheets of water flowing over rock faces, and although it is hardly a running-water habitat in the ordinary sense it often occurs very close to one.

As would be expected, many of the inhabitants are peculiarly modified for maintaining their position, remaining submerged to the right extent

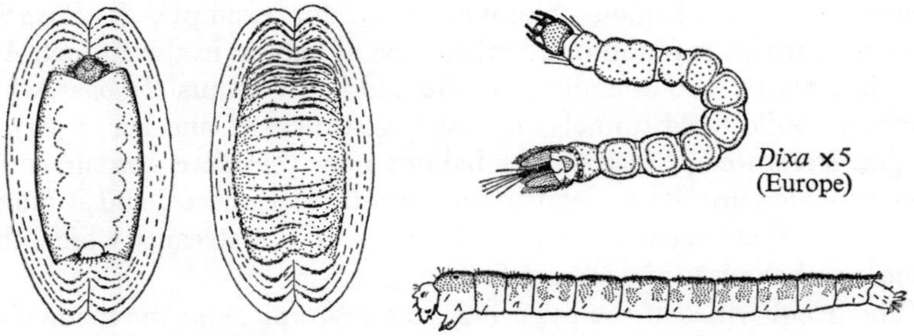

FIG. XXI, 1. Examples of madicolous animals which may occur in collections made in streams where there are wet rock-surfaces nearby.

in the water film, being able to move readily as conditions change, etc. For a discussion of these and many other points the reader is referred to the excellent and extensive work of Vaillant (1956, 1961). The habitat is mentioned here primarily because, at the edges of stony streams, at the sides of waterfalls and on rocky chutes, running water frequently produces such habitats. They are often colonized by typical madicoles and they occur even beside lowland streams (Marlier, 1951). Similarly, in shallow mossy streams, the moss forms a habitat, at any rate at times of low water, hardly distinguishable from that on steep, wet, moss-covered rock faces. There is, therefore, much similarity and intermingling of the faunas and floras of the small stream or stream-edge and the rock-face habitats, and many animals which are often collected in stony and rocky streams are really madicoles from the edge (Fig. XXI, 1). These include some Lumbricidae, small species of Hydrophilidae, some small caddis-worms such as *Stactobia*, and a variety of Diptera including several types of cranefly larvae, Psychodidae, Stratiomyidae, and the peculiar U-shaped Dixidae. The Diptera are the most numerous madicoles, and in contrast to the truly stream-dwelling families they are all air-breathing.

There is also a limited amount of movement to the edge habitat from the stream bed. For example, the Japanese mayfly *Bleptus* often moves into thin films of water at the edges of swift streams (Uéno, 1931), and the common little hydraenid beetles, the adults of which occur in stony rivers and streams, spend their larval life at the edge as air-breathers. Beier and Pomeisl (1959) describe how *Octhebius exsculpus* of this family breeds in rocky streams in the Vienna woods in a fringe of wet algae which grow in summertime just above the water level. Pupation occurs a little further up the rock and the adults then move out into the stream bed where they pass the winter. At least some species seem to have moved out of the normal stream habitat and taken up permanent residence in the madicolous biotope. Such, for example, is *Simulium auricoma*, which is found in no other type of habitat in the French Alps (Dorier, 1961), and members of the caddis-fly genus *Tinodes* which normally build fixed tunnels on submerged rocks (Danecker, 1961).

The stream-edge madicolous habitat may therefore contain a few species which are not present in its more typical phases, and it may at times lose specimens into the main body of the streams which then appear to belong to the stream fauna.

The peculiarities of the hyporheic, the psammic, and the madicolous faunas indeed indicate how very important are the microhabitats in running water, and how necessary it is to examine all types of habitat to obtain unequivocal data on the stream biota as a whole.

CHAPTER XXII

The ecosystem

The traditional way of considering an ecosystem is to regard it as a self-contained complex to which energy is supplied in the form of radiation. This it passes through a series of trophic levels—primary producer, primary consumer, secondary consumer, and so on—and it returns basic materials to their original form and, as it were, rolls on cycles of carbon, nitrogen, phosphorus, sulphur, and other elements. This model applies fairly well to terrestrial situations and to lakes, although even the latter are no longer regarded as microcosms and the concept of the trophic level is undoubtedly a considerable over-simplification (Ivlev, 1945; Darnell, 1961). The model clearly, however, does not apply to the ecosystem in running water, where everything released into solution by metabolism tends to go on downstream and has little opportunity of being recycled on the spot. Any cycling which does occur is continually displaced in a downstream direction (Shadin, 1956; Schmitz, 1961).

There is, moreover, much evidence to show that allochthonous organic material is of great importance in running water and that by no means all the energy which drives the ecosystem is derived from radiation which it itself receives. In this respect it is, of course, not unique as the same must apply to all deep-water systems, but it is certainly unusual for one which normally lies within reach of the sunshine.

Thienemann (1912) very early made the observation that stream-dwellers include many species which feed on detritus, and Jewell (1927) stressed that prairie streams are poor, probably because they receive little organic matter from plants in their valleys. Since that time many workers have pointed out, or stressed, the importance of allochthonous material as forming a large part of the base of the food nexe (Coker, 1954; Pleskot, 1954; Scott, D.C., 1958; Ivlev and Ivassik, 1961; Hynes, 1963). Most of these authors have been concerned with the temperate zone and they have considered primarily the autumnal contribution of fallen leaves, but Fittkau (1964) pointed out that the same general principle applies in the Amazonian rain forest, and that blossoms, fruit, and pollen may be even more important than dead leaves, since they are probably better food for animals.

Much of the information on detrital feeding is, however, only qualitative, as we have seen in Chapters X and XVIII, and this cannot lead to any deep understanding of the system. There is also undoubtedly some local photosynthesis and hence autochthonous primary production, so we will first discuss this and then consider such, very inadequate, data as we have on the other trophic levels.

PRIMARY PRODUCTION

Odum (1956) argued that it should theoretically be easy to determine the gross primary production, i.e. the amount of carbon assimilated by photosynthesis, over a length of stream, by the simple expedient of monitoring, continuously or at frequent intervals, the oxygen content of the water entering and leaving it, and that the same concept can be applied to measurements made at a single point. Then, because

$$Q = P - R + D + A$$

where Q = rate of change of dissolved oxygen, P = rate of gross primary production, R = rate of respiration, D = rate of uptake from (or loss to) the atmosphere, and A = rate of drainage accrual—all in terms of oxygen per unit area or volume—it should be possible to work out P.

This can be done by working on short intervals of time throughout the period of twenty-four hours and assuming that R is constant and so can be determined at night, that A can be measured or is negligible, and that D is determinable, or that it can be inferred from the type of stream and the oxygen deficit during any particular calculation interval.

Using this idea, and equating one atom of carbon to one molecule of oxygen, Odum worked out the weight of carbon fixed per unit area in a number of streams for which diurnal oxygen curves were available, and he showed that in many of them, particularly in those which were polluted, primary production rates were high, higher even than on good agricultural land or in cultures of *Chlorella*.

A major weakness of this scheme, however, is the assumption made about D, the rate of exchange with the atmosphere, and other workers have since tried to determine this directly, either by experimentally lowering the oxygen content of the water or by direct measurement. Edwards *et al.* (1961) lowered the oxygen content, by the addition of nitrite and a cobalt catalyst to an experimental reach of a small river from which the light was excluded, and then measured its rate of recovery. They thus obtained a figure for the coefficient of oxygen exchange through unit surface area; and Copeland and Duffer (1964) floated an inverted plastic dome on the water surface and withdrew gas samples at intervals to determine gain or loss of oxygen to the atmosphere. The reader is referred to Owens (1965) for a thoughtful discussion of the dissolved-oxygen method of determining the primary production of streams and its various possible sources of error.

Odum's original method was successfully used to study stream microcosms in glass tubes, which, being full of water, presented no problems in respect of D, and these experiments showed that the ratio P/R was often less than one (Odum and Hoskin, 1957). In effect then the plant community, of which the microcosms were almost entirely composed, was actually producing rather little which it was not itself respiring. Field studies using the method, with or without the refinement of direct measurement of the rate of exchange of oxygen through the water surface, have produced rather similar results.

Hoskin (1959) found that in the Neuse River, North Carolina, primary production varied from 0·3 to 9·8 g. oxygen per square metre per day, but that respiration used 0·7–21·5 g./sq. m./day in such a way that respiration was always the greater at all stations. In the Middle Ocanee River, Georgia, Nelson and Scott (1962) found that respiration was greater than photosynthesis on the four occasions on which measurements were made. In the River Ivel, England, Edwards (1962; Edwards and Owens, 1962) found that, although rather exceptionally large amounts of plants were present, their respiration was so high that they were, in effect, contributing little in the way of production to the ecosystem. Woods (1965*b*) was unable to apply the method in the upper Ohio River because of the irregularity of change of the oxygen levels, and McConnell and Sigler (1959) had a similar difficulty in the Logan River, Utah, because the turbulent flow kept the oxygen content of the water in equilibrium with the atmosphere. This problem was to some extent surmounted by Hufford (1965) who found that he could detect photosynthetic activity in early spring, in the Maumee River, by a fall in the carbon dioxide content of the water, even when oxygen levels remained stable. Finally Duffer and Dorris (1966), found, during their extensive study of the Blue River, Oklahoma, that respiration always exceeded photosynthesis at all stations except one. This one was a granite outcrop which, in summertime, produced more than it consumed.

From this brief survey of the results of gross estimations made on entire ecosystems it seems fair to conclude that only in limited situations, or where pollution has materially altered the nutritive status of the water, does local primary production supply more than a proportion, possibly a very small one, of the energy used by the running-water ecosystem.

Another approach to the measurement of primary production is to measure it directly by more conventional means, and studies of this kind have been directed towards both the periphyton and the higher plants, and to a lesser extent towards the plankton.

Direct measurement of periphyton production *in situ* presents many difficulties. These are partly caused by the irregular distribution of the growths and partly by the fact that it is known that the current influences the metabolic rate of attached algae (Chapter IV). Obviously, because of the known differences between populations on natural and artificial substrata, it is unrealistic to attempt to apply to the natural situation results obtained with materials left in a stream for some while and then removed to chambers where measurements of photosynthetic rate may be made. Such data, although interesting, cannot be expected to relate very closely to natural substrata in the water.

King and Ball (1966) have, however, been able to arrive at some estimate of the net production of diatoms by observing their rate of increase, in terms of ash-free dry weight, on sheets of plastic placed in a stream. The weight increased in summertime for fifteen days and then levelled off because, they surmised, attrition then equalled production. They obtained the figure of 0·253 g./sq. m./day for the rate of increase during the fifteen-day period, but there is no means of knowing how this is related to actual production on the stream bed, nor how much of it represented colonization of vacant space by drifting cells and thus not local production at all. The presence of drifting cells, which presumably attach where they can, is an inevitable and possibly large distortion of results obtained by direct study based on initially bare surfaces.

Wetzel (1963, 1965) has described methods whereby the periphyton on lake shores may be covered with transparent and opaque chambers, inside which the usual measurements of change in oxygen, carbon dioxide, pH, or uptake of introduced C_{14} may be used to assess carbon assimilation, as is done for instance, in light and dark bottle experiments with phytoplankton in lakes. For work in a stream, however, the absence of flow would inevitably introduce an unknown factor, a limitation which applied to the work of McConnell and Sigler (1959) who placed alga-covered stones in light and dark bottles. Thomas and O'Connell (1966) describe a chamber in which the water can be kept circulating by a pump, and which has a flexible top and sides, allowing water samples to be removed without the introduction of air or new water. Such a device could be used for study of photosynthetic activity of portions of natural substratum *in situ*, although they used it only for the study of algae grown on artificial materials in a stream.

Another approach is to remove the periphyton, break it up and then treat it in light and dark bottles as if it were plankton. Kobayasi (1961*b*) did this with samples from the River Arakawa in Japan, and he was able to show that photosynthetic rate was affected by both light and temperature, and to calculate that production was higher in the warmer, less shaded, lower reaches of the river than in its cooler, heavily shaded, canyon reach. He worked out the amounts of carbon fixed as 810 and 210 g./sq. m./yr. respectively, but, while these figures may be correctly related from a comparative standpoint, one would hesitate to accept them in absolute terms in view of their foundation on the behaviour of homogenized periphyton maintained in still water.

It is indeed true to say that the only *direct* measurements on photosynthetic rates of stream periphyton which have so far been made have been in artificial streams by the Odum method, or by an extension of it using carbon dioxide or pH (McIntyre and Phinney, 1965; Kevern and Ball, 1965). These give efficiencies of energy usage (15·1 and 18·3 per cent) which are at least an order of magnitude higher than those which seem to occur in the field, and Kevern's study showed that there was poor agreement between the measurement of photosynthesis and the amount of algal material produced. There was, however, agreement between photosynthetic rate, as measured by oxygen change, and the instantaneous rate of growth (p. 326) of algae on artificial substrata during the logarithmic phase of their increase after colonization (Kevern *et al.*, 1966). It would seem then that our present knowledge is quantitatively rather doubtful, but that there are indications of ways in which it could be improved.

Another approach to the periphyton problem is to use the indirect method of measuring the amount of chlorophyll A present. It has, for instance, been shown that the quantities of this pigment, which can be extracted by steeping stones in solvent, vary seasonally more or less in concert with the expected rate of primary production (Waters, 1961*b*). Unfortunately, however, the assimilation number, the weight of oxygen produced per unit weight of chlorophyll per hour, is by no means constant. Odum *et al.* (1958) found that it varies from 0·4 to 4·0 and that it is highest in thin communities in bright light, and Woods (1965*b*) quotes similar ranges of figures for phytoplankton. It appears, therefore, that only direct measurements of undisturbed periphyton in chambers in which the water is kept moving, offer any real hope of determining the primary production by periphyton, and extensive field measurements of this type have still to be made.

For higher plants, Wetzel (1964, 1965) suggests that a suitable method for determination of primary production is to use chambers similar to those recommended for periphyton, and to measure the uptake of radiocarbon. He suggests that measurement of oxygen change is not meaningful because of the storage of gases in aerenchyma. Obviously also, for running-water plants, there must be provision for water movement. Odum (1957) attempted to measure the production of *Sagittaria* in Silver Springs, Florida, by covering plants with light and dark bell-jars and measuring the change in oxygen. He had some problems with oxygen stored in the tissues, and he noted that the concentration of this gas in the water continued to rise for half an hour after a black bell-jar had been put into position. This, and the lack of current in the jars, probably made his estimates inaccurate, and this seems to have been so in that he records over 6,000 g. carbon fixed/sq. m./yr. and an efficiency of use of light of 5·3 per cent. Such high figures, even if they are more or less correct for that particular spring, are probably outside the normal range of running waters.

Direct measurements of the net production of macrophytes in some southern English rivers have been made by cropping and weighing plants from known areas at intervals during the growing season (Edwards and Owens, 1960; Owens and Edwards, 1961, 1962). The technique was to cut and dry the plants in a spin-dryer, used with a battery at the stream side, and to convert the weights to dry weight as determined on small samples carried to the laboratory. From changes in standing biomass in these very thickly vegetated streams it could be calculated that the net fixation of carbon was of the order of one to two or more g./sq. m./day during the growing season. This high figure represents, nevertheless, an efficiency of less than 1 per cent conversion of energy, but it should be noted here that this is a measure of production less any material that had been respired or eaten by animals during the intervals between croppings. A rather similar method involving frequent sampling was used by Nelson and Scott (1962) to determine the production of *Podostemon* on rock (see p. 424), and they obtained the very high figure of 930·3 g. ash-free dry weight/sq. m./yr. This is probably equivalent to several times that amount in terms of wet weight. It is clear therefore that in some situations rooted plants can produce great amounts of material in running water, and a standing biomass measurable in kilograms of wet weight per square metre. The River Ivel was reported as supporting 7,300 g./sq. m., and Greze (1953) records 11,900 g./sq. m. in parts of the River Angara. These are, however, rather exceptional situations in rivers with shallow water, slow flow, and even discharges, and it seems that light rather than nutrients puts an upper limit to production, as growths in the Ivel above and below a sewage outfall were similar. Indeed in a polluted stream near the River Ivel, Westlake (1961) recorded only 1,460 g. of *Potamogeton pectinatus* per square metre, which is in contrast to the expectation that increased nutrients increase primary production. Doubtless the interception of light by suspended matter in sewage is a factor of great importance. Even the brown coloration of the water of the Rio Negro causes the compensation point, where respiration equals photosynthesis, to lie at only 1·5 m. (Hammer, 1965), so only particularly clear, shallow, and unshaded reaches which are not subjected to periodic wash-out can support high levels of primary production by higher plants. In this connection it is perhaps important to stress that higher plants build up biomass in one season and lose it at others. Owens and Edwards (1961) found that in the River Ivel amounts rose by an average of 0·85 g. carbon/sq. m./day from March to June and declined by an average of 0·36 g./sq. m./day from June to December. There is, therefore, a seasonal accrual which is presumably used up by primary consumers throughout the year.

Westlake (1966) has recently suggested that it may become possible to work out the rates of photosynthesis in stands of higher plants in running water by combining knowledge, obtained in the laboratory, of their photosynthetic characteristics with measurements of density, depth distribution, and self-shading in the field. This idea clearly has considerable possibilities, but his present model is too crude to give useful results, and it would be an enormous task to attempt to apply it to mixed populations of plants in a varied watercourse.

Published accounts of studies on primary production by plankton in running water are few and scattered, and they have all been based on the light and dark bottle technique, which would seem on *a priori* grounds to be quite acceptable despite the lack of current. A planktonic alga moves with the water, so in contrast to a sessile plant it would seem that it is irrelevant whether or not the water itself is moving. It should not be overlooked that, even in still water, plankton is either slowly sinking through the water or floating up through it, so the possibility of cells forming a depleted layer of water around themselves does not arise—except, of course, in experimental bottles!

Duffer and Dorris (1966) found that the amounts of suspended chlorophyll A were low in the Blue River, Oklahoma, and they assumed, as one would expect from what is known of the amounts of plankton in small rivers, that planktonic primary production is low. Similarly, Hammer (1965) recorded only about 0·1 g. carbon fixed/sq. m./day in the brown water of the Rio Negro. On the other hand Knöpp (1960) records 1·6 g. oxygen produced/sq. m./day in the Rhine at Koblenz (equivalent to

0·6 g. carbon) and Woods (1965*b*) obtained intermediate values from the Upper Ohio River. Woods was also able to show a general increase in production in the downstream direction, which, of course, fits with what is known of the distribution of phytoplankton in rivers.

Knöpp collected water samples at all seasons from the rivers Rhine and Main, and incubated them in the laboratory in order to determine their potential ability to produce oxygen. This was almost zero in the winter and rose to maxima in the summer, and he was able to show that the potential production lagged about a month behind the increase in daylength and was lower during the dull cool summer of 1958 than it was during the warm sunny one of 1959. Clearly, therefore, plankton needs time to build up in a river, and he was also able to show that this occurs almost exclusively in water warmer than about 12 °C. This is an important point as it shows that the only time at which local plankton can contribute much energy to the ecosystem is during the summer, and, in contrast to the higher plants and the periphyton, any material that it builds up then is either used up at once or is passed on downstream.

SECONDARY PRODUCTION

The fact that it is usually necessary to consider secondary production, which in streams is in effect production of benthic invertebrates, as a unit is, of course, an admission of failure to disentangle the several trophic levels. It is an admission that in studies of the ecosystem as a whole we cannot distinguish between primary consumers, secondary consumers, and the many species with multiple trophic spectra. Its study has mostly, and for obvious reasons, been carried out with special reference to the amount of food available for fish. It is perhaps of interest, and salutary, to note that while the biologists who study fishes appreciate the great differences in metabolic rate and efficiency of conversion of foodstuffs into flesh, even between closely related species, e.g. Mann's (1965) careful distinction between the cyprinids *Rutilus rutilus* and *Alburnus alburnus*, there has been a marked tendency to lump the invertebrates together and to treat them as a single ecological entity. This is said not in criticism of this kind of work but to emphasize the magnitude of the problem of working out the production rates of a fauna consisting of mixed and very varied animals. In reality each species should be treated separately, and only when this is done will there be any true indication of what is occurring. However, to do this for all species, or even for the most important of them, in any one habitat would be an enormous task, and it has not yet been attempted.

To date there have been two rather different approaches to the problem. The first is back calculation from the measured production of fishes which feed on invertebrates, and the second is through direct study of the invertebrates themselves. It should perhaps be stressed at this point that the term production is here being used in the sense of Ivlev, to mean all the organic matter built up by the organisms under consideration, other than that used by them for metabolism, whether or not it is all present at one time.

Back calculation from the known, or rather from the calculated, production of fishes was first performed by Allen (1951) during his monumental and pioneer study of brown trout in the Horokiwi Stream, New Zealand. He, by methods discussed in the next section, worked out the production of trout flesh per unit area of stream, and then, on the basis of published data on the amounts of food needed by laboratory-reared fish to maintain themselves and to grow, it was possible to calculate the amount of food which must have been eaten. After subtraction of the amount which stomach analyses indicated would have been of terrestrial origin, this was found to be many times both the biomass and the numbers of the invertebrates in the benthos at any one time. This factor, of about 17 in terms of numbers of the fauna as a whole, and as much as 50 for the mayfly *Deleatidium*, or about 100 in terms of weight for the whole fauna, came to be known as the 'Allen paradox', because it was difficult to explain when it was known that most of the invertebrates involved have an annual cycle. The individuals, it would seem, should all be there at the start of their growing season even if they are eaten as they grow. The average standing biomass in the stream varied from 2·62 to 7·19 g. wet weight/sq. m. at various points in the stream.

A partial explanation of the paradox was provided by Gerking (1962), who pointed out some faulty assumptions in Allen's calculation of fish production. The details do not concern us here, but the correction would divide the paradox factor by only about 2 or 3. A further partial explanation is provided by the fact that Allen's benthic samples did not include very small animals, below 2 mm. long. He therefore did not count the more numerous young specimens, and we now know that most stream insects go on hatching from eggs for a long period (Chapter XIV). It is therefore not surprising that more specimens are eaten than are there at any one time even though many of the prey species are univoltine. This point, which is discussed by Macan (1958*a*) and Hynes (1961), does not, however, alter the fact that Allen's work showed that the production of the benthos in terms of both numbers and weight is many times that of the amount present at any one time, and that the paradox is a real phenomenon. This fact was further emphasized when it was later shown that the population of eels in the Horokiwi Stream has twice the biomass of the trout (Burnet, 1952*b*), and they also must eat considerable numbers of invertebrates.

The same paradox has been revealed by rather similar methods in still waters, with the added finding that if the fishes are removed the biomass of the invertebrates increases, but its rate of production falls (Hayne and Ball, 1956). The last point probably applies to streams also, and it almost certainly results from intraspecific competition. When one specimen is removed another grows to replace it, whereas if the first one is not eaten it alone grows up. In the rhithron, where places to shelter are probably important in controlling numbers, one might expect this to be particularly obvious, but it has not been studied.

Mann (1964, 1965), after a somewhat similar study of the fishes in a reach of the River Thames, calculated, on the basis of oxygen-uptake figures and population studies, that the annual energy requirements of the fish population were over 700 kcal./sq. m. One kilocalorie is about equivalent to one gram of fish flesh, wet weight (Winberg, 1956), and presumably this is more or less true of most invertebrates also; figures in kilocalories and grams of wet weight are thus at any rate roughly comparable. The invertebrate biomass present at any one time was, however, only about 68 kcal./sq. m. of which all but 5·4 were present as large bivalves which are not eaten by the fishes. Here again, even allowing for the fact that much of the food of the fishes was plant material and allochthonous insects, the paradox is once more observed. In this instance, as many of the invertebrates, which were mostly worms, leeches, and molluscs, would not have been missed in the benthic samples, the high rate of turnover must have been provided by the continuous production of young by many invertebrates, e.g. oligochaetes, snails and *Sphaerium*, the multivoltine cycles of the Chironomidae, and possibly the zooplankton of which no account was taken. Nevertheless, it seems that here also there is an annual production which is many times the energy content of the standing biomass of the food organisms.

Horton (1961) also estimated the quantities of animals eaten by trout in an acid stream in south-western England and showed that the weight was 8·7 to 25·9 times the observed average biomass at different stations and in different years, and the number 19 to 65 times. She interpreted the discrepancy between weight and numbers as being because she did not find the very small animals in her benthic samples, and she believed that some of the excess of consumption over standing crop would be accounted for by continuous hatching of eggs, a point already alluded to. So once again we see that production is many times the standing biomass, but there is still the uncertainty as to how many times. In view though of the inefficiency of our quantitative benthic sampling methods (Chapter XII) and the fact that in none of these studies was the sampling method outstandingly good, it would seem that all the turnover ratios (production/biomass) which have so far been calculated are probably too high. One might hazard a guess that the true value is perhaps about 10 for populations dominated by arthropods, but it may be much less.

	Protonemura meyeri		*Amphinemura sulcicollis*	
Number of:	Population at the start of each stage	Number consumed by trout	Population at the start of each stage	Number consumed by trout
Eggs hatched (minimum)	179,843 ↓	→ 107,613	92,610 ↓	→ 5,220
Half-grown nymphs	72,230 ↓	→ 65,518		
Full-grown nymphs	6,712 ↓	→ 2,530	87,390 ↓	→ 71,900
Emerging imagines	4,182 ↓	→ 575	15,490 ↓	→ 5,800
Eggs laid (maximum)	2,880,000 ↓		484,500 ↓	
Eggs hatched	68,879	→ 22,521	156,405 ↓	→ 142,890
Half-grown nymphs	46,358 ↓	→ 23,424		
Full-grown nymphs	22,934 ↓	→ 15,968	13,517 ↓	→ 3,814
Emerging imagines	8,966	→ 0?	9,703 ↓	→ 6,565
Eggs laid	7,190,000		156,900	

TABLE XXII, 1. The estimated numbers at different stages in their life histories and the fates of two species of stonefly during two years (1956–8) in a reach of Walla Brook, Devonshire, England; based upon benthic samples, trout stomach contents, and estimates of numbers of trout present. From Horton (1961).

Horton has calculated the loss of individuals of two species of stonefly caused by predation by trout, on the basis of growth rates, estimated numbers of each size group eaten by the fish, the number of adults emerging and the number of eggs per female. Her table is reproduced here (Table XXII, 1), as it is an almost unique attempt to draw up this sort of life table for stream insects. Unfortunately she gave no information on the area involved in her study and it is easy to find fault with some of the basic assumptions. For example, the numbers of eggs estimated to be laid by *Amphinemura sulcicollis* are undoubtedly far too low, and the apparently very low percentage-hatch of the eggs reflects the fact that her benthic samples did not contain small specimens in proportion to their numbers; also, of course, no account is taken of deaths from other causes. The last point tends to reduce the apparent rate of mortality during nymphal life, as can be seen in Table XXII, 2, where Horton's results are compared with some similar figures calculated from data given by Hynes (1961) who collected the insects repeatedly during a year from very small areas with a very fine-meshed sampler. It can be seen that the percentages surviving from newly hatched to emerging as calculated from Horton's figures are greater than those recalculated from the data of Hynes except for *Protonemura meyeri* in one year, and it is perhaps relevant to note that Hynes commented on the fact that this species seemed to have a particularly low death rate among the species which he studied. Probably the survival rates determined by both studies are too high because of failure to find a large proportion of the very young specimens, but undoubtedly further attempts at this kind of investigation would lead to a better understanding of benthic production.

	Protonemura meyeri			*Amphinemura sulcicollis*		
	Walla Brook		Afon Hirnant	Walla Brook		Afon Hirnant
	1956–7	1957–8	1955–6	1956–7	1957–8	1955–6
Number of newly hatched	179,843	68,879	586	92,610	156,405	1,612
% *survivors*						
Half-grown	40·0	67·2	33·8	—	—	12·5
Full-grown	3·7	33·3	13·8	94·5	8·8	3·3
Emerging	2·3	13·0	4·1	16·7	6·2	1·0

TABLE XXII, 2. The original numbers and percentage survivors of two species of stonefly at various stages in their growth. Calculated from data given by Horton (1961) for Walla Brook, Dartmoor, and by Hynes (1961) for Afon Hirnant, Wales. No area of study is given for Walla Brook; that for Afon Hirnant was 1,800 sq. cm.

A point which is worth noticing in Table XXII, 1 is the way in which the figures vary from year to year, sometimes by factors of three or more. It is clear therefore that data from one year are not sufficient, and Hunter (1961) who studied a population of *Ancylus* in Craigton Burn in Scotland during nine consecutive breeding seasons found that the numbers varied sevenfold.

It is often assumed that the proportional composition of stream ecosystems is fairly constant, even though the total biomass may be affected by spates, and some workers (e.g. Hynes, 1961) have found this to be true of two consecutive years. That there can, however, be large variations in species composition for no apparent reason is illustrated in Table XXII, 3. The collections on which this table is based were made at the same season at the same three points along about a kilometre of a stony trout stream every year for nine years. There seemed to be no obvious change in the river, which drains open moorland, and the collections were made so early in the year that the fauna could have been little changed by possible variations in the emergence dates of early insects. There is also no doubt that the collections, which were made by netting at random over the river bed for a fixed time, were comparable from year to year at least in respect of the proportions of the various

	1955	1956	1957	1958	1959	1960	1961	1962	1963
Polycelis felina	1·2	0·5	0·1	0·8	0·9	0·9	0·6	0·9	0·9
Protonemura meyeri	+	0·2	0·1	0·3	0·1	0	0	0·5	3·7
Nemoura 3 spp.	0·1	0·4	0·2	0·3	0·8	0·1	0·2	0·1	2·6
Amphinemura sulcicollis	11·1	25·3	17·9	19·2	3·8	9·6	0·2	0·1	9·6
Brachypteri risi	0·1	+	+	0·1	+	0	0·3	0·2	1·5
Leuctra 4 spp.	3·0	0·8	5·0	1·2	1·7	0·2	+	0	0·4
Chloroperla 2 spp.	2·8	3·1	1·2	2·4	1·1	1·8	0	0·1	0·2
Isoperla grammatica	5·1	13·6	3·5	6·8	3·1	9·3	0·1	0	0
Perlodes microcephala	0·7	0·2	0·1	1·3	0·1	0·9	0	0·1	0·1
Rhithrogena semicolorata	34·5	21·8	15·8	40·1	7·0	13·7	+	0·1	0·2
Hepagenia lateralis	0	0·2	0	0·1	0·1	0	+	0	0·4
Ecdyonurus venosus	0·2	0·5	0·1	0·8	0·2	0·5	0	0	0
Baetis 2 spp.	7·8	5·1	35·6	5·6	13·8	3·9	84·0	65·7	44·3
Hydropsyche 2 spp.	0	0·8	3·6	9·6	7·1	17·3	0	0	0·6
Rhyacophila dorsalis	0·5	1·0	1·2	0·8	0·3	0·2	0·3	0	0
Polycentropus sp.	0·3	1·5	0·1	0·2	0·1	3·8	0	0·1	0·1
Sericostoma sp.	0·4	0·8	0·3	0·3	0·1	0·6	0	0	0
Stenophylax sp.	0·1	0	0	1·0	0·6	0·5	0	0·2	0·3
Esolus parallelopipedus	11·8	2·4	3·3	1·5	12·3	14·1	9·1	17·8	10·7
Limnius volkmari	0·4	0·7	0·2	0·2	2·4	1·2	0·8	4·9	3·8
Elmis maugetti	0·3	0	0·2	0·2	0·5	1·9	0·7	1·4	2·6
Ceratopogonidae 1 sp.	0·2	3·3	0·6	0·3	3·0	3·2	0·7	3·9	1·1
Orthocladiinae spp.	15·6	14·8	3·4	6·3	25·7	10·9	0·5	2·7	9·7
Tanypodinae spp.	+	0·2	0·1	0·1	0·2	1·0	0·2	0·3	1·0
Simulium spp.	0	0·4	2·6	0	13·3	0·4	0	0	0·4
Empidae 1 sp.	0	0	0·1	0	0·4	0·1	0·1	0·1	0
Limnaea pereger	0	0·2	0	0·3	0	0·4	0	0·1	0·1
Total number of animals	2,154	2,264	3,239	1,034	2,901	1,167	5,295	1,974	1,245

TABLE XXII, 3. The percentage composition of the important members of the invertebrate fauna of the upper River Derwent, England, in the spring of nine successive years, as revealed by standardized coarse-net collections made at the end of March or the beginning of April in each year. The results of collections at three stations have been pooled in constructing the table as the stations gave similar results. + = <0·05 per cent.

species caught even though the total numbers taken may have been affected by differences in discharge.

It can be seen from the table that the percentage composition of the fauna varied considerably from year to year, and that in 1961 there was a major change in that *Baetis* became very abundant and several other animals quite scarce. Moreover, this persisted for the next two years. No explanation could be found for this on the basis of discharge or temperature. The two exceptionally dry summers of 1955 and 1959 which might have been expected to affect the fauna in the springs of 1956 and 1960 had apparently no marked effect, and the climate during 1960, which might have been suspected of causing the changes registered in 1961, was in no way remarkable except perhaps that the winter was a fairly wet one. The point, however, is that for no obviously catastrophic reason there can be great changes in the faunal composition from year to year and this fact has not often been stressed, primarily because few long series of observations have been made. The existence of this variability should not be ignored in our considerations of production and our picture of the ecosystem as a whole.

The direct approach to determination of secondary production of the invertebrate community has been taken by Nelson and Scott (1962) and by Hynes (1961). The former sampled a rocky *Podostemon*-covered outcrop in the sandy Ocanee River, and they determined the numbers and weights of the organisms present in a certain area each month. These, of course, fluctuated, but by adding together all the increases which occurred during the year they arrived at a minimum estimate of production. Decreases were regarded as losses downstream and by death, and there seemed little chance of disturbance caused by drift on to the rock as it was isolated below a long barren sandy reach. This procedure produced the following figures in grams of ash-free dry weight per square metre for net production:

Podostemon	930·3
Filter feeders	20·2
Herbivores	7·9
Detritus feeders	2·0
Mixed herbivore and detritus feeders	3·3
Carnivores	4·3

This is a total of 33·4 g./sq. m. for primary consumers and 37·7 for all invertebrates. They also calculated, on the basis of calories, to which they converted their figures according to published data, that the turnover ratios of production to standing biomass were 11 and 12 respectively. This is in good general agreement with the figures estimated from the fish-food approach discussed above, and this would suggest that both are of about the right order of magnitude. In this study a very fine mesh was used so there is some confidence that small specimens were not lost, and the rather special habitat, on rock, also leaves no doubt that the whole biota was sampled.

Hynes (1961) sampled a small area of a mountain stream in Wales, on thirteen occasions spread over a year, using a very fine mesh. All the animals were identified and measured, and ten taxa were found to constitute 71 per cent of the invertebrate biomass. By summing the total number of specimens of each of the taxa in each millimetre size-group it was possible to work out the loss of specimens as the animals grew, and to convert this to an estimate of the loss of weight through the year. On the assumption then that the ten taxa were typical of the whole fauna, a correction applied to data worked out on 71 per cent of it gave an annual production of almost 3 g./sq. m./yr. *wet* weight. This is a very low figure, much lower than the figures for the Thames, the Horokiwi, and Ocanee rivers above, and for Walla Brook for which Horton's figures range from 17·8 to 41·6 g./sq. m./*dry* weight. The reason for this discrepancy is quite unknown, and it may be that the Welsh stream really is very unproductive.

It seems likely then that production in the rhithron can range from about 3 g./sq. m./yr. wet weight to tens of grams dry weight, which represents about one-quarter to one-third of wet weight (Allen, 1951). And if one accepts Allen's figure of about 100 times the weight of the standing biomass as the production, the Horokiwi yielded an average of over 400 g./sq. m. of invertebrates in 1940. We know, however, that this estimate is certainly too high, and clearly we need more work to establish if in fact such a wide range of rates of production occurs. At the present time it is easy to be critical of all the approaches that have been made to this problem, and so far the direct and indirect methods have not been checked against one another in the same stream.

Two studies have been made of individual species by the method of calculating the 'instantaneous growth rate', which was originally devised for fishes and which has been briefly described earlier (p. 326). This statistic (i) is such that

$$P = iB$$

where P is production rate in weight per unit area, B is the mean biomass per unit area, and i is the difference between the natural logarithms of the mean weights, or sizes, of the individuals in the population on two successive days. In practice i is determined over a longer period than one day, by sampling at intervals and dividing the difference ($\log_e$ weight on day $1 - \log_e$ weight on day n) by the number of days (n) elapsing between samples; and B is the mean of the standing biomasses found at the two sampling periods.

It will be clear that the above equation will provide, by calculation or graphically, an estimate of production, as long as there are no selective additions to or subtractions from the population of particular size-groups during the interval between samplings. It is obvious also that the more frequent the samplings during the growth period the more accurate the assessment. Unfortunately, as we have seen, many stream-dwelling insects hatch over a long period, so the mean sizes of the population tend to be continuously lowered. They also come to full size and emerge, thus disappearing from samples, more or less as a group. To use the method for insects therefore requires a good knowledge of the life history, and it is also necessary, at any rate at times, to calculate i from maximum rather than mean sizes, or from the mean size of the largest quartile or decile. Moreover, in multivoltine species, each generation has to be treated separately so that small specimens of generations $x + 1$ which appear while some individuals of generation x are still present do not interfere with the calculations.

These problems do not arise with fishes, where the generations can be distinguished by study of scales or other annular structures, and the same applies to large pelecypods in which age can also be calculated from annual rings on the shell. Negus (1966) studied the Unionidae in a reach of the River Thames and was able to establish the truly annual nature of the rings and the fact that growth occurs only during the period April to October and that it varies with temperature. By determining the density of the population and using the method of instantaneous growth rate she determined for the three common species the following figures in g./sq. m.

	Standing biomass	Annual production	Turnover ratio
Anodonta anatina	64·82	13·22	0·20
Unio pictorum	38·21	5·29	0·14
Unio tumidus	15·61	2·00	0·13

The turnover ratios have been calculated from her data and are much lower than have been found for the arthropods, or for faunas consisting mostly of arthropods which have already been discussed. It may also be noted here, as this point is referred to later in the chapter, that over 90 per cent of the benthic biomass was contributed by four species of Unionidae (120·7 g./sq. m.) and that this was nearly twice the biomass of the fishes (65·9 g./sq. m. Mann, 1965).

Waters (1966), similarly using the method of instantaneous growth rate, determined the production rate of *Baetis vagans* on a riffle in a stream in Minnesota which he sampled at monthly intervals for a year. He obtained the figures of 7·6 and 5·0 g./sq. m. for the summer and winter generations respectively. The mean standing biomass was 1·3 g./sq. m., so the turnover ratio was 9·7, which is once more in good agreement with values for this ratio discussed above. It was somewhat unclear during his study whether the summer population was single or multiple, and he suggests that the ratio may be about 3 per generation and cites some other work on arthropods in other environments where this ratio appears to be about three. This is, however, much less than the value inferred from studies based on back calculation and derived from populations composed mainly of univoltine invertebrates. It may therefore be that the value of about 10 is associated with the habitat rather than with the number of generations.

Waters had noted earlier (1961*a*) that far larger numbers of *B. vagans* drifted over an area than were present on it, a point also stressed by Horton (1961) about the invertebrates in Walla Brook, and he had suggested that the drift may represent excess production (p. 266). As there appeared to be few predators on the riffle which would cause mortality he argued that the following relationship should hold:

$$P = \Delta B + (Do - Di) + E$$

where P = production, ΔB = rate of change of biomass per unit area, Do and Di = rate of drift out of and into the area, and E = emergence of adults. By installing drift nets at the two ends of a riffle of which the area was known, he obtained values for Do and Di; and some idea of E was provided by floating adults. By sampling at intervals over periods of twenty-four hours, and calculating ΔB on a twenty-four-hour basis from his other data, he was able, by planimetry of curves of P plotted against time, to obtain values of P for the summer and winter generations of 6·7 and 2·4 g./sq. m. These are lower than the values obtained by the instantaneous growth method, but they take no account of losses occurring on the riffle, where there was a large population of *Gammarus* which would eat dead or moribund specimens. Also E was undoubtedly under estimated. Allowing for these points the agreement is remarkably good, and, while the method could not be applied to a more normal situation where predation is important, it does lend credibility to the suggestion that the amount of drift bears some relation to production.

On the other hand, figures for drift and estimated consumption of aquatic insects by fishes in the artificially controlled stream studied by Warren *et al.* (1964) do not agree at all closely. These figures are shown in Table XXII, 5, where it can be seen that drift was about four times the amount eaten in two untreated sections of the stream and rather less than the amounts eaten in two sections enriched with sucrose. The drift figures refer, however, only to riffle areas whereas the others include pool areas as well, but the differences between the two types of section illustrate the fact that the relationships between production, standing biomass and drift are by no means simple.

In concluding this section on production of the benthos it is perhaps worth stressing that this area of investigation is in its infancy, and that, despite its difficulty, it is of considerable importance for our understanding of the ecosystem. It would seem that the most fruitful approach is to work species by species, and that the best way to do this at present is by the instantaneous growth rate method coupled with a clear understanding of the life history. It would also seem important to determine the turnover ratios for as many species as possible under a wide variety of conditions, because if these prove to be reasonably constant for particular life forms, or groups of taxa, they would provide a very simple way of estimating production from relatively few samples. Great accuracy will probably not be obtainable because of the problems of sampling discussed in Chapter XII, but at the present time, as the above discussion shows, we do not know within two or three orders of magnitude what are likely to be the rates of production. Walla Brook and the Afon Hirnant, for instance, are both small soft-water trout streams in western Britain with very similar faunas, but they seem, as judged by our present methods, to differ greatly in benthic production.

PRODUCTION OF FISHES

Methods of studying the production of fishes are far in advance of those available for other parts of the biota, and a thorough discussion of them is beyond the scope of this book. For details the reader is referred to Allen (1951), Ricker (1958), Mann (1965), and Williams (1965), where further references are also available.

For our purposes only an outline is needed. Firstly it is necessary to determine the population present at various times of the year. To do this effectively one must measure the efficiency of fishing gear, a point we have discussed in Chapter XV, and, because every fishing method is selective for or against certain size-groups, different correction factors have to be applied to each group. Thus even the initial statistic is quite difficult to obtain.

The population is sampled repeatedly during at least a year, the ages of the fishes are determined as well as the mortality rates of each year class between samplings. Estimates are also made of the number of eggs laid, by direct count of ovaries often supplemented by counts of the number of nests, and of the percentage hatch of the eggs. Then, knowing the growth rate of each age-class between samplings, the instantaneous growth rate can be calculated, and this, combined with the death rate, can be used to calculate production, by mathematical or graphical methods. At this point various refinements can be introduced into the calculations, such as allowance for non-linear death-rates and estimated movements of fishes into and out of the area under study.

Using essentially this method on brown trout in the Horokiwi River Allen (1951) worked out that, in a normal year, the average standing biomass of trout was 26·8 g./sq. m. and that the average turnover ratio was about 1·7. Both biomass and production varied somewhat from point to point in the stream, but the ratio remained within the range 1·6 to 2·3, except in a year of severe flood damage and disruption where it became less than one at most points. Horton (1961) made a similar study of brown trout in an English stream, and a summary of her results is given in Table XXII,4. It is clear not only that her fish were much less abundant than Allen's, but that they were at least five times as productive in that the turnover ratio varied from 8 to 15. This would seem to indicate that this ratio can vary greatly from place to place even for one species under more or less natural conditions, and once again it is

Section		Average density of fish	Total production for year	Turnover ratio
I	1956–7	1·49	15·0	10·0
	1957–8	1·31	18·5	14·1
II	1956–7	1·14	11·8	10·3
	1957–8	0·82	12·3	15·0
III	1956–7	1·34	10·7	8·0
	1957–8	0·85	10·8	12·7

TABLE XXII,4. The average standing biomass of brown trout and the total production in grams per square metre at three stations and in two years in Walla Brook, Dartmoor, England, together with turnover ratios. Calculated from data given by Horton (1961).

desirable to know whether this really is so and what are the controlling factors. That it is not simply a matter of numbers controlling growth rate is shown by the lowered ratio found by Allen during the period following a severe flood which cut down the population, and also by the experiment of removing a large number of brook trout from a stream, which did not alter the rate of growth of those that remained (Cooper *et al.*, 1962). The whole environment would appear to be involved in differences of this kind, or, perhaps, genetic differences between populations of fishes are important.

Mann (1965) also measured the rate of production of a population of cyprinids and perch in a potamon reach of the River Thames, and he has produced figures showing a standing biomass of 65·9 g./sq. m. and production of 42·6 g. The turnover ratio is here only 0·65; for the species *Rutilus rutilus*, *Leuciscus leuciscus*, and *Perca fluviatilis* it was less than 0·5, but for *Alburnus alburnus* it was 0·83. During this study Mann also worked out the energy requirements of the population based upon measured respiratory rates adjusted to temperature. The details of this interesting study do not concern us here, but his results indicate that 16·5 times as much energy was used as was produced in the form of fish flesh. It is this large figure which contributes to the paradox mentioned on p. 419.

Lastly, two studies made under more or less artificial conditions, where fishes were confined and could be measured at intervals, have thrown some light on factors which may affect production.

Warren *et al.* (1964) confined cutthroat trout, *Salmo clarkii*, in reaches of a stream and examined them at intervals, when they were weighed and measured and food was removed from their stomachs. The data thus obtained, together with parallel studies on the food consumption and growth of captive fish, enabled the calculation of production rates. Data from their one complete year of observations, together with some ratios calculated from them, are given in Table XXII, 5.

The stream was controlled by a bypass so that the discharge could be kept fairly constant, and the four sections were isolated by screens. As shown in the table, two were partially cleared of trees and two were enriched by addition of sucrose to the water to the point where it caused growths of *Sphaerotilus natans*. This is the commonest 'sewage-fungus' bacterium, and it obviously increased the production in the sections in which it occurred. The trout were all yearlings, and it can be seen from the table that their production was low or even negative in the upper two sections, but it was intermediate between the findings of Allen and Horton on wild brown trout in the two lower ones. It will be recalled that 1 kcal. of fish is about equal to 1 g. wet weight, so the figures can be compared fairly directly.

Warren *et al.* conclude that sections I and II were probably overstocked in 1960–1, as they obtained higher productions—between half and 1 kcal./sq. m.—in 1962 and 1963 with lower stocking rates. The figures for all years cannot, however, be compared directly as the later experiments were conducted during only four months, which did not include the whole growing season nor the winter period when losses of weight had occurred in the first experiment. The later experiments nevertheless continued to show higher productions in the sucrose-enriched sections such that, although the estimated amount of food consumed there was only doubled, the production of fish was increased at least sevenfold. It seems undoubtedly true, therefore, that the *capacity to produce*, the productivity, bears no direct relationship to the *actual production*, which is a difference between the amount of food eaten and the amount used in metabolism. This difference is the amount of energy available for growth, and if the fish population is large it may even be a negative quantity. The ratio of fish production to fish biomass therefore

			Enriched with sucrose	
		Partially cleared of trees		
Section no. (in order downstream)	I	II	III	IV
Average standing biomass of trout	4·61	3·96	7·17	4·01
Production of trout during the year	0·79	−0·30	5·92	5·40
Ratio of production to biomass	0·17	—	0·83	1·34
Estimated total consumption of food	25·15	20·13	46·58	34·17
% of food in stomachs of aquatic origin	67·5	54·1	65·1	70·9
% of food which was aquatic insects	24·6	39·6	58·6	60·2
Estimated consumption of aquatic insects	6·19	7·95	27·3	20·6
Annual mean biomass of aquatic insects on riffles	2·19	5·03	20·40	12·20
Ratio of insects eaten to insect biomass	2·8	1·6	1·3	1·7
Annual drift of aquatic insects from riffles	24·11	30·04	17·00	14·03
Ratio of insect drift to insect biomass	11·0	5·98	0·83	1·16

TABLE XXII, 5. The biomass and production of *Salmo clarkii* in four sections of a controlled stream, Berry Creek, Oregon, during the year 1960–1; with estimates of food consumed and measurements of standing biomass and drift of aquatic insects and some calculated ratios. Figures, other than percentages and ratios, in kcal./sq. m. Data from Warren *et al.* (1964).

will vary, although one would expect that in a natural situation it would settle down at some equilibrium level. We have seen that in field studies it has been found to vary from 0·65 in the Thames to double figures in Walla Brook.

Similarly the paradox ratio, benthic insects estimated as eaten divided by standing biomass of insects, ranged from 1·3 to 2·8 in the one year for which complete data are available, and this is considerably lower than has been found under normal field conditions. This is perhaps surprising as one might have expected that with the apparent overstocking in sections I and II it would have been at an unusually high level. There are, however, so many uncertainties connected with the field data that it is difficult to be sure of the realities of differences, and it would seem that the experimental approach initiated by Dr. Warren and his colleagues promises to go a long way towards resolving these problems of production.

An important contribution that they have already made is their demonstration of the reality of the difference between productivity and production. This they have further elaborated by experiments in laboratory streams in which the sculpin *Cottus perplexus* was kept at different population levels (Davis and Warren, 1965). The streams resembled one another in having a simple algal and chironomid community, and the *Cottus* fed on the Chironomidae. When there were few fishes they grew well but they did not fully exploit the food available, and when there were many fishes they over-exploited the food and grew little or not at all. Maximum *production* therefore was obtained at intermediate populations of fishes even though the productivity presumably remained constant. These experiments also showed that the carnivorous stonefly *Acroneuria pacifica* was more efficient than *Cottus* in producing biomass under the conditions of the experiment, and that it competed with the fishes for the food and reduced their rate of production when both species were present together.

Such results are, of course, what one would expect, but they illustrate very well the fallacy of treating the benthos as a whole as if it were a unit quite separate from the fishes. This point becomes even more apparent when it is remembered that many fish species feed only partly, or not at all, on benthic invertebrates. Which brings us back to the point made earlier that complete understanding of production can only be obtained by study at the specific level. The results also add to that point the necessity of being aware of reactions between taxonomically remote species.

HETEROTROPHY AND MICRO-ORGANISMS

We have already noted in Chapters X and XVIII that many invertebrates and some species of fish feed on dead organic matter which is mainly of terrestrial origin. Chapman (1966; Chapman and Demory, 1963) has calculated that, because of this, well over 50 per cent of the energy obtained by young coho salmon in wooded coastal streams in Oregon is derived ultimately from primary production which did not occur in the water. Quite a proportion of this energy comes, however, directly from the terrestrial insects which these fish eat very readily. Other fish feed primarily on the benthos, and Chapman's study showed that 51 per cent of the food in the guts of the benthos was allochthonous material. In other types of stream the dependence upon detritus is also great. Nelson and Scott (1962) estimated that 66 per cent of primary consumer energy on a rocky outcrop in the middle Ocanee River, was derived from allochthonous material, and Cummins *et al.* (1966) record about 30 per cent of the food of invertebrates in Linesville Creek, Pennsylvania, as being detritus. Also in many situations a large proportion of the fishes feed directly on detritus, and dense populations of stoneflies or crustaceans occur in small woodland streams and play an important role in the degradation of fallen leaves (Bick, 1959). So dense are the populations of insects in such places that even their parasites may become spectacularly abundant. Efford (1965) records densities of the water mite *Feltria*, a parasite of *Tanytarsus*, which are so great that they resemble those of soil mites, a most unusual situation in fresh water. That the quantity of leaves really is great is shown by the finding of Szczepański (1965) that about 500 g. of dry leaves per year fall directly on to the surface from each metre of wooded shore of Lake Mikołajki in Poland. This figure is certainly minimal as it takes no account of leaves which fall on to the ground and are subsequently washed or blown into the water. A stream in a wooded valley has, of course, woodland on both banks, and so probably receives at least a kilogram per metre of length per year.

It seems likely that allochthonous organic matter is also an important source of energy to the plankton. It has fairly recently been shown that the production of zooplankton in lakes is too great to be fully accounted for by local primary production, and the conclusion is that bacteria and detritus must be involved and are possibly the major source of energy (Nauwerck, 1963). River water contains more detritus than lake water, so it seems probable that it may be even more important to plankton in running water (Bursche, 1959). Maciolek (1966) found an average of 0·67 mg./l. of fine suspended organic matter in a clear mountain stream in California, equivalent to 3·24 calories, and of this between 40 and 80 per cent was detrital material derived from outside the stream. Presumably in the more murky waters of lowland streams the quantities are much greater, and, as we shall see, the amounts are increased by riparian vegetation.

In temperate latitudes most of the larger allochthonous material finds its way into the rhithron during the autumn and winter, and species which feed directly on it are most abundant during the winter and early spring (Jones, 1950; Hynes, 1961). Hynes (1963) has pointed out that the dependence of the benthos in the rhithron on allochthonous matter results in the highest biomass at that season, and several workers have indeed found denser populations in winter than in summer (Hynes, 1961, 1963; Komatsu, 1964*b*; Tanaka, 1966). This is not, however, always true as has been shown by studies on streams in Scotland, many of which were found to have higher populations in summer than in spring (Maitland, 1964; Morgan and Egglishaw, 1965). Nevertheless, the fact that there *are* high wintertime populations of benthos is almost certainly attributable to allochthonous materials.

Autumn-shed leaves placed in water lose weight slowly, but for a long period. Figures of up to 30 per cent in six months are given by Levanidov (1949) for deciduous species, and 75 per cent for maple leaves and 50 per cent for oak and pine have been recorded after one year in water (Chase, 1957); during all this time the leaves exert an oxygen demand which must be attributed to the activity of heterotrophic micro-organisms. Mosevich (1947) conducted a series of experiments which showed that fallen deciduous and coniferous leaves fairly rapidly reduced the dissolved oxygen in closed containers, and that the deciduous leaves were the more effective. He also showed that water extracts of the leaves exerted a great oxygen demand, and that here the difference between the deciduous and coniferous leaves was even more marked. Nykvist (1963) similarly experimented with the extraction of soluble materials from fallen leaves, and he gives the following figures for the percentages of dry weight extracted from the litter of various European forest trees in one day at 25 °C.

Fraxinus excelsior (ash)	16·5
Alnus glutinosa (alder)	12·0
Betula verrucosa (birch)	10·7
Quercus robur (oak)	7·1
Fagus silvatica (beech)	3·8
Picea abies (spruce)	1·1
Pinus sylvestris (pine)	0·9

From these it can be seen that coniferous needles, and to a lesser extent beech leaves, produce dissolved material much less readily than do most of the broad-leaved species. They continue, however, to release a small amount for much longer, and in all the species the rate of release of the soluble material is increased if the leaves are broken up.

We shall consider the microbial breakdown of the dissolved material shortly, but we should note here that the organic fraction of river sediments is largely composed of broken-up plant material and of animal faeces derived from it. Such sediments consume much oxygen, especially when they are inhabited by benthic animals which carry currents downwards into their burrows or turn the material over by their activities (Edwards, 1958, 1962; Owens and Edwards, 1963; Schumacher, 1963).

Here also there is evidence of microbial activity, but we know very little about the micro-organisms themselves.

In recent years a whole complex of hyphomycete fungi has been described from dead leaves in streams in many parts of the world (Ingold, 1959, 1960, 1961; Petersen, 1962, 1963*a*, *b*; Nilsson, 1964). These have peculiar tetraradiate spores which appear to have been convergently evolved in many genera and are almost certainly an adaptation to the running-water environment (Ingold, 1966). The spores are so abundant in streams that the fungi undoubtedly play an important role in the ecosystem. Dead vegetation also provides a medium for a great variety of other fungi, most of which are normal inhabitants of forest litter. Cooke (1961) stresses that these have been much neglected, and it has been confirmed by Mr. N. Kaushik in my laboratory that many species grow densely on and in leaves kept in well-oxygenated stream water. Undoubtedly therefore, fungi are important in the breakdown of allochthonous organic matter and in providing food for the animals which feed on it, but we lack any really good quantitative information. Teal (1957) calculated that micro-organisms in a small leaf-filled spring in Massachusetts respired 350 kcal./sq. m./yr. of which only 55 kcal. could be accounted for by algae. He could not distinguish the contributions by bacteria, fungi, and small animals, but it seems probable that a large proportion of this metabolic activity was contributed by fungi.

It should also not be overlooked that fallen leaves are quite a good foodstuff in their own right. Levanodov's (1949) analyses show that dead leaves are not inferior to hay, and Hynes (1963) reports that elm leaves contain 7·5 per cent of their air-dried weight in the form of protein. Other species of tree leaf have since been found to contain similar amounts. Percentages of insoluble protein in the autumn-shed leaves of elm, alder, oak, maple, and beech have been found to range from 3·5 to 10·0 per cent of the air-dried weight (Kaushik, unpublished), and Melin (1930) reports that the crude water-insoluble protein in the autumn-shed leaves of eleven species ranged from 1·99 to 11·00 per cent. Similarly, the studies of woodland ecologists reveal that such leaves normally contain at least 2 per cent of total nitrogen. Much of this is probably present in the form of protein, so we can conclude that, for animals which can digest them, dead leaves are well worth eating even without the assistance of fungi.

Little mention has so far been made of bacteria which might be supposed to be even more important than fungi in the breakdown of the solid organic matter on the stream bed. Our, as yet unpublished, experiments indicate, however, that whatever their importance elsewhere, and it may be great in the once or twice eaten or otherwise fragmented detritus, it is not great on leaves while they remain whole.

Another pointer to the probable importance of heterotrophic organisms to animals which feed on allochthonous matter was provided by Nelson and Scott (1962). They found, as is usual, that the ratio of weight of carnivores to weight of primary consumers was, in all their samples, higher than the ratio of weight of herbivores to weight of plant material—in this instance *Podostemon*. This is because, as is well known, the production efficiency of secondary consumers is higher than that of herbivores. The ratio of detritus feeders to detritus was, however, intermediate, and if the animals had been feeding directly on the detritus there would seem no reason to expect it to differ from the herbivore to plant ratio. They therefore suggest that this indicates that part of the food is being derived from micro-organisms and that detritus feeders are, in effect, partially secondary consumers. This approach to the problem of the role of micro-organisms in streams might lead to some means of quantitative assessment of their importance.

It may be noted, in passing, that it is well established that many cave-dwelling crustaceans feed on the bacteria which occur in clay. These are mostly chemotrophic species of *Leptothrix* and *Perabacterium* which oxidize ferrous iron (Husson, 1962). In Lake Michigan the distribution of *Pontoporeia affinis* is related to the abundance of bacteria in the sediments and not to particle size or organic matter (Marzolf, 1965, 1966) and, as we have seen (p. 194), it is possible to rear Chironomidae on a diet of bacteria grown upon filter paper. It seems, therefore, that simple bacteria/animal situations can occur in aquatic habitats, and that these would provide a starting point for studies on production ratios in which detritus is also involved.

The water itself is full of dissolved or colloidal organic matter, often greatly in excess of the amount in suspension. This is the material which, through micro-organisms, exerts the 'biochemical oxygen demand' which is so frequently used as a measure of organic pollution. In unpolluted streams much of it is presumably derived from dead leaves and other materials, and it supports great numbers of bacteria. Nykvist (1963) found that, under aerobic conditions, large populations developed within a day in his leaf extracts, and within four days they had eliminated all the amino-acids which were originally present. They also, together with some fungi, formed small clots which precipitated out of the water. This, in the stream environment, would, of course, carry high-grade food material to the benthos.

Many so-called 'plate-counts' have been made on stream waters, mostly in connection with pollution. In these the water is placed on nutrient agar and the number of colonies of bacteria which develop is counted. It seems though that this technique greatly underestimates the numbers present in the water as many species do not grow under these conditions. Recently more accurate methods of counting the cells directly on membrane filters or after centrifugation have been employed (e.g. Beling, 1953; Rodina, 1964). Details of such bacteriological methods do not concern us here, but they give much higher figures. Beling records the following counts from the Fulda River, Germany, in numbers of bacteria per ml. of water.

	Plate method	Membrane filter method
Upper reach	535	30,124
Upper middle reach	8,533	227,509
Lower middle reach	3,128	173,840
Lower reach	325	18,450

Even higher numbers, ranging from 0·1 to 25×10^6 cells/ml., have been recorded from streams in Saskatchewan, and these enormous numbers provide food directly for such animals as *Simulium*. They also undoubtedly contribute larger particles of food to other benthic filter feeders such as the great populations of mussels, snails, and insects in the potamon (Richardson, 1921). We have seen that they spontaneously aggregate into clots in the laboratory and sink, and, as it is known that, in the less concentrated situation in the sea, aeration causes the aggregation of bacteria and organic matter (p. 50), it seems highly probable that the turbulent flow of running water enhances this tendency to precipitation. We can take it then that bacteria continually return material to the stream bed, which was originally derived by solution from solid organic matter there and on the land. Moreover, this microbial material is particulate, and thus edible, and it is undoubtedly high-grade food.

Unfortunately we know very little about the bacteria which are actually involved in these processes. For a brief review of the literature and some preliminary information the reader is referred to Botan *et al.* (1960), who suggest that, as with the fungi, many soil-dwelling species may be important. We also do not know if different situations encourage different micro-organisms; but it would seem, on the basis of oxygen demands and the amounts of weight lost when they are soaked in water, that different types of forest litter are likely to provide very different quantities of those organisms to the benthos. Variations of this kind could well be responsible for the observed differences in rates of production between apparently similar streams (p. 428) and between acid and alkaline streams. In general acid landscapes tend to favour coniferous trees and alkaline ones deciduous species. They may also account for the apparently deleterious effects on streams of afforestation of their valleys with conifers (p. 231).

THE ENERGY RELATIONS OF THE ECOSYSTEM

With this information on the sources of energy to the ecosystem and our previous considerations of the biomasses of various groups of organisms and their turnover rates, we can, inadequate though it be, construct a crude model of the ecosystem as it exists in streams and rivers. This has been done in Figs. XXII, 1 and XXII, 2, in which the size of the boxes bears some relation to the biomass of the organisms shown in them and the thickness of the arrows the supposed importance of the energy pathways. It should not, however, be forgotten that the amount of the biomass is not related to the rate of energy turnover. Some organisms, such as Unionidae, have low turnover rates (p. 426), and even small plants, such as Cyanophyta, may lock up energy for long periods as they are unpalatable (Margalef, 1958; Gajevskaja, 1958). In lakes it seems probable that blue-green algae become available as food for animals primarily after their death (Darnell, 1961), and the same seems to apply to many angiosperms in running water, in contrast to the situation in still water (Frohne, 1956). These points cannot be adequately shown on a diagram, and, similarly, the diagrams cannot give any indication of the respiratory loss of energy without becoming extremely complex.

In the *rhithron* (Fig. XXII, 1) the main source of energy is usually allochthonous particulate matter which is repeatedly passed through the guts of detritus feeders. This point is emphasized by the repetition of this cycle in the diagram. It is known, for instance that *Asellus* defecates 30 per cent of the calorific value of leaf tissue which it eats (Levanidov, 1949), and this probably applies to other detritus feeders also. The faeces, the broken-up plant tissue and most of the material produced by any rooted plants present, go to form the detritus on which the same and other species feed, presumably aided by micro-organisms of which we know very little. The allochthonous matter, the detritus, and the land surface provide dissolved organic matter for bacteria, which add material to the detritus but also support Simuliidae which seem to be the only important members of the fauna which can here make direct use of suspended bacteria. This may account for their almost universal occurrence in rhithron situations. Attached algae and mosses also contribute to the detritus, but they also support some insects and snails directly, and the algae are most important in this respect during the spring and autumn.

Carnivorous invertebrates and fishes are the secondary consumers, although many of both groups are in part primary consumers also. Metabolism releases plant nutrients and dissolved organic matter and more allochthonous matter is added from the land, so the cycle is continuously repeated downstream, but with changes of species as the

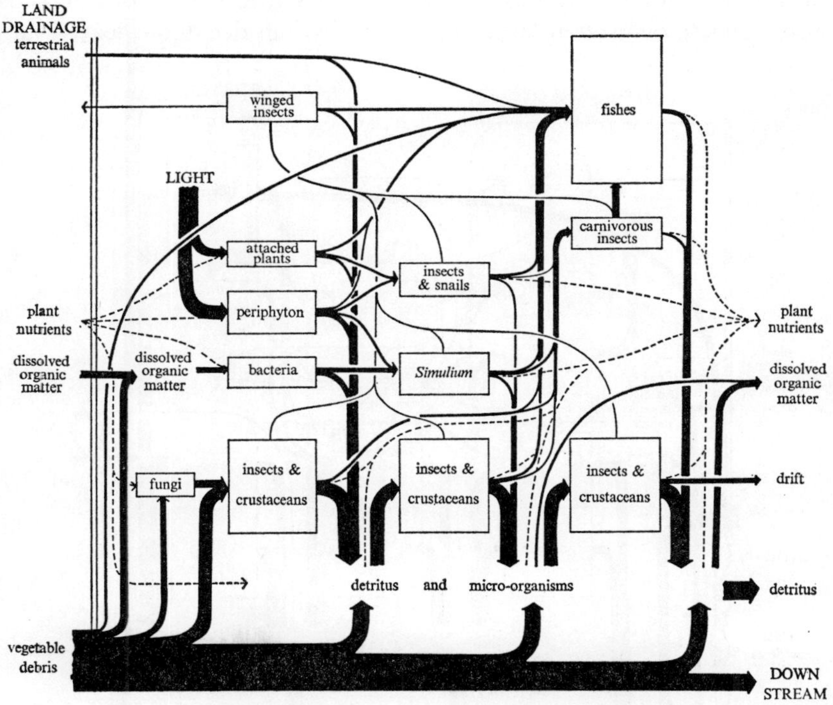

FIG. XXII, 1. A diagrammatic representation of the trophic relationships of the ecosystem in the rhithron. The relative sizes of the boxes are an indication of the biomass and the width of the arrows is proportional to the supposed relative importance of the energy pathways. Dotted lines represent salts in solution.

conditions change. A pool and riffle channel may impose some alternation on the system, since riffles support more algae and pools collect and oxidize more detritus (Neel, 1951), and the general level of activity depends upon the fertility of the valley in respect of its ability to produce both nutrient salts and allochthonous vegetable matter. Some studies have indicated that headwaters are particularly rich (e.g. Müller, 1955*a*), and this may be because they are narrow and so receive more organic matter per unit area of stream bed. There are also suggestions that the rate of consumption of organic matter is higher in hard water. Egglishaw

(1965), for instance, found that the ratio of the weight of bottom fauna to the weight of detritus increased linearly with the calcium content of the water in Scottish streams. This is similar to the situation in soils, where organic matter tends to accumulate in the absence of calcium, and it is probably related to micro-organismal activity. Soft water therefore probably reduces the rate of turnover of the detritus cycle, and this may be why soft waters produce fewer or smaller fish. Coniferous detritus also seems to be used up much more slowly than deciduous leaves, and

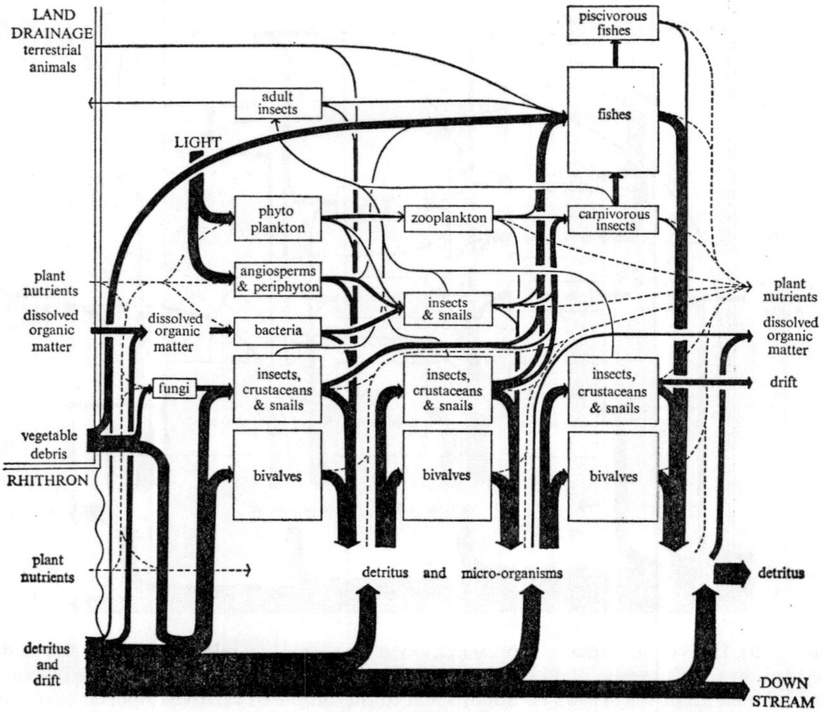

FIG. XXII, 2. A diagrammatic representation of the trophic relationships of the eco-system in the potamon. Details as in Fig. XXII, 1.

this is probably at least in part because of micro-organisms. We are clearly in great need of a much better understanding of the role of micro-organisms in the detrital cycle.

In the *potamon* (Fig. XXII, 2) the main source of energy is fine detritus from the rhithron, but dissolved organic matter supports suspended bacteria and, at any rate in summertime, nutrient salts from the rhithron and the land allow the development of plankton. All the finely divided material provides food for bivalves and, when it settles, for annelid worms, prosobranch snails, and insects, chiefly Chironomidae and burrowing mayflies in many situations. We can assume that here also the detritus cycle is repeated and that the material is consumed many times. We have no direct observations on species from the potamon, but Newell (1965) has shown that the estuarine prosobranch snail, *Hydrobia ulvae*, which feeds on mud like most of its relatives in rivers, digests nitrogenous matter from the mud, but not much carbon, and that the nitrogenous matter builds up again on the faeces, which can then again serve as food. Presumably the nitrogenous material is bacterial protein, to which the energy of the mud and nitrate from the water is converted and is thus made available to the snail. It is interesting though to note that in this system, which may be similar to that occurring in rivers, the nitrogen is apparently derived from the water and not from the mud. Adequate supplies of dissolved nitrate may therefore be important in the rate of turnover, and possibly the fact that a common riparian tree, the alder *Alnus*, is a nitrogen-fixer, and sheds leaves which contain four times as much nitrogen as other species, is important. Alder trees in the drainage basin have been shown to increase the productivity of a lake (Goldman, 1961), and they may be even more important to rivers.

In some potamon reaches the attached and rooted plants may play roles similar to those that they have in the rhithron, but they are probably less important because of the relatively small area of shallow water. Their influence is also less obvious because the supply of detritus, although locally augmented in the autumn, is more evenly spread throughout the year. There is thus much less obvious seasonal change in the potamon than in the rhithron. Only temperature exerts its seasonal effect so that plankton develops and insects emerge primarily in summer, but the main bulk of the biomass is composed of perennial (non-seasonal, p. 295) species of worms, snails, bivalves, and fishes. Also, possibly simply because of its more completely year-round availability, a larger proportion of the fishes in many parts of the world tend to feed directly on allochthonous organic matter. The whole system, although similar in principle to that of the rhithron, is thus probably more stable and less subject to seasonal change of energy sources, but like the rhithron it also rolls on repetitively downstream.

Lastly we may note, as we have discussed in Chapter XX, that although the type of cycle changes little in the downstream direction the species do change. They also differ from one region to another and from one type of river to another. It seems, however, that the *total number* of species remains about the same. Patrick (1961, 1964) carried out very thorough surveys of a number of watercourses in the south-eastern United States and in the headwater regions of the Amazon, and she identified as far as possible all the species collected. She found that the total *numbers of species* of algae, protozoans, insects, and fishes varied within only about ±33 per cent, but the *lists of species* differed widely. In the south-eastern

	Total no. of taxa	Percentages of taxa which occur in different numbers of rivers								
		1	2	3	4	5	6	7	8	9
Algae	354	55·7	17·2	10·7	7·1	4·5	2·3	1·7	0	0·9
Protozoa[1]	299	62·9	13·4	10·7	7·7	2·7	1·7	0·1	0·7	
Insects[1]	283	73·9	10·9	8·5	3·2	2·1	1·1	0·1	0	
Fish	132	56·8	16·7	16·7	6·8	0	2·3	0	0	0

1. Studied in eight rivers only.

TABLE XXII, 6. The percentages of taxa which occur in different numbers of nine rivers in the south-eastern United States (Potomac, Ottawa, Rock Creek, North Fork, Holston, North Anna, White Clay, Savannah, and Escambia). Data from Patrick (1961).

United States about 73 per cent of the taxa were found in only one or two rivers and very few were found in all; a summary of her data is given in Table XXII, 6. From this we can conclude, with Dr. Patrick, that there are many more species available than there are niches for them in any one river. In other words, the competition for places in the stream or river ecosystem must be very severe.

CHAPTER XXIII

Effects of man on watercourses

Human activity has profoundly affected rivers and streams in all parts of the world, to such an extent that it is now extremely difficult to find any stream which has not been in some way altered, and probably quite impossible to find any such river. The effects range from pollution to changes in the pattern of flow, and they have become increasingly marked during the past two or three centuries.

We shall not here consider pollution nor the direct or indirect addition of biocides, for a general discussion of which topics the reader is referred to Hynes (1960), but rather we shall deal with a range of practices which produce less immediately obvious, but none the less profound, effects on the biota of running water. These are usually cumulative and many are so insidious that they occur without being noticed. It is probably true to say that even stream biologists have not been fully aware of their magnitude until quite recent times.

At several points earlier in this book we have noted that man has produced alterations by clearing stream banks, substituting agricultural crops for native vegetation, and impounding streams and rivers, and it is the effects of these activities, together with others, such as canalization for navigation and drainage, abstraction of water, and the introduction of species from other parts of the world, which will be considered here.

Removal of vegetation from the banks has immediate and presumably direct effects on the fish fauna. For instance, trout require some cover, but they tend to be scarce in densely shaded water (Hobbs, 1948; Boussu, 1954). Similarly, removal of forest from a watershed may have direct and rapid effects, especially where improper practices lead to drastic soil erosion from logging roads. Serious silting up of streams and reductions of fish and invertebrate populations have resulted from such activities in both eastern and western North America (Tebo, 1955; Chapman, 1962) and the same has doubtless also occurred elsewhere.

In the same way any engineering work carried out near a stream, without adequate provision for the prevention of erosion, can rapidly cause enormous changes through silting up and scour by mineral particles. King and Ball (1964) recorded great reductions in the invertebrate populations, periphyton, and general biological activity in the Red Cedar River caused by the building of an interstate highway in Michigan. Construction of any kind almost inevitably results in some erosion, and even small amounts of mineral matter cause changes in the fauna. Hamilton (1961), for example, observed in a Scottish stream that quite small amounts of fine washings from quarried sand eliminated some mayflies and stoneflies and encouraged oligochaetes.

More subtle effects result from the less spectacular sheet-erosion from cleared land, the changes in the patterns of run-off caused by agriculture, and the elimination, for land use, of marshes, swamps, and flood-plain lakes. All these activities tend to make the water more silty, the minimal flows lower and the maximal temperatures higher. The results appear to have passed unnoticed in Europe, where presumably they occurred during the Middle Ages, but in North America they have attracted considerable notice because they have occurred, and are occurring increasingly, in modern times. Cordone and Kelley (1961) have reviewed the effects of inorganic silt on streams, and they conclude that almost all the trout streams in North America will be seriously affected unless steps are taken to reduce erosion. This is because the eggs and fry of Salmonidae are eliminated by suffocation when interstices in the gravel are clogged, and so are many of the invertebrates on which the fish feed.

Adult fishes are not apparently directly affected by reasonable amounts of silt (Wallen, 1951), so there are no spectacular mortalities such as occasionally occur in polluted waters, but the ultimate results are similar. The same applies to mussels, of which the adults survive silting but the young do not (Ellis, 1936), and this also is causing concern because of its economic implications. At one time North American freshwater clams supported a flourishing button industry (Carlander, 1954), and when that declined, and buttons came to be made of plastic, a new use for the shells was found as seed material for the Japanese artificial-pearl industry. Great quantities of shells are now shipped to Japan, and, as with fishing, money is involved, so people are conscious of the decline in a source of income.

They are less aware that many other changes have occurred—that many clear-water species of fish have been reduced in numbers or eliminated in Indiana, Wisconsin, Ohio, and Illinois and have been replaced during the past half century by warm-water species (Gerking, 1945; Black, 1949; Lachner, 1956; Larimore and Smith, 1963; Mills *et al.*,

1966); the topeka shiner, *Notropis topeka*, and the crayfish, *Orconectes neglectus*, have much more restricted distributions in Kansas and neighbouring states than in former times (Minckley and Cross, 1959; Williams, 1954). These changes are all probably directly attributable to the effects of agricultural development, and they are certainly only a very small proportion of those that have occurred.

In other parts of the world some species have been recorded as having been reduced or eliminated in recent times in the same manner. The large carnivorous mayfly *Centroptiloides* is now absent from some places in South Africa, presumably because of erosion silting (Agnew, 1962), and silting has changed the fauna of streams in New Zealand (Allen, 1960*b*). It may indeed have caused the extinction of the New Zealand grayling *Prototroctes oxyrhynchus* which vanished between 1870 and 1925 (Allen, 1961). These, however, are only the changes that have been recorded. It is quite certain that many more have occurred among less spectacular organisms, and that the distribution of many, possibly most, running-water species has been altered in some way. Probably plants and animals which are now characteristic of cool shaded headwaters once had enormously greater distributions.

Another effect of land use is the enrichment of the water with plant nutrients. Quite apart from the enormous amounts of nitrate and phosphate which are added in effluents, especially sewage, the mere fact that the run-off pattern is altered changes the amounts of nutrient entering streams, and we are continually increasing this by the use of fertilizers. In 1850 the United States used about 3,000 tons of nitrogenous fertilizer; by 1950 this had risen to about 1 million and by 1960 to about 3 million tons. Similarly, in 1940 only about 390,000 tons of phosphate fertilizer were applied to agricultural land in that country, but by 1960 the tonnage was about 1·3 million (McVicker *et al.*, 1963). The rate of increase is therefore enormous, and it is increasing year by year. Presumably not all of this fertilizer reaches the watercourses, because phosphate is firmly held by soil, but nitrate leaches out fairly readily and phosphate is carried on eroded particulate matter. Recent studies in Ontario have shown that large amounts, ranging from 290 to 600 lb./sq. mile/yr., of phosphate are carried by streams draining agricultural areas, and that they are increased manyfold when urban areas occur in the watershed (Owen and Johnson, 1966). Some of this urban influence is undoubtedly caused by sewage, but it has been shown, during study of run-off water from an urban area in Cincinnati, Ohio, that 9 per cent of the phosphate and 11 per cent of the nitrate that would be expected in the sewage of the population inhabiting the area is actually present in the surface drainage (Weibel *et al.*, 1964). As our urban areas sprawl at an ever-increasing rate it would seem that, whatever we do with our sewage and industrial wastes, we shall continuously increase the fertility of stream and river water.

The effects of enrichment on lakes are fairly well documented, and, as almost all of them are bad from a human point of view, they are attracting a great deal of attention. We know far less about the effects on running water. It is certain that enrichment causes changes in the flora and encourages Chlorophyceae and rooted plants (Chapters IV and V), but the extent to which this occurs depends upon local circumstances. If, however, enrichment is accompanied by reductions in flow during warm weather, it must, especially if riparian trees have been cleared, lead to increased primary production. There are few 'before and after' studies which are free of the complication of organic pollution, but changes in the attached algae have been noted during twenty years in the Horokiwi River, New Zealand (Allen, 1960*a*), and they have probably occurred almost everywhere.

Presumably enrichment also encourages planktonic algae in large rivers, but there appears to be little information on this point. Perhaps, however, this is not surprising because many factors limit the development of plankton in rivers and one of the most important is turbidity (Chapter VI). Enrichment is usually accompanied by soil erosion and increased turbidity, especially where there is shipping (Mills *et al.*, 1966), so presumably the two effects tend to cancel one another out. It is therefore usually only where the rate of flow of a river has been reduced by barrages that dense blooms of planktonic algae develop. Without human intervention in the flow of the water it seems that these occur only in the lacustrine conditions of side arms and flood-plain lakes.

Direct alteration of channels for drainage, water extraction, operation of mills, and navigation also produce profound effects. We have seen that barrages encourage phytoplankton; they also, of course, interfere with the migration of fishes, and in many parts of the world alterations in the fish fauna of large rivers have been attributed to this cause. It is possible to arrange passes which allow the upstream movement of active fishes, such as Salmonidae which can leap, but they do not solve the problems which dams present to descending young fish (p. 354), and it seems that it is difficult or impossible to cater for the more sluggish species typical of larger rivers, many of which have become very scarce, or even extinct, within quite recent times.

Dredging and straightening of channels have the two-fold effect of eliminating littoral areas and rooted plants, and of producing great uniformity along the watercourse. Several studies have shown that dredging greatly reduces the abundance and variety of the flora and invertebrate fauna, and that newly cut channels develop very little of their own if they are swift and canal-like (Engelhardt, 1951; Gaufin, 1959; Hynes, 1960). Mussels are often eliminated by direct removal (Clark, 1944), and the larger species of fishes move out because of the absence of pools, shelter, and suitable breeding sites (Trautman, 1939; Einsele, 1957; Stuart, 1959; Whitney and Bailey, 1959). Regulation of the water of the upper Rhine has eliminated the turtle *Emys orbicularis* (Thienemann, 1950), which is a sad loss from the already rather limited European fauna, and could undoubtedly have been prevented by more understanding engineering.

Canalization of small streams also seems to encourage *Simulium* because of its need for swift fairly laminar flow over a stable substratum (Müller, 1953). Canalization, therefore, particularly if it is accompanied by increases in suspended bacteria caused by pollution, could lead to outbreaks of undesirable blackflies (Freeden, 1960). And while we are on the subject of nuisance flies it should be noted that all solid stable structures erected in rivers form excellent substrata for net-spinning Trichoptera, several species of which can, as a result, occur in troublesome swarms (Fremling, 1960*b*). Man therefore creates pests for himself and his livestock by altering watercourses in ignorance of the consequences.

In recent times great numbers of streams and rivers have been dammed for various purposes all over the world, and this has many effects apart from the obvious one of changing a watercourse into a lake. For general discussions of the ecology of dam-lakes the reader is referred to Neel (1963), and, especially for tropical conditions, to the symposium edited by Lowe-McConnell (1966). A dam-lake differs from most natural ones in a number of ways. Perhaps the most important of these are that the retention-time of the water is usually relatively short, the outflow may be from deep water, and hence hypolimnial, and in many the water level is allowed to fluctuate much more widely than it does in a natural lake.

We are not here primarily concerned with conditions within artificial lakes, but it may be noted that a number of studies has shown that after the dam is closed the typical running-water benthic fauna is eliminated and is replaced by a lacustrine one. At first the new fauna is unstable and some animals, especially *Chironomus*, occur in great abundance, particularly on the newly flooded areas, and several years are needed before some stability is attained (Beling, 1949; O'Connell and Campbell, 1953; Morduchai-Boltovskoi, 1961*a*, *c*; Sokolowa, 1961; Miron, 1962). As we have seen in Chapter VI, plankton of a lacustrine type develops, and in large rivers it becomes increasingly lacustrine as the water approaches the dam. This indicates that current and its associated phenomenon of turbidity is fairly certainly the main factor militating against the development of lake-plankters, particularly crustaceans, in rivers (Vyushkova, 1962; Uspenskii, 1963). Similarly, the fishes change as slow-water species largely or entirely replace the running-water ones (Martin and Campbell, 1953; Swingle, 1954; Hall, 1955). In the tropics it has often happened that floating plants, e.g. *Eichhornia*, *Salvinia*, or *Pistia*, or in shallow water mats of littoral vegetation, develop on impoundments and lead to de-oxygenation of the water (Damas, 1961; Lowe-McConnell, 1966), and this, of course, has far-reaching ecological effects.

These are, however, all effects on the lakes themselves and so do not greatly concern us, except in so far as they affect the river. Similarly the fact that dam lakes, because they become stratified, offer cold-water hypolimnial refuges is of little importance in river biology. It is, however, interesting to note that in Russia, where several large rivers which flow southward have been dammed, many species of fishes and crustaceans now occur in reservoirs, and at times in the rivers below them, far south of their original range (Romadina, 1959; Dzyuban, 1962). This indicates that drift must previously have carried these species out of areas where they could survive permanently, and it emphasizes the role of rivers in the distribution of aquatic organisms.

Although the physical influence of an impoundment does not stretch far above the altitude of the crest of the dam, the biological effects may be felt for a long way upstream. Migratory fish may be eliminated from the upper reaches, and this causes changes in the ecosystem. For example, dams in New Zealand have eliminated predatory eels from some watersheds with corresponding benefits to the trout (Hobbs, 1948). Or the presence of a lake may result in the migration into tributary streams of fish species which did not previously occur there, but which develop large populations in the lake from which they wander (Ruhr, 1957).

The effects on downstream reaches are more marked, and they vary greatly according to the way in which the water in the impoundment is used (Neel, 1963). Usually the presence of a dam increases the plankton content of the water, with the effects on the benthic fauna which we have discussed in Chapter XIII. As we have seen in Chapter VI lake plankton is fairly rapidly eliminated from a river, but sometimes, as in the Missouri, many of the characteristics of the plankton originating in reservoirs may persist for long distances (Neel *et al.*, 1963). Usually, also, reservoirs act as settling basins and reduce the turbidity of the water; but some management regimes may increase it by the discharge of turbid layers from the lake. Indeed, it is occasionally the practice to discharge accumulated sediments from behind dams with disastrous effects on the river. Nisbet (1961) describes how periodic discharges of tons of sediment from the dam at Verbois turn the upper Rhone into a stream of mud. Clearly this is an activity which should be discouraged if we place any value on the biota of the river.

When the discharge is from the upper layers of the reservoir the effects downstream are similar to those of a natural lake, with slower warming in the spring and later cooling in the autumn, and also less fluctuation of discharge than in the undammed river. The temperature regime alters the fauna, and the relatively stable discharge encourages plants. It has often been observed that dense growths of filamentous algae occur below dams, probably because they are less subject to wash-out and attrition than in an unregulated river (e.g. Neel, 1963; Stober, 1964). This, of course, also produces changes in the fauna.

When the discharge is from below the thermocline of a stratified reservoir the summer temperature of the river is kept very low, and a cold-water area may be produced far away from any which occur naturally (Tarzwell, 1939). The water may also be chemically unusual and contain ferrous, manganous, and sulphide ions and, at first, little oxygen (Symons *et al.*, 1964). It may also contain great quantities of bacteria as well as moribund algae from the hypolimnion (Coutant, 1963). Thus, at any rate for some distance below the dam, conditions in the river are most unusual, and they are reflected in the biota.

Occasionally, and especially where dams are used for irrigation, the dry-weather discharge is greatly reduced and the level of the river fluctuates much more than it did before. Neel (1953) describes the effects of this in the North Platte River, Wyoming, and points out that conditions there now vary from those of a large river to those of a small head-water, with the result that organisms characteristic of neither can persist.

The effects of reservoirs therefore vary very much with management, and dams could quite often be operated to man's advantage when their biological effects are clearly understood. For example, migratory fishes are often encouraged to move upstream by rising discharge and they do not move much when it is uniformly low. It is possible therefore to manipulate discharges so as to maintain populations of valuable salmonids in rivers from which differently managed impoundments would eliminate them (Huntsman, 1945; Baxter, 1961). It should also be possible, by suitable management, to prevent the losses of fishes caused by their stranding in irrigation canals (Clothier, 1953) and so to have both crops and fish. Many of the riverine problems which stem from the impoundment of water could be solved by symbiosis between biologists and engineers.

Indeed, in a relatively small way, a number of biologists in the United States have tried applying their knowledge of the ecology of trout to the improvement of sport fishing, by the provision of shelter, pools, and small barriers to diversify stream beds and so to encourage suitable food organisms (Tarzwell, 1935, 1938*a*, *b*; Morofsky, 1936, Hunter *et al.*, 1941; Shetter *et al.*, 1946). This sort of activity, however, although it has continued to be given honourable mention in textbooks on fishery biology (Lagler, 1952; Rounsefell and Everhart, 1953) seems to have gone to some extent out of fashion. This is a pity, as we have there the germ of the idea of the management of running water to biological rather than purely engineering ends. If we had more experience of the former we might produce less devastating effects with the latter.

Finally man has altered the biota of running water in many parts of the world by moving species between continents. Much of this activity is quite recent, and almost all of it has occurred within the last century, but despite this we often have little solid information on its effects. *Eichhornia* has produced marked but biologically rather unspecified changes along the banks of the Nile in the Sudan, carp are accused of the removal of weed beds and the muddying up of rivers in the New World, trout may have reduced the populations of some invertebrates in New Zealand (Allen, 1961), and rainbow trout and brown trout are probably the cause of the decline of the dolly varden, *Salvelinus malma*, and the cutthroat trout in the Canadian Rocky Mountains (Nelson, 1965). There are many other examples of effects which *may* have resulted from introductions, and our inability to make any more definite statements than these point up how little we know about the ecology of rivers and streams and how much they have probably been altered before we even started to study them seriously.

Bibliography and index of citations

The abbreviations used for the titles of periodicals are those from *The World List of Scientific Periodicals*, 4th ed., Butterworth, London, 1965.

Some titles of articles are given in a language other than that of the body of the paper. This is indicated in brackets following the title, which is often that supplied by the author as a heading to a summary. Where papers are not in English, French, German, or Italian the presence of a summary in one of these four languages is indicated. Sources of translations into English of some Russian papers follow the references to them.

In order to conserve space the terms 'idem' and 'ibid.' have been used liberally, papers with three or more authors have been listed by the name of the first followed by *et al.* and the numbers of pages in substantial books have not been given.

Numbers in *italic* type following individual references are those of pages in this book where the papers are cited.

ABDIN, G., 1948. Seasonal distribution of phytoplankton and sessile algae in the River Nile, Cairo. *Bull, Inst. Égypte*, **29**, 369–82. *70, 99, 106.*

ABELL, D. A., 1959. Observations on mosquito populations of an intermittent foothills stream in California. *Ecology*, **40**, 186–93. *405.*

ABELL, D. L., 1961. The role of drainage analysis in biological work on streams. *Verh. int. Verein. theor. angew. Limnol.* **14**, 533–7. *14.*

ACKENHEIL, H. V., 1946. Rheon aus dem Flusse Lagan bei Ågård. Ein Beitrag zur Kenntnis der Mikroorganismentrift in Fliessgewässern. *Meddn telmat. Stn Åfard*, **4**, 1–34. *101.*

ADAM, W., 1942. Sur la répartition et la biologie de *Hydrobia jenkinsi* Smith en Belgique. *Bull. Mus. r. Hist. nat. Belg.* **18, 23,** 18 pp. *153.*

ADAMS, C. C., 1915. The variations and ecological distribution of the snails of the genus *Io. Mem. natn Acad. Sci.* **12**, 1–92. *143.*

AGNEW, J. D., 1962. The distribution of *Centroptiloides bifasciata* (E. P.) (Baëtidae: Ephem.) in Southern Africa, with ecological observations on the nymphs. *Hydrobiologia*, **20**, 367–72. *195, 219, 445.*

ALABASTER, J. S., and HARTLEY, W. G., 1962. The efficiency of a direct current electric fishing method in trout streams. *J. Anim. Ecol.* **31**, 385–8. *315.*

ALBRECHT, M.-L., 1953*a*. Ergebnisse quantitativer Untersuchungen an fliessender Gewässer. *Ber. limnol. Flussstn Freudenthal*, **4**, 10–11. *247–8.*

ALBRECHT, M.-L., 1953*b*. Die Plane und andere Flämingbäche. *Z. Fisch.* N.F. **1**, 389–476. *199, 207–8, 215, 218, 249, 390.*

—— 1959. Die quantitative Untersuchung der Bodenfauna fliessender Gewässer (Untersuchungsmethoden und Arbeitsergebnisse). Ibid. **8**, 481–550. *236, 248–51.*

—— 1961. Ein Vergleich quantitativer Methoden zur Untersuchung der Makrofauna fliessender Gewässer. *Verh. int. Verein. theor. angew. Limnol.* **14**, 486–90. *245.*

—— and BURSCHE, E.-M., 1957. Fischereibiologische Untersuchungen an Fliessgewässer. I. Physiographisch-biologische Studien an der Polenz. *Z. Fisch.* N.F. **6**, 209–40. *229, 239, 390.*

ALLAN, I. R. H., 1952. A hand-operated grab for sampling river beds. *J. Anim. Ecol.* **21**, 159–60. *237.*

—— *et al.*, 1958. A field and laboratory investigation of fish in a sewage effluent. *Fishery Invest., Lond.* **1**, **6**, **2**, 76 pp. *317.*

ALLANSON, B. R., 1961. The physical, chemical and biological conditions in the Jukskei–Crocodile River System. *Hydrobiologia,* **18**, 1–76. *208, 227.*

—— and KERRICH, J. E., 1961. A statistical method for estimating the number of animals found in field samples drawn from polluted rivers. *Verh. int. Verein. theor. angew. Limnol.* 14, 491–4. *244.*

ALLEN, K. R., 1940. Studies on the biology of the early stages of the salmon (*Salmo salar.*). 1. Growth in the River Eden. *J. Anim. Ecol.* **9**, 1–23. *237, 371.*

—— 1941*a*. Idem, 2. Feeding habits. Ibid., **10**, 47–76. *367, 373.*

—— 1941*b*. Idem, 3. Growth in the Thurso River system, Caithness. Ibid. **10**, 273–95. *371.*

—— 1942. Comparison of bottom faunas as sources of available fish food. *Trans. Am. Fish. Soc.* **71**, 275–83. *373.*

—— 1944. Studies on the biology of the early stages of the salmon (*Salmo salar*). 4. The smolt migration in the Thurso River in 1938. *J. Anim. Ecol.* **13**, 63–85. *353.*

—— 1951. The Horokiwi stream. A study of a trout population. *Fish. Bull. N.Z.* **10**, 231 pp. *198, 226, 237, 313, 326, 339, 343, 353, 367, 391, 419, 425, 428–9.*

—— 1956. The geography of New Zealand's freshwater fish. *N.Z. Sci. Rev.* **14**, 3–9. *302, 312, 346, 387.*

—— 1959. The distribution of stream bottom fauna. *Proc. N.Z. ecol. Soc.* **6**, 5–8. *200, 226.*

—— 1960*a*. Effect of land development on stream bottom faunas. Ibid. **7**, 20–1. *446.*

—— 1960*b*. Freshwater fauna of Wellington. *Science in Wellington,* Roy. Soc N.Z. Handbook, pp. 23–4. *337, 387, 445.*

—— 1961. Relations between Salmonidae and the native freshwater fauna in New Zealand. *Proc. N.Z. ecol. Soc.* **8**, 66–70. *445, 450.*

ALLEN, W. E., 1921. A quantitative and statistical study of the plankton of the San Joaquin River and its tributaries in and near Stockton, California, in 1913. *Univ. Calif. Publs Zool.* **22**, 1–292. *99.*

ALLEN, W. R., and CLARK, M. E., 1943. Bottom preferences of fishes of northeastern Kentucky streams. *Trans. Ky Acad. Sci.* **11**, 26–30. *335.*

AMBÜHL, H., 1959. Die Bedeutung der Strömung als ökologischer Faktor. *Schweiz. Z. Hydrol.* **21**, 133–264. *7, 148, 163, 166, 197.*

AMBÜHL, H., 1961. Die Strömung als physiologischer und ökologischer Faktor. Experimentelle Untersuchungen an Bachtieren. *Verh. int. Verein. theor. angew. Limnol.* **14**, 390–5. *7, 163.*

—— 1962. Die Besonderheiten der Wasserströmung in physikalischer, chemischer und biologischer Hinsicht. *Schweiz. Z. Hydrol.* **24**, 367–82. *7, 163.*

AN DER LAN, H., 1962. Zur Turbellarian-Fauna der Donau. *Arch. Hydrobiol.* Suppl. **27**, 3–27. *387.*

ANDERSON, J. R., and DICKE, R. J., 1960. Ecology of the immature stages of some Wisconsin black flies (Simuliidae: Diptera). *Ann. ent. Soc. Am.* **53**, 386–404. *193, 224, 257, 273, 275.*

—— and VOSKUIL, G. H., 1963. A reduction in milk production caused by the feeding of blackflies (Diptera: Simuliidae) on dairy cattle in California, with notes on the feeding activity on other animals. *Mosquito News*, **23**, 126–31. *186.*

ANDERSON, N. H., 1966. Depressant effect of moonlight on activity of aquatic insects. *Nature, Lond.* **209**, 319–20. *263–5.*

ANDERSON, R. O., 1959. A modified flotation technique for sorting bottom fauna samples. *Limnol. Oceanogr.* **4**, 223–5. *243.*

ANGELIER, E., 1952. Le peuplement en Hydracariens des eaux courantes françaises et son origine. *Fasc. hors-sér. Soc. Biogéogr.* pp. 67–74. *389, 408.*

—— 1953. Recherches écologiques et biogéographiques sur la faune des sables submergés. *Archs. Zool. exp. gén.* **90**, 37–162. *408.*

—— 1962. Remarques sur la répartition de la faune dans le milieu interstitiel hyporhéique. *Zool. Anz.* **168**, 351–6. *407.*

—— *et al.*, 1963. Les Hydracariens du Céret; étude systématique et écologique. *Bull. Soc. Hist. nat. Toulouse*, **98**, 459–500. *202, 215, 273, 275.*

ANNANDALE, H., and HORA, S. L., 1922. Parallel evolution in the fish and tadpoles of mountain torrents. *Rec. Indian Mus.* **24**, 505–10. *306.*

APPLEGATE, V. C., and BRYNILDSON, C. I., 1952. Downstream movement of recently transformed sea lampreys, *Petromyzon marinus*, in the Carp Lake River, Michigan. *Trans. Am. Fish. Soc.* **81**, 275–90. *353.*

ARMITAGE, K. B., 1958. Ecology of riffle insects of the Firehole River, Wyoming. *Ecology*, **39**, 571–80. *251, 255, 292, 391.*

—— 1961. Distribution of riffle insects of the Firehole River, Wyoming. *Hydrobiologia*, **17**, 152–74. *205, 391.*

ATZ, J. W., 1940. Reproductive behaviour in the Eastern Johnny Darter *Boleosoma nigrum olmstedi* (Storer). *Copeia*, **1940**, 100–6. *356.*

AUBERT, J., 1945. Le microptérisme chez les Plécoptères (Perlariés). *Revue suisse Zool.* **52**, 395–9. *145.*

—— 1963. Les Plécoptères des cours d'eau temporaires de la péninsule ibérique. *Mitt. Schweiz. ent. Ges.* **35**, 301–15. *403.*

BACKIEL, T., 1964. On the fish populations in small streams. *Verh. int. Verein. theor. angew. Limnol.* **15**, 529–34. *338, 385.*

BADCOCK, R. M., 1949. Studies on stream life in tributaries of the Welsh Dee. *J. Anim. Ecol.* **18**, 193–208. *114, 256, 292.*

—— 1953*a*. Studies of the benthic fauna in tributaries of the Kävlinge River, Southern Sweden. *Rep. Inst. Freshwat. Res. Drottningholm*, **35**, 21–34. *293, 390.*

—— 1953*b*. Comparative studies in the populations of streams. Ibid. **35**, 38–50. *226–7, 239.*

BADCOCK, R. M., 1953*c*. Observation of oviposition under water of the aerial insect *Hydropsyche angustipennis* (Curtis) (Trichoptera). *Hydrobiologia*, **5,** 222–5. *158.*

BAILEY, J. E., 1952. Life history and ecology of the sculpin *Cottus bairdi punctulatus* in Southwestern Montana. *Copeia*, **1952,** 243–55. *340.*

BAILEY, R. G., 1966. Observations on the nature and importance of organic drift in a Devon river. *Hydrobiologia*, **27,** 353–67. *261.*

BAILEY, R. M., and HARRISON, H. M., 1945. Food habits of the southern channel catfish (*Ictalurus lacustris punctulatus*) in the Des Moines River, Iowa. *Trans. Am. Fish. Sos.* **75,** 110–38. *367.*

—— *et al.*, 1954. Fishes from the Escambia River, Alabama and Florida, with ecologic and taxonomic notes. *Proc. Pa. Acad. Sci.* **106,** 109–64. *346, 350.*

BAINBRIDGE, R., 1960. Speed and stamina in three fish. *J. exp. Biol.* **37,** 129–53. *308–9, 355.*

BALDWIN, N. S., 1956. Food consumption and growth of brook trout at different temperatures. *Trans. Am. Fish. Soc.* **86,** 323–8. *372.*

BALL, R. C., *et al.*, 1963. Upstream dispersion of radiophosphorus in a Michigan trout stream. *Pap. Mich. Acad. Sci.* **48,** 57–64. *269.*

BAMFORTH, S. S., 1962. Diurnal changes in shallow aquatic habitats. *Limnol. Oceanogr.* **7,** 348–53. *42.*

BĂNĂRESCU, P., 1961. Tiergeographische Betrachtungen über die Fischfauna des Donaubeckens. *Verh. int. Verein. theor. angew. Limnol.* **14,** 386–9. *338.*

BANGHAM, R. V., and BENNINGTON, N. L., 1938. Movement of fish in streams. *Trans. Am. Fish. Soc.* **68,** 256–62. *343.*

BARDACH, J., 1964. *Downstream: A Natural History of the River.* Harper & Row, New York. *vii.*

BARNARD, K. H., 1932. South African may-flies (Ephemeroptera). *Trans. R. Soc. S. Afr.* **20,** 201–59. *184.*

BATTLE, H. I., and SPRULES, W. M., 1960. A description of the semi-buoyant eggs and early developmental stages of the goldeye, *Hiodon alosoides* (Rafinesque). *J. Fish. Res. Bd Can.* **17,** 245–66. *361.*

BAXTER, G., 1961. River utilisation and the preservation of migratory fish life. *Minut. Proc. Instn civ. Engrs,* **18,** 225–44. *27, 354, 449.*

BEACH, N. W., 1960. A study of the planktonic rotifers of the Ocqueoc River System, Presque Isle County, Michigan. *Ecol. Monogr.* **30,** 339–57. *103–4.*

BEADLE, L. C., 1957. Comparative physiology; osmotic and ionic regulation in aquatic animals. *A. Rev. Physiol.* **19,** 329–58. *218, 220.*

BEAK, T. W., 1938*a*. Investigations of the fauna of a water meadow carrier. *Rep. Avon biol. Res.* **5,** 29–41. *294.*

—— 1938*b*. Methods of making and sorting collections for an ecological study of a stream. Ibid. **5,** 42–6. *243.*

BEAUCHAMP, R. S. A., 1933. Rheotaxis in *Planaria alpina. J. exp. Biol.* **10,** 113–29. *151–2.*

—— 1935. The rate of movement of *Planaria alpina.* Ibid. **12,** 271–85. *151.*

—— 1937. Rate of movement and rheotaxis in *Planaria alpina.* Ibid. **14,** 104–16. *151.*

—— and ULLYOT, P., 1932. Competitive relationships between certain species of freshwater triclads. *J. Ecol.* **20,** 200–8. *387.*

BECK, W. M., 1965. The streams of Florida. *Bull. Fla State Mus.* **10,** 91–126. *30, 207, 216, 219, 221, 391.*

BEETON, A. M., 1956. Food habits of the burbot (*Lota lota lacustris*) in the White River, a Michigan trout stream. *Copeia*, **1956**, 58–60. *370*.

BEHMER, D. J., 1964. Movement and angler harvest of fishes in the Des Moines River, Boone County, Iowa. *Proc. Iowa Acad. Sci.* **71**, 259–63. *343*.

BEHNEY, W. H., 1937. Food organisms of some New Hampshire trout streams. *Surv. Rep. N.H. Fish. Comm.* **2**, 76–80. *247, 255*.

BEHNING, A., 1924. Einige Ergebnisse qualitativer und quantitativer Untersuchungen der Bodenfauna der Wolga. *Verh. int. Verein. theor. angew. Limnol.* **2**, 71–94. *207, 250*.

—— 1928. Das Leben der Wolga, zugleich eine Einführung in die Flussbiologie. *Die Binnengewässer*, Stuttgart, **5**, 162 pp. *26, 74, 105, 110, 112, 119, 122, 346, 351, 361, 367, 389, 391, 396*.

—— 1929*a*. Über das Plankton der Wolga. *Verh. int. Verein. theor. angew. Limnol.* **4**, 192–212. *98–9, 108*.

—— 1929*b*. Das Leben der Wolga. in Vogt O. *Die Naturwissenschaft der Sowjet-Union*, Berlin, 25 pp. *63, 74, 109, 346–7, 391, 396*.

BEHRE, K., and WEHRLE, E., 1942. Welche Faktoren entscheiden über die Zuzammensetzung von Algengesellschaften? Zur Kritik algenökologischer Fragenstellungen. *Arch. Hydrobiol.* **39**, 1–23. *72*.

BEIER, M., 1948. Zur Kenntnis von Körperbau und Lebensweise der Helminen (Col. Dryopidae). *Eos, Wien.* **24**, 123–211. *132, 172, 222–3, 230, 277*.

—— and POMEISL, E., 1959. Einiges über Körperbau und Lebensweise von *Octhebius exsculptus* Germ. und seiner Larve (Col. Hydroph. Hydraen.). *Z. Morph. Ökol. Tiere*, **48**, 72–88. *410*.

BELING, A., 1953. Anwendung von Membranfiltern für Fliesswasseruntersuchungen. *Ber. limnol. Flussstn Freudenthal*, **4**, 14–16. *437*.

BELING, D., 1929. La faune aquatique des fleuves méridianaux de l'Ukraine en rapport avec le question de son origine. *Verh. int. Verein. theor. angew. Limnol.* **4**, 213–39. *105, 346*.

—— 1949. Der Einfluss des Wasserbaues auf die Tierwelt des Dnjepr-Strömschnellen-gebietes. *Zool. Anz.* Suppl. **13**, 172–80. *346, 349, 447*.

BENGTSSON, S., 1924. La nutrition des larvae des éphémerès. *Annls Biol. lacustre*, **13**, 215–17. *194*.

BENNIN, E., 1926. Das Plankton der Warthe in den Jahren 1920–24. *Arch. Hydrobiol.* **17**, 545–93. *98, 109*.

BENSON, N. G., 1953. The importance of ground water to trout populations in the Pigeon River, Michigan. *Trans. N. Am. Wildl. Conf.* **18**, 269–81. *324*.

—— 1954. Seasonal fluctuations in the feeding of brook trout in the Pigeon River, Michigan, *Trans. Am. Fish. Soc.* **83**, 8 pp. *372*.

—— 1955. Observations on anchor ice in a Michigan trout stream. *Ecology*, **36**, 529–30. *32, 206*.

BERG, K., 1938. Studies on the bottom animals of Esrom Lake, *K. danske Vidensk. Selsk. Skr.* **9**, **8**, 247 pp. *237*.

—— 1943. Physiographical studies on the River Susaa. *Folia limnol. Scand.* **1**, 174 pp. *27, 39*.

—— 1952. On the oxygen consumption of Ancylidae (Gastropoda) from an ecological point of view. *Hydrobiologia*, **4**, 225–67. *162*.

—— *et al.*, 1948. Biological studies on the River Susaa. *Folia limnol. Scand.* **4**, 318 pp. *196, 207, 248, 278, 390, 396*.

—— *et al.*, 1958. Seasonal and experimental variations of the oxygen consumption of the limpet *Ancylus fluviatilis* (O. F. Müller). *J. exp. Biol.* **35,** 43–73. *162.*

Berner, L., 1950. The mayflies of Florida. *Univ. Fla Publ. biol. Sci. Ser.* **4, 4,** 267 pp. *128, 405.*

—— 1954. Phoretic association between a species of *Simulium* and a mayfly nymph, with a description of the nymph. *Ann. Mag. nat. Hist.* **12, 7,** 116–21. *181, 184.*

—— and Sloan, W. C., 1954. The occurrence of a mayfly nymph in brackish water. *Ecology,* **35,** 98. *218.*

Berner, L. M., 1951. Limnology of the lower Missouri River. Ibid. **32,** 1–12. *98–9, 259–60, 366.*

Bernet, B., 1960. Recherches biologiques sur les populations de *Gobio gobio* (Linné 1758) de la Nivelle (fleuve côtier du Pays Basque). *Annls Stn cent. Hydrobiol. appl.* **8,** 127–80. *323, 335.*

Berry, P. Y., 1966. The food and feeding habits of the torrent frog, *Amolops larutensis. J. Zool. Lond.* **149,** 204–14. *378.*

Berthélemy, C., 1966. Recherches écologiques et biogéographiques sur les plécoptères et coléoptères d'eau courante (*Hydraena* et Elminthidae) des Pyrénées. *Ann. Limnologie,* **2,** 227–458. *144, 202–3, 222, 387, 390, 393, 402.*

Bertin, L., 1956. *Eels: A Biological Study.* Cleaver-Hume Press, London. *308, 313, 345, 353.*

Besch, W., 1966. Driftnetzmethode und biologische Fliesswasseruntersuchung. *Verh. int. Verein. theor. angew. Limnol.* **16,** 669–78. *260.*

Beyer, H., 1932. Die Tierwelt der Quellen und Bäche des Baumberggebietes. *Abh. westf. ProvMus. Naturk.* **3,** 9–187. *400.*

Bick, G. H., *et al.,* 1953. An ecological reconnaisance of a naturally acid stream in Southern Louisiana. *J. Tenn. Acad. Sci.* **28,** 221–31. *210, 391.*

Bick, H., 1959. *Gammarus pulex fossarum* Koch 1835 als Fallaubersetzer. *Z. Fisch.* N.F. **8,** 635–38. *433.*

—— 1963. A review of Central European methods for the biological estimation of water pollution levels. *Bull. Wld Hlth Org.* **29,** 401–13. *241.*

Bílý, *et al.,* 1952. *Hydrobiologia Hnilca a Hornadu.* (Czech: German summary) *Nakl. Slov. Akad.* Bratislava, 183 pp. *391.*

Bishai, H. M., 1960. The effect of water currents on the survival and distribution of fish larvae. *J. Cons. perm. int. Explor. Mer.* **25,** 134–46. *363.*

Björk, S., 1962. Investigations on *Margaritifera margaritifera* and *Unio crassus. Acta Limnol.* **4,** 109 pp. *221, 276.*

Black, J. B., 1963. Observations on the home range of stream-dwelling crawfishes. *Ecology,* **44,** 592–5. *153.*

Black, J. D., 1949. Changing fish populations as an index of pollution and soil erosion. *Trans. Ill. St. Acad. Sci.* **42,** 145–8. *444.*

Blum, J. L., 1954*a*. Evidence for a diurnal pulse in stream phytoplankton. *Science N.Y.* **119,** 732–4. *96.*

—— 1954*b*. Two winter diatom communities of Michigan streams. *Pap. Mich. Acad. Sci.* **39,** 3–7. *71.*

—— 1956. The ecology of river algae. *Bot. Rev.* **22,** 291–341. *53–4, 60, 94.*

—— 1957. An ecological study of the algae of the Saline River, Michigan. *Hydrobiologia,* **9,** 361–408. *58, 67–8, 70–1, 77.*

—— 1960. Algal populations in flowing waters. *Spec. publs. Pymatuning Lab. Fld Biol.* **2,** 11–21. *53, 56–7, 60, 63, 66–7, 72–5, 101, 107.*

BOCCARDY, J. A., and COOPER, E. L., 1963. The use of rotenone and electrofishing in surveying small streams. *Trans. Am. Fish. Soc.* **92,** 307–10. *316.*

BORNHAUSER, K., 1913. Die Tierwelt der Quellen in der Umgebung Basels. *Int. Revue ges. Hydrobiol. Hydrogr.* Suppl. **5, 3,** 90 pp. *398.*

BOTAN, E. A., *et al.*, 1960. A study of the microbiol decomposition of nitrogenous matter in fresh water. *Arch. Hydrobiol.* **56,** 334–54. *437.*

BOTOSANEANU, L., 1960. Sur quelques regularités observées dans le domaine de l'écologie des insectes aquatiques. Ibid. **56,** 370–7. *204.*

—— and NEGREA, ST., 1961. Une oasis aquatique à faune relique dans la plaine du Danube inférieur. *Hydrobiologia,* **18,** 199–218. *400.*

—— *et al.*, 1959. Contributions à l'étude hydrobiologique d'une deereá de Dobrogea: la Casimcea. *Arch. Hydrobiol.* **55,** 30–51. *391, 400.*

BOURNAUD, M., 1963. Le courant, facteur écologique et éthologique de la vie aquatique. *Hydrobiologia,* **21,** 125–65. *121, 125.*

BOUSFIELD, E. L., 1958. Fresh-water amphipod crustaceans of glaciated North America. *Can. Fld Nat.* **72,** 55–113. *275.*

BOUSSU, M. F., 1954. Relationship between trout populations and cover on a small stream. *J. Wildl. Mgmt,* **18,** 229–39. *313, 443.*

BOVBJERG, R. V., 1952*a*. Comparative ecology and physiology of the crayfish *Orconectes propinquus* and *Cambarus fodiens. Physiol. Zoöl.* **25,** 34–56. *207.*

—— 1952*b*. Ecological aspects of dispersal of the snail *Campeloma decisum. Ecology,* **33,** 169–76. *153.*

—— 1964. Dispersal of aquatic animals relative to density. *Verh. int. Verein. theor. angew. Limnol.* **15,** 879–84. *153.*

BRADEN, G. E., 1951. Turbulence, diffusion and sedimentation in stream channel expansions and contractions. *Proc. Okla. Acad. Sci.* **31,** 73–7. *21.*

—— 1958. The hydraulic jump in natural streams. Ibid. **38,** 78–9. *9.*

BREHM, V., 1911. Beobachtungen über die Entstehung des Potamoplanktons. *Int. Revue ges. Hydrobiol. Hydrogr.* **4,** 311–14. *94.*

—— and RUTTNER, F., 1926. Die Biocönosen der Lunzer Gewässer. Ibid. **16,** 281–391. *48, 256, 390.*

BRICKENSTEIN, C., 1955. Über den Netzbau der Larve von *Neureclepsis bimaculata* (Trichopt., Polycentropidae). *Abh. bayer. Akad. Wiss.* N.F. **69,** 44 pp. *158, 187–8.*

BRIGGS, J. C., 1948. The quantitative effects of a dam upon the bottom fauna of a small California stream. *Trans. Am. Fish. Soc.* **78,** 70–81. *257.*

BRINCK, P., 1949. Studies on Swedish stoneflies. *Opusc. ent.* Suppl. **11,** 250 pp. *140, 145–6, 159, 203, 234, 278–9, 286–7.*

—— and WINGSTRAND, K. G., 1949. The mountain fauna of the Varihaure area in Swedish Lapland. I. General account. *K. fysiogr. Sällsk. Lund Förh.* N.F. **60, 2,** 69 pp. *390, 402.*

—— —— 1951. Idem. II. Special account. Ibid. **61, 2,** 173 pp. *390.*

BRINKHURST, R. O., 1965. Observations on the recovery of a British river from gross organic pollution. *Hydrobiologia,* **25,** 9–51. *269.*

BRINLEY, F. J., and KATZIN, L. J., 1942. Distribution of stream plankton in the Ohio River System. *Am. Midl. Nat.* **27,** 177–90. *99, 109.*

BRITT, N. W., 1955. New methods of collecting bottom fauna from shoals or rubble bottoms of lakes and streams. *Ecology,* **36,** 524–5. *239, 242.*

BROCK, T. D., and BROCK, M. L., 1966. Temperature optima for algal development in Yellowstone and Iceland hot springs. *Nature, Lond.* **209,** 733–4. *62.*

BROOK, A. J., 1955. The aquatic fauna as an ecological factor in studies of the occurrence of freshwater algae. *Revue algol.* N.S. **1**, 142–5. *193*.

—— and RZÓSKA, J., 1954. The influence of the Gebel Aulyia Dam on the development of Nile plankton. *J. Anim. Ecol.* **23**, 101–14. *108*.

—— and WOODWARD, W. B., 1956. Some observations on the effects of water inflow and outflow on the plankton of small lakes. Ibid. **25**, 22–35. *101–2*.

BROWN, C. J. D., 1952. Spawning habits and early development of the mountain whitefish, *Prosopium williamsoni*, in Montana. *Copeia*, **1952**, 109–13. *343, 348, 359*.

—— *et al.*, 1953. Observations on ice conditions in the West Gallatin River, Montana. *Proc. Mont. Acad. Sci.* **13**, 21–7. *32, 205*.

BROWN, D. S., 1961. The food of the larvae of *Chloeon dipterum* L. and *Baetis rhodani* (Pictet) (Insecta, Ephemeroptera). *J. Anim. Ecol.* **30**, 55–75. *194*.

BROWN, M. E., 1946*a*. The growth of the brown trout (*Salmo trutta* Linn.). I. Factors influencing the growth of trout fry. *J. exp. Biol.* **22**, 118–29. *311*.

—— 1946*b*. Idem. II. The growth of two-year-old trout at a constant temperature of 11·5 degrees. Ibid. **22**, 130–44. *311, 325*.

—— 1946*c*. Idem. III. The effect of temperature on the growth of two-year-old trout. Ibid. **22**, 145–55. *372*.

BROWZIN, B. S., 1962. On the classification of rivers in the Great Lakes–St. Lawrence Basin. *Publs gt Lakes Res. Inst.* **9**, 86–92. *27*.

BRUES, C. T., 1927. Animal life in hot springs. *Q. Rev. Biol.* **2**, 181–203. *399*.

—— 1928. Studies on the fauna of hot springs in the western United States and the biology of thermophilous animals. *Proc. Am. Acad. Arts Sci.* **63**, 139–228. *399*.

BUCHHOLZ, M., 1957. Age and growth of river carpsucker in Des Moines River, Iowa. *Proc. Iowa Acad. Sci.* **64**, 589–600. *366, 368*.

BUDDE, H., 1928. Die Algenflora des Sauerländischen Gebirgsbaches. *Arch. Hydrobiol.* **19**, 433–520. *70, 72–3*.

—— 1930. Die Algenflora der Ruhr. Ibid. **21**, 559–648. *70*.

—— 1932. Limnologische Untersuchungen niederrheinischer und westfälischer Gewässer. Die Algenflora der Lippe und ihrer Zuflüsse. Ibid. **24**, 188–252. *70*.

BURKS, B. D., 1953. The mayflies, or Ephemeroptera, of Illinois. *Bull. Ill. St. nat. Hist. Surv.* **26**, 1–216. *176, 279, 404*.

BURNET, A. M. R., 1952*a*. Studies on the ecology of the New Zealand long-finned eel, *Anguilla dieffenbachii* Gray. *Aust. J. mar. Freshwat. Res.* **3**, 32–63. *367–8*.

—— 1952*b*. Studies on the ecology of the New Zealand freshwater eels I. Design and use of an electric fishing machine. Ibid. 3, 111–25. *315, 420*.

—— 1965. Observations on the spawning migrations of *Galaxias attenuatus* (Jenyns). *N.Z. Jl Sci. Technol.* **8**, 79–87. *346*.

BURSCHE, E.-M., 1959. Untersuchungen über den Detritusgehalt in Seen und Flüssen. *Int. Revue ges. Hydrobiol. Hydrogr.* **44**, 438–48. *433*.

—— 1962. Untersuchungen über die in der Polenz vorherrschenden Bewuchsdiatomeen. *Arch. Hydrobiol.* **58**, 474–89. *70*.

BURTON, G. W., and ODUM, E. P., 1945. The distribution of stream fish in the vicinity of Mountain Lake, Virginia. *Ecology*, **26**, 182–94. *306, 324, 385–6*.

BUSCEMI, P. A., 1958. Littoral oxygen depletion produced by a cover of *Elodea canadiensis*. *Oikos*, **9**, 239–45. *39*.

BUSCEMI, P. A., 1966. The importance of sedimentary organics in the distribution of benthic organisms. *Spec. Publs Pymatuning Lab. Fld Biol.* **4,** 79–86. *48.*

BUSH, S. F., 1933. Note on hydrogen-ion concentration of the waters of certain Natal trout streams. *S. Afr. J. Sci.* **30,** 388–93. *332.*

BUȘNITĂ, T., 1961. Die Wandlungen der Fischfauna der unteren Donau. *Verh. int. Verein. theor. angew. Limnol.* **14,** 381–5. *346–7.*

[1]BUTCHER, A. D., 1946. Freshwater fish of Victoria and their food. *Publn Victoria Dept Fish Game,* Melbourne. *367.*

BUTCHER, R. W., 1932. Studies on the ecology of rivers II. The microflora of rivers with special reference to the algae on the river bed. *Ann. Bot.* **46,** 813–61. *57, 68, 94–5, 109.*

—— 1933. Idem. I. On the distribution of macrophytic vegetation in the rivers of Britain. *J. Ecol.* **21,** 58–91. *83–4, 87–8, 90.*

—— 1938. The algae of the river. *Rep. Avon biol. Res.* **5,** 47–52. *66, 72.*

—— 1940. Studies on the ecology of rivers IV. Observations on the growth and distribution of the sessile algae in the River Hull, Yorkshire. *J. Ecol.* **28,** 210–23. *63, 72.*

[1]—— 1946. Idem. VI. The algal growth in certain highly calcareous streams. Ibid. **33,** 268–83. *63, 68, 76.*

—— 1947. Idem. VII. The algae of organically enriched waters. Ibid. **35,** 186–91. *58, 68–9.*

—— *et al.*, 1930. Variations in composition of river waters. *Int. Revue ges. Hydrobiol. Hydrogr.* **24,** 47–80. *41.*

—— *et al.*, 1931. A biological investigation of the River Lark. *Fishery Invest., Lond.* **1, 3, 3,** 112 pp. *85, 87, 390.*

—— *et al.*, 1937. Survey of the River Tees. Part III.—The non-tidal reaches—chemical and biological. *Tech. Pap. Wat. Pollut. Res. D.S.I.R.* **6,** 189 pp. *41, 106, 390.*

CAINES, L. A., 1965. The phosphorus content of some aquatic macrophytes with special reference to seasonal fluctuations and applications of phosphate fertilisers. *Hydrobiologia.* **25,** 289–301. *87.*

CAIRNS, D., 1941. Life history of the two species of New Zealand fresh-water eel. I. Taxonomy, age and growth, migration, and distribution. *N.Z. Jl Sci. Technol.* **23,** 53–72. *345.*

—— 1942*a*. Idem. II. Food and inter-relationships with trout. Ibid. **23,** 132–48. *345.*

—— 1942*b*. Idem. III. Development of sex. Campaign of eel destruction. Ibid. **23,** 173–8. *345.*

CAMPBELL, N., 1961. The growth of brown trout in acid and alkaline waters. *Salm. Trout Mag.* **1961,** 47–51. *333.*

CARLANDER, H. B., 1954. *History of Fish and Fishing in the Upper Mississippi River.* Upper Mississippi River Conservation Committee, 96 pp. *444.*

CARLANDER, K. D., 1950. *Handbook of Fresh-water Fishery Biology.* Wm C. Brown, Dubuque. *339.*

—— 1953. *First Supplement to the Handbook of Fresh-water Fishery Biology.* Wm. C. Brown, Dubuque. *339.*

1. The 1946 papers of Butcher, A. D., and Butcher, R. W., are distinguished as Butcher, A. D. (1946) and Butcher (1946) in the text.

CARPENTER, K. E., 1927. Faunistic ecology of some Cardiganshire streams. *J. Ecol.* **15,** 33–54. *390.*

—— 1928*a*. On the distribution of freshwater Turbellaria in the Aberystwyth district, with especial reference to two ice-age relicts. Ibid. **16,** 105–22. *204, 219, 387, 400.*

—— 1928*b*. *Life in Inland Waters*. Sidgwick and Jackson, London. *vii, 219, 384.*

—— 1940. The feeding of salmon parr in the Cheshire Dee. *Proc. zool. Soc. Lond.* **110,** 81–96. *367, 372.*

CARRÉ, C., 1957. Contribution à l'étude de la biologie d'*Agrion virgo* (Insecte, Odonate, Zygoptère). *Bull. Soc. scient. Bretagne,* **32,** 81–102. *276.*

CARTER, B. T., 1950. The movement of fishes through navigation lock chambers in the Kentucky River. *Trans. Ky Acad. Sci.* **15,** 48–56. *349.*

CARTER, G. S., undated. *The Papyrus Swamps of Uganda*. Heffer, Cambridge. *92.*

CHACKO, P. I., and GANAPATI, S. V., 1952. Hydrobiological survey of the Suruli River in the Highwavys, Mudurai District, India, to determining its suitability for the introduction of the rainbow trout. *Arch. Hydrobiol.* **46,** 128–41. *324, 391.*

CHANDLER, C. M., 1966. Environmental factors affecting the local distribution and abundance of four species of stream-dwelling triclads. *Invest. Indiana Lakes Streams,* **7,** 1–56. *204, 387, 400.*

CHANDLER, D. C., 1937. Fate of typical lake plankton in streams. *Ecol. Monogr.* **7,** 445–79. *103.*

—— 1939. Plankton entering the Huron River from Portage and Base Line Lakes, Michigan. *Trans. Am. microsc. Soc.* **58,** 24–41. *102.*

CHAPMAN, D. W., 1962. Effects of logging upon fish resources of the west coast. *J. For.* **60,** 533–7. *443.*

—— 1966. The relative contribution of aquatic and terrestrial primary producers to the trophic relations of stream organisms. *Spec. Publs Pymatuning Lab. Fld Biol.* **4,** 116–30. *432.*

—— and DEMORY, R., 1963. Seasonal changes in the food ingested by aquatic insect larvae and nymphs in two Oregon streams. *Ecology,* **44,** 140–6. *192, 194, 229, 432.*

CHASE, E. S., 1957. Oxygen demand exerted by leaves stored under water. *J. New Engl. Wat. Wks Ass.* **71,** 307–12. *434.*

CHUTTER, F. M., 1963. Hydrobiological studies on the Vaal River in the Vereeniging Area. Part I; Introduction, water chemistry and biological studies on the fauna of habitats other than muddy bottom sediments. *Hydrobiologia,* **21,** 1–65. *228, 256.*

—— and NOBLE, R. G., 1966. The reliability of a method of sampling stream invertebrates. *Arch. Hydrobiol.* **62,** 95–103. *247.*

CIANFICCONI, F., and RIATTI, M., 1957. Impiego di pietre artificiali per l'analysi quantitativa delle colonizzazioni faunistiche dei fondi potamici. *Boll. Pesca Piscic. Idrobiol.* N.S. **12,** 299–334. *198, 210, 239.*

CLAASSEN, P. W., 1922. The larva of a Chironomid (*Trissocladius equitans* n. sp.) which is parasitic upon a may-fly nymph (*Rhithrogena* sp.). *Kans. Univ. Sci. Bull.* **14,** 395–405. *180.*

CLARK, C. C., 1944. The fresh-water naiades of Auglaize County, Ohio. *Ohio J. Sci.* **54,** 167–76. *119, 389, 447.*

CLARK, W. J., 1960. The phytoplankton of the Logan River, Utah, a mountain stream. *Diss. Abstr.* **21, 6,** 1339. *94, 109.*

CLARKE, A. H., and BERG, C. O., 1959. *The Freshwater Mussels of Central New York*. Cornell Univ. Agr. Exp. Sta., Ithaca. *220, 389*.

CLAUS, G., 1961. Monthly ecological studies on the flora of the Danube at Vienna in 1957–8. *Verh. int. Verein. theor. angew. Limnol.* **14,** 458–65. *107*.

CLAUSEN, R. G., 1931. Orientation in fresh water fishes. *Ecology,* **12,** 541–6. *326*.

CLEARY, R. E., and GREENBANK, J., 1954. An analysis of techniques used in estimating fish populations in streams, with particular reference to large non-trout streams. *J. Wildl. Mgmt,* **18,** 461–77. *313–14, 317, 335*.

CLELAND, D. M., 1954. A study of the habits of *Valvata piscinalis* (Müller) and the structure and function of the alimentary canal and reproductive system. *Proc. malac. Soc. Lond.* **1954,** 167–203. *194*.

CLEMENS, W. A., 1917. An ecological study of the mayfly *Chirotenetes*. *Univ. Toronto Stud. biol. Ser.* **17,** 5–43. *185*.

CLIFFORD, H. F., 1966*a*. The ecology of invertebrates in an intermittent stream. *Invest. Indiana Lakes Streams,* **7,** 57–98. *278–9, 404–6*.

—— 1966*b*. Some limnological characteristics of six Ozark streams. *Missouri Dept Conservation, Divn Fish.* D-J Series **4,** 55 pp. *94, 292*.

CLOTHIER, W. D., 1953. Fish loss and movement in irrigation diversions from the West Gallatin River, Montana. *J. Wildl. Mgmt,* **17,** 144–58. *350, 449*.

CLUGSTON, J. P., and COOPER, E. L., 1960. Growth of the common eastern madtom, *Noturus insignis*, in Central Pennsylvania. *Copeia,* **1960,** 9–16. *305*.

CODREANU, R., 1939. Recherches biologiques sur un chironomide *Symbocladius rhitrogenae* (Zavr.) ectoparasite 'cancérigène' des éphémères torrenticoles. *Archs Zool. exp. gén.* **81,** 1–283. *180*.

COKER, R. E., 1929. Keokuk dam and the fisheries of the upper Mississippi River. *Bull. Bur. Fish., Wash.* **45,** 87–139. *349*.

—— 1930. Studies of the common fishes of the Mississippi River at Keokuk. Ibid. **45,** 141–225. *349*.

—— 1954. *Streams, Lakes, Ponds*. Univ. North Carolina Press. *vii, 32, 411*.

—— *et al.*, 1936. Swimming plume and claws of the broad-shouldered water-strider *Rhagovelia flavocincta* Bueno (Hemiptera). *Bull. Brooklyn ent. Soc.* **31,** 81–4. *175*.

COLE, G. A., and MINCKLEY, W. L., 1961. A new species of amphipod crustacean (genus *Gammarus*) from Kentucky. *Trans. Am. microsc. Soc.* **8,** 391–8. *401*.

COLE, L. C., 1957. A surprising case of survival. *Ecology,* **38,** 357. *168*.

CONAWAY, C. H., 1952. Life history of the water shrew (*Sorex palustris navigator*). *Am. Midl. Nat.* **48,** 219–48. *380*.

CONRAD, W., 1942. Sur la faune et la flore d'un ruisseau de l'Ardenne belge. *Mém. Mus. r. Hist. nat. Belg.* **99,** 1–177. *390*.

COOKE, W. B., 1956. Colonisation of artificial bare areas by microorganisms. *Bot. Rev.* **22,** 613–38. *53, 56*.

—— 1961. Pollution effects on the fungus population of a stream. *Ecology,* **42,** 1–18. *435*.

COOPER, E. L., 1953. Periodicity of growth and change of condition of brook trout (*Salvelinus fontinalis*) in three Michigan trout streams. *Copeia,* **1953,** 107–14. *326*.

—— *et al.*, 1962. Growth rate of brook trout at different population densities in a small infertile stream. *Progve FishCult.* April 1962, 74–80. *333, 429*.

COPELAND, B. J., and DUFFER, W. R., 1964. Use of a clear plastic dome to measure gaseous exchange diffusion rates in natural waters. *Limnol. Oceanogr.* **9,** 494–9. *413.*

CORBET, P. S., 1956*a*. Larvae of East African Odonata. Introduction and 1. *Ictinogomphus ferox* Rambur. *Entomologist,* **89,** 97–100. *178.*

—— 1956*b*. Idem. 4–5. 4. *Lestinogomphus angustus* Martin. 5. *Phyllogomphus aethiops* Selys. Ibid. **89,** 216–19. *178.*

—— 1957*a*. Idem. 6–8. 6. *Brachythemis lacustris* Kirby. 7. *Brachythemis leucosticla* (Burmeister). 8. *Zyxomma flavicans* (Martin). Ibid. **90,** 28–34. *178.*

—— 1957*b*. Idem. 12–14. Ibid. **90,** 143–7. *178.*

—— 1958. Some effects of DDT on the fauna of the Victoria Nile. *Revue Zool. Bot. afr.* **57,** 73–95. *256.*

—— 1959. Notes on the insect food of the Nile crocodile in Uganda. *Proc. R. ent. Soc. Lond.* **A34,** 17–22. *378.*

—— 1960*a*. A new species of *Afronurus* (Ephemeroptera) and its association with *Simulium* in Uganda. Ibid. **B29,** 68–72. *181.*

—— 1960*b*. The food of a sample of crocodiles (*Crocodrilus niloticus* L.) from Lake Victoria. *Proc. zool. Soc. Lond.* **133,** 561–72. *378.*

—— 1961. The biological significance of the attachment of immature stages of *Simulium* to mayflies and crabs. *Bull. ent. Res.* **52,** 695–9. *182.*

CORDONE, A. J., and KELLEY, D. W., 1961. The influence of inorganic sediment on the aquatic life of streams. *Calif. Fish Game,* **47,** 189–228. *444.*

COUTANT, C. C., 1962. The effect of a heated water effluent upon the macro-invertebrate riffle fauna of the Delaware River. *Proc. Pa Acad. Sci.* **36,** 58–71. *205, 267.*

—— 1963. Stream plankton above and below Green Lane Reservoir. Ibid. **37,** 122–6. *449.*

CREASER, C. W., 1930. Relative importance of hydrogen-ion concentration, temperature, dissolved oxygen, and carbon dioxide tension, on habitat selection by brook trout. *Ecology,* **11,** 246–62. *332.*

CRIDLAND, C. C., 1957. Ecological factors affecting the numbers of snails in temporary bodies of water. *J. trop. Med. Hyg.* **60,** 1–7. *404, 406.*

—— 1958. Ecological factors affecting the numbers of snails in a permanent stream. Ibid. **61,** 16–20. *291.*

CRISP, G., 1956. An ephemeral fauna of torrents in the N. Territories of the Gold Coast, with special reference to the enemies of *Simulium*. *Ann. trop. Med. Parasit.* **50,** 260–7. *195, 226, 403.*

CUMMINS, K. W., 1962. An evaluation of some techniques for the collection and analysis of benthic samples with special emphasis on lotic waters. *Am. Midl. Nat.* **67,** 477–504. *23–4, 236.*

—— 1964. Factors limiting the microdistribution of larvae of the caddisflies *Pycnopsyche lepida* (Hagen) and *Pycnopsyche guttifer* (Walker) in a Michigan stream. *Ecol. Monogr.* **34,** 271–95. *128, 272, 280, 283.*

—— 1966. A review of stream ecology with special emphasis on organism-substrate relationships. *Spec. Publs Pymatuning Lab. Fld Biol.* **4,** 2–51. *vii.*

—— *et al.*, 1966. Trophic relations in a small woodland stream. *Verh. int. Verein. theor. angew. Limnol.* **16,** 627–38. *433.*

CUNNINGHAM, B., 1937. River flow around bends. *Nature, Lond.* **140,** 728–9. *21.*

CURRY, L. V. L., 1954. Notes on the ecology of the midge fauna (Diptera; Tendipedidae) of Hunt Creek, Montmorency County, Michigan. *Ecology,* **35,** 541–50. *381.*

CUSHING, C. E., 1963. Filter-feeding insect distribution and planktonic food in the Montreal River. *Trans. Am. Fish. Soc.* **92,** 216–19. *257–8.*

—— 1964. Plankton and water chemistry in the Montreal River lake-stream system, Saskatchewan. *Ecology,* **45,** 306–13. *101–2.*

CZERNIN-CHUDENITZ, C. W., 1966. Das Plankton der österreichischen Donau und seine Bedeutung für die Selbstreinigung. *Arch. Hydrobiol.* Suppl. **30,** 194–217. *98, 109.*

DAHM, A. G., 1949. *Phagocata* (= *Fonticola*) from South Sweden. (Turbellaria Tricladida Paludicola). Taxonomical, ecological and chlorological studies. *Acta Univ. Lund,* N.F. **45, 7,** 32 pp. *233, 400.*

DAIBER, F. C., 1956. A comparative analysis of winter feeding habits of two benthic stream fishes. *Copeia,* **1956,** 141–51. *369, 371.*

DAMAS, H., 1961. L'évolution d'un lac barrage du Katanga. *Verh. int. Verein. theor. angew. Limnol.* **14,** 661–4. *448.*

D'ANCONA, U., and MERLO, S., 1959. La speciazone nelle trote italiane ed in particolare in quelle del Lago di Garda. *Atti Ist. veneto Sci.* **117,** 21–6. *301.*

DANECKER, E., 1961. Studien zur hygropetrischen Fauna. Biologie und Ökologie von *Stactobia* und *Tinodes* (Insect., Trichopt.). *Int. Revue ges. Hydrobiol. Hydrogr.* **46,** 214–54. *410.*

DARNELL, R. M., 1961. Trophic spectrum of an estuarine community, based on studies of Lake Pontchartrain, Louisiana. *Ecology,* **42,** 553–68. *368, 411, 438.*

—— 1964. Organic detritus in relation to secondary production in aquatic communities. *Verh. int. Verein. theor. angew. Limnol.* **15,** 462–70. *368.*

DAVIDSON, F. A., and WILDING, J. L., 1943. A quantitative faunal investigation of a cold spring community. *Am. Midl. Nat.* **29,** 200–9. *238, 400.*

DAVIES, D. M., 1950. A study of the black fly populations of a stream in Algonquin Park, Ontario. *Trans. R. Can. Inst.* **28,** 121–59. *275, 277.*

—— *et al.,* 1962. The black flies (Diptera: Simuliidae) of Ontario. Part I. Adult identification and distribution with descriptions of six new species. *Proc. ent. Soc. Ont.* **92,** 70–154. *140, 155, 159.*

DAVIES, L., 1954. Observations on *Prosimulium ursinum* Edw. at Holandsfjord, Norway. *Oikos,* **5,** 94–8, *286, 402.*

—— 1960. The first instar larva of a species of *Prosimulium* (Diptera: Simuliidae). *Can. Ent.* **92,** 81–4. *186.*

—— 1961. Ecology of two *Prosimulium* species (Diptera) with reference to their ovarian cycles. Ibid. **93,** 1113–40. *286, 289.*

—— 1965. The structure of certain atypical Simuliidae (Diptera) in relation to evolution within the family, and the erection of a new genus for the Crozet Island black fly. *Proc. Linn. Soc. Lond.* **176,** 159–80. *186.*

—— 1966. The taxonomy of British black-flies (Diptera: Simuliidae). *Trans. R. ent. Soc. Lond.* **118,** 413–506. *201, 257.*

—— and SMITH, C. D., 1958. The distribution and growth of *Prosimulium* larvae (Diptera: Simuliidae) in hill streams in northern England. *J. Anim. Ecol.* **27,** 335–48. *387.*

DAVIS, G. E., and WARREN, C. E., 1965. Trophic relations of a sculpin in laboratory stream communities. *J. Wildl. Mgmt,* **29,** 846–71. *432.*

—— *et al.,* 1963. The influence of oxygen concentration on the swimming performance of juvenile Pacific salmon at various temperatures. *Trans. Am. Fish. Soc.* **92,** 111–24. *321–2.*

DAVIS, H. S., 1938. Instructions for conducting stream and lake surveys. *Fishery Circ. Bur. Fish.* **26,** 55 pp. *4, 247.*

DAZO, B. C., 1965. The morphology and natural history of *Pleurocera acuta* and *Goniobasis livescens* (Gastropoda: Cerithiacea: Pleuroceridae). *Malacologia,* **3,** 1–80. *276, 281.*

DEFFEYES, K. S., 1965. Carbonate equilibria: a graphic and algebraic approach. *Limnol. Oceanogr.* **10,** 412–26. *43.*

DEGRANGE, C., 1960. Recherches sur la reproduction des Ephéméroptères. *Trav. Lab. Hydrobiol. Piscic. Univ. Grenoble,* **50/1,** 7–194. *140.*

DEMEL, C., 1923. La faune hivernale des sources du lac de Wigry. *Annls Biol. lacustre,* **11,** 187–95. *400–1.*

DENDY, J. S., 1944. The fate of animals in stream drift when carried into lakes. *Ecol. Monogr.* **14,** 333–57. *268–9.*

—— and SCOTT, D. S., 1953. Distribution, life history, and morphological variations of the southern lamprey, *Ichthyomyzon gagei*. *Copeia,* **1953,** 152–62. *357.*

DENHAM, S. C., 1938. A limnological investigation of the West Fork and Common Branch of White River. *Invest. Indiana Lakes Streams,* **1,** 17–71. *99, 249, 259, 261.*

DES CILLEULS, J., 1928*a*. Revue générale des études sur le plancton des grandes fleuves ou rivières. *Int. Revue ges. Hydrobiol. Hydrogr.* **20,** 174–206. *105.*

—— 1928*b*. Le phytoplancton de la Loire dans la région Saumuroise. Ibid. **21,** 217–67. *101.*

—— 1932. Contribution nouvelle à l'étude du phytoplancton de la Loire. Ibid. **27,** 120–9. *105.*

DIMHIRN, I., 1953. Über die Strahlungsvorgänge in Fliessgewässer. *Wett. Leben Sonderh.* **2,** 52–63. *34.*

DINSMORE, J. J., 1962. Life history of the creek chub, with emphasis on growth. *Proc. Iowa Acad. Sci.* **69,** 296–301. *372.*

DITTMAR, H., 1951. Anwendungsmöglichkeiten der zerhackten Gleichstroms für ökologische Arbeiten. *Arch. Hydrobiol.* **45,** 217–23. *242.*

—— 1953. Die Bedeutung des Ca- und Mg-Gehaltes für die Fauna fliessender Gewässer. *Ber. limnol. Flussstn Freudenthal,* **4,** 20–3. *222.*

—— 1955*a*. Ein Sauerlandbach. Untersuchungen an einem Wiesen-Mittelgebirgsbach. *Arch. Hydrobiol.* **50,** 305–552. *149, 219, 221, 287, 390, 400.*

—— 1955*b*. Die quantitative Analyse des Fliesswasser-Benthos. Anregungen zu ihrer methodischen Anwendung und ihre praktische Bedeutung. Ibid. Suppl. **22,** 295–300. *237.*

DODDS, G. S., and HISAW, F. L., 1924*a*. Ecological studies of aquatic insects. I. Adaptations of mayfly nymphs to swift streams. *Ecology,* **5,** 137–48. *122, 128, 134.*

—— —— 1924*b*. Idem. II. Size of respiratory organs in relation to environmental conditions. Ibid. **5,** 262–71. *166.*

—— —— 1925*a*. Idem. III. Adaptations of caddisfly larvae to swift streams. Ibid. **6,** 123–37. *391.*

—— —— 1925*b*. Idem. IV. Altitudinal range and zonation of mayflies, stoneflies and caddisflies in the Colorado Rockies. Ibid. **6,** 380–90. *387, 391.*

DOLGOFF, G. I., 1929. Über die Ungleichartigkeit des Flusswassers. *Int. Revue ges. Hydrobiol. Hydrogr.* **22,** 371–412. *38.*

DORIER, A., 1937. La faune des eaux courantes alpines. *Verh. int. Verein. theor. angew. Limnol.* **8,** 33–41. *80, 201, 390, 402.*

DORIER, A., 1961. Sur la répartition des Simuliidae (Dipt.) des Alpes françaises. Ibid. **14,** 369–71. *257, 387, 410.*

—— and VAILLANT, F., 1954. Observations et expériences relatives à la resistance au courant de divers invertébrés aquatiques. *Trav. Lab. Hydrobiol. Piscic. Univ. Grenoble,* **45/6,** 9–31. *130, 132.*

—— —— 1955. Sur le facteur vitesse du courant. *Verh. int. Verein. theor. angew. Limnol.* **12,** 593–7. *149–50, 152.*

DORRIS, T. C., and COPELAND, B. J., 1962. Limnology of the middle Mississippi River. III. Mayfly populations in relation to navigation water-level control. *Limnol. Oceanogr.* **7,** 240–7. *261.*

—— *et al.,* 1963. Idem. IV. Physical and chemical limnology of river and chute. Ibid. **8,** 79–88. *29, 33, 40.*

DOUGLAS, B., 1958. The ecology of the attached diatoms and other algae in a stony stream. *J. Ecol.* **46,** 295–322. *57, 66, 69, 70, 193.*

DUDICH, E., 1958. Die Grundlagen der Fauna eines Karpaten-Flusses. *Acta zool. hung.* **3,** 179–200. *391.*

DUFFER, W. R., and DORRIS, T. C., 1966. Primary productivity in a southern Great Plains stream. *Limnol. Oceanogr.* **11,** 143–51. *413, 417.*

DUFFIELD, J. E., 1933. Fluctuations in numbers among freshwater crayfish, *Potamobius pallipes* Lereboullet. *J. Anim. Ecol.* **2,** 184–96. *231.*

DUNCAN, C. J., 1959. The life cycle and ecology of the freshwater snail *Physa fontinalis* (L.). Ibid. **28,** 97–117. *287.*

DZYUBAN, N. A., 1962. Reservoirs as a zoogeographical factor (Russian). *Trudy Zon. Sov. po. Tipol. i Biol. Obosn. Ryb. Ispol. Vnut. Vod. Yuzh. Zony S.S.S.R.* (Transactions of the zonal conference on the typology and biological basis of the use of inland bodies of water in commercial fishing in the southern zone of the U.S.S.R.) **1962,** 105–10 (Nat. lend. Libr., Boston Spa, Yorks, Engl. R.T.S. 2936). *448.*

EASTHAM, L. E. S., 1932. Currents produced by the gills of mayfly nymphs. *Nature, Lond.* **130,** 58. *166, 177–8.*

—— 1937. The gill movements of nymphal *Ecdyonurus venosus* (Ephemeroptera) and the currents produced by them in the water. *J. exp. Biol.* **14,** 219–28. *177.*

—— 1939. Gill movements of nymphal *Ephemera danica* (Ephemeroptera) and the water currents caused by them. Ibid. **16,** 18–33. *177.*

ECKEL, O., 1953. Zur Thermik der Fliesgewässer: Über die Änderung der Wassertemperatur entlang des Flusslaufes. *Wett. Leben Sonderh.* **2,** 41–47. *28–30.*

EDDY, S., 1931. The plankton of the Sangamon River in the summer of 1929. *Bull. Ill. St. nat. Hist. Surv.* **19,** 469–86. *99, 100.*

—— 1934. A study of fresh-water plankton communities. *Illinois biol. Monogr.* **12,** 4, 93 pp. *98–9.*

EDINGTON, J. M., 1965. The effect of water flow on populations of net-spinning Trichoptera. *Mitt. int. Verein. theor. angew. Limnol.* **13,** 40–8. *188, 200.*

—— 1966. Some observations on stream temperature. *Oikos.* **15,** 265–73. *28, 30, 204.*

—— and MOLYNEUX, L., 1960. Portable water velocity meter. *J. scient. Instrum.* **35,** 455–7. *5.*

EDMUNDS, G. F. 1956. Exuviation of subimaginal Ephemeroptera in flight. *Ent. News*, **67**, 91–3. *145*.

—— 1957. The predaceous mayfly nymphs of North America. *Proc. Utah Acad. Sci.* **34**, 23–4. *195*.

—— and TRAVER, J. R., 1959. The classification of the Ephemeroptera I. Ephemeroidea: Behningiidae. *Ann. ent. Soc. Am.* **52**, 43–51. *176*.

EDWARDS, R. W., 1958. The effect of larvae of *Chironomus riparius* Meigen on the redox potentials of settled activated sludge. *Ann. appl. Biol.* **46**, 457–64. *434*.

—— 1962. Some effects of plants and animals on the conditions in fresh-water streams with particular reference to their oxygen balance. *Int. J. Air Wat. Pollut.* **6**, 505–20. *51, 413, 434*.

—— and BROWN, M. W., 1960. An aerial photographic method for studying the distribution of aquatic macrophytes in shallow waters. *J. Ecol.* **48**, 161–3. *88*.

—— and OWENS, M., 1960. The effect of plants on river conditions. I. Summer crops and estimates of net productivity of macrophytes in a chalk stream. Ibid. **48**, 151–60. *416*.

—— —— 1962. Idem. IV. The oxygen balance of a chalk stream. Ibid. **50**, 207–20. *413*.

—— and ROLLEY, H. L. J., 1965. Oxygen consumption of river muds. Ibid. **53**, 1–19. *51*.

—— *et al.*, 1961. Estimates of surface aeration in two streams. *J. Instn Wat. Engrs*, **15**, 395–405. *412*.

EFFORD, I. E., 1960. A method of studying the vertical distribution of the bottom fauna in shallow waters. *Hydrobiologia*, **16**, 288–92. *237*.

—— 1965. Ecology of the watermite *Feltria romijni* Besseling. *J. Anim. Ecol.* **34**, 233–51. *433*.

EGGLISHAW, H. J., 1964. The distributional relationship between the bottom fauna and plant detritus in streams. Ibid. **33**, 463–76. *211–12, 239*.

—— 1965. Estimating and accounting for the quantity of bottom fauna in Highland streams. *Rep. Challenger Soc.* **3**, 18. *439–40*.

—— and MORGAN, N. C., 1965. A survey of the bottom fauna of streams in the Scottish Highlands Part II. The relationship of the fauna to the chemical and geological conditions. *Hydrobiologia*, **26**, 173–83. *251–2*.

EICHER, G. J., 1953. Aerial methods of assessing red salmon populations in Western Alaska. *J. Wildl. Mgmt*, **17**, 521–7. *358*.

EIDEL, K., 1933. Beiträge zur Biologie einiger Bäche des Schwarzwaldes, mit besonderer Berüchsichtigung der Insektenfauna der Elz und Kinzig. *Arch. Hydrobiol.* **25**, 543–615. *390, 400*.

EINSELE, W., 1957. Flussbiologie, Kraftwerke und Fischerei. *Schr. err.Öst Fischereiverb.* **1**, 1–63. *447*.

—— 1960. Die Strömungsgeschwindigkeit als beherrschender Faktor bei den limnologischen Gestaltung der Gewässer. *Österreichs Fischerei*, Suppl. **1**, **2**, 40 pp. *8, 12, 18–19, 49, 50, 108, 126, 146, 208*.

ELIAN, L., and PRUNESCU-ARION, E., 1964. Biocoenotische Untersuchungen im Felsenbereich der unteren Donau. *Arch. Hydrobiol.* Suppl. **27**, 457–63. *207*.

ELLIOTT, J. M., 1965*a*. Daily fluctuations of drift invertebrates in a Dartmoor stream. *Nature, Lond.* **205**, 1127–9. *262–3*.

—— 1965*b*. Invertebrate drift in a mountain stream in Norway. *Norsk ent. Tidsskr.* **13**, 97–9. *261*.

ELLIOTT, J. M., 1966. Downstream movements of trout fry (*Salmo trutta*) in a Dartmoor stream. *J. Fish. Res. Bd Can.* **23,** 157–9. *355.*

ELLIS, M. M., 1936. Erosion silt as a factor in aquatic environments. *Ecology,* **17,** 29–42. *444.*

ELLIS, R. J., 1961. A life history study of *Asellus intermedius* Forbes. *Trans. Am. microsc. Soc.* **80,** 80–102. *287.*

—— and GOWING, H., 1957. Relationship between food supply and condition of wild brown trout, *Salmo trutta* Linnaeus, in a Michigan stream. *Limnol. Oceanogr.* **2,** 299–308. *292, 367.*

EMBODY, G. C., 1928. Stocking policy for the streams, smaller lakes and ponds of the Oswego watershed. *Rep. N.Y. Conserv. Dep.* Suppl. **17,** 17–39. *247.*

ENĂCEANU, V., 1964. Das Donauplankton auf rumänischen Gebiet. *Arch. Hydrobiol.* Suppl. **27,** 442–56. *98–9, 109.*

ENGELHARDT, W., 1951. Faunistisch-ökologische Untersuchungen über Wasserinsekten an den südlichen Zuflüssen des Ammersees. *Mitt. münch. ent. Ges.* **41,** 1–135. *155, 223, 226, 390, 404, 447.*

—— 1957. Faunistisch-ökologische Untersuchungen an lappländischen Fliessgewässern. *Arch. Hydrobiol.* **53,** 499–519. *401.*

ENTZ, B., 1961. Die Beziehungen der Wasserchemie und primären Produktion in kleinen Bächen der Balatonsee-Gegend. *Verh. int. Verein. theor. angew. Limnol.* **14,** 495–8. *42, 51, 79, 387.*

ERIKSEN, C. H., 1964. The influence of respiration and substrate upon the distribution of burrowing mayfly naiads. Ibid. **15,** 903–11. *163, 166.*

—— 1966. Benthic invertebrates and some substrate-current-oxygen interrelationships. *Spec. Publs Pymatuning Lab. Fld Biol.* **4,** 98–115. *406.*

EYSTER, C., 1964. Micronutrient requirements for green plants, especially algae. *Algae and Man* (ed. D. F. Jackson). Plenum Press, New York, pp. 77–85. *46.*

FABRICIUS, E., 1951. The topography of the spawning bottom as a factor influencing the size of the territory in some species of fish. *Rep. Inst. Freshwat. Res. Drottningholm,* **32,** 43–9. *360.*

—— 1954. Aquarium observations on the spawning behaviour of the burbot, *Lota vulgaris* L. Ibid. **35,** 51–7. *360, 362.*

—— and LINDROTH, A., 1954. Experimental observations on the spawning of whitefish, *Coregonus laveratus* L., in the stream aquarium of the Hölle Laboratory at River Indalsälven. Ibid. **35,** 105–12. *348, 359, 361.*

FAHY, W. E., 1954. The life history of the northern greenside darter *Etheostoma blenniodes blenniodes* Rafinesque. *J. Elisha Mitchell scient. Soc.* **70,** 139–205. *360.*

FINDENEGG, I., 1959. *Die Gewässer Österreichs. Ein limnologischer Überblick.* Biologische Station, Lunz. *60, 80, 390, 400.*

FISHER, K., 1932. *Agriotypus armatus* (Walk.) (Hymenoptera) and its relations with its hosts. *Proc. zool. Soc. Lond.* **1932,** 451–61. *158, 172.*

FITTKAU, E. J., 1953. Odonaten aus der Fulda. *Ber. limnol. Flussstn Freudenthal,* **5,** 29–36. *387.*

—— 1964. Remarks on limnology of central-Amazon rain-forest streams. *Verh. int. Verein. theor. angew. Limnol.* **15,** 1092–6. *200, 221, 391, 411.*

FJERDINGSTAD, E., 1950. The microflora of the River Mølleaa with special reference to the relation of the benthal algae to pollution. *Folia limnol. Scand.* **5,** 124 pp. *105.*

FLEENER, G. C., 1951. Life history of the cut-throat trout, *Salmo clarki* Richardson, in Logan River, Utah. *Trans. Am. Fish. Soc.* **81**, 235–48. *362*.

[1]FONTAINE, J., 1964. Commensalisme et parasitisme chez les larves d'éphéméroptères. *Bull. mens. Soc. linn. Lyon*, **33**, 163–74. *180–2*.

FONTAINE, M., 1961. Action des eaux courantes sur certains aspects de la physiologie téléostéens. *Verh. int. Verein. theor. angew, Limnol.* **14**, 400–3. *355*.

[1]—— 1964. Les mécanismes physiologiques du comportement migratoire amphibiotique catadrome des poissons téléostéens. Ibid. **15**, 959–67. *355*.

FORD, J. B., 1962. The vertical distribution of larval Chironomidae (Dipt.) in the mud of a stream. *Hydrobiologia*, **19**, 262–72. *175*.

—— and HALL, R. E., 1958. A grab for quantitative sampling in stream muds. Ibid. **11**, 198–205. *237*.

FOX, H. M., and SIDNEY, J., 1953. The influence of dissolved oxygen on the respiratory movements of caddis larvae. *J. exp. Biol.* **30**, 235–7. *166*.

—— and SIMMONDS, B. G., 1933. Metabolic rates of aquatic arthropods from different habitats. Ibid. **10**, 67–74. *161*.

—— *et al.*, 1934. Metabolic rates of Ephemerid nymphs from swiftly flowing and from still waters. Ibid. **12**, 179–84. *161*.

—— *et al.*, 1937. The oxygen consumption of Ephemerid nymphs from flowing and from still water in relation to the concentration of oxygen in the water. Ibid. **14**, 210–18. *161–2*.

FREDEEN, F. J. H., 1958. Black flies (Diptera: Simuliidae) of the agricultural areas of Manitoba, Saskatchewan, and Alberta. *Int. Congr. Ent.* **10, 3**, 819–23. *273, 275, 405*.

—— 1959*a*. Rearing black flies in the laboratory (Diptera: Simuliidae). *Can. Ent.* **91**, 73–83. *277*.

—— 1959*b*. Collection, extraction, sterilization and low-temperature storage of black-fly eggs (Diptera: Simuliidae). Ibid. **91**, 450–3. *277*.

—— 1960. Bacteria as a source of food for black-fly larvae. *Nature, Lond.* **187**, 963. *193, 447*.

—— 1962. DDT and Heptachlor as black-fly larvicides in clear and turbid water. *Can. Ent.* **94**, 875–80. *186*.

—— 1964. Bacteria as food for blackfly larvae (Diptera: Simuliidae) in laboratory cultures and in natural streams. *Can. J. Zool.* **42**, 527–38. *193*.

—— *et al.*, 1951. Egg-laying habits, overwintering stages, and life-cycle of *Simulium arcticum* Mall. (Diptera: Simuliidae). *Can. Ent.* **83**, 73–6. *273, 277, 289*.

—— *et al.*, 1953. Adsorption of DDT on suspended solids in river water and its role in black-fly control. *Nature, Lond.* **171**, 700. *186*.

FREMLING, C. R., 1960*a*. Biology of a large mayfly, *Hexagenia bilineata* (Say), of the upper Mississippi River. *Res. Bull. Iowa agric. Exp. Stn.* **482**, 842–52. *286*.

—— 1960*b*. Biology and possible control of nuisance caddisflies of the upper Mississippi River. Ibid. **483**, 856–79. *158, 188, 447*.

—— 1964. Rhythmic *Hexagenia* mayfly emergences and the environmental factors which influence them. *Verh. int. Verein. theor. angew. Limnol.* **15**, 912–16. *286*.

FRISON, T. H., 1929. Fall and winter stoneflies, or Plecoptera, of Illinois. *Bull. Ill. St. nat. Hist. Surv.* **18**, 345–409. *169, 272, 278*.

1. The 1964 papers of Fontaine, J., and Fontaine, M., are distinguished as Fontaine, J. (1964) and Fontaine (1964) in the text.

FRITSCH, F. E., 1902. Preliminary report on the phytoplankton of the River Thames. *Ann. Bot.* **16,** 576–84. *101.*

—— 1903. Further observations on the phytoplankton of the River Thames. Ibid. **17,** 631–47. *101.*

—— 1905. The plankton of some English rivers. Ibid. **19,** 163–7. *101.*

—— 1929. The encrusting algal communities of certain fast flowing streams. *New Phytol.* **28,** 165–96. *61, 74.*

—— 1949. The lime-encrusting *Phormidium* community of British streams. *Verh. int. Verein. theor. angew. Limnol.* **10,** 141–4. *76.*

—— 1950. *Phormidium incrustatum* (Naeg.) Gom., an important member of the lime-encrusted communities of flowing water. *Biol. Jaarb.* **17,** 27–39. *76.*

FROHNE, W. C., 1956. The provendering role of the larger aquatic plants. *Ecology,* **37,** 387–8. *438.*

FROST, W. E., 1939. River Liffey survey II—The food consumed by the brown trout (*Salmo trutta* Linn.) in acid and alkaline waters. *Proc. R. Ir. Acad.* **45B,** 139–206. *332–3.*

—— 1942. Idem, IV—The fauna of the submerged 'mosses' in an acid and an alkaline water. Ibid. **47B,** 293–369. *241, 271, 276, 292.*

—— 1943. The natural history of the minnow, *Phoxinus phoxinus. J. Anim. Ecol.* **12,** 139–62. *310, 349, 359, 366–7.*

—— 1945*a*. River Liffey survey VI—Discussion on the results obtained from investigations on the food and growth of brown trout (*Salmo trutta* L.) in alkaline and acid waters. *Proc. R. Ir. Acad.* **50B,** 321–42. *332–3.*

—— 1945*b*. The age and growth of eels (*Anguilla anguilla*) from the Windermere catchment area. Parts I and II. *J. Anim. Ecol.* **14,** 23–36 and 106–24. *345, 355.*

—— 1946. Observations on the food of eels (*Anguilla anguilla*) from the Windermere catchment area. Ibid. **15,** 43–53. *367.*

—— 1950. The growth and food of young salmon (*Salmo salar*) and trout (*S. trutta*) in the River Forss, Caithness. Ibid. **19,** 147–58. *373.*

—— 1952. Predators on the eggs of char in Windermere. *Salm. Trout Mag.* **1952,** 193–7. *348.*

—— and WENT, A. E. J., 1940. River Liffey survey III—The growth and food of young salmon. *Proc. R. Ir. Acad.* **46B,** 53–80. *355.*

FRY, F. E. J., *et al.,* 1946. Lethal temperature relations for a sample of young speckled trout, *Salvelinus fontinalis. Publs Ont. Fish. Res. Lab.* **66,** 1–35. *322–3.*

FRYER, G., 1959. The trophic interrelationship and ecology of some littoral communities of Lake Nyasa with especial reference to the fishes, and a discussion of the evolution of a group of rock-frequenting Cichlidae. *Proc. zool. Soc. Lond.* **132,** 153–281. *112.*

FUNK, J. L., 1949. Wider application of the electrical method of collecting fish. *Trans Am. Fish. Soc.* **77,** 49–60. *316.*

—— 1957*a*. Movement of stream fishes in Missouri. Ibid. **85,** 39–57. *343.*

—— 1957*b*. The Missouri Conservation Commission's electric seine. *Symposium on evaluation of fish populations in warm-water streams.* Iowa St. Coll, Ames. mimeo. rep. *316.*

—— 1958. Relative efficiency and selectivity of gear used in the study of stream fish populations. *Trans. N. Am. Wildl. Conf.* **23,** 236–48. *317.*

—— and CAMPBELL, R. S., 1953. The population of larger fishes in Black River, Missouri. *Univ. Mo Stud.* **26, 2,** 69–82. *317, 385.*

GAJEVSKAJA, N. S., 1958. Le rôle des groups principaux de la flore aquatique dans les cycles trophiques des différents bassins d'eau douce. *Verh. int. Verein. theor. angew. Limnol.* **13**, 350–62. *70, 192, 438.*

GALTSOFF, P. S., 1924. Limnological observations in the Upper Missouri. *Bull. Bur. Fish., Wash.* **39**, 347–438. *100.*

GAMESON, A. L. H., 1957. Weirs and the aeration of rivers. *J. Instn Wat. Engrs,* **11**, 477–90. *42.*

GARD, R., 1961. Effects of beaver on trout in Sagehen Creek, California. *J. Wildl. Mgmt,* **25**, 221–42. *353, 381–2.*

—— 1963. Insulation of a Sierra stream by snow cover. *Ecology,* **44**, 194–7. *31.*

GAUFIN, A. R., 1959. Production of bottom fauna in the Provo River, Utah. *Iowa St. Coll. J. Sci.* **33**, 395–419. *205, 227, 255, 292, 391, 447.*

—— *et al.*, 1956. A statistical evaluation of stream bottom sampling data obtained from three standard samplers. *Ecology,* **37**, 643–8. *249.*

GAUFIN, R. F., and GAUFIN, A. R., 1961. The effect of low oxygen concentrations on stoneflies. *Proc. Utah Acad. Sci.* **38**, 57–64. *167.*

GAY, P. A., 1960. Ecological studies of *Eichhornia crassipes* Solms. in the Sudan I. Analysis of spread in the Nile. *J. Ecol.* **48**, 183–91. *93.*

—— and BERRY, L., 1959. The water hyacinth: a new problem on the Nile. *Geogrl J.* **125**, 89–91. *93.*

GEE, J. H., and NORTHCOTE, T. G., 1963. Comparative ecology of two sympatric species of dace (*Rhinicthys*) in the Fraser River system, British Columbia. *J. Fish. Res. Bd Can.* **20**, 105–18. *306.*

GEIJSKES, D. C., 1935. Faunistisch-ökologische Untersuchungen am Röserenbach bei Leistal im Basler Tafeljura. *Tijdschr. Ent.* **78**, 249–382. *222, 233, 387, 390, 400.*

—— 1942. Observations on temperature in a tropical river. *Ecology,* **23**, 106–10. *29.*

GEITLER, L., 1927. Über Vegetationsfärbungen in Bächen. *Biologia gen.* **3**, 791–814. *57.*

GELDIAY, R., 1956. Studies on local populations of the freshwater limpet *Ancylus fluviatilis Müller. J. Anim. Ecol.* **25**, 389–402. *180, 206, 275.*

GERKING, S. D., 1945. Distribution of the fishes of Indiana. *Invest. Indiana Lakes Streams,* **3**, 1–137. *328, 335, 385, 444.*

—— 1949. Characteristics of stream fish populations. Ibid. **3**, 285–309. *338–9.*

—— 1950. Stability of a stream fish population. *J. Wildl. Mgmt,* **14**, 193–202. *343.*

—— 1953. Evidence for the concepts of home range and territory in stream fishes. *Ecology,* **34**, 347–65. *343–4.*

—— 1957. A method of sampling the littoral macrofauna and its application. Ibid. **38**, 219–26. *238.*

—— 1959. The restricted movements of fish populations. *Biol. Rev.* **34**, 221–42. *311, 343–4, 351.*

—— 1962. Production and food utilisation in a population of blue gill sunfish. *Ecol. Monogr.* **32**, 31–78. *244, 419.*

GERSBACHER, W. M., 1937. The development of stream bottom communities in central Illinois. *Ecology,* **18**, 359–90. *328, 391, 404.*

GESSNER F., 1937. Untersuchungen über die Assimilation und Atmung submerser Wasserpflanzen. *Jb. wiss. Bot.* **85**, 267–328. *82.*

GESSNER, F., 1950. Die ökologische Bedeutung der Strömungsgeschwindigkeit fliessender Gewässer und ihre Messung auf kleinstem Raum. *Arch. Hydrobiol.* **43**, 159–65. *5, 89.*

—— 1955*a*. Die limnologischen Verhältnisse in den Seen und Flüssen von Venezuela. *Verh. int. Verein. theor. angew. Limnol.* **12**, 284–95. *60, 61, 81, 99.*

—— 1955*b*. *Hydrobotanik I. Energiehaushalt.* Veb. Deutsch. Ver. Wissensch., Berlin. *5, 6, 53–4, 61, 70, 75, 81–3, 87–8, 90–92, 99.*

—— 1959. Idem *II. Stoffhaushalt.* Veb. Deutsch. Ver. Wissensch., Berlin. *53, 75–7, 79, 86, 92.*

—— 1960. Limnologische Untersuchungen am Zusammenfluss des Rio Negro und des Amazonas (Solimoes). *Int. Revue ges. Hydrobiol. Hydrogr.* **45**, 55–79. *37–8, 44.*

—— 1961. Der Sauerstoffhaushalt des Amazonas. Ibid. **46**, 542–61. *23, 26, 28, 33, 40, 50.*

—— 1963. *Hydrobotanik III. Biozonotik.* Veb. Deutsch, Ver. Wissensch., Berlin. *53.*

—— 1965. Zur Limnoligie des unteren Orinoco. *Int. Revue ges. Hydrobiol. Hydrogr.* **50**, 305–33. *33, 38.*

—— and PANNIER, F., 1958. Der Sauerstoffverbrauch der Wasserpflanzen bei verschiedenen Sauerstoffspannungen. *Hydrobiologia*, **10**, 323–51. *65, 82, 162.*

GILLIES, M. T., 1954. The adult stages of *Prosopistoma* Latreille (Ephemeroptera), with descriptions of two new species from Africa. *Trans. R. ent. Soc. Lond.* **105**, 355–72. *123, 145, 291.*

GIUDICELLI, J., 1964. L'oviposition chez les Blépharocérides (Diptera). *Revue fr. Ent.* **31**, 116–19. *159.*

GLEDHILL, T., 1959. The life-history of *Ameletus inopinatus* (Siphlonuridae, Ephemeroptera). *Hydrobiologia*, **14**, 85–90. *289.*

—— 1960. The Ephemeroptera, Plecoptera and Trichoptera caught by emergence traps in two streams during 1958. Ibid. **15**, 179–88. *282–3.*

GOLDMAN, C. R., 1961. The contribution of alder trees (*Alnus tenuifolia*) to the primary productivity of Castle Lake, California. *Ecology*, **42**, 282–8. *441.*

GOODRICH, C., 1937. Studies of the gastropod family Pleuroceridae—VI. *Occ. Pap. Mus. Zool. Univ. Mich.* **347**, 1–12. *142.*

—— 1941. Distribution of the gastropods of the Cahaba River, Alabama. Ibid. **428**, 1–30. *143.*

—— and VAN DER SCHALIE, H., 1944. A revision of the Mollusca of Indiana. *Am. Midl. Nat.* **32**, 257–326. *389, 394.*

GORBUNOV, K. V., 1955. Dynamics of overgrowth in the belts of the lower zone of the Volga Delta and their role in the nutrition of young carp (Russian). *Trudy vses. gidrobiol. Obshch.* **6**, 80–103. *62.*

GORHAM, E., 1955. On the acidity and salinity of rain. *Geochim. cosmochim. Acta*, **7**, 231–9. *36.*

—— 1961. Factors influencing supply of major ions to inland waters, with special reference to the atmosphere. *Bull. geol. Soc. Am.* **72**, 795–840. *36.*

GRAHAM, J. M., 1949. Some effects of temperature and oxygen pressure on the metabolism and activity of the speckled trout, *Salvelinus fontinalis. Can. J. Res.* **27**, 270–88. *320.*

GRAY, E., 1952. The ecology of the ciliate fauna of Hobson's Brook, a Cambridgeshire chalk stream. *J. gen. Microbiol.* **6**, 108–22. *114.*

GREEN, J., 1960. Zooplankton of the River Sokoto. The Rotifera. *Proc. zool. Soc. Lond.* **135**, 491–523. *106.*

—— 1962. Idem. The Crustacea. Ibid. **138**, 415–53. *106.*

—— 1963. Idem. The Rhizopoda Testacea. Ibid. **141**, 497–514. *106.*

GREENBERG, A. E., 1964. Plankton of the Sacramento River. *Ecology*, **45**, 40–9. *97, 99.*

GRENIER, P., 1949. Contribution à l'étude biologique des Simuliides de France. *Physiologia comp. Oecol.* **1**, 165–330. *5, 138, 152, 186, 198, 200, 202, 220, 275, 286–7, 289, 387.*

GREZE, I. I., 1953. Hydrobiology of the lower part of the River Angara (Russian). *Trudy vses. gidrobiol. Obshch.* **5**, 203–11. *98, 117, 215–16, 367, 396, 416.*

GRIFFITH, R. E., and VOORHEES, D. J., 1960. Ingestion of available phytoplankton by forage fish from Catoctin Creek, Maryland. *Trans. Am. microsc. Soc.* **79**, 309–15. *366.*

GRIMÅS, U., 1963. Reflections on the availability and utilization degree of bottom animals as fish food. *Zool. Bidr. Upps.* **35**, 497–503. *248.*

GROVER, N. C., and HARRINGTON, W. H., 1943. *Stream Flow Measurements, Records and their Uses.* Wiley, New York. *7.*

GRZENDA, A. R., and BREHMER, M. L., 1960. A quantitative method for the collection and measurement of stream periphyton. *Limnol. Oceanogr.* **5**, 190–4. *58.*

GUMTOW, R. B., 1955. An investigation of the periphyton in a riffle of the West Gallatin River, Montana. *Trans. Am. microsc. Soc.* **74**, 278–92. *58, 70, 71.*

GUNNING, G. E., 1959. The sensory basis for homing in the longear sunfish *Lepomis megalotis megalotis* (Rafinesque). *Invest. Indiana Lakes Streams,* **5**, 103–30. *344, 352.*

—— and LEWIS, W. M., 1956. Age and growth of two important bait species in a cold-water stream in southern Illinois. *Am. Midl. Nat.* **55**, 118–20. *340.*

—— and SHOOP, C. R., 1962. Restricted movements of the American eel, *Anguilla rostrata* (Le Sueur), in freshwater streams, with comments on growth rate. *Tulane Stud. Zool.* **9**, 265–72. *343.*

HAGELIN, L. O., and STEFFNER, N., 1958. Notes on the spawning habits of the river lamprey (*Petromyzon fluviatilis*). *Oikos*, **9**, 221–38. *357.*

HALL, G. E., 1955. Preliminary observations on the presence of stream-inhabiting fishes in Tenkiller Reservoir, a new Oklahoma impoundment. *Proc. Okla. Acad. Sci.* **34**, 34–40. *448.*

HALL, R. E., 1951. Comparative observations on the chironomid fauna of a chalk stream and a system of acid streams. *J. Soc. Br. Ent.* **3**, 253–62. *219.*

—— 1960. The Chironomidae of three chalk streams in southern England. *Int. Congr. Ent.* **11**, **1**, 178–81. *222, 273, 275.*

HALLAM, J. C., 1959. Habitat and associated fauna of four species of fish in Ontario streams. *J. Fish. Res. Bd Can.* **16**, 147–73. *338, 385.*

HAMILTON, A., 1940. The New Zealand Dobson fly (*Archichauliodes diversus* Walk); life history and bionomics. *N.Z. Jl Sci. Technol.* **22**, 45–55. *160, 276.*

HAMILTON, J. D., 1961. The effect of sand-pit washings on a stream fauna. *Verh. int. Verein. theor. angew. Limnol.* **14**, 435–9. *444.*

HAMILTON, W. J., 1932. The food and feeding habits of some eastern salamanders. *Copeia*, **1932**, 83–6. *377–8.*

HAMMER, L., 1965. Phytosynthese und Primärproduktion im Rio Negro. *Int. Revue ges. Hydrobiol. Hydrogr.* **50**, 335–9. *417*.

HANKINSON, T. L., 1910. An ecological study of the fish of a small stream. *Trans. Ill. St. Acad. Sci.* **3**, 23–31. *368*.

—— 1919. Notes on life-histories of Illinois fish. Ibid. **12**, 132–50. *356–9*.

—— 1930. Breeding behavior of the silverfin minnow, *Notropis whipplii spilopterus* (Cope). *Copeia*, **1930**, 73–4. *360*.

—— 1932. Observations on the breeding behavior and habitats of fishes in southern Michigan. *Pap. Mich. Acad. Sci.* **15**, 411–25. *358–9*.

HARDISTY, M. W., 1961*a*. Studies on an isolated spawning population of the brook lamprey (*Lampetra planeri*). *J. Anim. Ecol.* **30**, 339–55. *362*.

—— 1961*b*. The growth of larval lampreys. Ibid. **30**, 357–71. *363*.

HARKER, J. E., 1952. A study of the life cycles and growth-rates of four species of mayflies. *Proc. R. ent. Soc. Lond.* **A 27**, 77–85. *271, 284, 286, 288*.

—— 1953. An investigation of the distribution of the mayfly fauna of a Lancashire stream. *J. Anim. Ecol.* **22**, 1–13. *152, 226, 229*.

—— 1954. The Ephemeroptera of eastern Australia. *Trans. R. ent. Soc. Lond.* **105**, 241–68. *123, 133*.

HARRINGTON, R. W., 1947. The breeding behavior of the bridled shiner, *Notropis bifrenatus. Copeia*, **1947**, 186–92. *360*.

—— 1957. Sexual periodicity of the cyprinid fish *Notropis bifrenatus* (Cope), in relation to the phases in its annual reproductive cycle. *J. exp. Zool.* **135**, 529–55. *362*.

HARRIS, E. K., 1957. Further results in the statistical analysis of stream sampling. *Ecology*, **38**, 463–8. *236*.

HARRISON, A. D., 1958. Hydrobiological studies on the Great Berg River, Western Cape Province. Part 2. Quantitative studies on sandy bottoms, notes on tributaries and further information on the fauna, arranged systematically. *Trans. R. Soc. S. Afr.* **35**, 227–76. *227, 405*.

—— 1963. An ecological survey of the Great Berg River. *Monographiae biol.* **14**, 143–58. *391*.

—— 1965*a*. Geographical distribution of riverine invertebrates in Southern Africa. *Arch. Hydrobiol.* **61**, 387–94. *117*.

—— 1965*b*. River zonation in Southern Africa. Ibid. **61**, 380–6. *391, 393, 402*.

—— 1966. Recolonisation of a Rhodesian stream after drought. Ibid. **62**, 405–21. *405, 406*.

—— and AGNEW, J. D., 1962. The distribution of invertebrates endemic to acid streams in the Western and Southern Cape Province. *Ann. Cape Prov. Mus.* **2**, 273–91. *119*.

—— and ELSWORTH, J. F., 1958. Hydrobiological studies on the Great Berg River, Western Cape Province. Part I. General description, chemical studies and main features of the flora and fauna. *Trans. R. Soc. S. Afr.* **35**, 125–226. *44, 80, 86, 208, 227, 293, 391*.

—— and FARINA, T. D. W., 1965. A naturally turbid water with deleterious effects on the egg capsules of planorbid snails. *Ann. trop. Med. Parasit.* **59**, 327–30. *231*.

—— *et al.*, 1963. Hydrobiological studies on the Vaal River in the Vereeniging Area. Part II. The chemistry, bacteriology and invertebrates of the bottom muds. *Hydrobiologia*, **21**, 66–89. *210*.

HARROD, J. J., 1964. The distribution of invertebrates on submerged aquatic plants in a chalk stream. *J. Anim. Ecol.* **33,** 335–48. *215.*

HARTLAND-ROWE, R., 1964. Factors influencing the life histories of some stream insects in Alberta. *Verh. int. Verein. theor. angew. Limnol.* **15,** 917–25. *231, 272, 283–4, 292, 296.*

HARTLEY, J. C., 1962. The life history of *Trocheta subviridis* Dutrochet. *J. Anim. Ecol.* **31,** 519–24. *276.*

HARTLEY, P. H. T., 1947. The natural history of some British freshwater fishes. *Proc. zool. Soc. Lond.* **117,** 129–206. *366–7, 373.*

—— 1948. Food and feeding relationships in a community of fresh-water fishes. *J. Anim. Ecol.* **17,** 1–14. *373.*

HARTMAN, C. F., 1963. Observations on behaviour of juvenile brown trout on a stream aquarium during winter and spring. *J. Fish. Res. Bd. Can.* **20,** 769–87. *335.*

HARTMAN, R. T., 1960. Algae and metabolites of natural waters. *Spec. Publs Pymatuning Lab. Fld. Biol.* **2,** 38–55. *60.*

—— 1965. Composition and distribution of phytoplankton communities in the Upper Ohio River. Ibid. **3,** 45–65. *99, 105, 109.*

—— and HIMES, C. L., 1961. Phytoplankton from the Pymatuning Reservoir in downstream areas of the Shenango River. *Ecology,* **42,** 180–3. *102.*

HASLER, A. D., 1954. Odour perception and orientation in fishes. *J. Fish Res. Bd Can.* **11,** 107–29. *352.*

—— 1960. Guideposts of migrating fishes. *Science N.Y.* **132,** 785–92. *351–2.*

—— and WISBY, W. J., 1958. The return of displaced largemouth bass and green sun-fish to a 'home' area. *Ecology,* **39,** 289–93. *344.*

—— *et al.*, 1958. Sun-orientation and homing in fishes. *Limnol. Oceanogr.* **3,** 353–61. *352.*

HAYNE, D. W., and BALL, R. C., 1956. Benthic productivity as influenced by fish predation. Ibid. **1,** 162–75. *420.*

HAZARD, T. P., and EDDY, S., 1951. Modification of the seasonal cycle in brook trout (*Salvelinus frontinalis*) by control of light. *Trans. Am. Fish. Soc.* **80,** 158–62. *362.*

HAZZARD, A. S., 1932. Some phases of the life history of the eastern brook trout, *Salvelinus fontinalis* Mitchill. Ibid. **62,** 344–50. *357.*

—— 1933. Low water temperature, a limiting factor in the successful production of trout in natural waters. Ibid. **63,** 204–7. *324.*

—— 1935. Instructions for lake and stream survey work. *Mimeo. Circular, U.S. Bur. Fish.* 35 pp. *247.*

HENDELBERG, J., 1960. The freshwater pearl mussel, *Margaritifera margaritifera* (L.). *Rep. Inst. Freshwat. Res. Drottningholm,* **41,** 149–71. *221, 276.*

HESS, A. D., and SWARTZ, A., 1941. The forage ratio and its use in determining the food grade of streams. *Trans. N. Am. Wildl. Conf.* **5,** 162–4. *246, 373.*

HESTER, F. E., and DENDY, J. S., 1962. A multiple plate sampler for aquatic macroinvertebrates. *Trans. Am. Fish. Soc.* **91,** 420–1. *239.*

HEUSS, K., 1966. Beitrag zur Fauna der Werra, einem salinaren Binnengewässer. *Gewäss. Abwäss.* **43,** 48–64. *218.*

HEYWOOD, J., and EDWARDS, R. W., 1962. Some aspects of the ecology of *Potamopyrgus jenkinsi* Smith. *J. Anim. Ecol.* **31,** 239–50. *153.*

HILMY, A. M., 1962. Experimente zur Atmungsphysiologie von Ephemeropteren-larven. *Int. Congr. Ent.* **11,** 3, 254–8. *167.*

HINTON, H. E., 1947. The gills of some aquatic beetle pupae (Coleoptera, Psephenidae). *Proc. R. ent. Soc. Lond.* **A 22,** 52–60. *172.*

—— 1952. Survival of a chironomid larva after 20 months dehydration. *Int. Congr. Ent.* **9, 1,** 478–82. *279.*

—— 1953. Some adaptations of insects to environments that are alternately dry and flooded, with some notes on the habits of the Stratiomyidae. *Trans. Soc. Br. Ent.* **11,** 209–27. *172.*

—— 1958. The spiracular gills of insects. *Int. Congr. Ent.* **10, 1,** 543–8. *172.*

—— 1962. The structure and function of the spiracular gills of *Deuterophlebia* (Deuterophlebiidae) in relation to those of other Diptera. *Proc. zool. Soc. Lond.* **138,** 111–22. *172.*

—— 1964. The respiratory efficiency of the spiracular gill of *Simulium. J. Insect Physiol.* **10,** 73–80. *172.*

HIRSCH, A., 1958. Biological evaluation of organic pollution of New Zealand streams. *N.Z. Jl Sci. Technol.* **1,** 550–3. *119.*

HOAR, W. S., 1942. Diurnal variations in feeding activity of young salmon and trout. *J. Fish. Res. Bd Can.* **6,** 90–101. *313.*

—— 1951. The behaviour of chum, pink and coho salmon in relation to their seaward migration. Ibid. **8,** 241–63. *309, 313, 346, 354.*

HOBBS, D. F., 1940. Natural reproduction of trout in New Zealand and its relation to density of population in certain New Zealand waters. *Fish. Bull. N.Z.* **8,** 93 pp. *359.*

—— 1948. Trout fisheries in New Zealand, their development and management. Ibid. **9,** 159 pp. *443, 448.*

HOESTLANDT, H., 1948. Recherches sur la biologie de l'*Eriocheir sinensis,* H. Milne-Edwards (Crustacé Brachyoure). *Annls Inst. océonogr., Monaco,* **24,** 1–116. *153, 276.*

HOLDEN, M. J., and GREEN, J., 1960. The hydrology and plankton of the River Sokoto. *J. Anim. Ecol.* **29,** 65–84. *106, 110.*

HOLDSWORTH, R. P., 1941. The life history and growth of *Pteronarcys proteus* Newman. *Ann. ent. Soc. Am.* **34,** 495–502. *276, 278.*

HÖLL, K., 1955. Chemische Untersuchungen an kleinen Fliessgewässern. *Verh. int. Verein. theor. angew. Limnol.* **12,** 360–72. *37, 40, 49.*

HOLTON, G. D., 1953. A trout population study on a small creek in Gallatin County, Montana. *J. Wildl. Mgmt,* **17,** 62–82. *387.*

HOOPES, D. T., 1961. Utilization of mayflies and caddisflies by some Mississippi River fishes. *Trans. Am. Fish. Soc.* **89,** 32–3. *367.*

HOOVER, E. E., 1936. The spawning activities of fresh water smelt, with special reference to the sex ratio. *Copeia,* **1936,** 85–91. *362.*

—— 1938. Fish populations of primitive brook trout streams in New Hampshire. *Trans. N. Am. Wildl. Conf.* **3,** 486–96. *385.*

—— and HUBBARD, H. E., 1937. Modification of the sexual cycle in trout by control of light. *Copeia,* **1937,** 206–10. *362.*

HORA, S. L., 1922. Structural modifications in fish of mountain torrents. *Rec. Indian Mus.* **24,** 31–61. *304–6.*

—— 1923. Observations on the fauna of certain torrential streams in the Khasi Hills. Ibid. **25,** 579–600. *117, 146, 306, 349, 391.*

—— 1928. Animal life in torrential streams. *J. Bombay nat. Hist. Soc.* **32,** 111–26. *122, 131, 137, 178, 193.*

—— 1930. Ecology, bionomics and evolution of the torrential fauna, with special reference to the organs of attachment. *Phil. Trans. R. Soc.* **B218,** 171–282. *81, 122, 129–30, 132, 134, 137, 158, 304–7, 378, 391.*

——1936. Nature of the substratum as an important factor in the ecology of torrential fauna. *Proc. natn. Inst. Sci. India,* **2,** 45–7. *131.*

HORNUFF, L. E., 1957. A survey of four Oklahoma streams with reference to production. *Rep. Okla. Fish. Res. Lab.* **62,** 1–22. *200.*

HORNUNG, H., 1959. Floristisch-ökologische Untersuchungen an der Echaz unter besonderer Berüchsichtung der Verunreinigung durch Abwässer. *Arch. Hydrobiol.* **55,** 52–126. *71.*

HORTON, P. A., 1961. The bionomics of brown trout in a Dartmoor stream. *J. Anim. Ecol.* **30,** 311–38. *260–1, 420–2, 425, 427, 429.*

HOSKIN, C. M., 1959. Studies of oxygen metabolism of streams of North Carolina. *Publs Inst. mar. Sci. Univ. Tex.* **6,** 186–92. *413.*

HOWARD, A. D., 1951. A river mussel parasitic on a salamander. *Chicago Acad. Sci. nat. Hist. Misc.* **77,** 6 pp. *337, 377, 389.*

HUBAULT, E., 1927. Contribution à l'étude des invertébrés torrenticoles. *Bull. biol. Fr. Belg.* Suppl. **9,** 390 pp. *7, 40, 122, 127, 130, 136, 152, 156–7, 159, 273, 390, 402.*

HUBBS, C., 1964. Effects of thermal fluctuations on the relative survival of greenthroat darter young from stenothermal and eurythermal waters. *Ecology* **45,** 376–79. *323.*

—— and STRAWN, K., 1957. The effects of light and temperature on the fecundity of the greenthroat darter, *Etheostoma lepidum.* Ibid. **38,** 596–602. *360, 362.*

HUBBS, C. L., 1941. The relation of hydrological conditions to speciation in fishes. *A Symposium on Hydrobiology.* Univ. Wisconsin Press, Madison, pp. 182–95. *302, 304–5.*

——and LAGLER, K. F., 1958. Fishes of the Great Lakes region. *Bull. Cranbrook Inst. Sci.* **26,** 213 pp. *328, 346.*

—— and WALKER, B. W., 1942. Habitat and breeding behavior of the American cyprinid fish *Notropis longirostris. Copeia,* **1942,** 101–4. *305, 335, 360.*

HUET, M., 1938. Hydrobiologie piscicole du bassin moyen de la Lesse. *Mém. Mus. r. Hist. nat. Belg.* **82,** 1–128. *390.*

——1949. Aperçu des relations entre la pente et les populations piscicoles des eaux courantes. *Schweiz. Z. Hydrol.* **11,** 333–51. *383.*

—— 1951. Nocivité des boisements en Epicéas (*Picea excelsa* Link.) pour certains cours d'eau de l'Ardenne Belge. *Verh. int. Verein. theor. angew. Limnol.* **11,** 189–200. *231.*

—— 1954. Biologie, profils en long et en travers des eaux courantes. *Bull. fr. Piscic.* **175,** 41–53. *338, 383.*

—— 1959. Profiles and biology of Western European streams as related to fish management. *Trans. Am. Fish. Soc.* **88,** 155–63. *383–4.*

—— 1961. Reproduction et migrations de la truite commune (*Salmo trutta fario* L.) dans un ruisselet salmonicole de l'Ardenne belge. *Verh. int. Verein. theor. angew. Limnol.* **14,** 757–62. *349.*

—— 1962. Influence du courant sur la distribution des poissons dans les eaux courantes. *Schweiz. Z. Hydrol.* **24,** 413–32. *322, 324, 327, 383.*

HUFFORD, T. L., 1965. A comparison of photosynthetic yields in the Maumee River, Steidtmann's pond, and Urschel's quarry under natural conditions. *Ohio J. Sci.* **65,** 176–82. *413.*

HUGHES, D. A., 1966*a*. Mountain streams of the Barberton area, eastern Transvaal. Part I. A survey of the fauna. *Hydrobiologia*, **27**, 401–38. *230*.

—— 1966*b*. Idem. Part II. The effect of vegetational shading and direct illumination on the distribution of stream fauna. Ibid. **27**, 439–59. *230*.

HUMPHRIES, C. F., and FROST, W. E., 1937. River Liffey Survey. The Chironomid fauna of the submerged mosses. *Proc. R. Ir. Acad.* **43 B**, 161–81. *292*.

HUNT, B. P., 1953. The life history and economic importance of a burrowing mayfly, *Hexagenia limbata*, in southern Michigan lakes. *Bull. Inst. Fish. Res. Univ. Mich.* **4**, 1–151. *288*.

HUNTER, G. W., *et al.*, 1941. An attempt to evaluate the effect of stream improvement in Connecticut. *Trans. N. Am. Wildl. Conf.* **5**, 276–91. *253, 450*.

HUNTER, W. R., 1953*a*. The condition of the mantle cavity in two pulmonate snails living in Loch Lomond. *Proc. R. Soc. Edinb.* **B65**, 143–65. *167*.

—— 1953*b*. On the growth of the fresh-water limpet, *Ancylus fluviatilis* Müller. *Proc. zool. Soc. Lond.* **123**, 623–36. *132, 275*.

—— 1961. Annual variations in growth and density in natural populations of freshwater snails in the west of Scotland. Ibid. **136**, 219–53. *422*.

HUNTSMAN, A. G., 1945. Freshets and fish. *Trans. Am. Fish. Soc.* **75**, 257–66. *349, 449*.

—— 1948. Fertility and fertilisation of streams. *J. Fish. Res. Bd Can.* **7**, 248–53. *69*.

HUSSON, R., 1962. Les ressources alimentaires des animaux cavernicoles. *Cah. Etud. biol. Lyon*, **8–9**, 103–16. *436*.

HUTCHINSON, G. E., 1957. *A Treatise on Limnology*, vol. i. Wiley, New York. *vii, 28, 31, 36, 43, 46, 68*.

HUTCHINSON, L., 1939. Some factors influencing zooplankton distribution in the Hocking River. *Ohio J. Sci.* **39**, 259–73. *98*.

HYNES, H. B. N., 1941. The taxonomy and ecology of the nymphs of British Plecoptera, with notes on the adults and eggs. *Trans. R. ent. Soc. Lond.* **91**, 459–557. *118, 140, 145, 147, 159, 167, 177, 183, 192–4, 207, 223, 234, 272, 276–9, 286–7*.

—— 1948. Notes on the aquatic Hemiptera-Heteroptera of Trinidad and Tobago, B.W.I., with a description of a new species of *Martarega* B. White (Notonectidae). Ibid. **99**, 341–60. *116–17, 175, 387*.

—— 1950. The food of fresh-water sticklebacks (*Gasterosteus aculeatus* and *Pygosteus pungitius*) with a review of methods used in studies of the food of fishes. *J. Anim. Ecol.* **19**, 36–58. *357, 365, 373–4*.

—— 1952*a*. The Plecoptera of the Isle of Man. *Proc. R. ent. Soc. Lond.* **A27**, 71–6. *233*.

—— 1952*b*. The Neoperlinae of the Ethiopian Region (Plecoptera: Perlidae). *Trans. R. ent. Soc. Lond.* **103**, 85–108. *117*.

—— 1952*c*. Perlidae (Plecoptera). *Mission G. F. de Witte. Parc National de l'Upemba*, Brussels, **8**, 3–11. *291*.

—— 1953. The Plecoptera of some small streams near Silkeborg, Jutland. *Ent. Meddr*, **26**, 489–94. *219*.

—— 1954. The ecology of *Gammarus duebeni* Lilljeborg and its occurrence in fresh water in western Britain. *J. Anim. Ecol.* **23**, 38–84. *147, 169, 193, 195, 221, 233*.

—— 1955*a*. Biological notes on some East African aquatic Heteroptera. *Proc. R. ent. Soc. Lond.* **A 30**, 43–54. *116, 152, 387*.

HYNES, H. B. N., 1955*b*. The reproductive cycle of some British freshwater Gammaridae. *J. Anim. Ecol.* **24,** 352–87. *275, 287.*
—— 1955*c*. Distribution of some freshwater Amphipoda in Britain. *Verh. int. Verein. theor. angew. Limnol.* **12,** 620–8. *218.*
—— 1958. The effect of drought on the fauna of a small mountain stream in Wales. Ibid. **13,** 826–33. *223–4, 279, 404.*
—— 1960. *The Biology of Polluted Waters*. Liverpool University Press. *vii, ix, 153, 216, 254, 269, 443, 447.*
—— 1961. The invertebrate fauna of a Welsh mountain stream. *Arch. Hydrobiol.* **57,** 344–88. *51, 79, 80, 144, 167, 215, 223–4, 237, 241, 244, 255, 269, 271–2, 277, 283–6, 289, 292, 294, 298, 390, 393–4, 419, 421, 424, 433–4.*
—— 1962. The hatching and growth of the nymphs of several species of Plecoptera. *Int. Congr. Ent.* **11, 3,** 271–3. *272, 282, 284, 289.*
—— 1963. Imported organic matter and secondary productivity in streams. *Int. Congr. Zool.* **16, 4,** 324–9. *168, 194, 411, 433–5.*
—— 1964*a*. The interpretation of biological data with reference to water quality. *Publ. U.S. Pub. Hlth Serv.* **999–AP–15,** 289–98. *241.*
——1964*b*. Some Australian Plecopteran nymphs. *Gewäss. Abwäss.* **34/5,** 17–22. *117, 272, 276.*
—— 1965. The significance of macroinvertebrates in the study of mild river pollution. *Publ. U.S. Pub. Hlth Serv.* **999–WP–25,** 235–40. *269.*
—— and WILLIAMS, T. R., 1962. The effect of DDT on the fauna of a Central African stream. *Ann. trop. Med. Parasit.* **56,** 78–91. *29, 61, 79, 117, 284, 290–1, 391, 403.*
—— *et al.*, 1961. Freshwater crabs and *Simulium neavei* I. Preliminary observations made on the slopes of Mount Elgon in December, 1960, and January, 1961. Ibid. **55,** 197–201. *387.*

IDE, F. P., 1935. The effect of temperature on the distribution of the mayfly fauna of a stream. *Publs Ont. Fish. Res. Lab.* **50,** 1–76. *203, 387.*
—— 1940. Quantitative determination of the insect fauna of rapid water. Ibid. **59,** 1–24. *203, 241, 283, 391.*
IDYLL, C., 1942. Food of rainbow, cutthroat and brown trout in the Cowichan River system, B.C. *J. Fish. Res. Bd Can.* **5,** 448–58. *367, 373.*
—— 1943. Bottom fauna of portions of the Cowichan River. B.C. Ibid. **6,** 133–9. *238.*
ILLIES, J., 1952*a*. Die Plecopteren und das Monardsche Prinzip. *Ber. limnol. Flussstn Freudenthal,* **3,** 53–69. *282–3.*
—— 1952*b*. Die Mölle. Faunistisch-ökologische Untersuchungen an einem Forellenbach in Lipper Bergland. *Arch. Hydrobiol.* **46,** 424–612. *79, 204, 220–2, 233, 390.*
—— 1954. Die Besiedlung der Fulda (insbes. das Benthos der Salmonidenregion) nach dem jetzigen Stand der Untersuchung. *Ber. limnol. Flussstn Freudenthal,* **5,** 1–28. *390, 392.*
—— 1955. Der biologische Aspekt der limnologischen Fliesswassertypisierung. *Arch. Hydrobiol.* Suppl. **22,** 337–46. *392–3.*
—— 1956. Seeausfluss-Biozönosen lappländischer Waldbäche. *Ent. Tidskr.* **77,** 138–53. *258–9.*
—— 1958. Die Barbenregion mitteleuropäischer Fliessgewässer. *Verh. int. Verein. theor. angew. Limnol.* **13,** 834–44. *98, 208, 390, 392, 396.*

Illies, J., 1959. Retardierte Schlupfzeit von *Baetis*-Gelegen (Ins., Ephem.). *Naturwissenschaften*, **46**, 119–20. *278*.

—— 1960. Zur Frage der Realität des Biozönose-Begriffes. *Z. angew. Ent.* **47**, 95–101. *392*.

—— 1961*a*. Gebirgsbäche in Europa und in Südamerika—ein limnologischer Vergleich. *Verh. int. Verein. theor. angew. Limnol.* **14**, 517–23. *117, 123, 144*.

—— 1961*b*. Versuch einer allgemein biozönotischen Gleiderung der Fliessgewässer. *Int. Revue ges. Hyrobiol. Hydrogr.* **46**, 205–13. *392*.

—— 1961*c*. *Die Lebensgemeinschaft des Bergbaches*. Neue Brehm-Bücherei, Wittenberg Lutherstadt. *vii*.

—— 1962. Die Bedeutung der Strömung für die Biozönose in Rhithron und Potamon. *Schweiz. Z. Hydrol.* **24**, 433–5. *392*.

—— 1963. Revision der südamerikanischen Gripopterygidae (Plecoptera). *Mitt. Schweiz. ent. Ges.* **36**, 145–248. *117–18, 144*.

—— 1964. The invertebrate fauna of the Huallaga, a Peruvian tributary of the Amazon River, from the sources down to Tingo Maria. *Verh. int. Verein. theor. angew. Limnol.* **15**, 1077–83. *391, 393*.

—— and Botosaneanu, L., 1963. Problèmes et méthodes de la classification et de la zonation écologique des eaux courantes, considerées surtout du point de vue faunistique. *Mitt. int. Verein. theor. angew. Limnol.* **12**, 57 pp. *14, 383, 392–3, 396*.

Imanishi, K., 1941. Mayflies from Japanese torrents X. Life forms and life zones of mayfly nymphs II. Ecological structure illustrated by life zone arrangement. *Mem. Coll. Sci. Engng Kyoto imp. Univ.* **B16**, 1–35. *387*.

Ingold, C. T., 1959. Submerged aquatic Hyphomycetes. *J. Queckett microsc. Club*, **4, 5**, 115–30. *435*.

—— 1960. Aquatic hyphomycetes from Canada. *Can. J. Bot.* **38**, 803–6. *435*.

—— 1961. Another aquatic spore type with clamp connections. *Trans. Br. mycol. Soc.* **44**, 27–30. *435*.

—— 1966. The tetraradiate aquatic fungal spore. *Mycologia*, **58**, 43–56. *435*.

Israelson, G., 1942. The freshwater Florideae of Sweden. Studies on their taxonomy, ecology and distribution. *Symb. bot. upsal.* **6, 1**, 134 pp. *61*.

Istenič, L., 1963. Rate of oxygen consumption of larvae of *Perla marginata* Pz. in relation to body size and temperature (Czech: English summary). *Raspr. slov. Akad. Znan. Umet.* **4, 7**, 201–36. *162, 167*.

Ivlev, V. S., 1945. The biological productivity of waters. *Uspekhi Sovremennoi Biologii*, **19, 1**, 98–120 (Russian, transl. W. E. Ricker, 1966. *J. Fish. Res. Bd Can.* **23**, 1727–59). *194, 411*.

—— and Ivassik, V. M., 1961. Materials to the biology of mountain rivers in the Soviet Transcarpathians (Russian). *Trudy vses. gidrobiol Obshch.* **11**, 171–88. *204, 226–7, 411*.

Jaag, O., 1938. Die Kryptogamenflora des Rheinfalls und des Hochrheins von Stein bis Eglisau. *Mitt. naturf. Ges. Schaffhausen*, **14**, 1–158. *72–3, 80, 102, 105*.

—— and Ambühl, H., 1964. The effect of the current on the composition of biocoenoses in flowing water streams. *Int. Conf. Wat. Pollut. Res. Lond.*, Pergamon Press, Oxford, pp. 31–49. *8, 163–5*.

Jackson, D. F., *et al.*, 1964. Ecological investigations of the Oswego River drainage basin. I, the outlet. *Publs gt Lakes Res. Inst.* **11**, 88–99. *69*.

JACKSON, D. J., 1956*a*. Observations on water beetles during drought. *Entomologist's mon. Mag.* **92**, 154–5. *404*.

—— 1956*b*. The capacity for flight of certain water beetles and its bearing on their origin in the Western Scottish Isles. *Proc. Linn. Soc. Lond.* **167**, 76–96. *145*.

—— 1958. Observations on *Hydroporus ferrugineus* Steph. (Col. Dytiscidae), and some further evidence indicating incapacity for flight. *Entomologist's Gaz.* **9**, 55–9. *145, 401*.

JACKSON, P. B. N., 1961. *The Fishes of Northern Rhodesia*. Government Printer, Lusaka. *302, 306, 329, 348–9, 360, 366*.

JAFFA, B. B., 1917. Notes on the breeding and incubation periods of the Iowa darter, *Etheostoma iowae* Jordan and Meek. *Copeia*, **1917**, 71–2. *360*.

JANKOVIĆ, D., 1958. Ecological research on Danubian sterlet (*Acipenser ruthenus* L.) (Yugoslavian: English summary). *Posebna Izd. srp. biol. Društ.* **2**, 145 pp. *359, 367*.

—— 1960. The systematics and ecology of graylings (*Thymallus thymallus* L.) in Yugoslavia (Yugoslavian: English summary). Ibid. **7**, 144 pp. *359, 362, 372*.

JENKINS, M. F., 1960. On the method by which *Stenus* and *Dianous* (Coleoptera: Staphylinidae) return to the banks of a pool. *Trans. R. ent. Soc. Lond.* **112**, 1–14. *175*.

JEWELL, M. E., 1922. The fauna of an acid stream. *Ecology*, **3**, 22–8. *219*.

—— 1927. Aquatic biology of the prairie. Ibid. **7**, 289–98. *254, 411*.

—— 1935. An ecological study of the fresh-water sponges of Northern Wisconsin. *Ecol. Monogr.* **5**, 461–504. *121*.

—— 1939. An ecological study of the fresh-water sponges of Wisconsin II. The influence of calcium. *Ecology*, **20**, 11–28. *220, 222*.

JOHNSON, D. S., 1957. A survey of Malayan freshwater life. *Malay. Nat. J.* **12**, 57–65. *117, 306, 391*.

JOHNSON, R. P., 1963. Studies on the life history and ecology of the bigmouth buffalo *Ictiobus cyprinellus* (Valenciennes). *J. Fish. Res. Bd Can.* **20**, 1397–429. *328*.

JÓNASSON, P. M., 1955. The efficiency of sieving techniques for sampling freshwater bottom fauna. *Oikos*, **6**, 183–207. *239*.

—— 1958. The mesh factor in sieving techniques. *Verh. int. Verein. theor. angew. Limnol.* **13**, 860–6. *239*.

JONES, J. R. E., 1941. The fauna of the River Dovey, West Wales. *J. Anim. Ecol.* **10**, 12–24. *253, 390*.

—— 1943. The fauna of the River Teifi, West Wales. Ibid. **12**, 115–23. *199, 390*.

—— 1948. The fauna of four streams in the 'Black Mountain' district of South Wales. Ibid. **17**, 51–65. *226, 253, 390*.

—— 1949*a*. An ecological study of the River Rheidol, North Cardiganshire, Wales. Ibid. **18**, 67–88. *207, 332, 390*.

—— 1949*b*. A further study of calcareous streams in the 'Black Mountain' district of South Wales. Ibid. **18**, 142–59. *194–5*.

—— 1950. A further ecological study of the River Rheidol: the food of the common insects of the main-stream. Ibid. **19**, 159–74. *194, 433*.

—— 1951. An ecological study of the River Towy. Ibid. **20**, 68–86. *63, 65, 226, 390*.

JONES, J. W., 1959. *The Salmon*. Collins, London. *309, 313, 346, 351, 353, 357, 364*.

JONES, J. W., and BALL, J. N., 1954. The spawning behaviour of brown trout and salmon. *Br. J. Anim. Behav.* **2,** 103–14. *357.*

—— and KING, G. M., 1949. Experimental observations on the spawning behaviour of the Atlantic salmon (*Salmo salar* L.). *Proc. zool. Soc. Lond.* **119,** 33–48. *357.*

—— —— 1951. The underwater activity of the dipper. *Br. Birds,* **45,** 400–1. *380.*

—— —— 1952. The spawning of the male salmon parr (*Salmo salar* Linn. juv.). *Proc. zool. Soc. Lond.* **122,** 615–19. *357.*

—— and ORTON, J. H., 1940. The paedogenetic male cycle in *Salmo salar* L. *Proc. R. Soc.* **B128,** 485–99. *357.*

JÜRGENSEN, C., 1935. Die Mainalgen bei Würzburg. *Arch. Hydrobiol.* **28,** 361–414. *98.*

KAISER, E. W., 1947. Commensale og parasitiske Chironomidelarver. *Flora Fauna,* **1947,** 14–16. *180.*

—— 1949. Fysisk, kemisk, hydrometrisk og biologisk undersøgelser af Mølleåen fra Lyngby Sø til Øresund 1946 og 1947. *Skr. dansk Ingen.-foren, Spildevandskom.* **1,** 39 pp. *390.*

—— 1950. *Sialis nigripes* Ed. Pict., ny for Danmark, og udbredelsen af *S. lutaria* L. og *S. fuliginosa* Pict. i Danmark (English summary). *Flora Fauna,* **56,** 17–36. *276, 387.*

—— 1951. Biologiske, biokemiske, bakteriologiske samt hydrometriske undersøgelser af Pøleåen 1946 og 1947. *Skr. dansk Ingen.-foren. Spildevandskom.* **3,** 53 pp. *390.*

—— 1961. Studier over de danske *Sialis*-arter II. Biologien hos *S. fuliginosa Pict.* og *S. nigripes* Ed. Pict. (English summary). *Flora Fauna,* **67,** 74–96. *160, 276.*

—— 1965. Über Netzbau und Strömungssinn bei den Larven der Gattung *Hydropsyche* Pict. (Ins., Trichoptera). *Int. Revue ges. Hydrobiol. Hydrogr.* **50,** 169–224. *188–9.*

KAJAK, Z., 1958. An attempt at interpreting the quantitative dynamics of benthic fauna in a chosen environment in the 'Konfederatka' Pool (old river bed) adjoining the Vistula. (Polish: English summary). *Ekol. pol.* **A6, 7,** 205–91. *273, 275.*

KAJEWSKI, G., 1955. Einige Beobachtungen zur Frage der Ökologie der Bachfauna in strengen Wintern. *Z. Fisch.* N.F. **4,** 143–58. *371.*

KALLEBERG, H., 1958. Observations in a stream tank of territoriality and competition in juvenile salmon and trout (*Salmo salar* L. and *S. trutta* L.). *Rep. Inst. Freshwat. Res. Drottningholm,* **39,** 55–98. *305, 310–11, 313, 327.*

KALMAN, L., 1966. Beitrag zur Messung von Fliessgeschwindigkeiten in Klärbecken mit Thermosonden. *Schweiz. Z. Hydrol.* **28,** 69–87. *5.*

KAMLER, E., 1960. Notes on the Ephemeroptera fauna of Tatra streams. *Polskie Archwm Hydrobiol.* **8 (21),** 107–27. *199.*

—— 1962. La faune des éphémères de deux torrents des Tatras. Ibid. **10 (23),** 15–38. *199, 215.*

—— 1965. Thermal conditions in mountain waters and their influence on the distribution of Plecoptera and Ephemeroptera larvae. *Ekol. pol.* **A13,** 377–414. *30–1, 201.*

—— 1966. L'influence du degré d'astatisme de certains facteurs du milieu sur la répartition des larves d'Ephémères et des Plécoptères dans les eaux des montagnes. *Verh. int. Verein. theor. angew. Limnol.* **16,** 663–8. *30, 199, 201, 208, 223.*

KAMLER, E., and RIEDEL, W., 1960*a*. The effect of drought on the fauna, Ephemeroptera, Plecoptera and Trichoptera, of a mountain stream. *Polskie Archwm Hydrobiol.* **8 (21),** 87–94. *238, 404.*

—— —— 1960*b*. A method for quantitative study of the bottom fauna of Tatra streams. Ibid. **8 (21),** 95–105. *237, 241.*

KANN, E., 1966. Der Algenaufwuchs in einigen Bächen Österreichs. *Verh. int. Verein. theor. angew. Limnol.* **16,** 464–54. *63, 68.*

KARR, J. R., 1963. Age, growth and food habits of Johnny, Slenderhead and Blacksided darters of Boone County, Iowa. *Proc. Iowa Acad. Sci.* **70,** 228–36. *367.*

—— 1964. Age, growth, fecundity and food habits of fantail darters in Boone County, Iowa. Ibid. **71,** 274–80. *337, 367.*

KAWAI, T., 1959. Foods of *Salvelinus pluvius* of the streams connecting with Lake Ôtori-ike (Japanese: English summary). *Jap. J. Limnol.* **20,** 167–73. *367.*

KAWANABE, H., 1958. On the significance of the social structure for the mode of density effect in a salmon-like fish, 'Ayu', *Plecoglossus altivelis* Temminck and Schlegel. *Mem. Coll. Sci. Engng Kyoto imp. Univ.* **B25,** 171–80. *310, 311, 346.*

—— 1959. Food competition on some rivers of Kyoto Prefecture, Japan. Ibid. **B26,** 253–68. *366–7, 376.*

KEAST, A., 1966. Trophic interrelationships in the fish fauna of a small stream. *Publs gt Lakes Res. Inst.* **15,** 51–79. *368–9, 371, 374.*

KEENLEYSIDE, M. H. A., 1962. Skin-diving observations of Atlantic salmon and brook trout in the Miramichi River, New Brunswick. *J. Fish. Res. Bd Can.* **16,** 625–34. *305, 309, 327, 364.*

KEETCH, D. P., and MORAN, V. C., 1966. Observations on the biology of nymphs of *Paragomphus cognatus* (Rambur) (Odonata: Gomphidae). I. Habitat selection in relation to substrate particle size. *Proc. R. ent. Soc. Lond.* **A 41,** 116–22. *206.*

KENNEDY, H. D., 1958. Biology and life history of a new species of mountain midge, *Deuterophlebia nielsoni*, from Eastern California (Diptera, Deuterophlebiidae). *Trans. Am. microsc. Soc.* **77,** 201–28. *136, 145, 267.*

—— 1960. *Deuterophlebia inyoensis*, a new species of mountain midge from the alpine zone of the Sierra Nevada range, California (Diptera: Deuterophlebiidae). Ibid. **79,** 191–210. *273, 275.*

KEVERN, N. R., and BALL, R. C., 1965. Primary productivity and energy relationships in artificial streams. *Limnol. Oceanogr.* **10,** 74–87. *415.*

—— *et al.*, 1966. Use of artificial substrata to estimate the productivity of periphyton. Ibid. **11,** 499–502. *415.*

KHOO, S. G. 1964*a*. Studies on the biology of stoneflies. Unpublished thesis, University of Liverpool. *278, 280, 284–5.*

—— 1964*b*. Studies on the biology of *Capnia bifrons* (Newman) and notes on the diapause in the nymphs of this species. *Gewäss. Abwäss.* **34/5,** 23–30. *278, 280–1, 284–5.*

KING, D. L., and BALL, R. C., 1964. The influence of highway construction on a stream. *Res. Rep. Mich. St. agric. Exp. Stn,* **19,** 4 pp. *444.*

—— —— 1966. A qualitative and quantitative measure of aufwuchs production. *Trans. Am. microsc. Soc.* **85,** 232–40. *414.*

KING, H. W., *et al.*, 1948. *Hydraulics.* Wiley, New York. 7.

KLEEREKOPER, H., 1955. Limnological observations in Northeastern Rio Grande do Sul, Brazil. *Arch. Hydrobiol.* **50,** 553–67. *92, 306, 338, 360, 387.*

KLIMOWICZ, H., 1961. Differentiation of rotifers in various zones of Nile near Cairo. *Polskie Archwm Hydrobiol.* **9 (22),** 223–42. *99.*

KLYUCHAREVA, O. A., 1963. Downstream and diurnal vertical migrations of benthic invertebrates of the Amur River (Russian). *Russk. zool. Zh.* **42,** 1601–12. *259.*

KNIGHT, A. W., and GAUFIN, A. R., 1963. The effect of water flow, temperature, and oxygen concentration on the Plecoptera nymph, *Acroneuria pacifica* Banks. *Proc. Utah Acad. Sci.* **40,** 175–84. *163.*

—— —— 1964. Relative importance of varying oxygen concentration, temperature, and water flow on the mechanical activity and survival of the Plecoptera nymph, *Pteronarcys californica* Newport. Ibid. **41,** 14–28. *163, 166, 168.*

—— ——1966. Oxygen consumption of several species of stoneflies (Plecoptera). *J. Insect Physiol.* **12,** 347–55. *167.*

—— *et al.*, 1965*a*. Description of the eggs of common Plecoptera of western United States. *Ent. News,* **76,** 105–11. *140,*

—— *et al.*, 1965*b*. Further descriptions of the eggs of Plecoptera of western United States. Ibid. **76,** 233–9. *140.*

KNÖPP, H., 1952. Studien zur Statik und Dynamik der Biozönose eines Teichausflusses. *Arch. Hydrobiol.* **46,** 15–102. *257.*

—— 1960. Untersuchungen über das Sauerstoff-Produktions-Potential von Flussplankton. *Schweiz. Z. Hydrol.* **22,** 152–66. *107, 417–18.*

KOBAYASI, H., 1961*a*. Chlorophyll content in sessile algal community of Japanese mountain river. *Bot. Mag., Tokyo,* **74,** 228–35. *66, 70.*

—— 1961*b*. Productivity in sessile algal community of Japanese mountain river. Ibid. **74,** 331–41. *415.*

KOFOID, C. A., 1903. The plankton of the Illinois River, 1894–1899, with introductory notes upon the hydrography of the Illinois River and its basin. Part I. Quantitative investigations and general results. *Bull. Ill. St. Lab. nat. Hist.* **6,** 95–629. *98, 108–9.*

—— 1908. Idem. Part II. Constituent organisms and their seasonal distribution. Ibid. **8,** 2–360. *108.*

KOMATSU, T., 1964*a*. Faunistic studies of aquatic insects in the Taru River system, Nagano Prefecture (Japanese: English summary). *New Ent., Ueda,* **13,** 1–12. *391, 400.*

—— 1964*b*. Aquatic insect communities in winter and the biotic index of the rivers which flow into Lake Biwa (Japanese: English summary). *Jap. J. Ecol.* **14,** 217–33. *391, 434.*

KOPECKÝ, K., 1965. Einfluss der Ufer- und Wassermakrophyten-Vegetation auf die Morphologie des Flussbettes einige tschechoslowakischer Flüsse. *Arch. Hydrobiol.* **61,** 137–60. *89.*

KOTHÉ, P., 1961. Hydrobiologie der Oderelbe. Natürliche, industrielle und wasserwirtschaftliche Faktoren in ihrer Auswirkung auf das Benthos des Stromgebietes oberhalb Hamburgs, Ibid. Suppl. **24,** 221–343. *390.*

KOZHOV, M., 1963. Lake Baikal and its life. *Monographiae biol.* **11,** 352 pp. *112, 146.*

KRAATZ, W. C., 1923. A study of the food of the minnow (*Campostoma anomalum*). *Ohio J. Sci.* **23,** 265–83. *366, 368.*

KRAWANY, H., 1930. Trichopterenstudien im Gebiet der Lunzer Seen. II and III. *Int. Revue ges. Hydrobiol. Hydrogr.* **23,** 417–27. *188–9, 256.*

KRECKER, F. H., 1939. A comparative study of the animal populations of certain submerged aquatic plants. *Ecology,* **20,** 553–62. *215.*

KRESSER, W., 1961. Hydrographische Betrachtung der österreichischen Gewässer. *Verh. int. Verein. theor. angew. Limnol.* **14,** 417–21. *26.*

KRUMHOLZ, L. A., 1950. Some practical considerations in the use of rotenone in fisheries research. *J. Wildl. Mgmt,* **14,** 413–24. *316.*

KUEHN, J. H., 1949. A study of a population of longnose dace (*Rhinicthys c. cataractae*). *Proc. Minn. Acad. Sci.* **17,** 81–7. *369.*

KUEHNE, R. A., 1962. A classification of streams, illustrated by fish distribution in an Eastern Kentucky creek. *Ecology,* **43,** 608–14. *14, 385.*

KÜHN, G., 1940. Zur Ökologie und Biologie der Gewässer (Quellen und Abflüsse) des Wassersprengs bei Wien. *Arch. Hydrobiol.* **36,** 157–262. *390, 400.*

KÜHTREIBER, J., 1934. Die Plekopterenfauna Nordtirols. *Ber. naturw.-med. Ver. Innsbruck,* **44,** 1–219. *145, 286.*

KURECK, A., 1966. Schlüpfrhythmus von *Diamesa artica* (Diptera Chironomidae) auf Spitzbergen. *Oikos,* **17,** 276–7. *402.*

KURENKOV, I. I., 1957. The effect of a volcano on a river fauna (Russian). *Priroda,* **12,** 49–54 (Fish. Res. Bd Canada, Ottawa, transl. serv. No. 184). *219, 328.*

LAAKSO, M., 1951. Food habits of the Yellowstone whitefish *Prosopium williamsoni cismontanus* (Jordan). *Trans. Am. Fish. Soc.* **80,** 99–109. *367, 372.*

LACHNER, E. A., 1946. Studies on the biology of the chubs (Genus *Nocomis,* Family Cyprinidae) of north-eastern United States. *Cornell Univ. Abstracts of Thesis,* **1946,** 207–10. *358, 367.*

—— 1950. Food, growth and habits of fingerling northern smallmouth bass, *Micropterus dolomieu dolomieu* Lacépède, in trout waters of western New York. *J. Wildl. Mgmt,* **14,** 50–6. *369.*

—— 1952. Studies of the biology of the Cyprinid fishes of the chub genus *Nocomis* of Northeastern United States. *Am. Midl. Nat.* **48,** 435–66. *358–9.*

—— 1956. The changing fish fauna of the Upper Ohio basin. *Spec. Publs Pymatuning Lab. Fld Biol.* **1,** 64–78. *328, 335, 359, 385, 444.*

—— *et al.,* 1950. Studies on the biology of some Percid fishes from Western Pennsylvania. *Am. Midl. Nat.* **43,** 92–111. *311, 340.*

LACKEY, J. B., 1942. The plankton algae and protozoa of two Tennessee rivers. Ibid. **27,** 191–202. *99.*

—— *et al.,* 1943. Some plankton relationships in a small unpolluted stream. Ibid. **30,** 403–25. *98.*

LAGLER, K. F., 1943. Food habits and economic relations of the turtles of Michigan with special reference to fish management. Ibid. **29,** 257–312. *379.*

—— 1952. *Freshwater Fishery Biology.* Wm C. Brown, Dubuque. *313, 339, 450.*

—— and SALYER, J. C., 1947. Food and habits of the common watersnake, *Natrix s. sipedon,* in Michigan. *Pap. Mich. Acad. Sci.* **31,** 169–80. *379.*

—— *et al.,* 1962. *Icthyology.* Wiley, New York. *301.*

LAIRD, M., 1956. Studies of mosquitoes and freshwater ecology in the South Pacific. *Bull. R. Soc. N.Z.* **6,** 213 pp. *119.*

LAKE, C. T., 1936. The life history of the fan-tailed darter *Catanotus flabellaris flabellaris* (Rafinesque). *Am. Midl. Nat.* **17,** 816–30. *356.*

LAKE, J. S., 1957. Trout populations and habitats in New South Wales. *Aust. J. mar. Freshwat. Res.* **8,** 1–38. *372.*

—— 1959. The freshwater fishes of New South Wales. *Res. Bull. St. Fish. N.S.W.* **5,** 19 pp. *346, 352, 360–1.*

LAKSHMINARAYANA, J. S. S., 1965. Studies on the phytoplankton of the River Ganges, Varanasi, India. Parts I–II. *Hydrobiologia,* **25,** 119–65. *98.*

LARIMORE, R. W., 1952. Home pools and homing behavior of smallmouth black bass in Jordan Creek. *Biol. Notes nat. Hist. Surv. Div. St. Ill.* **28,** 1–12. *344.*

—— 1954. Dispersal, growth and influence of smallmouth bass stocked in a warm-water stream. *J. Wildl. Mgmt.,* **18,** 207–16. *343.*

—— 1955. Minnow productivity in a small Illinois stream. *Trans. Am. Fish. Soc.* **84,** 110–16. *350.*

—— 1961. Fish population and electrofishing success in a warm-water stream. *J. Wildl. Mgmt,* **25,** 1–12. *317.*

—— and SMITH, P. W., 1963. The fishes of Champaign County, Illinois, as affected by 60 years of stream changes. *Bull. Ill. St. nat. Hist. Surv.* **28,** 299–382. *338, 444.*

—— *et al.,* 1959. Destruction and re-establishment of stream fish and invertebrates affected by drought. *Trans. Am. Fish. Soc.* **88,** 261–85. *216, 223, 329.*

LARSEN, K., 1961. Fish populations in small Danish streams. *Verh. int. Verein. theor. angew. Limnol.* **14,** 769–72. *339.*

LARSEN, L. O., 1962. Weight and length in the river lamprey. *Nature, Lond.* **194,** 1093. *308.*

LARSÉN, O., 1932. Beiträge zur Ökologie und Biologie von *Aphelocheirus aestivalis* Fabr. *Int. Revue ges. Hydrobiol. Hydrogr.* **26,** 1–19. *284.*

LASTOCHKIN, D. A., 1943. A plain river subdivision into geomorphological and biological districts on the basis of its structural and biological unity. *Dokl. Acad. Nauk. S.S.S.R.* **41,** 347–50. *26, 385, 396.*

LAUFF, G. H., and CUMMINS, K. W., 1964. A model stream for studies in lotic ecology. *Ecology,* **45,** 188–91. *151.*

—— *et al.,* 1961. A method for sorting bottom fauna samples by elutriation. *Limnol. Oceanogr.* **6,** 426–66. *242.*

LAURIE, E. M. O., 1942. The dissolved oxygen of an upland pond and its inflowing stream, at Ystumtuen, North Cardiganshire, Wales. *J. Ecol.* **30,** 357–82. *42.*

LAUTERBORN, R., 1893. Beiträge zur Rotatorienfauna des Rheins und seiner Altwässer. *Zool. Jb. Abt. Syst.* **7,** 254–73. *94.*

—— 1902. Beiträge zur Mikrofauna und -Flora der Mosel. *Z. Fisch.* **9,** 1–25. *98.*

—— 1910. Die Vegetation des Oberrheins. *Verh. naturh.-med. Ver. Heidelb.* **10,** 450–502. *76, 105.*

—— 1916. Die geographische und biologische Gliederung des Rheinstroms. I Teil. *Sber. heidelb. Akad. Wiss.* **B 7, 4,** 61 pp. *105.*

—— 1917. Idem. II Teil. Ibid. **B 8, 5,** 70 pp. *105.*

—— 1918. Idem. III Teil. Ibid. **B 9, 1,** 87 pp. *105.*

LEACH, W. J., 1940. Occurrence and life history of the northern brook lamprey, *Ichthyomyzon fossor,* in Indiana. *Copeia,* **1940,** 21–34. *363.*

LE CREN, E. D., 1958. Preliminary observations on populations of *Salmo trutta* in becks in Northern England. *Verh. int. Verein. theor. angew. Limnol.* **13,** 754–7. *348–9.*

LE GLEN, E. D., 1965. Some factors regulating the size of populations of freshwater fish. *Mitt. int. Verein. theor. angew. Limnol.* **13,** 88–105. *311–12.*

LEMMERMAN, E., 1907*a*. Das Plankton der Weser bei Bremen. *Arch. Hydrobiol.* **2,** 393–447. *108.*

—— 1907*b*. Das Plankton der Jang-tse-Kiang (China). Ibid. **2,** 534–44. *98.*

LENNON, R. E., and PARKER, P. S., 1960. The stoneroller *Campostoma anomalum* (Rafinesque) in Great Smoky Mountains National Park. *Trans. Am. Fish. Soc.* **89,** 263–70. *357.*

LEONARD, J. W., 1939. Comments on the adequacy of accepted stream bottom sampling technique. *Trans. N. Am. Wildl. Conf.* **4,** 288–95. *239, 246.*

—— 1942. Some observations on the winter feeding habits of brook trout fingerlings in relation to natural food organisms present. *Trans. Am. Fish. Soc.* **71,** 219–27. *266.*

—— and LEONARD, F. A., 1962. Mayflies of Michigan trout streams. *Bull. Cranbrook Inst. Sci.* **43,** 139 pp. *147, 152, 158–9, 207.*

LEOPOLD, L. B., 1962. Rivers. *Am. Scient.* **1962,** 511–37. *2.*

—— and LANGBEIN, W. B., 1966. River meanders. *Scient. Am.* **214,** 60–70. *22.*

—— *et al.*, 1964. *Fluvial Processes in Geomorphology*. Freeman, San Francisco. *1, 3, 8–10, 12–23, 25–6, 37–8.*

LEVANIDOV, U. YA., 1949. Significance of allochthonous material as a food resource in a water body as exemplified by the nutrition of the water louse (*Asellus aquaticus* L.) (Russian). *Trudy vses. gidrobiol Obshch.* **1,** 100–17. *434–5, 438.*

LEWIS, D. J., 1957. Observations on Chironomidae at Khartoum. *Bull. ent. Res.* **48,** 155–84. *259.*

—— 1960. Observations on the *Simulium neavei* complex at Amani in Tanganyika. Ibid. **51,** 95–113. *182.*

—— 1961. The *Simulium neavei* complex (Diptera, Simuliidae) in Nyasaland. *J. Anim. Ecol.* **30,** 303–10. *182.*

—— *et al.*, 1960. The attachment of immature Simuliidae to other arthropods. *Nature, Lond.* **187,** 618–9. *182.*

LEWIS, W. M., and ELDER, D. 1953. The fish population of the headwaters of a spotted bass stream in southern Illinois. *Trans. Am. Fish. Soc.* **82,** 193–202. *338.*

LIBOSVÁRSKY, J., 1966. Successive removals with electrical fishing gear—a suitable method for making populations estimates in small streams. *Verh. int. Verein. theor. angew. Limnol.* **16,** 1212–16. *318.*

LIEFTINCK, M. A., 1962. On the problem of intrinsic and adaptive characters in odonate larvae. *Int. Congr. Ent.* **11, 3,** 274–8. *166.*

LIEPOLT, R., 1961. Limnologische Forschungen im österreichischen Donauström. *Verh. int. Verein. theor. angew. Limnol.* **14,** 422–9. *98.*

LILLY, M. L., 1953. The mode of life and the structure and functioning of the reproductive ducts of *Bithynia tentaculata* (L.). *Proc. malac. Soc. Lond.* **30,** 87–110. *194.*

LINDUSKA, J. P., 1942. Bottom type as a factor influencing the local distribution of mayfly nymphs. *Can. Ent.* **74,** 26–30. *199, 207.*

LINSENMAIR, K. E., and JANDER, R., 1963. Das 'Entspannungschwimmen' von *Velia* und *Stenus*. *Naturwissenschaften,* **6,** 231. *174–5.*

LIVINGSTONE, D. A., 1963. Chemical composition of rivers and lakes. *Prof. Pap. U.S. Geol. Surv.* **440 G,** 64 pp. *44.*

LLOYD, J. T., 1921. The biology of North American caddis fly larvae. *Bull. Lloyd Libr. Ent. Ser.* **1,** 124 pp. *137–8, 185, 193.*

LOFTUS, K. H., 1958. Studies on river-spawning populations of lake trout in eastern Lake Superior. *Trans. Am. Fish. Soc.* **87,** 259–77. *348, 351.*

LOGAN, S. W., 1963. Winter observations on bottom organisms and trout in Badger Creek, Montana. Ibid. **92,** 140–5. *206, 255, 261, 343.*

LOHAMMAR, G. L., 1954. The distribution and ecology of *Fissidens julianus* in northern Europe. *Svensk bot. Tidskr.* **48,** 162–73. *80.*

LONGHURST, A. R., 1959. The sampling problem in benthic ecology. *Proc. N.Z. ecol. Soc.* **6,** 8–12. *236.*

LOWE, R. H., 1951. Factors influencing the runs of elvers in the River Bann, Northern Ireland. *J. Cons. perm. int. Explor. Mer,* **17,** 299–315. *353.*

—— 1952. The influence of light and other factors on the seaward migration of the silver eel (*Anguilla anguilla* L.). *J. Anim. Ecol.* **21,** 275–309. *308, 313.*

LOWE-MCCONNELL, R. H. (ed.), 1966. Man-made lakes. *Symp. Inst. Biol. Lond.* **15,** 217 pp. *447–8.*

LOWRY, G. R., 1965. Movement of cutthroat trout, *Salmo clarki clarki* (Richardson), in three Oregon coastal streams. *Trans. Am. Fish. Soc.* **94,** 334–8. *346.*

LUCE, W. M., 1933. A survey of the fishery of the Kaskaskia River. *Bull. Ill. St. nat. Hist. Surv.* **20,** 71–123. *338.*

LUDWIG, W. B., 1932. The bottom invertebrates of the Hocking River. *Bull. Ohio biol. Surv.* **26,** 223–49. *132, 206, 391.*

LUND, J. W. G., and TALLING, J. F., 1957. Botanical limnological methods with special reference to the algae. *Bot. Rev.* **23,** 489–583. *56–8.*

LUTHER, H., 1954. Über Krustenbewuchs an Steinen fliessender Gewässer, speziell in Südfinnland. *Acta bot. fenn.* **55,** 1–61. *61, 66.*

LYMAN, F. E., 1943. A pre-impoundment bottom-fauna study of the Watts Bar Reservoir area (Tennessee), *Trans. Am. Fish. Soc.* **72,** 52–62. *272, 284.*

—— 1944. Effect of temperature on the emergence of mayfly imagos from the subimago stage. *Ent. News,* **55,** 113–15. *145.*

—— 1945. Reactions of certain nymphs of *Stenonema* (Ephemeroptera) to light as related to habitat preference. *Ann. ent. Soc. Am.* **38,** 234–6. *152.*

—— 1955. Seasonal distribution and life cycles of Ephemeroptera. Ibid. **48,** 380–91. *288.*

—— and DENDY, J. S., 1943. A pre-impoundment bottom-fauna study of Cherokee Reservoir area (Tennessee). *Trans. Am. Fish. Soc.* **73,** 194–208. *251, 254.*

MACAN, T. T., 1957*a*. The Ephemeroptera of a stony stream. *J. Anim. Ecol.* **26,** 317–42. *147, 222, 387, 389.*

—— 1957*b*. The life histories and migrations of the Ephemeroptera in a stony stream. *Trans. Soc. Br. Ent.* **12,** 129–56. *152, 261, 275, 278, 284, 286, 288–9.*

—— 1958*a*. Causes and effects of short emergence periods in insects. *Verh. int. Verein. theor. angew. Limnol.* **13,** 845–9. *419.*

—— 1958*b*. Methods of sampling the bottom fauna in stony streams. *Mitt. int. Verein. theor. angew. Limnol.* **8,** 1–21. *236–7, 244.*

—— 1958*c*. The temperature of a small stony stream. *Hydrobiologia,* **12,** 89–106. *28–31.*

—— 1960*a*. The occurrence of *Heptagenia lateralis* (Ephem.) in streams in the English Lake District. *Wett. Leben,* **12,** 231–4. *203.*

MACAN, T. T., 1960*b*. The effect of temperature on *Rhithrogena semicolorata* (Ephem.). *Int. Revue ges. Hydrobiol. Hydrogr.* **45,** 197–201. *284–5.*

—— 1961*a*. Factors that limit the range of freshwater animals. *Biol. Rev.* **36,** 151–98. *157–9, 222, 230, 336.*

—— 1961*b*. A review of running water studies. *Verh. int. Verein. theor. angew. Limnol.* **14,** 587–602. *393.*

—— 1962. Biotic factors in running water. *Schweiz. Z. Hydrol.* **24,** 386–407. *232.*

—— 1963. *Freshwater Ecology*. Longmans, London. *169, 232.*

—— 1964. Emergence traps and the investigation of stream faunas. *Riv. Idrobiol.* **3,** 75–92. *241, 261.*

—— and WORTHINGTON, E. B., 1951. *Life in Lakes and Rivers*. Collins, London. *vii.*

MACIOLEK, J. A., 1966. Abundance and character of microseston in a California mountain stream. *Verh. int. Verein. theor. angew. Limnol.* **16,** 639–45. *49, 433.*

—— and NEEDHAM, P. R., 1951. Ecological effects of winter conditions on trout and trout foods in Convict Creek, California. *Trans. Am. Fish. Soc.* **81,** 202–17. *32–3, 262, 371.*

MACK, B., 1953. Zur Algen und Pilzflora des Liesingsbaches. *Wett. Leben Sonderh.* **2,** 136–49. *70, 73.*

MACKENTHUN, K. M., *et al.*, 1964. Limnological aspects of recreational lakes. *Publ. U.S. Pub. Hlth Serv.* **1167,** 176 pp. *47.*

MACKERETH, J. C., 1957. Notes on the Plecoptera from a stony stream. *J. Anim. Ecol.* **26,** 343–51. *195.*

—— 1960. Notes on the Trichoptera of a stony stream. *Proc. R. ent. Soc. Lond.* **A35,** 17–23. *276, 284.*

MADALIŃSKI, K., 1961. Moss dwelling rotifers of Tatra streams. *Polskie Archwm Hydrobiol.* **9 (22),** 243–63. *114.*

MÄDLER, K., 1961. Untersuchungen über den Phosphorgehalt in Bächen. *Int. Revue ges. Hydrobiol. Hydrogr.* **46,** 75–83. *48.*

MADSEN, B. L., 1962. Økologiske undersøgelser i nogle østjyske vandløb. l. Fysiske og kemike forhold (English summary). *Flora Fauna,* **68,** 185–95. *28.*

—— 1963. Idem. 2. Planarier og igler (English summary). Ibid. **69,** 113–25. *201.*

—— 1966. Om rytmisk aktivitet hos døgnfluennymfer (English summary). Ibid. **72,** 148–50. *264.*

MAGUIRE, B., 1963. The passive dispersal of small aquatic organisms and their colonisation of isolated bodies of water. *Ecol. Monogr.* **33,** 161–85. *401.*

MAITLAND, P. S., 1964. Quantitative studies on the invertebrate fauna of sandy and stony substrates in the River Endrick, Scotland. *Proc. R. Soc. Edinb.* **B68,** 277–301. *226, 434.*

—— 1965*a*. The feeding relationships of salmon, trout, minnows, stone loach and three-spined stickleback in the River Endrick, Scotland. *J. Anim. Ecol.* **34,** 109–33. *366–7, 369–70, 373–4.*

—— 1965*b*. Notes on the biology of *Ancylus fluviatilis* in the River Endrick, Scotland. *Proc. malac. Soc. Lond.* **36,** 339–47. *180, 221.*

—— 1966. The fauna of the River Endrick. *Studies on Loch Lomond,* **2,** 194 pp., Glasgow. *114, 367, 390, 394, 396.*

MANN, K. H., 1953. The life history of *Erpobdella octoculata* (Linnaeus 1758). *J. Anim. Ecol.* **22,** 199–207. *275, 282.*

—— 1955. The ecology of British freshwater leeches. Ibid. **24,** 98–119. *220.*

Mann, K. H., 1957. A study of a population of the leech *Glossiphonia complanata* (L). Ibid. **26**, 99–111. *275–6.*

—— 1959. On *Trocheta bykowskii* Gedroyć, 1913, a leech new to the British fauna, with notes on the taxonomy and ecology of other Erpobdellidae. *Proc. zool. Soc. Lond.* **132**, 369–79. *201, 223.*

—— 1961*a*. The oxygen requirements of leeches considered in relation to their habitats. *Verh. int. Verein. theor. angew. Limnol.* **14**, 1009–13. *216–7.*

—— 1961*b*. The life history of the leech *Erpobdella testacea Sav.* and its adaptive significance. *Oikos*, **12**, 164–9. *216, 275, 282.*

—— 1964. The pattern of energy flow in the fish and invertebrate fauna of the River Thames. *Verh. int. Verein. theor. angew. Limnol.* **15**, 485–95. *420.*

—— 1965. Energy transformations by a population of fish in the River Thames. *J. Anim. Ecol.* **34**, 253–75. *418, 420, 426, 428–9.*

Margalef, R., 1949. A new limnological method for the investigation of thin-layered epilithic communities. *Hydrobiologia*, **1**, 215–16. *57.*

—— 1958. 'Trophic' versus biotic typology, as exemplified in the regional limnology of norther Spain. *Verh. int. Verein. theor. angew. Limnol.* **13**, 339–49. *438.*

—— 1960. Ideas for a synthetic approach to the ecology of running waters. *Int. Revue ges. Hydrobiol. Hydrogr.* **45**, 133–53. *72, 100.*

Markus, H. C., 1934. Life history of the blackhead minnow (*Pimephales promelas*). *Copeia*, **1934**, 116–22. *356.*

Marlier, G., 1950. Sur deux larves de *Simulium* commensales de nymphes Ephémères. *Revue Zool. Bot. afr.* **43**, 135–44. *181.*

—— 1951. La biologie d'un ruisseau de plaine. Le Smohain. *Mém. Inst. r. Sci. nat. Belg.* **114**, 1–98. *207, 390, 409.*

—— 1952. Fish feeding on *Simulium* larvae. *Nature, Lond.* **170**, 496. *348.*

—— 1953. Etude biogéographique du bassin de la Ruzizi, basée sur la distribution des poissons. *Annls Soc. r. zool. Belg.* **84**, 175–224. *306, 348, 387.*

—— 1954. Recherches hydrobiologiques dans les rivières du Congo oriental II. Etude écologique. *Hydrobiologia*, **6**, 225–64. *200, 208, 210, 387, 391.*

Marshall, N., 1947*a*. Studies on the life history and ecology of *Notropis chalybaeus*. *Q. Jl Fla Acad. Sci.* **9**, 163–88. *360, 366.*

—— 1947*b*. The spring run and cave habitats of *Erimystax harperi* (Fowler). *Ecology*, **28**, 68–75. *336.*

Martin, R. G., and Campbell, R. S., 1953. The fishes of Black River and Clearwater Lake, Missouri. *Univ. Mo Stud.* **26**, 45–66. *328, 338, 448.*

Martof, B. S., 1962. Some aspects of the life history and ecology of the salamander *Leurognathus*. *Am. Midl. Nat.* **67**, 1–35. *378.*

—— and Scott, D. C., 1957. The food of the salamander *Leurognathus*. *Ecology*, **38**, 494–501. *378.*

Marzolf, G. R., 1965. Substrate relations of the burrowing amphipod *Pontoporeia affinis* in Lake Michigan. Ibid. **46**, 579–92. *436.*

—— 1966. The trophic position of bacteria and their relation to the distribution of invertebrates. *Spec. Publs Pymatuning Lab. Fld Biol.* **4**, 131–5. *436.*

Mason, J. L., 1939. Studies on the fauna of an Algerian hot spring. *J. exp. Biol.* **16**, 487–98. *322–3, 399.*

Matheson, D. H., 1951. Inorganic nitrogen in precipitation and atmospheric sediments. *Can. J. Technol.* **29**, 406–12. *36.*

Matoničkin, I., and Pavletić, Z., 1959. A contribution to the knowledge of the biocenotic relations in the River Pliva in Bosnia (Yugoslavian: English summary). *Arh. biol. Nauka*, **11**, 1–12. *391*.

MATONIČKIN, I., and PAVLETIĆ, Z., 1960. Biological characteristics of the erosive cataracts of the River Bosna (Yugoslavian: English summary). *Biol. Glasn.* **13**, 295–305. *400*.

—— —— 1961. Contributo alla conoscenza dell'ecologia delle biocenosi sulle cascade travertinose nella regione Carstica, Jugoslava. *Hydrobiologia*, **18**, 225–44. *76–7*, *80*.

—— —— 1962*a*. Die Lebensbedingungen an Kalktuffwasserfällen der Jugoslawischen Karstgewässer (Yugoslavian: German summary). *Acta bot. croat.* **20/1**, 175–98. *76–7*.

—— —— 1962*b*. Entwicklung der Lebensgemeinschaften und ihre Bedeutung für die Bildung und Erhaltung der Kalkstuff-Wasserfälle. *Arch. Hydrobiol.* **58**, 467–73. *76–7*, *80*.

MAUCH, E., 1963. Untersuchungen über das Benthos der deutschen Mosel unter besonderer Breüchsichtigung der Wassergute. *Mitt. zool. Mus. Berl.* **39**, 1–172. *198*, *223*, *240*.

MAXWELL, G. R., and BENSON, A., 1963. Wing pad and tergite growth of mayfly nymphs in winter. *Am. Midl. Nat.* **69**, 224–30. *284*.

McCONNELL, W. J., and SIGLER, W. F., 1959. Chlorophyll and productivity in a mountain river. *Limnol. Oceanogr.* **4**, 335–51. *5*, *58*, *413–14*.

McCORMACK, J. C., 1962. The food of young trout (*Salmo trutta*) in two different becks. *J. Anim. Ecol.* **31**, 305–16. *267*, *369*.

McCRIMMON, H. R., and BIDGOOD, B., 1965. Abnormal vertebrae in rainbow trout with particular reference to electrofishing. *Trans. Am. Fish. Soc.* **94**, 84–8. *316*.

—— and KWAIN, W.-H., 1966. Use of overhead cover by rainbow trout exposed to a series of light intensities. *J. Fish. Res. Bd Can.* **23**, 983–90. *313*.

McDOWALL, R. M., 1965*a*. Studies on the biology of the red-finned bully *Gobiomorphus huttoni* (Ogilby). II. Breeding and life history. *Trans. R. Soc. N.Z.* (*Zool.*), **5**, 177–96. *356*, *361*.

—— 1965*b*. Idem. III. Food studies. Ibid. **5**, 233–54. *367*, *369*.

McFADDEN, J. T., 1961. A population study of the brook trout *Salvelinus fontinalis*. *Wildl. Monogr. Chestertown*, **7**, 73 pp. *315*, *325–6*, *343*.

—— and COOPER, E. L., 1962. An ecological comparison of six populations of brown trout (*Salmo trutta*). *Trans. Am. Fish. Soc.* **91**, 53–62. *328*, *332–3*, *339*.

—— *et al.*, 1965. Some effects of environment on egg production in brown trout (*Salmo trutta*). *Limnol. Oceanogr.* **10**, 88–95. *333*.

McINTYRE, C. D., 1966*a*. Some effects of current velocity on periphyton communities in laboratory streams. *Hydrobiologia*, **27**, 559–70. *65*.

—— 1966*b*. Some factors affecting respiration in lotic environments. *Ecology*, **47**, 918–30. *65*.

—— and PHINNEY, H. K., 1965. Laboratory studies of periphyton production and community metabolism in lotic environments. *Ecol. Monogr.* **35**, 237–58. *415*.

McLANE, W. M., 1948. The seasonal food of the largemouth black bass, *Micropterus salmoides floridianus* (Lacépède), in the St. Johns River, Welaka, Florida. *Q. Jl Fla Acad. Sci.* **10**, 103–38. *368–9*.

McMAHON, J., 1952. Phoretic association between Simuliidae and crabs. *Nature, Lond.* **169**, 1018. *182*.

McNeil, W. J., 1962. Variations in the dissolved oxygen content of intragravel water in four spawning streams of southeastern Alaska. *Spec. scient. Rep. U.S. Fish. Wildl. Serv.* **402,** 15 pp. *358.*

McVicker, M., *et al.*, 1963. *Fertiliser Technology and Usage.* Soil Science Society of America, Madison, Wisconsin, 464 pp. *445.*

Melin, E., 1930. Biological decomposition of some types of litter from North American forests. *Ecology,* **11,** 72–101. *435.*

Metcalf, A. L., 1959. Fishes of Chautaugua, Cowley and Elk counties, Kansas. *Univ. Kans. Publs Mus. nat. Hist.* **11,** 345–400. *329, 335.*

Meyer, W. H., 1962. Life history of three species of redhorse (*Moxostoma*) in the Des Moines River, Iowa. *Trans. Am. Fish. Soc.* **91,** 412–19. *337.*

Mikulski, J. S., 1961. Ecological studies upon bottom communities in the River Wisła (Vistula). *Verh. int. Verein. theor. angew. Limnol.* **14,** 372–5. *250.*

Miller, R. B., 1952. Survival of hatchery-reared cutthroat trout in an Alberta stream. *Trans. Am. Fish. Soc.* **81,** 35–42. *312.*

—— 1954*a*. Comparative survival of wild and hatchery-reared cutthroat trout in a stream. Ibid. **83,** 120–30. *312.*

—— 1954*b*. Movements of cutthroat trout after different periods of retention upstream and downstream from their homes. *J. Fish. Res Bd Can.* **11,** 550–8. *344, 351.*

—— 1958. The role of competition in the mortality of hatchery trout. Ibid. **15,** 27–45. *312.*

—— and Kennedy, W. A., 1946. Color change in a sculpin. *Copeia,* **1946,** 100. *308.*

Miller, R. J., 1962. Reproductive behavior of the stoneroller minnow, *Campostoma anomalum pullum.* Ibid. **1962,** 407–17. *357.*

Miller, R. R., 1960. Systematics and biology of the gizzard shad (*Dorosoma cepedianum*) and related species. *Fishery Bull. Fish Wildl. Serv. U.S.* **60,** 371–92. *360, 366, 371.*

Mills, D. H., 1964. The ecology of the young stages of the Atlantic salmon in the River Bran, Ross-shire. *Freshwat. Salm. Fish. Res.* **32,** 58 pp. *354.*

Mills, H. B., *et al.*, 1966. Man's effect on the fish and wildlife of the Illinois River. *Biol. Notes nat. Hist. Surv. Div. St. Ill.* **57,** 24 pp. *444–6.*

Minckley, W. L., 1961. Occurrence of subterranean isopods in the epigean environment. *Am. Midl. Nat.* **63,** 452–5. *114, 400.*

—— 1963. The ecology of a spring stream Doe Run, Meade County, Kentucky. *Wildl. Monogr. Chestertown,* **11,** 124 pp. *29, 35, 42, 48, 73, 76, 80, 89, 92, 112, 195, 213–14, 226, 234, 237, 241, 328, 330–1, 368, 391, 400.*

—— 1964. Upstream movements of *Gammarus* (Amphipoda) in Doe Run, Meade County, Kentucky. *Ecology,* **45,** 195–7. *154.*

—— and Cross, F. B., 1959. Distribution, habitat, and abundance of the Topeka shiner, *Notropis topeka* (Gilbert) in Kansas. *Am. Midl. Nat.* **61,** 210–17. *445.*

—— and Deacon, J. E., 1959. Biology of the flathead catfish in Kansas. Ibid. **88,** 344–55. *335, 369.*

—— and Tindall, D. R., 1963. Ecology of *Batrachospermum* sp (Rhodophyta) in Doe Run, Meade County, Kentucky. *Bull. Torrey bot. Club,* **90,** 391–400. *61, 63, 74.*

Minshall, G. W., and Minshall, J. N., 1966. Notes on the life history and ecology of *Isoperla clio* (Newman) and *Isogenus decisus* Walker (Plecoptera: Perlodidae). *Am. Midl. Nat.* **76,** 340–50. *159, 272.*

MIRON, I., 1962. Cìteva date privind efectele inundării asupra unor animale de current (French summary). *Anal. ştiinţ. Univ. Al. I. Cuza II*, N.S. **2**, 269–78. *447*.

MOFFETT, J. W., 1936. A quantitative study of the bottom fauna in some Utah streams variously affected by erosion. *Bull. Univ. Utah. biol. Ser.* **26, 9,** 33 pp. *224–5*.

MOKEEVA, N. P., 1964. The algoflora of the Oka River (Russian). *Trudy zool. Inst. Leningr.* **32,** 92–105. *67, 107*.

MONAKOV, A. V., 1964. Zooplankton of the Oka River (Russian). Ibid. **32,** 106–12. *98*.

MOON, H. P., 1935. Methods and apparatus suitable for an investigation of the littoral region of oligotrophic lakes. *Int. Revue ges. Hydrobiol. Hydrogr.* **32,** 319–33. *239, 242*.

—— 1939. Aspects of the ecology of aquatic insects. *Trans. Soc. Br. Ent.* **6,** 39–49. *194*.

—— 1940. An investigation of the movements of freshwater invertebrate faunas. *J. Anim. Ecol.* **9,** 76–83. *234*.

—— 1956. Observations on a small portion of a drying chalk stream. *Proc. zool. Soc. Lond.* **126,** 327–33. *406*.

MOORE, G. A., 1950. The cutaneous sense organs of barbeled minnows adapted to life in the muddy waters of the Great Plains Region. *Trans. Am. microsc. Soc.* **69,** 69–95. *308*.

—— and PADEN, A., 1950. Fishes of the Illinois River in Oklahoma and Arkansas. *Am. Midl. Nat.* **44,** 76–95. *338*.

MOORE, I. J., 1964. Effects of water current on fresh-water snails *Stagnicola palustris* and *Physa propinqua*. *Ecology*, **45,** 558–64. *131, 147*.

MORDUCHAI-BOLTOVSKOI, PH. D., 1961*a*. The development of the bottom-fauna in the Gorbovskoie and Kuibyshevskoie Reservoirs (Russian). *Trudy Inst. Biol. Vodokhran.* **4, 7,** 49–177. *447*.

—— 1961*b*. Bottom fauna of the deltas of the Pontocaspian Basin (Russian). *Trudy vses. gidrobiol. Obshch.* **11,** 136–49. *268*.

—— 1961*c*. Die Entwicklung der Bodenfauna in den Stauseen der Wolga. *Verh. int. Verein. theor. angew. Limnol.* **14,** 647–51. *447*.

MORDUKHAY-BOLTOVSKOY, PH. D., 1964. Caspian fauna in fresh waters outside the Ponto-Caspian Basin. *Hydrobiologia*, **23,** 159–64. *117*.

MORETTI, G. P., 1949. Valuttizone biologica del fiume Potenza come esponente delle acque fluviali delle Marche. *Verh. int. Verein. theor. angew. Limnol.* **10,** 335–8. *391*.

—— 1953. I fattori ecologici che regolano la vite nelle acque correnti delle Alpi e degli Appenini. *Memorie Ist. ital. Idrobiol.* **7,** 229–316. *391*.

—— and GIANOTTI, F. S., 1962. Der Einfluss der Strömung auf die Verteilung der Trichopteren *Agapetus* gr. *fuscipes* Curt. und *Silo* gr. *nigricornis* Pict. *Schweiz. Z. Hydrol.* **24,** 467–84. *197*.

MORGAN, N. C., and EGGLISHAW, H. J., 1965. A survey of the bottom fauna of streams in the Scottish Highlands Part I. Composition of the fauna. *Hydrobiologia*, **25,** 181–211. *240, 434*.

MOROFSKY, W. F., 1936. Survey of insect fauna of some Michigan trout streams in connection with improved and unimproved streams. *J. econ. Ent.* **29,** 749–54. *450*.

MOSEVICH, N. A., 1947. Winter ice conditions in the rivers of the Ob-Irtysh basin (Russian). *Izv. vses. Inst. ozern. rechn. ryb. Khoz.* **25,** 1–56. *32, 319, 434*.

MOSSEWITSCH, N. A., 1961. Sauerstoffdefizit in den Flussen des Westsibirischen Tieflandes, seine Ursachen und Einflüsse auf die aquatische Fauna. *Verh. int. Verein. theor. angew. Limnol.* **14**, 447–50. *319–20.*

MOTAS, C., *et al.*, 1962. *Cercetări asupra biologiei izvoarelor şi apelor freatice din partea centrala a Cîmpiei Romîne.* Edit. Acad. Repub. Pop. Romîne, Bucharest. *400.*

MOTTLEY, *et al.*, 1939. The determination of the food grade of streams. *Trans. Am. Fish. Soc.* **68**, 336–43. *225, 246, 248.*

MÜLLER, K., 1952. Über das Wachstum verschiedener Forellenpopulationen in Mittelgebirgsbächen. *Ber. limnol Flussstn Freudenthal,* **3**, 47–53. *303.*

—— 1953. Produktionsbiologische Untersuchungen in nordschwedischen Fliessgewässern. Teil I. Der Einfluss der Flössereiregulierungen auf den quantitativen und qualitativen Bestand der Bodenfauna. *Rep. Inst. Freshwat. Res. Drottningholm,* **34**, 90–121. *447.*

—— 1954*a*. Die Fischbesiedlung und die regionale Einstufung der Fliessgewässer der nordschwedischen Waldregion. *Ber. limnol. Flussstn Freudenthal,* **6**, 51–6. *385.*

—— 1954*b*. Wachstum und Ernährung des Gründlings (*Gobio fluviatilis Cuv.*) in der Fulda. Ibid. **6**, 61–4. *367.*

—— 1954*c*. Faunistisch-ökologische Untersuchungen in nordschwedischen Waldbächen. *Oikos,* **5**, 77–93. *256, 390.*

—— 1954*d*. Investigations on the organic drift in North Swedish streams. *Rep. Inst. Freshwat. Res. Drottningholm.* **35**, 133–48. *251, 259, 268.*

—— 1954*e*. Produktionsbiologische Untersuchungen in nordschwedischen Fliessgewässern 2. Untersuchungen über Verbreitung, Bestandsdichte, Wachstum und Ernährung der Fische der nordschwedischen Waldregion. Ibid. **35**, 149–83. *366–7, 373.*

—— 1954*f*. Die Drift in fliessenden Gewässern. *Arch. Hydrobiol.* **49**, 539–45. *155, 267–8.*

—— 1955*a*. Qualitative und quantitative Untersuchungen an Fischen der Fulda. *Hydrobiologia,* **7**, 230–44. *338, 439.*

—— 1955*b*. Produktionsbiologische Untersuchungen in nordschwedischen Fliesgewässern 3. Die Bedeutung der Seen und Stillwasserzonen für die Produktion in Fliessgewässern. *Rep. Inst. Freshwat. Res. Drottningholm,* **36**, 148–62. *256–7.*

—— 1956. Das produktionsbiologische Zusammenspiel zwischen See und Fluss. *Ber. limnol. Flussstn Freudenthal,* **7**, 1–8. *256, 258–9.*

—— 1958. Beitrag zur Methodik der Untersuchung fliessender Gewässer. *Arch. Hydrobiol.* **54**, 567–70. *239, 261.*

—— 1961. Die Biologie der Äscher (*Thymallus thymallus* L.) im Lule Älv (Schwedisch Lappland). *Z. Fisch.* N.F. **10**, 173–201. *359, 367.*

—— 1963. Diurnal rhythm in 'organic drift' of *Gammarus pulex. Nature, Lond.* **198**, 806–7. *263.*

—— 1965. An automatic stream drift sampler. *Limnol. Oceanogr.* **10**, 483–5. *262–3.*

—— 1966*a*. Die Tagesperiodik von Fliesswasserorganismen. *Z. Morph. Ökol. Tiere,* **56**, 93–142. *261–5.*

—— 1966*b*. Zur Periodik von *Gammarus pulex. Oikos,* **17**, 207–11. *263.*

MÜLLER-HAECKEL, A., 1965. Tagesperiodik des Siliziumgehaltes in einem Fliessgewässer. Ibid. **16**, 232–3. *67, 96.*

MÜLLER-HAECKEL, A., 1966. Diatomeendrift in Fliessgewässern. *Hydrobiologia,* **28,** 73–87. *67.*

MUNCY, R. J., 1958. Movements of channel catfish in Des Moines River, Boone County, Iowa. *Iowa St. Coll. J. Sci.* **32,** 563–71. *343.*

MUNDIE, J. H., 1956. Emergence traps for aquatic insects. *Mitt. int. Verein. theor. angew. Limnol.* **7,** 1–13. *241.*

—— 1964. A sampler for catching emerging insects and drifting materials in streams. *Limnol. Oceanogr.* **9,** 456–9. *262.*

MUNRO, W. R., and BALMAIN, K. H., 1956. Observations on the spawning runs of brown trout in the South Queich, Loch Leven. *Freshwat. Salm. Fish. Res.* **13,** 17 pp. *349.*

MURPHY, G., 1942. Relationship of the fresh-water mussel to trout in the Truchee River. *Calif. Fish Game,* **28,** 89–102. *221.*

MURPHY, H. E., 1922. Notes on the biology of some of our North American species of mayflies II. Notes on the biology of mayflies of the genus *Baetis. Bull. Lloyd Libr.* **22,** 40–6. *288.*

MURRAY, M. J., 1938. An ecological study of the invertebrate fauna of some northern Indiana streams. *Invest. Indiana Lakes Streams,* **1,** 101–10. *249.*

MUTTKOWSKI, R. A., 1929. The ecology of trout streams in Yellowstone National Park. *Roosevelt wild Life Ann.* **2,** 155–240. *155, 160, 391.*

—— and SMITH, G. M., 1929. The food of trout stream insects in Yellowstone National Park. Ibid. **2,** 241–63. *193, 195.*

NAKAMURA, K. *et al.,* 1954. Qualitative study on benthos, periphyton and potamoplankton referred to hydrography in Chikuma River (Japanese: English summary). *Bull. Freshwat. Fish. Res. Lab., Tokyo,* **3, 1,** 1–25. *391.*

NARAYANAYYA, D. V., 1928. The aquatic weeds in Deccan irrigation canals. *J. Ecol.* **16,** 123–33. *92.*

NAUMANN, E., 1925. Die höhere Vegetation des Bach- und Teichgebietes bei Aneboda. *Ark. Bot.* **19, 2,** 31 pp. *85.*

NAUWERCK, A., 1963. Die Beziehungen zwischen Zooplankton und Phytoplankton im See Erken. *Symb. bot. upsal.* **17, 5,** 163 pp. *433.*

NEAVE, F., 1930. Migratory habits of the mayfly, *Blasturus cupidus* Say. *Ecology,* **11,** 568–76. *155, 405.*

—— 1943*a*. Racial characteristics and migratory habits in *Salmo gairdneri. J. Fish. Res. Bd Can.* **6,** 245–51. *348.*

—— 1943*b*. Diurnal fluctuations in the upstream migration of coho and spring salmon. Ibid. **6,** 158–63. *355.*

NEEDHAM, J. G., *et al.,* 1935. *The Biology of Mayflies.* Comstock, Ithaca, N.Y. *159.*

NEEDHAM, P. R., 1928. A net for the capture of stream drift organisms. *Ecology,* **9,** 339–42. *261.*

—— 1934. Quantitative studies of streams bottom foods. *Trans. Am. Fish. Soc.* **64,** 238–47. *238, 248, 294.*

—— 1938. *Trout Streams.* Comstock, Ithaca, N.Y. *vii.*

—— 1961. Observations on the natural spawning of eastern brook trout. *Calif. Fish Game,* **47,** 27–40. *357.*

—— and CRAMER, F. K., 1943. Movement of trout in Convict Creek, California. *J. Wildl. Mgmt,* **7,** 142–8. *354.*

—— and JONES, A. C., 1959. Flow, temperature, solar radiation, and ice in relation to activities of fishes in Sagehen Creek, California. *Ecology*, **40**, 465–74. *32, 310–12, 322.*

NEEDHAM, P. R., and TAFT, A. C., 1934. Observations on the spawning of steelhead trout. *Trans. Am. Fish. Soc.* **64**, 332–8. *357–8.*

—— and USINGER, R. L., 1956. Variability in the macrofauna of a single riffle in Prosser Creek, California, as indicated by the Surber sampler. *Hilgardia*, **24, 14**, 383–409. *246.*

—— and WELSH, J. P., 1953. Rainbow trout (*Salmo gairdneri* Richardson) in the Hawaiian Islands. *J. Wildl. Mgmt*, **17**, 233–55. *119–20, 137, 302.*

NEEL, J. K., 1951. Interrelations of certain physical and chemical features in a head-water limestone stream. *Ecology*, **32**, 368–91. *30, 40, 42, 48, 439.*

—— 1953. Certain limnological features of a polluted irrigation stream. *Trans. Am. Fish. Soc.* **72**, 119–35. *449.*

—— 1963. Impact of reservoirs. *Limnology in North America.* (ed. D. G. Frey). Madison, Wisconsin, pp. 575–93. *447, 449.*

—— *et al.*, 1963. Main stem reservoir effects on water quality in the Central Missouri River. *Mimeo. Rep. U.S. Dept Hlth*, Kansas City, 111 pp. *34, 108, 448.*

NEFF, D. J., 1957. Ecological effects of beaver habitat abandonment in the Colorado Rockies. *J. Wildl. Mgmt*, **21**, 80–4. *382.*

NEGUS, C. L., 1966. A quantitative study of growth and production of unionid mussels in the River Thames at Reading. *J. Anim. Ecol.* **35**, 513–32. *276, 281, 426.*

NEIL, J. H., and OWEN, G. E., 1964. Distribution, environmental requirements and significance of *Cladophora* in the Great Lakes. *Publs gt Lakes Res. Inst.* **11**, 113–21. *68–9.*

NEILL, R. M., 1938. The food and feeding of the brown trout (*Salmo trutta L.*) in relation to the organic environment. *Trans. R. Soc. Edinb.* **59**, 481–520. *238, 372.*

NEISWESTNOWA-SHADINA, K., 1935. Zur Kenntniss der rheophilen Mikrobenthos. *Arch. Hydrobiol.* **28**, 555–82. *114, 176.*

—— 1937. Verteilung und Saisondynamik der Biozönosen des Flussbettes und ihre Studiummethoden (Russian: German summary). *Dokl. Akad. Nauk S.S.S.R.* biol. ser. 1247–72. *176, 250.*

NELSON, D. J., and SCOTT, D. C., 1962. Role of detritus in the productivity of a rock-outcrop community in a Piedmont stream. *Limnol. Oceanogr.* **7**, 396–413. *255, 272, 292, 413, 416, 424, 433, 436.*

NELSON, E. M., 1961. The comparative morphology of the definitive swim bladder in the Catastomidae. *Am. Midl. Nat.* **65**, 101–10. *306.*

NELSON, J. S., 1965. Effects of fish introductions and hydroelectric development on fishes in the Kananaskis River system. *J. Fish. Res. Bd Can.* **22**, 721–53. *450.*

NEVIN, F. R., 1936. A study of the larger invertebrate forage organisms in selected areas of the Delaware and Susquehanna watersheds. *Rep. N.Y. St. Conserv. Dep.* **25** Suppl., 195–204. *226.*

NEWELL, A. E., 1957. Two-year study of movements of stocked brook trout and rainbow trout in a mountain stream. *Progve FishCult.* **19**, 76–80. *343.*

NEWELL, R., 1965. The role of detritus in the nutrition of two marine deposit feeders, the Prosobranch *Hydrobia ulvae* and the bivalve *Macoma balthiea. Proc. Zool. Soc. Lond.* **144**, 25–45. *441.*

NICOLAU-GUILLAMET, P., 1959. Recherches faunistiques et écologiques sur la rivière 'La Massane'. *Vie Milieu*, **10**, 217–66. *79, 80, 276, 280, 390.*

NIELSEN, A., 1942. Über die Entwicklung und Biologie der Trichopteren. *Arch. Hydrobiol.* Suppl. **17**, 255–631. *128, 135, 141, 157, 159, 166–7, 184, 187, 193, 272, 275, 277, 280, 387.*

—— 1943. Postembryonale Entwicklung und Biologie der rheophilen Köcherfliege *Oligoplectrum maculatum* Fowicroy. *K. danske Vidensk. Selsk. Skr.* **19, 2,** 86 pp. *137, 156, 159, 185.*

—— 1950*a*. On the zoogeography of springs. *Hydrobiologia*, **2,** 313–21. *287, 400.*

—— 1950*b*. The torrential invertebrate fauna. *Oikos*, **2,** 177–96. *8, 11, 134, 137, 146, 156, 178.*

—— 1950*c*. Notes on the genus *Apatidea* MacLachlan, with descriptions of two new and possibly endemic species from the springs of Himmerland. *Ent. Meddr*, **25,** 384–404. *287.*

—— 1951*a*. Is dorsoventral flattening of the body an adaptation to torrentia life? *Verh. int. Verein. theor. angew. Limnol.* **11,** 264–7. *122, 134.*

—— 1951*b*. Spring fauna and speciation. Ibid. **11,** 261–3. *287.*

NIETZKE, G., 1937. Die Kossau. Hydrobiologische-faunistische Untersuchungen an schleswig-holsteinischen Fliessgewässern. *Arch. Hydrobiol.* **32,** 1–74. *199, 390.*

NIKOLSKI, G. V., 1933. On the influence of the rate of flow on the fish fauna of the rivers of Central Asia. *J. Anim. Ecol.* **2,** 266–81. *302–3, 387.*

—— 1937. On the distribution of fishes according to the nature of their food in rivers flowing from the mountains of middle Asia. *Verh. int. Verein. theor. angew. Limnol.* **8, 2,** 169–76. *371.*

—— 1963. *The Ecology of Fishes.* Academic Press, London. *310, 325, 347, 351, 361, 368.*

NILSSON, N., 1957. On the feeding habits of trout in a stream of northern Sweden. *Rep. Inst. Freshwat. Res. Drottningholm*, **38,** 154–66. *267, 367.*

NILSSON, S., 1964. Fresh water Hyphomycetes. Taxonomy, morphology and ecology. *Symb. bot. upsal.* **18,** 5–130. *435.*

NISBET, M., 1961. Un example de pollution de rivière par vidage d'une retenue hydroélectrique. *Verh. int. Verein. theor. angew. Limnol.* **14,** 678–80. *449.*

NOBLE, G. K., 1931. *The Biology of the Amphibia.* McGraw-Hill, New York. *377–8.*

NOEL, M. S., 1954. Animal ecology of a New Mexico spring brook. *Hydrobiologia*, **6,** 120–35. *153–4, 391, 400–1.*

NOEL-BUXTON, M. B., 1956. Field experiments with DDT in association with finely divided organic material for the destruction of the immature stages of the genus *Simulium* in the Gold Coast. *Jl W. Afr. Sci. Ass.* **2,** 36–40. *186.*

NORTHCOTE, T. G., 1958. Effect of photoperiodism on response of juvenile trout to water currents. *Nature, Lond.* **181,** 1283–4. *354.*

—— 1962. Migratory behaviour of juvenile rainbow trout, *Salmo gairdneri*, in outlet and inlet streams of Loon Lake, British Columbia. *J. Fish. Res. Bd Can.* **19,** 201–70. *354.*

NOYES, A. A., 1914. Biology of net-spinning Trichoptera of Cascadilla Creek. *Ann. ent. Soc. Am.* **7,** 250–71. *188.*

NYKVIST, N., 1963. Leaching and decomposition of water-soluble organic substances from different types of leaf and needle litter. *Studia Forestalia Suecica*, **3,** 31 pp. *434, 436.*

O'CONNELL, T. R., and CAMPBELL, R. S., 1953. The benthos of the Black River and Clearwater Lake, Missouri. *Univ. Mo Stud.* **26, 2,** 25–41. *208, 211, 447.*

ODUM, H. T., 1956. Primary production in flowing waters. *Limnol. Oceanogr.* **1,** 102–17. *48, 50, 412.*

—— 1957. Trophic structure and productivity of Silver Springs. *Ecol. Monogr.* **27,** 55–112. *401, 416.*

—— and CALDWELL, D. K., 1955. Fish respiration in the natural oxygen gradient of an anaerobic spring in Florida. *Copeia,* **1955,** 104–6. *42, 320, 398.*

—— and HOSKIN, C. M., 1957. Metabolism of a laboratory stream microcosm. *Publs Inst. mar. Sci. Univ. Tex.* **4,** 115–33. *413.*

—— *et al.*, 1958. The chlorophyl 'A' of communities. Ibid. **5,** 65–96. *415.*

O'FARRELL, A. F., 1949. The blue hole expedition, 1949. a short term limnological study on the Gara River, New South Wales. *J. New Engl. Univ. Coll. Sci.* **5,** 46–82. *117.*

OHLMER, W., and SCHWARTZKOPFF, J., 1959. Schwimmgeschwindigkeiten von Fischen aus stehenden Binnengewässern. *Naturwissenschaften,* **46,** 362–3. *302, 310.*

ØKLAND, J., 1963. A review of the quantity of bottom fauna in Norwegian lakes and rivers (Norwegian: English summary). *Fauna, Oslo,* **16,** Suppl., 1–67. *254, 257.*

OLIFF, W. D., 1960. Hydrobiological studies on the Tugela River system. Part I. The main Tugela River. *Hydrobiologia,* **16,** 281–385. *80, 227, 293, 387, 391.*

—— 1963. Idem. Part III. The Buffalo River. Ibid. **21,** 355–79. *227, 391.*

—— and KING, J. L., 1964. Idem. Part IV. The Mooi River. Ibid. **24,** 567–83. *213, 391.*

—— *et al.*, 1965. Idem. Part V. The Sundays River. Ibid. **26,** 189–202. *68, 391.*

ORGAN, J. A., 1961. Studies on the local distribution, life history, and population dynamics of the salamander genus *Desmognathus* in Virginia. *Ecol. Monogr.* **31,** 189–220. *378.*

ORGHIDAN, T., 1959. Ein neuer Lebensraum des unterirdischen Wassers: Der hyporheische Biotop. *Arch. Hydrobiol.* **55,** 392–414. *407.*

ORTMAN, A. E., 1920. Correlation of shape and station in fresh-water mussels (Naiades). *Proc. Am. phil. Soc.* **59,** 269–312. *143.*

OUTTEN, L. M., 1957. A study of the life history of the cyprinid fish *Notropis coccogenis*. *J. Elisha Mitchell scient. Soc.* **73,** 68–84. *359.*

—— 1958. Studies of the life history of the cyprinid fishes *Notropis galacturus* and *rubricroceus*. Ibid. **74,** 122–34. *336, 359.*

OWEN, G. E., and JOHNSON, M. G., 1966. Significance of some factors affecting yields of phosphorus from several Lake Ontario watersheds. *Publs gt Lakes Res. Inst.,* **15,** 400–10. *47, 445.*

OWENS, M., 1965. Some factors involved in the use of dissolved-oxygen distributions in streams to determine productivity. *Mem. Ist. Ital. Idrobiol.* **18** Suppl., 209–24. *413.*

—— and EDWARDS, R. W., 1961. The effect of plants on river conditions II. Further crop studies and estimates of net productivity of macrophytes in a chalk stream. *J. Ecol.* **49,** 119–26. *51, 416–17.*

—— —— 1962. Idem III. Crop studies and estimates of net productivity in four streams in southern England. Ibid. **50,** 157–62. *416.*

—— —— 1963. Some oxygen studies in the River Lark. *Proc. Soc. Wat. Treat. Exam.* **12,** 126–46. *51, 434.*

OWENS, M., and EDWARDS, R. W., 1964. A chemical survey of some English rivers. Ibid. **13,** 134–44. *41.*

—— and MARIS, P. J., 1964. Some factors affecting the respiration of some aquatic plants. *Hydrobiologia,* **23,** 533–43. *82.*

PAGAST, F., 1934. Über die Metamorphose von *Chironomus xenolabis* K. eines Schwammparasiten. *Zool. Anz.* **105,** 155–8. *122, 180.*

PALOUMPIS, A. A., 1958*a*. Measurement of some factors affecting the catch in a minnow seine. *Proc. Iowa Acad. Sci.* **65,** 580–6. *314.*

—— 1958*b*. Responses of some minnows to flood and drought conditions in an intermittent stream. *Iowa St. Coll. J. Sci.* **32,** 547–61. *329.*

PANKNIN, W., 1947. Zur Entwicklung der Algensoziologie und zum Problem der 'echten' und 'zügehörigen' Algengesellschaften. *Arch. Hydrobiol.* **41,** 92–111. *72.*

PATE, V. S. L., 1932. Studies on the fish food supply in selected areas. *Rep. N.Y. St. Conserv. Dep.* **21** Suppl., 133–49. *248.*

—— 1933. Idem. Ibid. **22** Suppl., 130–56. *248.*

—— 1934. Idem. Ibid. **23** Suppl., 136–57. *248.*

PATRICK, R., 1959. Aquatic life in a new stream. *Wat. Sewage Wks,* Dec. 1959, 5 pp. *266.*

—— 1961. A study of the numbers and kinds of species found in rivers in eastern United States. *Proc. Acad. nat. Sci. Philad.* **113,** 215–58. *441–2.*

—— 1964. A discussion of the results of the Catherwood Expedition to the Peruvian headwaters of the Amazon. *Verh. int. Verein. theor. angew. Limnol.* **15,** 1084–90. *441.*

—— *et al.*, 1954. A new method for determining the pattern of the diatom flora. *Notul. Nat.* **259,** 1–12. *58, 73.*

PAWŁOWSKI, L. K., 1948. Contribution à la systématique des sangsues du genre *Erpobdella* de Blainville. *Pr. łódź. Tow. Nauk,* **3, 8,** 54 pp. *201, 387.*

—— 1956. Première liste des Rotifères trouvés dans la riviére Grabia. *Bull. Soc. Sci. Lett. Łódź,* **3, 7,** 1–54. *114.*

—— 1959*a*. Remarques sur la répartition de la faune torrenticole des Carpathes. Ibid. **3, 57,** 7–84. *391.*

—— 1959*b*. Aperçu sur la faune des hirudinées des Carpathes. *Pr. łódź. Tow. Nauk,* **3, 55,** 49 pp. *201.*

PEARSALL, W. H., 1923. A theory of diatom periodicity. *J. Ecol.* **11,** 165. *67.*

PECKHAM, R. S., and DINEEN, C. F., 1957. Ecology of the central mudminnow, *Umbra limi* (Kirtland). *Am. Midl. Nat.* **58,** 222–31. *335, 349.*

PENNAK, R. W., 1943. Limnological variables in a Colorado mountain stream. Ibid. **28,** 189–99. *101.*

—— and VAN GERPEN, E. D., 1947. Bottom fauna production and physical nature of the substrate in a northern Colorado trout stream. *Ecology,* **28,** 42–8. *249, 391.*

PENTELOW, F. T. K., *et al.*, 1938. An investigation of the effects of milk wastes on the Bristol Avon. *Fishery Invest., Lond.* **1, 4, 1,** 80 pp. *249, 292.*

PERCIVAL, E., and WHITEHEAD, H., 1926. Observations on the biology of *Ephemera danica*, Müll. *Proc. Leeds phil. lit. Soc.* **1,** 136–48. *206.*

—— —— 1928. Observations on the ova and oviposition of certain Ephemeroptera and Plecoptera. Ibid. **1,** 271–88. *140, 278.*

PERCIVAL, E., and WHITEHEAD, H., 1929. A quantitative study of some types of stream-bed. *J. Ecol.* **17,** 282–314. *208–9, 213, 249, 390.*

—— —— 1930. Biological survey of the River Wharfe II. Report on the invertebrate fauna. Ibid. **18,** 286–302. *207, 286, 390.*

PETERSEN, R. H., 1962. Aquatic Hyphomycetes from North America. I. Aleuriosporae (Part I), and a key to the genera. *Mycologia,* **54,** 117–51. *435.*

—— 1963*a*. Idem. II. Aleuriosporae (Part II), and Blastosporae. Ibid. **55,** 18–29. *435.*

—— 1963*b*. Idem. III. Phialosporae and miscellaneous species. Ibid. **55,** 570–81. *435.*

PETERSON, D. G., and WOLFE, L. S., 1958. The biology and control of black flies (Diptera: Simuliidae) in Canada. *Int. Congr. Ent.* **10, 3,** 551–64. *159, 186, 273, 275.*

PETR, T., 1961. Quantitative Studie des Makrozoobenthos der Zuflüsse des Staubeckens Sedlice mit Rüchsicht auf ihre Verunreinigung. *Sb. vys. Šk. chem-technol. Praze, Technol. vody,* **5,** 367–494. *283, 287.*

PETRAVICZ, J. J., 1936. The breeding habits of the least darter, *Microperca punctulata* Putnam. *Copeia,* **1936,** 77–82. *360.*

—— 1938. The breeding habits of the blacksided darter, *Hadropterus maculatus* Girard. Ibid. **1938,** 40–4. *359.*

PETROPAVLOVSKAYA, V. N., 1951. Nutrition of young sturgeon in the River Don in the period of their hatching (Russian). *Trudy vses. gidrobiol. Obshch.* **3,** 58–71. *367, 369, 373.*

PHILIPSON, G. N., 1953. The larva and pupa of *Hydropsyche instabilis* Curtis (Trichoptera: Hydropsychidae). *Proc. R. ent. Soc. Lond.* **A28,** 17–23. *188–9.*

—— 1954. The effect of water flow and oxygen concentration on six species of caddis fly (Trichoptera). *Proc. zool. Soc. Lond.* **124,** 547–64. *163, 166, 188.*

PHILLIPS, J. S., 1929. A report on the food of trout and other conditions affecting their well-being in the Wellington district. *Fish. Bull. N.Z.* **2,** 31 pp. *119.*

—— 1931. A further report on the conditions affecting the well-being of trout in New Zealand. Ibid **3,** 27 pp. *253.*

PHILLIPPS, R. W., and CLAIRE, E. W., 1966. Intragravel movement of the reticulate sculpin, *Cottus perplexus,* and its potential as a predator on salmon embryos. *Trans. Am. Fish. Soc.* **95,** 210–12. *336.*

PHILLIPSON, J., 1956. A study of factors determining the distribution of the blackfly *Simulium ornatum* Mg. *Bull. ent. Res.* **47,** 227–38. *148, 197.*

PHINNEY, H. K., and MCINTYRE, C. D., 1965. Effect of temperature on metabolism of periphyton communities developed in laboratory streams. *Limnol. Oceanogr.* **10,** 341–4. *62.*

PICKEN, L. E. R., 1937. The structure of some protozoan communities. *J. Ecol.* **25,** 368–84. *114.*

PINET, J.-M., 1962. Observations sur la biologie et l'écologie d'*Oligoneuriella rhenana* Imhoff (Ephemeroptera). *Annls Stn cent. Hydrobiol, appl.* **9,** 301–31. *184, 231, 271.*

PIROZHNIKOV, P. L., and SHULGA, Y. E., 1957. Basic characteristics of the zooplankton of the lower River Lena (Russian). *Trudy vses. gidrobiol. Obshch.* **8,** 219–30. *98–9.*

PLESKOT, G., 1951. Wassertemperatur und Leben im Bach. *Wett. Leben,* **3,** 129–43. *202, 402.*

PLESKOT, G., 1953. Zur Ökologie der Leptophlebiiden (Ins., Ephemeroptera). *Öst. zool. Z.* **4,** 45–107. *147–8, 159, 167, 195, 271.*

—— 1954. Die Lebensgemeinschaften der Bäche, Flüsse und Ströme. *Protokoll der 2 österreichischen Naturschutztagung*, Vienna, pp. 39–58. *411.*

—— 1958. Die Periodizität einiger Ephemeropteren der Schwechat. *Wasser und Abwasser*, **1958,** 1–32. *271, 275, 279, 286–7.*

—— 1960. Beobachtungen über diapausen in der Entwicklung der Ephemeropteren. *Int. Congr. Ent.* **11, 1,** 363–6. *271, 275, 280, 284–5, 287, 289.*

—— 1961. Die Periodizität der Ephemeroptern-Fauna einiger österreichischer Fliessgewässer. *Verh. int. Verein. theor. angew. Limnol.* **14,** 410–16. *278, 280, 283, 289.*

POLLARD, R. A., 1955. Measuring seepage through salmon spawning gravel. *J. Fish. Res. Bd Can.* **12,** 706–41. *358.*

POMEISL, E., 1953*a*. Der Mauerbach. *Wett. Leben Sonderh.* **2,** 103–21. *32, 197, 225, 390.*

—— 1953*b*. Studien an Dipterenlarven des Mauerbaches. Ibid. **2,** 165–75. *389.*

—— 1961. Ökologische-biologische Untersuchungen an Plecopteren in Gebiet der niederösterreichischen Kalkalpen und des diesem vorgelagerten Flyschgebietes. *Verh. int. Verein. theor. angew. Limnol.* **14,** 351–4. *168, 219.*

PONYI, E., 1962. Beiträge zur Kenntnis des Crustaceen-Planktons der ungarischen Donau. *Opusc. zool. Bpest*, **4,** 127–32. *99.*

POPOVICI-BAZNOSANU, A., 1928. Sur la prétendue adaptation morphologique des larves à la vie rhéophile. *Bull. biol. Fr. Belg.* **62,** 126–48. *122.*

POWER, G., 1958. The evolution of freshwater races of the Atlantic salmon (*Salmo salar* L.). *Arctic*, **11,** 86–92. *301.*

—— 1959. Field measurements of the basal oxygen consumption of Atlantic salmon parr and smolts. Ibid. **12,** 195–202. *321.*

PROWSE, G. A., and TALLING, J. F., 1958. The seasonal growth and succession of plankton algae in the White Nile. *Limnol. Oceanogr.* **3,** 222–38. *108.*

PRUTHI, H. S., 1928. Observations on the biology and morphology of the immature stages of *Aulacodes peribocalis* Wlk (Hydrocampinae—Lepidoptera). *Rec. Indian Mus.* **30,** 353–6. *137, 156.*

PURKETT, C. A., 1958. Growth of the fishes of the Salt River, Missouri. *Trans. Am. Fish. Soc.* **87,** 116–31. *343.*

—— 1961. Reproduction and early development of the paddlefish. Ibid. **90,** 125–9. *359, 361.*

RAABE, H., 1951. Die Diatomenflora der ostholsteinischer Fliessgewässer. *Arch. Hydrobiol.* **44,** 521–638. *71.*

RADFORTH, I., 1940. The food of the grayling (*Thymallus thymallus*), flounder (*Platychthys flesus*), roach (*Rutilus rutilus*), and gudgeon (*Gobio fluviatilis*), with special reference to the Tweed watershed. *J. Anim. Ecol.* **9,** 302–18. *366–7.*

RANEY, E. C., 1939. The breeding habits of *Ichthyomyzon greeleyi* Hubbs and Trautman. *Copeia*, **1939,** 111–12. *357.*

—— 1940. The breeding behavior of the common shiner, *Notropis cornutus* (Mitchill). *Zoologica N.Y.* **25,** 1–14. *356, 359.*

—— 1947. *Nocomis* nests used by other breeding Cyprinid fishes in Virginia. Ibid. **32,** 125–32. *358–9.*

—— and LACHNER, E. A., 1939. Observations on the life history of the spotted darter, *Poecilichthys maculatus* (Kirtland). *Copeia*, **1939,** 157–65. *356.*

RANEY, E. C., and LACHNER, E. A., 1946. Age, growth, and habits of the hog sucker *Hypentelium nigricans* (Le Sueur) in New York. *Am. Midl. Nat.* **36,** 76–86. *359, 365, 368.*

—— and MASSMAN, W. H., 1953. The fishes of the tidewater section of the Pamunkey River, Virginia. *J. Wash. Acad. Sci.* **43,** 424–32. *301–2, 346.*

—— and ROECKER, R. M., 1947. Food and growth of two species of watersnakes from Western New York. *Copeia,* **1947,** 171–4. *379.*

—— and WEBSTER, D. A., 1942. The spring migration of the common white sucker, *Catastomus c. commersonnii* (Lacépède), in Skaneatales Lake Inlet, New York. Ibid. **1942,** 139–48. *359.*

RASMUSSEN, D. I., 1940. Beaver–trout relationship in the Rocky Mountain region. *Trans. N. Am. Wildl. Conf.* **5,** 256–63. *381.*

RAUŠER, J., 1962. Zur Verbreitungsgeschichte einer Insektendauergruppe (Plecoptera) in Europa. *Pr. brn. Zakl. čsl Akad. Věd.* **34,** 281–383. *284.*

RAWLINSON, R., 1939. Studies on the life history and breeding of *Ecdyonurus venosus* (Ephemeroptera). *Proc. zool. Soc. Lond.* **B 109,** 377–450. *125, 152.*

REED, E. B., and BEAR, G., 1966. Benthic animals and food eaten by brook trout in Archuleta Creek, Colorado. *Hydrobiologia,* **27,** 227–37. *367.*

REEVES, C. D., 1907. The breeding habits of the rainbow darter (*Etheostoma coeruleum* Storer), a study in sexual selection. *Biol. Bull. mar. biol. Lab., Woods Hole,* **4,** 23–59. *359.*

REHDER, D. R., 1959. Some aspects of the life history of the carp, *Cyprinus carpio,* in the Des Moines River, Boone County, Iowa. *Iowa St. Coll. J. Sci.* **34,** 11–26. *349, 366–8.*

REID, G. K., 1961. *Ecology of Inland Waters and Estuaries.* Reinhold, New York. *vii, 2, 23.*

REIF, C. B., 1939. The effect of stream conditions on lake plankton. *Trans. Am. microsc. Soc.* **58,** 398–403. *102–3.*

REIGHARD, J., 1913. The breeding habits of the log perch (*Percina caprodes*). *Rep. Mich. Acad. Sci.* **15,** 104–5. *360.*

—— 1920. The breeding behavior of the suckers and minnows. *Biol. Bull. mar. biol. Lab., Woods Hole,* **38,** 1–32. *359.*

—— 1943. The breeding habits of the river chub, *Nocomis micropogon* (Cope). *Pap. Mich. Acad. Sci.* **28,** 397–423. *358–9.*

REINHARD, E. G., 1931. The plankton ecology of the Upper Mississippi, Minneapolis to Winona. *Ecol. Monogr.* **1,** 395–464. *99, 109.*

REYNOLDS, J. B., 1965. Life history of smallmouth bass, *Micropterus dolomieui* Lacépède, in the Des Moines River, Boone County, Iowa. *Iowa St. Coll. J. Sci.* **39,** 417–36. *335, 343.*

REYNOLDSON, T. B., 1961. Observations on the occurrence of *Asellus* (Isopoda, Crustacea) in some lakes of northern Britain. *Verh. int. Verein. theor. angew. Limnol.* **14,** 988–94. *220.*

RHOADES, R., 1962. Further studies on Ohio crayfishes. Cases of sympatry of stream species in southern Ohio. *Ohio J. Sci.* **62,** 27–33. *207.*

RICE, C. H., 1938. Studies in the phytoplankton of the River Thames (1928–1932). *Ann. Bot.* N.S. **2,** 539–81. *98, 101.*

RICHARD, G., 1960. Contribution à l'étude ethologique des odonates. *Int. Congr. Ent.* **11, 1,** 604–7. *192.*

RICHARDSON, R. E., 1921. The small bottom and shore fauna of the Middle and Lower Illinois River and its connecting lakes, Chillicothe to Grafton: its valuation: its sources of food supply: and its relation to the fishery. *Bull. Ill. St. nat. Hist. Surv.* **13,** 363–522. *391, 437.*

RICHARDSON, R. E., 1928. The bottom fauna of the Middle Illinois River, 1913–1925. Its distribution, abundance, valuation, and index value in the study of stream pollution. Ibid. **17,** 387–475. *391.*

RICKER, W. E., 1934. An ecological classification of certain Ontario streams. *Univ. Toronto Stud. biol. Ser.* **37,** 114 pp. *196, 324, 385, 391, 395.*

—— 1958. Handbook of computations for biological statistics of fish populations. *Bull. Fish. Res. Bd Can.* **119,** 300 pp. *339, 428.*

RILEY, C. F. C., 1921. Distribution of the large waterstrider *Gerris remigis* Say throughout a river system. *Ecology,* **2,** 32–6. *116.*

ROBERT, A., 1960. Les Trichoptères des eaux agitées de la rivière du Diable. *Int. Congr. Ent.* **11, 1,** 565–8. *283.*

ROBINS, C. R., and CRAWFORD, R. W., 1954. A short accurate method for estimating the volume of stream flow. *J. Wildl. Mgmt,* **18,** 366–9. *4.*

RODINA, A. G., 1964. Microbiological investigations of the River Oka (Russian). *Trudy zool. Inst. Leningr.* **32,** 52–80. *38–9, 437.*

ROLL, H., 1938. Die Pflanzengesellschaften Östholsteinischer Fliessgewässer. *Arch. Hydrobiol.* **34,** 159–305. *85.*

ROMADINA, E. S., 1959. A study of the zooplankton of the Volga River, using samples collected at the construction site of the Gorkii Power Station in 1953–1955 (Russian). *Trudy VI Sov. po Probl. Biol. Vnut. Vod., Leningrad,* **1959,** 347–51 (Nat. lend. Libr., Boston Spa, Yorks, Engl. R.T.S. 2930). *98–9, 109, 448.*

ROOS, T., 1957. Studies on upstream migration in adult stream-dwelling insects. *Rep. Inst. Freshwat. Res. Drottningholm,* **38,** 167–93. *155.*

ROSS, H. H., 1963. Stream communities and terrestrial biomes. *Arch. Hydrobiol.* **59,** 235–42. *113.*

ROUND, F. E., 1964. The ecology of benthic algae. *Algae and Man* (ed. D. F. Jackson). Plenum Press, New York, pp. 138–84. *53, 55, 57, 59, 60, 65, 67.*

—— 1965. *The Biology of Algae.* Arnold, London. *53, 75, 399.*

ROUNSEFELL, G. A., and EVERHART, W. H., 1953. *Fishery Science, Its Methods and Applications.* Wiley, New York. *339, 450.*

ROY, H. K., 1955. Plankton ecology of the River Hooghly at Palta, West Bengal. *Ecology,* **36,** 169–75. *107.*

RUGGLES, C. P., 1959. Salmon populations and bottom fauna in the Wenatchee River, Washington. *Trans. Am. Fish. Soc.* **88,** 186–90. *254.*

RUHR, C. E., 1957. Effect of stream impoundment in Tennessee on the fish populations of tributary streams. Ibid. **86,** 144–57. *448.*

RUPP, R., 1955. Beaver–trout relationships in the headwaters of the Sunkhaze Stream, Maine. Ibid. **84,** 75–85. *381.*

RUTTER, A. J., 1958. The effects of afforestation on rainfall and run-off. *J. Instn Publ. Hlth Engrs,* 119–38. *2, 3.*

RUTTNER, F., 1926. Bermerkungen über den Sauerstoffgehalt der Gewässer und dessen respiratorischen Wert. *Naturwissenschaften,* **14,** 1237–9. *48.*

—— 1956. Einige Beobachtungen über das Verhalten des Planktons in Seeabflüssen. *Öst. bot. Z.* **103,** 98–109. *102–3.*

—— 1963. *Fundamentals of Limnology.* Transl. D. G. Frey and F. E. J. Fry, University of Toronto Press. *vii, 5, 28, 30, 36, 43, 46, 79.*

RUTTNER-KOLISKO, A. 1961. Biotop und Biozönose des Sandufers einiger österreichischer Flüsse. *Verh. int. Verein. theor. angew. Limnol.* **14**, 362–8. *408.*

—— 1962. Porenraum und kapillare Wasserströmung im Limnopsammal, ein Beispiel für die Bedeutung verlangsamter Strömung. *Schweiz. Z. Hydrol.* **24**, 444–58. *408.*

RYLOV, V. M., 1940. On the negative effect of mineral seston on the nutrition of some planktonic Entomostraca under conditions of river flow (Russian). *Dokl. Acad. Nauk S.S.S.R.* **29**, 7. *104.*

RZÓSKA, J., 1961. Some aspects of the hydrobiology of the River Nile. *Verh. int. Verein. theor. angew. Limnol.* **14**, 505–7. *104.*

—— *et al.*, 1955. Seasonal plankton development in the White and Blue Nile near Khartoum. Ibid. **12**, 327–34. *99, 110.*

SABANEEFF, P., 1952. Das Zooplankton der Fulda-Expedition 1948. *Ber. limnol. Flussstn Freudenthal*, **3**, 1–4. *98.*

—— 1956. Über das Zooplankton der Weser. Ibid. **7**, 28–42. *107.*

SATTLER, W., 1955. Über den Netzbau der Larve von *Hydropsyche angustipennis* Curt. *Naturwissenschaften*, **42**, 186–7. *188.*

—— 1958. Beiträge zur Kenntnis von Lebensweise und Körperbau der Larve und Puppe von *Hydropsyche* Pict. (Trichoptera) mit besonderen Berüchsichtung des Netzbaues. *Z. Morph. Ökol. Tiere*, **47**, 115–92. *188.*

—— 1962. Über einen Fall von hygropetrischen Lebensweise einer Philopotomide (*Chimarrha*, Trichoptera) aus dem brasilianischen Amazonasgebiet. *Arch. Hydrobiol.* **58**, 125–35. *187.*

—— 1963*a*. Über den Körperbau und Ethologie der Larve und Puppe von *Macronema* Pict. (Hydropsychidae), ein als Larve sich von 'Mikro-Drift' ernährendes Trichopter aus dem Amazongebiet. Ibid. **59**, 26–60. *190.*

—— 1963*b*. Die Larven- und Puppenbauten von *Diplectrona felix* McLach. (Trichoptera). *Zool. Anz.* **170**, 53–5. *188–9.*

—— 1965. Über Messvorgänge bei der Bautätigkeit von Trichopteren-Larven. Ibid. **175**, 378–86. *189.*

—— and KRACHT, A., 1963. Drift-Fang einer Trichopterenlarve unter Ausnutzung der Differenz von Gesamtdruck und statischem Druck des fliessenden Wassers. *Naturwissenschaften*, **50**, 362–3. *190.*

SAUNDERS, J. W., and SMITH, M. W., 1962. Physical alteration of stream habitat to improve trout production. *Trans. Am. Fish. Soc.* **91**, 185–8. *332.*

SAWYER, C. N., 1947. Fertilisation of lakes by agricultural and urban drainage. *J. New Engl. Wat. Wks Ass.* **61**, 109–27. *47.*

SAXENA, S. C., 1961. Adhesive apparatus of an Indian hill-stream Sisorid fish, *Pseudechineis sulcatus*. *Copeia*, **1961**, 471–3. *306.*

—— 1962. On the pelvic girdle and fin of a hill-stream Sisorid fish, *Pseudechineis sulcatus*. Ibid. **1962**, 656–7. *306.*

—— and CHANDY, M., 1966. Adhesive apparatus in certain Indian hill stream fishes. *J. Zool. Lond.* **148**, 315–40. *306.*

SCHALLGRUBER, F., 1944. Das Plankton des Donaustromes bei Wein, in qualitativer und quantitativer Hinsicht. *Arch. Hydrobiol.* **39**, 665–89. *98, 109.*

SCHEELE, M., 1952. Systematisch-ökologische Untersuchungen über die Diatomenflora der Fulda. Ibid. **46**, 305–424. *65, 73.*

SCHLIEPER, C., 1952. Versuch einer physiologischen Analyse der besonderen Eigenschaften einiger eurythermer Wassertiere. *Biol. Zbl.* **71**, 449–61. *319.*

SCHLIEPER, C., and BLÄSING, J., 1952. Über Unterscheide in dem individuellen und ökologischen Temperatubereich von *Planaria alpina* Dana. *Arch. Hydrobiol.* **47,** 288–94. *204.*

—— *et al.*, 1952. Experimentelle Veränderungen der Temperatutoleranz bei stenothermen und eurythermen Wassertieren. *Zool. Anz.* **149,** 163–9. *220, 334.*

SCHMASSMANN, H. J., 1951. Untersuchungen über Stoffhaushalt fliessender Gewässer. *Schweiz. Z. Hydrol.* **13,** 300–35. *41.*

SCHMITZ, W., 1954. Phytoplankton-Massenentwicklung in Staubecken und Fliessgewässern. *Verh. int. Verein. theor. angew. Limnol.* **12,** 241–52. *107.*

—— 1955. Physiographische Aspekte der limnologischen Fliessgewässertypen. *Arch. Hydrobiol.* Suppl. **22,** 510–23. *392.*

—— 1956. Salzgehaltschwankungen in der Werra und ihre fischereilichen Auswirkungen. *Vom Wass.* **23,** 113–36. *334.*

—— 1957. Die Bergbach-Zoozönosen und ihre Abgrenzung, dargestellt am Beispiel der oberen Fulda. *Arch. Hydrobiol.* **53,** 465–98. *392.*

—— 1960. Die Einbürgerung von *Gammarus tigrinus* Sexton auf dem europäischen Kontinent. Ibid. **57,** 223–5. *218.*

—— 1961. Fliesswässerforschung—Hydrographie und Botanik. *Verh. int. Verein. theor. angew. Limnol.* **14,** 541–86. *1, 7, 11, 26, 29, 34, 43, 50, 60, 73, 85, 88, 411.*

—— and VOLKERT, E., 1959. Die Messung von Mitteltemperaturen auf reaktionskinetischer Grundlage mit dem Kreispolarimeter und ihre Anwendung in Klimatologie und Bioökologie, speziell in Forst- und Gewåsserkunde. *Zeiss Mitt. Fortschr.* **1,** 300–37. *29–31.*

SCHNELLER, M. V., 1955. Oxygen depletion in Salt Creek, Indiana. *Invest. Indiana Lakes Streams,* **4,** 163–75. *41, 216, 329, 403–4.*

SCHORLER, R., 1907. Mitteilung über das Plankton der Elbe bei Dresden in Sommer 1904. *Arch. Hydrobiol.* **2,** 355–7. *105.*

SCHRÄDER, T., 1932. Über die Möglichkeit einer quantitativen Untersuchung der Boden- und Ufertierwelt fliessender Gewässer, Zugleich: Fischereibiologische Untersuchungen im Wesergebiet I. *Z. Fisch.* **30,** 105–25. *207, 237, 239.*

—— 1959. Zur Limnologie und Abwasserbiologie von Talsperren—Obere Saale. *Int. Revue ges. Hydrobiol. Hydrogr.* **44,** 486–619. *390.*

SCHROEDER, W. L., 1930. Biological Survey of the River Wharfe III. Algae present in the Wharfe plankton. *J. Ecol.* **18,** 303–5. *108.*

SCHROLL, F., 1959. Zur Ernährungsbiologie der steirischen Ammocöten *Lampetra planeri* (Bloch) und *Eudontomyzon danfordi* (Regan). *Int. Revue ges. Hydrobiol. Hydrogr.* **44,** 395–429. *366, 368.*

SCHUBART, O., 1946. Observaçôes sobre a productividade biológica das áquas de Monte Alegre. A fauna aquatica da regiäo. *Bolm. Ind. anim.* **8,** 22–54. *391.*

—— 1953. Über einen subtropischen Fluss Brasiliens, den Mogi-Guassu, insbesondere seine physikalischen Bedingungen wie Wasserstand, Temperatur uns Sichttiefe. *Arch. Hydrobiol.* **48,** 350–430. *391.*

SCHUCK, H. A., 1945. Survival, population density, growth, and movement of the wild brown trout in Crystal Creek. *Trans. Am. Fish. Soc.* **73,** 209–29. *344.*

SCHULZ, W., 1931. Die Orientierung des Rückenschwimmers zum Licht und zur Strömung. *Z. vergl. Physiol.* **14,** 392–404. *264.*

SCHUMACHER, A., 1963. Quantitative Aspekte der Beziehung zwischen Stärke der Tubificiden Besiedlung und Schichtdicke der Oxydationszone in den Susswasserwatten der Unterelbe. *Arch. FischWiss.* **14**, 48–50. *434.*

SCHWARTZ, F. J., 1958. The breeding behavior of the southern blacknose dace, *Rhinichthys atratulus obtusus* Agassiz. *Copeia*, **1958**, 141–3. *356.*

—— and NORVELL, J., 1958. Food, growth and sexual dimorphism of the redside dace *Clinostomus elongatus* (Kirtland) in Linesville Creek, Crawford County, Pennsylvania. *Ohio J. Sci.* **58**, 311–16. *367.*

SCHWOERBEL, J., 1959. Ökologische und tiergeographische Untersuchungen über die Milben (Acari, Hydrachnellae) der Quellen und Bäche des südlichen Schwarzwaldes und seiner Randgebiet. *Arch. Hydrobiol.* Suppl. **24**, 385–546. *154, 401.*

—— 1961*a*. Über die Lebensbedingungen und die Besiedlung des Hyporheischen Lebensraumes. Ibid. Suppl. **25**, 182–214. *114, 407.*

—— 1961*b*. Die Bedeutung der Wassermilben für die biozönotische Gleiderung. *Verh. int. Verein. theor. angew. Limnol.* **14**, 355–61. *136, 222, 389.*

—— 1964. Die Bedeutung des Hyporheals für die benthische Lebensgemeinschaft der Fliessgewässern. Ibid. **15**, 215–26. *407.*

—— and TILLIMANNS, G. C., 1964*a*. Untersuchungen über die Stoffwechseldynamik in Fliessgewässern. I. Die Rolle höherer Wasserpflanzen: *Callitriche hamulata* Kütz. *Arch. Hydrobiol.* Suppl. **28**, 245–58. *80, 87.*

—— —— 1964*b*. Idem. II. Experimentelle Untersuchungen über die Ammoniumaufnahme und pH-Anderung in Wasser durch *Callitriche hamulata* Kütz. und *Fontinalis antipyretica* L. Ibid. Suppl. **28**, 259–67. *80, 87.*

[1]SCOTT, D., 1958. Ecological studies on the Trichoptera of the River Dean, Cheshire. Ibid. **54**, 340–92. *150–1, 184, 197, 222.*

—— and RUSHFORTH, J. M., 1959. Cover on river bottoms. *Nature, Lond.* **183**, 836–7. *250.*

SCOTT, D. C., 1949. A study of a stream population of rock bass *Ambloplites rupestris*. *Invest. Indiana Lakes Streams*, **3**, 169–234. *343.*

—— 1951. Sampling fish populations in the Coosa River, Alabama. *Trans. Am. Fish. Soc.* **80**, 28–40. *317.*

[1]—— 1958. Biological balance in streams. *Sewage ind. Wastes*, **1958**, 1169–73. *210, 239, 411.*

—— *et al.*, 1959. The nymph of the mayfly genus *Tortopus* (Ephemeroptera: Polymitarcidae). *Ann. ent. Soc. Am.* **52**, 205–13. *119.*

[1]SCOTT, K. F. M., 1958. Hydrobiological studies on the Great Berg River. Part 4. The Chironomidae. *Trans. R. Soc. S. Afr.* **35**, 277–98. *389.*

SEELER, T., 1932. Das Metazoenplankton der Alster in quantitativer Betrachtung. *Int. Revue ges. Hydrobiol. Hydrogr.* **27**, 345–426. *105, 109.*

SEIFERT, M. F., 1963. Some aspects of territorial behavior of the fantail darter. *Proc. Iowa Acad. Sci.* **70**, 224–7. *309.*

SEVERSMITH, H. F., 1953. Distribution, morphology and life history of *Lampetra aepyptera*, a brook lamprey, in Maryland. *Copeia*, **1953**, 225–32. *357, 363.*

SHACKLETTE, H. T., 1965. A leafy liverwort hydrosere on Yakobi Island, Alaska. *Ecology*, **46**, 377–8. *79.*

SHADIN, V. I., 1956. Life in rivers (Russian). *Jizni presnih vod S.S.S.R.* **3**, 113–256. Moscow. *vii, 1, 2, 25–6, 29, 30, 32–3, 38, 41, 44–5, 47–9, 86, 94, 98, 100, 109–10, 117, 176, 193, 208, 210, 250, 255, 346–7, 353, 391, 396, 411.*

1. The 1958 papers of Scott, D., Scott, D. C., and Scott, K. F. M., are distinguished as Scott (1958), Scott, D. C. (1958), and Scott, K. F. M. (1958) in the text.

SHADIN, V. I., 1962. Wechsel der Biozönose beim Übergang von Fluss zu Stausee. *Schweiz. Z. Hydrol.* **24,** 459–66. *108, 391.*

—— 1964. Bottom biocenosis of the Oka River and their changes for 35 years (Russian). *Trudy zool. Inst. Leningr.* **32,** 226–88. *391.*

SHAPIRO, J., 1958. The core freezer—a new sampler for lake sediments. *Ecology,* **39,** 758. *237.*

SHAPOVALOV, L., and TAFT, A. C., 1954. The life histories of the steelhead rainbow trout (*Salmo gairdneri gairdneri*) and silver salmon (*Oncorhynchus kisutch*) with special reference to Waddell Creek, California, and recommendations regarding their management. *Bull. Dep. Fish Game St. Calif.* **98,** 375 pp. *346.*

SHELFORD, V. E., 1911. Ecological succession I. Stream fishes and the method of physiographic analysis. *Biol. Bull. mar. biol. Lab., Woods Hole,* **21,** 9–35. *383, 385.*

——1937. *Animal Communities in Temperate America.* University of Chicago Press. *153, 208, 404.*

SHELOMOV, I. K., and SPICHAK, M. K., 1960. The effect of reservoirs on the downstream hydrological and hydrochemical regimes of the river. *Dokl. Akad. Nauk S.S.S.R.* **133,** 457–8. *34, 49.*

SHERIDAN, W. L., 1961. Temperature relationships in a pink salmon stream in Alaska. *Ecology,* **42,** 91–8. *29, 31.*

SHETTER, D. S., 1937. Migration, growth rate, and population density of brook trout in the north branch of the Au Sable River, Michigan. *Trans. Am. Fish. Soc.* **66,** 203–10. *344.*

—— 1961. Survival of brook trout from egg to fingerling stage in two Michigan trout streams. Ibid. **90,** 252–8. *357.*

—— and HAZZARD, A. S., 1939. Species composition by age groups and stability of fish populations in sections of three Michigan trout streams during the summer of 1937. Ibid. **69,** 281–302. *314.*

—— and LEONARD, J. W., 1943. A population study of a limited area in a Michigan trout stream. September 1940. Ibid. **72,** 35–51. *314, 338–9.*

—— *et al.*, 1946. The effects of deflectors in a section of a Michigan trout stream. Ibid. **76,** 248–78. *450.*

SHOCKLEY, C. H., 1949. Fish and invertebrate populations of an Indiana bass stream. *Invest. Indiana Lakes Streams,* **3,** 247–70. *208, 379.*

SHOUP, C. S., 1943. Distribution of fresh-water gastropods in relation to total alkalinity of streams. *Nautilus,* **56,** 130–4. *220.*

—— 1947. Geochemical interpretation of water analyses from Tennessee streams. *Trans. Am. Fish. Soc.* **74,** 223–39. *43.*

—— 1948. Limnological observations on some streams of the New River watershed in the vicinity of Mountain Lake, Virginia. *J. Elisha Mitchell scient. Soc.* **64,** 1–12. *43, 247.*

SHUMWAY, D. L., *et al.*, 1964. Influence of oxygen concentration and water movement on the growth of steelhead trout and coho salmon embryos. *Trans. Am. Fish. Soc.* **93,** 342–56. *321.*

SIEMIŃSKA, J., 1956. The River Brynica from the point of view of hydrobiology and fishery (Polish: English summary). *Polskie Archwm Hydrobiol.* **3,** 69–160. *98.*

SIGLER, W. F., 1951. The life history and management of the mountain whitefish

Prosopium williamsoni (Girard) in Logan River, Utah. *Bull. Utah agric. Exp. Stn*, **347**, 21 pp. *313, 343, 348, 367*.

SILVER, S. J., *et al.*, 1963. Dissolved oxygen requirements of developing steelhead trout and chinook salmon embryos at different water velocities. *Trans. Am. Fish. Soc.* **92**, 327–43. *321*.

SINCLAIR, R. M., 1963. Effects of an introduced clam (*Corbicula*) on water quality in the Tennessee River valley. *Mimeo. Pap. presented at 2nd Ann. sanit. Engng Conf. Nashville, Tennessee*, 11 pp. *196*.

—— and ISOM, B. G., 1963. *Further Studies on the Introduced Clam* (Corbicula) *in Tennessee*. Tennessee Pollution Control Board, Nashville, 75 pp. *196*.

SINHA, V. R. P., and JONES, J. W., 1966. On the sex and distribution of the freshwater eel (*Anguilla anguilla*). *J. Zool. Lond.* **150**, 371–85. *345*.

SIOLI, H., 1963. Bietrage zur regionalen Limnologie des brazilianischen Amazonasgebietes. *Arch. Hydrobiol.* **59**, 311–50. *219, 221*.

—— 1964. General features of the limnology of Amazonia. *Verh. int. Verein. theor. angew. Limnol.* **15**, 1053–8. *221*.

—— 1965. Zur Morphologie des Flussbettes des Unteren Amazonas. *Naturwissenschaften*, **5**, 104. *16*.

SLACK, H. D., 1934. The winter food of brown trout (*Salmo trutta* L.) *J. Anim. Ecol.* **3**, 105–8. *371*.

—— 1936. The food of caddis fly (Trichoptera) larvae. Ibid. **5**, 105–15. *193–4*.

SLACK, K. V., 1955. A study of the factors affecting stream productivity by the comparative method. *Invest. Indiana Lakes Streams*, **4**, 3–47. *220*.

—— 1964. Effect of tree leaves on water quality in the Cacapon River, West Virginia. *Prof. Pap. U.S. Geol. Surv.* **475–D**, 181–5. *41*.

SLÁDEČKOVÁ, A., 1962. Limnological investigation methods for the periphyton ('Aufwuchs') community. *Bot. Rev.* **28**, 287–350. *56–8*.

SLOAN, W. C., 1956. The distribution of aquatic insects in two Florida springs. *Ecology*, **37**, 81–98. *398*.

SMART, G. C., 1962. The life history of the crayfish *Cambarus longulus longulus*. *Am. Midl. Nat.* **68**, 83–94. *276, 281*.

SMISSAERT, H. R., 1959. Limburgse Beken I, II & III. Faunistisch, oriënterend-oecologisch. *Natuurh. Maandbl.* **48**, 7–18, 35–46, 70–8. *390*.

SMITH, B. G., 1908. The spawning habits of *Chrosomus erythrogaster* Rafinesque. *Biol. Bull. mar. biol. Lab., Woods Hole*, **14**, 9–18. *359*.

SMITH, E. W., 1953. The life history of the crawfish *Orconectes* (*Faxonella*) *clypeatus* (Hay). *Tulane Stud. Zool.* **1**, 79–96. *276, 281, 404*.

SMITH, L. L., and MOYLE, J. B., 1944. A biological survey and fishery management plan for the streams of the Lake Superior North Shore watershed. *Tech. Bull. Minn. Dep. Conserv. Div. Fish Game*, **1**, 228 pp. *79, 86, 247, 324*.

—— *et al.*, 1949. Fish populations in some Minnesota trout streams. *Trans. Am. Fish. Soc.* **76**, 204–14. *339*.

SMITH, O. R., 1941. The spawning habits of cutthroat and eastern brook trouts. *J. Wildl. Mgmt*, **5**, 461–71. *357*.

SMITH, S. H., 1957. Evolution and distribution of the Coregonids. *J. Fish. Res. Bd Can.* **14**, 599–604. *301*.

SMYLY, W. J. P., 1955. On the biology of the stone loach *Nemacheilus barbatula* (L.). *J. Anim. Ecol.* **24**, 167–86. *313, 340, 359, 367*.

—— 1957. The life-history of the bull-head or Miller's thumb (*Cottus gobio* L.). *Proc. zool. Soc. Lond.* **128**, 431–53. *309, 311, 313, 340, 356, 367*.

SOKOLOWA, N., 1961. Die Entwicklung der Bodenfauna des Utscha-Wasserbeckens. *Verh. int. Verein. theor. angew. Limnol.* **14**, 640–2. *447*.

SOMMANI, E., 1953. Il concetto di 'zona ittica' e il suo reale sinificato ecologico. *Boll. Pesca Piscic. Idrobiol.* N.S. **7**, 61–71. *385*.

SOMMERMAN, K. M., *et al.*, 1955. Biology of Alaskan black flies (Simuliidae, Diptera). *Ecol. Monogr.* **25**, 345–85. *186*, *229*, *257*, *287*.

SÖRENSEN, I., 1951. An investigation of some factors affecting the upstream migration of the eel. *Rep. Inst. Freshwat. Res. Drottningholm*, **32**, 126–32. *308*, *353–4*.

SOURIE, R., 1962. Les tufs à chironomides dans quelques ruisseaux de la region de Campon (Hautes-Pyrénées). *Vie Milieu*, **13**, 741–6. *76*, *138*.

SOUTHERN, R., 1935. The food and growth of brown trout from Lough Derg and the River Shannon. *Proc. R. Ir. Acad.* **B 42**, 87–172. *367*.

—— and GARDINER, A. C., 1938. The phytoplankton of the River Shannon and Lough Derg. Ibid. **B 45**, 89–124. *98*, *100*.

SOWA, R., 1961. The bottom fauna of the River Bajerka (Polish: English summary). *Acta hydrobiol.* **3**, 1–32. *208*, *391*.

—— 1965. Ecological characteristics of the bottom fauna of the Wielka Puszcza Stream. Ibid. **7**, Suppl. 1, 61–92. *292*, *391*.

SPANOVSKAYA, V. D., 1963. Food of pike fingerlings (*Esox lucius*) (Russian: English summary). *Russk. zool. Zh.* **42**, 1071–9. *371*.

SPRULES, W. M., 1941. The effect of a beaver dam on the insect fauna of a trout stream. *Trans. Am. Fish. Soc.* **70**, 236–48. *209*, *381*.

—— 1947. An ecological investigation of stream insects in Algonquin Park, Ontario. *Univ. Toronto Stud. biol. ser.* **56**, 1–81. *202*, *208–9*, *223–4*, *226*, *234*, *241*, *275*, *283*, *391*, *403*.

STANKOVIĆ, S., 1960. The Balkan Lake Ohrid and its living world. *Monographiae biol.* **9**, 357 pp. *112*, *348*.

STARMACH, K., 1938. Untersuchungen über das Seston der oberen Wisła und Bioła Przemsza. *Polksa Akademia Umiejetności, Krakow. Komisja Fisjograficzna.* **73**, 1–146. *94*.

STARMÜHLNER, F., 1953. Die Molluskenfauna unsere Wienerwaldbächer. *Wett. Leben* Sonderh. **2**, 184–205. *143*, *198*, *205*.

—— 1961. Biologische Untersuchungen in isländischen, mitteleuropäischen und madagasischen Warmbächen. *Verh. int. Verein. theor. angew. Limnol.* **14**, 404–9. *399*.

STARRET, W. C., 1950*a*. Distribution of the fishes of Boone County, Iowa, with special reference to the minnows and darters. *Am. Midl. Nat.* **43**, 112–27. *313*, *329*, *337*.

—— 1950*b*. Food relationships of the minnows of the Des Moines River, Iowa. *Ecology*, **31**, 216–33. *368*.

—— 1951. Some factors affecting the abundance of minnows in the Des Moines River, Iowa. Ibid. **32**, 13–27. *328*.

STAUFFER, T. M., 1962. Duration of larval life of sea lampreys in Carp Lake River, Michigan. *Trans. Am. Fish. Soc.* **91**, 422–3. *363*.

STAVE, U., 1956. Wuchszonen auf wasserüberströmten Gestein. *Ber. limnol. Flussstn Freudenthal*, **7**, 19–20. *81*.

STEFANICH, F. A., 1952. The population and movement of fish in Prickley Pear Creek, Montana. *Trans. Am. Fish. Soc.* **81**, 260–74. *343*.

STEFFAN, A. W., 1960. Phylogenetische Trends in der Gattung *Dryops* (Dryopidae, Coleoptera). *Int. Congr. Ent.* **11**, **1**, 97–103. *387*.

—— 1963. Beziehungen zwischen Lebensraum und Körpergrösse bei mitteleuropäischen Elminthidae (Coleoptera: Dryopoidea). *Z. Morph. Ökol. Tiere*, **53**, 1–21. *168*.

—— 1964. Biozönotische Parallelität der Körpergrösse bei mitteleuropäischen Elminthidae (Coleoptera: Dryopoidea). *Naturwissenschaften*, **51**, 20–1. *168*.

—— 1965*a*. *Plecopteracoluthus downesi* gen. et sp. nov. (Diptera: Chironomidae), a species whose larvae live phoretically on larvae of Plecoptera. *Can. Ent.* **97**, 1323–44. *181*.

—— 1965*b*. Zur Statik und Dynamik im Ökosystem der Fliessgewässer und zu den Möglichkeiten ihrer Klassifizierung. *Biosoziologie*, Junk, The Hague, pp. 65–110. *392*.

STEGMAN, J. L., and MINCKLEY, W. L., 1959. Occurrence of three species of fishes in interstices of gravel in an area of subsurface flow. *Copeia*, **1959**, 341. *336*.

STEHR, W. C., and BRANSON, J. W., 1938. An ecological study of an intermittent stream. *Ecology*, **19**, 294–310. *329*, *404*.

STEINBÖCK, O., 1942. Das Verhalten von *Planaria alpina* Dana in der Natur und im Laboratoriumversuch. *Memorie Ist. ital. Idrobiol.* **1**, 63–75. *204*.

STEINMANN, P., 1907. Die Tierwelt der Gebirgsbäche. Eine faunistisch-biologische Studie. *Annls Biol. lacustre*, **2**, 30–150. *vii*, *122*, *383*, *390*.

—— 1908. Die Tierwelt der Gebirgsbäche. *Arch. Hydrobiol.* **3**, 266–73. *122*.

—— 1909. Die neuesten Arbeiten über Bachfauna. *Int. Revue ges. Hydrobiol. Hydrogr.* **2**, 241–6. *vii*.

STELLA, E., 1956. Le biocenosi del sistema sorgivo del Fiume Ninfa (Agro Romano). *Boll. Pesca Piscic. Idrobiol.* **31**, **10**, 5–52. *400*.

STEUSLOFF, U., 1939. Zusammenhänge zwischen Boden, Chemismus des Wassers und Phanerogamenflora in fliessenden Gewässern der Lüneberger Heide um Celle und Ülzen. *Arch. Hydrobiol.* **35**, 70–106. *221*.

—— 1943. Ein Beitrag zur Kenntnis der Verbreitung und der Lebensräume von *Gammarus*-Arten in Nordwestdeutschland. Ibid. **40**, 79–97. *193*.

STEWART, N. H., 1926. Development, growth, and food habits of the white sucker, *Catastomus commersonii* LeSueur. *Bull. Bur. Fish., Wash.* **42**, 147–84. *359*, *368*, *371*.

STILLER, J., 1957. Zur Biologie und Verbreitung der Protozoen- und Crustaceenfauna eines Mittelgebirgesbaches in Ungarn. *Arch. Hydrobiol.* **53**, 392–424. *114*.

STOBER, Q. J., 1964. Some limnological effects of Tiber Reservoir on the Marias River, Montana. *Proc. Mont. Acad. Sci.* **23**, 111–37. *68*, *449*.

STOTT, B., *et al.*, 1963. Homing behaviour in gudgeon (*Gobio gobio* (L.)). *Anim. Behav.* **11**, 93–6. *343–4*.

STRAŠKRABA, M., 1965. The effect of fish on the number of invertebrates in ponds and streams. *Mitt. int. Verein. theor. angew. Limnol.* **13**, 106–27. *232*.

—— 1966. On the distribution of the macrofauna and fish in two streams, Lucina and Morávka. *Arch. Hydrobiol.* **61**, 515–36. *391*.

—— *et al.*, 1966. Contribution to the problem of food competition among the sculpin, minnow and brown trout. *J. Anim. Ecol.* **35**, 303–11. *375*.

STRENGER, A., 1953. Zur Kopfmorphologie der Ephemeridenlarven. Erster Teil. *Ecdyonurus* und *Rhithrogena*. *Öst. zool. Z.* **4**, 191–228. *184*.

STUART, T. A., 1953*a*. Spawning migration, reproduction and young stages of loch trout (*Salmo trutta* L.). *Freshwat. Salm. Fish. Res.* **5,** 39 pp. *302, 310, 327, 348, 358, 363.*

—— 1953*b*. Water currents through permeable gravels and their significance to spawning Salmonids, etc. *Nature, Lond.* **172,** 407. *358.*

—— 1954. Spawning sites of trout. Ibid. **173,** 354. *358.*

—— 1957. The migrations and homing behaviour of brown trout. *Scient. Invest. Freshwat. Salm. Fish. Res. Scott. Home Dep.* **18,** 27 pp. *351.*

—— 1959. The influence of drainage works, levees, dykes, dredging, etc., on the aquatic environment and stocks. *Proc. I.U.C.N. Tech. Meeting, Athens,* **4,** 337–45. *447.*

—— 1962. The leaping behaviour of salmon and trout at falls and obstructions. *Freshwat. Salm. Fish. Res.* **28,** 46 pp. *10, 353–4.*

STUNDL, K., 1950. Zur Hydrographie und Biologie der österreichischer Donau. *Schweiz. Z. Hydrol.* **13,** 36–53. *94, 98, 109, 198.*

SUMNER, F. B., and SARGENT, M. C., 1940. Some observations on the physiology of warm spring fishes. *Ecology,* **21,** 45–54. *301, 322–3.*

SURBER, E. W., 1937. Rainbow trout and bottom fauna production in one mile of stream. *Trans. Am. Fish. Soc.* **66,** 193–202. *238.*

—— 1939. A comparison of four eastern smallmouth bass streams. Ibid. **68,** 322–35. *251.*

—— 1941. A quantitative study of the food of the smallmouth black bass, *Micopterus dolomieu,* in three eastern streams. Ibid. **70,** 311–34. *369.*

—— 1951. Bottom fauna and temperature conditions in relation to trout management in St. Mary's River, Augusta County, Virginia. *Va J. Sci.* N.S. **2,** 190–202. *255, 292.*

SVÄRDSON, G., 1949. Note on spawning habits of *Leuciscus eryopthalmus* (L.), *Abramis brama* (L.) and *Esox lucius* L. *Rep. Inst. Freshwat. Res. Drottningholm,* **29,** 102–7. *360.*

SWALE, E. M. F., 1964. A study of the phytoplankton of a calcareous river. *J. Ecol.* **52,** 433–46. *98, 106.*

SWINGLE, H. S., 1954. Fish populations in Alabama rivers and impoundments. *Trans. Am. Fish. Soc.* **83,** 47–57. *327, 339, 448.*

SYMOENS, J. J., 1949. Note sur des formations de tuf calcaire observées dans le bois d'Hautmont (Wauthier-Braine). *Bull. Soc. r. Bot. Belg.* **82,** 81–95. *76.*

—— 1951. Equisse d'un système des associations algale d'eau douce. *Verh. int. Verein. theor. angew. Limnol.* **11,** 395–408. *72.*

—— 1955. Découverte de tufs à Chironomides dans la région mosane. Ibid. **12**. 604–7. *138.*

—— 1957. Les eaux douces de l'Ardenne et des regions voisines: Les milieux et leur végetation algale. *Bull. Soc. r. Bot. Belg.* **89,** 111–314. *73–4, 94, 138.*

SYMONS, *et al.*, 1964. Influence of impoundments on water quality. *Publs U.S. Pub. Hlth Serv.* **999–WP–18,** 78 pp. *34, 46, 449.*

SZCZEPAŃSKI, A., Deciduous leaves as a source of organic matter in lakes. *Bull. Acad. pol. Sci. Cl.II, Sér. Sci. biol.* **13,** 215–17. *433.*

TACK, E., 1940. Die Elritze (*Phoxinus laevis* Ag.), ein monographische Bearbeitung. *Arch. Hydrobiol.* **37,** 321–425. *349, 351, 359, 369, 371.*

TAFT, A. C., and SHOPOVALOV, L., 1938. Homing instinct and straying among steelhead trout (*Salmo Gairdneri*) and silver salmon (*Onchorhynchus Kisutch*). *Calif. Fish. Game*, **24,** 118–25. *351.*

TAIT, J. S., 1960. The first filling of the swim bladder in salmonids. *Can. J. Zool.* **38,** 179–87. *305.*

TALLING, J. F., 1958. The longitudinal succession of the water characteristics in the White Nile. *Hydrobiologia*, **11,** 73–89. *42, 44, 48.*

TANAKA, H., 1960. On the daily change of the drifting of benthic animals in stream, especially on the types of daily change observed in taxonomic groups of insects. *Bull. Freshwat. Fish. Res. Lab., Tokyo*, **9,** 13–24. *263.*

—— 1966. Ecological studies on aquatic insects in upper reaches of the Kinu-Gawa River, Tchigi Prefecture, Japan (Japanese: English summary). Ibid. **15,** 123–47. *257, 272, 391, 434.*

TARZWELL, C. M., 1935. Progress in lake and stream improvement. *Trans. Am. Game Conf.* **21,** 119–34. *450.*

—— 1937. Experimental evidence on the values of trout stream improvement in Michigan. *Trans. Am. Fish. Soc.* **66,** 177–87. *253.*

—— 1938*a*. An evaluation of the methods and results of stream improvement in the northwest. *Trans. N. Am. Wildl. Conf.* **3,** 339–64. *450.*

—— 1938*b*. Factors influencing fish food and fish production in southwestern streams. *Trans. Am. Fish. Soc.* **67,** 246–55. *450.*

—— 1939. Changing the Clinch River into a trout stream. Ibid. **68,** 228–33. *205, 324, 449.*

TATE, W. H., 1949. Growth and food habit studies of smallmouth black bass in some Iowa streams. *Iowa St. Coll. J. Sci.* **23,** 343–54. *368.*

TAYLOR, C. B., 1941. Bacteriology of fresh water II. The distribution and types of coliform bacteria in lakes and streams. *J. Hyg., Camb.* **42,** 17–38. *106.*

TEAL, J. M., 1957. Community metabolism in a temperate cold spring. *Ecol. Monogr.* **27,** 283–302. *399, 400, 435.*

TEBO, L. B., 1955. Effects of siltation, resulting from improper logging, on the bottom fauna of a small trout stream in the Southern Appalachians. *Progve FishCult.* **17,** 64–70. *443.*

—— and HASSLER, W. W., 1961. Seasonal abundance of aquatic insects in western North Carolina trout streams. *J. Elisha Mitchell scient. Soc.* **77,** 249–59. *276, 292, 294, 367.*

—— —— 1963. Food of brook, brown and rainbow trout in streams in western North Carolina. Ibid. **79,** 44–53. *369, 373.*

TESCH, F. W., and ALBRECHT, M.-L., 1961. Über den Einfluss verschiedener Umweltfaktoren auf Wachstum und Bestand der Bachforelle (*Salmo trutta fario* L.) in Mittelgebirgsgewässer. *Verh. int. Verein. theor. angew. Limnol.* **14,** 763–8. *327, 332–3, 338.*

THIENEMANN, A., 1912. Der Bergbach des Sauerland. *Int. Revue ges. Hydrobiol. Hydrogr.* Suppl. **4,** 2, 1. 125 pp. *204, 383, 387, 390, 400, 411.*

—— 1925. Die Binnengewässer Mitteleuropas. *Die Binnengewässer*, Stuttgart, **1.** *vii, 390.*

—— 1926. Hydrobiologische Untersuchungen an den kalten Quellen und Bächen der Halbinsel Jasmund auf Rugen. *Arch. Hydrobiol.* **17,** 221–336. *390.*

—— 1949. Veränderungen in der Tierwelt unserer Quellen von 1918 bis 1948. *Die Heimat*, Kiel, **56,** 2–5. *400–1.*

—— 1950. Verbreitungsgeschichte der Süsswassertierwelt Europas. *Die Binnengewässer*, Stuttgart, **18.** *117, 120, 142, 196, 387, 400, 447.*

THIENEMANN, A., 1954*a*. *Chironomus*. Leben, Verbreitung und wirtschaftliche Bedeutung der Chironomiden. Ibid. **20.** *135, 137–8, 273–4, 389, 402.*

—— 1954*b*. Ein drittes biozönotisches Grundprinzip. *Arch. Hydrobiol.* **49,** 421–2. *234.*

THOMAS, E., 1966. Orientierung der Imagines von *Capnia atra* Morton (Plecoptera). *Oikos,* **17,** 278–80. *155.*

THOMAS, J. D., 1962. The food and growth of brown trout (*Salmo trutta* L.) and its feeding relationships with the salmon parr (*Salmo salar* L.) and the eel (*Anguilla anguilla* (L.)) in the River Teify, West Wales. *J. Anim. Ecol.* **31,** 175–205. *371, 373.*

—— 1964. Studies on the growth of trout, *Salmo trutta*, from contrasting habitats. *Proc. zool. Soc. Lond.* **142,** 459–509. *267, 337, 372.*

THOMAS, N. A., and O'CONNELL, R. L., 1966. A method for measuring primary production by stream benthos. *Limnol. Oceanogr.* **11,** 386–92. *414.*

THOMPSON, D. H., and HUNT, F. D., 1930. The fishes of Champaign County: A study of the distribution and abundance of fishes in small streams. *Bull. Ill. St. nat. Hist. Surv.* **19,** 5–101. *322, 338, 368.*

THORPE, W. H., 1950. Plastron respiration in aquatic insects. *Biol. Rev.* **25,** 344–90. *170, 172.*

THORUP, J., 1963. Growth and life-cycles of invertebrates from Danish springs. *Hydrobiologia,* **22,** 55–84. *285, 401.*

—— 1966. Substrate type and its value as a basis for the delimitation of bottom fauna communities in running waters. *Spec. Publs Pymatuning Lab. Fld Biol.* **4,** 59–74. *208, 210, 219, 230, 393, 401.*

TIMMERMANS, J. A., 1954. La pêche électrique en eau douce. *Trav. Stn Rech. Groenendaal,* **D 15,** 31 pp. *314–15.*

—— 1957. Estimation des populations piscicoles. Application aux eaux courantes rhéophiles. Ibid. **D 21,** 84 pp. *317.*

—— 1960. Observations concernant les populations de truite commune (*Salmo trutta fario* L.) dans les eaux courantes. Ibid. **D 28,** 36 pp. *343–4.*

—— 1961*a*. Essais en eaux courantes sur l'éfficacité de repeuplements en truites (*Salmo trutta fario*) et gardons (*Gardonus rutilus*) pêchables. Ibid. **D 32,** 51 pp. *312.*

—— 1961*b*. La population piscicole de l'Eau Blanche. Petite riviere du type supérieur de la zone à barbeau. Ibid. **D 30,** 16 pp. *339.*

TINDALL, D. R., and MINCKLEY, W. L., 1964. An integrated application of three kinds of sampling techniques to stream limnology. *Limnol. Oceanogr.* **9,** 270–2. *50.*

TRAUTMAN, M. B., 1939. The effects of man-made modifications on the fish fauna in Lost and Gordon Creeks, Ohio, between 1887–1938. *Ohio J. Sci.* **39,** 275–88. *447.*

TSUDA, M., 1961. Important role of net-spinning caddis-fly larvae in Japanese running water. *Verh. int. Verein. theor. angew. Limnol.* **14,** 376–7. *188.*

—— and KOMATSU, T., 1964*a*. Aquatic insect communities of the rivers of the Togakushi Highland (Japanese: English summary). *Jap. J. Ecol.* **14,** 14–18. *257.*

—— —— 1964*b*. Aquatic insect communities of Yoshino River, four years after the Ise-Wan typhoon. Ibid. **14,** 43–9. *253.*

TUXEN, S. L., 1944. The hot springs, their animal communities and their zoogeographical significance. *The Zoology of Iceland*, Copenhagen, **1, 2,** 206 pp. *399*.

Uéno, M., 1931. Contributions to the knowledge of Japanese Ephemeroptera. *Annotnes zool. jap.* **13,** 189–226. *128, 410*.
—— 1938. Bottom fauna of Lake Abasiri and the neighbouring waters in Hokkiado. *Trans. Sapporo nat. Hist. Soc.* **15,** 140–67. *391*.
—— 1952. Caddis fly larvae interfering with the flow in the waterway tunnels of a hydraulic power plant. *Kontyû,* **19,** 1–8. *188, 257*.
Uherkovich, G., 1965. Über das Potamo-Phytoplankton der Tisza (Theiss) in Ungarn. *Int. Revue ges. Hydrobiol. Hydrogr.* **50,** 269–80. *98, 109*.
Usinger, R. L., and Needham, P. R., 1956. A drag-type riffle-bottom sampler. *Progve FishCult.* **18,** 42–4. *237*.
Uspenskii, I. V., 1963. The zooplankton of the Mozhaisk Reservoir in the first year of its existence (Russian). *Uchinskoc i Mozhaiskoe Vodokranilishcha Mosk. Univ. Moscow,* **1963,** 375–88. (Nat. lend. Libr., Boston Spa, Yorks, Engl. R.T.S. 2942). *103, 105, 448*.

Vaillant, F., 1956. Recherches sur la faune madicole (hygropétrique *s.l.*) de France, de Corse et d'Afrique du Nord. *Mém. Mus. natn. Hist. nat., Paris,* N.S. **A11,** 1–258. *409*.
—— 1961. Fluctuations d'une population madicole au cours d'une année. *Verh. int. Verein. theor. angew. Limnol.* **14,** 513–16. *409*.
Van der Schalie, H., 1938. The naiad fauna of the Huron River, in south eastern Michigan. *Misc. Publs Mus. Zool. Univ. Mich.* **40,** 83 pp. *388–9*.
—— and Van der Schalie, A., 1950. The mussels of the Mississippi River. *Am. Midl. Nat.* **44,** 448–66. *209*.
Van Deusen, R. D., 1954. Maryland freshwater stream classification by water sheds. *Contr. Chesapeake biol. Lab.* **106,** 1–30. *385*.
Van Deventer, W. C., 1937. Studies on the biology of the crayfish, *Cambarus propinquus* Girard. *Illinois biol. Monogr.* **15, 3,** 67 pp. *287*.
Van Duzer, E. M., 1939. Observations on the breeding habits of the cut-lips minnow *Exoglossum maxillingua*. *Copeia,* **1939,** 65–75. *358*.
Van Oye, P., 1922. Zur Biologie des Potamoplanktons auf Java. *Int. Revue ges. Hydrobiol. Hydrogr.* **10,** 362–93. *110*.
—— 1926. Le potamoplankton du Ruki au Congo-Belge et des pays chaud en général. Ibid. **16,** 1–51. *100*.
Van Someren, V. D., 1952. *The Biology of Trout in Kenya Colony*. Government Printer, Nairobi, 114 pp. *228, 261, 290, 324–5, 362, 367, 391, 402*.
—— and McMahon, J., 1950. Phoretic association between *Afronurus* and *Simulium* species, and the discovery of the early stages of *Simulium neavei* on freshwater crabs. *Nature, Lond.* **166,** 350–1. *181*.
Vaux, W. G., 1962. Interchange of stream and intergravel water in a salmon spawning riffle. *Spec. scient. Rep. U.S. Fish. Wildl. Serv.* **405,** 11 pp. *10*.
Ventner, G. E., 1961. A new ephemeropteran record from Africa. *Hydrobiologia,* **18,** 327–31. *125, 133*.
Verrier, M.-L., 1948. La vitesse du courant et la répartition des larves d'Ephémères. *C. r. hebd. Séanc. Acad. Sci. Paris,* **227,** 1056–7. *126*.
—— 1953. Le rhéotropisme des larves d'éphémères. *Bull. biol. Fr. Belg.* **87,** 1–34. *155, 405*.

VERRIER, M.-L., 1956. *Biologie des Ephémères*. Collection Armand Colin, Paris, **306**, 316 pp. *134, 151–2, 157, 159.*

VIBERT, R., 1962. Quelques conséquences du courant d'eau et des champs électriques sur le comportement des poissons. *Schweiz. Z. Hydrol.* **24,** 436–43. *309, 311–12, 327.*

—— 1963. Neurophysiology of electric fishing. *Trans. Am. Fish. Soc.* **92,** 265–75. *314.*

—— *et al.*, 1960. Tests et indices d'efficacité des appareils de pêche électrique. *Annls Stn cent. Hydrobiol. appl.* **8,** 51–89. *314–15.*

VIVIER, P., 1965. La 'Peste', un facteur de régulation des populations d'écrevisses (*Astacus*). *Mitt. int. Verein. theor. angew. Limnol.* **13,** 49–62. *231–2.*

VOIGT, G. K., 1960. Alteration of the composition of rainwater by trees. *Am. Midl. Nat.* **63,** 321–6. *47.*

VON MITIS, H., 1938. Die Ybbs als Typus eines ostalpinen Kalkalpenfluss. Eine vorläufige Mitteilung. *Int. Revue ges. Hydrobiol. Hydrogr.* **37,** 425–44. *71, 229, 390.*

VONNEGUT, P., 1937. Die Barbenregion der Ems. *Arch. Hydrobiol.* **32,** 345–408. *98, 207, 390.*

VOO, E. E. VAN DER, and WESTHOFF, V., 1961. An autecological study of some limnophytes and helophytes in the area of the large rivers. *Wentia,* **5,** 163–258. *85.*

VYUSHKOVA, V. P., 1962. Zooplankton of the Volgograd Reservoir in the first year of its existence (Russian). *Tr. Zon. Sov. po Tipol. i Biol. Obz. Ispol. Vnut. Vod. Yuzh. Zony S.S.S.R.* **1962,** 81–4 (Nat. lend. Libr., Boston Spa, Yorks, Engl. R.T.S. 2939) (*see also* Dzyuban (1962) above: also published in *Vop. ekol.* **5,** 30). *448.*

WALKER, B. A., 1961. Studies on Doe Run, Meade County, Kentucky, IV. A new species of isopod crustacean (genus *Asellus*) from Kentucky. *Trans. Am. microsc. Soc.* **80,** 385–90. *401.*

WALLEN, I. E., 1951. The direct effect of turbidity on fishes. *Bull. Okla. agric. Exp. Stn,* **48, 2,** 1–27. *444.*

WALSHE, B. M., 1948. The oxygen requirements and thermal resistance of chironomid larvae from flowing and still waters. *J. exp. Biol.* **25,** 35–44. *161.*

—— 1950. Observations on the biology and behaviour of larvae of the midge *Rheotanytarsus. J. Queckett microsc. Club,* **3,** 171–8. *138, 187.*

—— 1951. The feeding habits of certain Chironomid larvae (subfamily Tendipedinae). *Proc. zool. Soc. Lond.* **121,** 63–79. *187.*

WALTER, G., 1961. Die aktive und passive Einwanderung tierischer und pflanzlicher Organismen in einem neuangelegten Durchflussgraben. *Verh. int. Verein. theor. angew. Limnol.* **14,** 508–12. *87, 91.*

WANGERSKY, P. J., 1965. The organic chemistry of sea water. *Am. Scient.* **53,** 358–74. *49, 50.*

WARREN, C. E., *et al.*, 1960. Progress report—Ecological studies of an experimental stream. *Mimeo. Report, Oregon St. Coll.*, Corvallis, Oregon. *61, 193–5.*

—— *et al.*, 1964. Trout production in an experimental stream enriched with sucrose. *J. Wildl. Mgmt,* **28,** 617–60. *262, 427, 430–1.*

WASER, E. and THOMAS, E. A., 1944. Untersuchungen an der Thur 1940/41. *Schweiz. Z. Hydrol.* **10,** 1–86. *94.*

—— *et al.*, 1943. Untersuchungen am Rhein von Schaffhausen bis Kaiserstuhl 1938–39 und 1940–41. Ibid. **9,** 225–309. *98.*

WATERS, T. F., 1961*a*. Standing crop and drift of stream bottom organisms. *Ecology*, **42**, 532–7. *268*, *427*.

—— 1961*b*. Notes on the chlorophyll method of estimating the photosynthetic capacity of stream periphyton. *Limnol. Oceanogr.* **6**, 486–8. *58*, *415*.

—— 1962*a*. Diurnal periodicity in the drift of stream invertebrates. *Ecology*, **43**, 316–20. *262–3*.

—— 1962*b*. A method to estimate the production rate of a stream bottom invertebrate. *Trans. Am. Fish. Soc.* **91**, 243–50. *268*.

—— 1964. Recolonisation of denuded stream bottom areas by drift. Ibid. **93**, 311–15. *267*.

—— 1965. Interpretation of invertebrate drift in streams. *Ecology*, **46**, 327–34. *262–3*, *266*.

—— 1966. Production rate, population density and drift of a stream invertebrate. Ibid., **47**, 595–604. *426*.

—— and KNAPP, R. J., 1961. An improved bottom fauna sampler. *Trans. Am. Fish. Soc.* **90**, 225–6. *238*.

WATSON, W., 1919. The bryophytes and lichens of fresh-water. *J. Ecol.* **7**, 71–83. *79*.

WATTS, R. L., *et al.*, 1942. Brook trout in Kettle Creek and tributaries. *Bull. Pa agric. Exp. Stn*, **437**, 41 pp. *344*, *351*.

WAUTIER, J., and PATTÉE, E., 1955. Expérience physiologique et expérience écologique. L'influence du substrat sur la consommation d'oxygène chez les larves d'Ephéméroptères. *Bull. mens. Soc. linn. Lyon*, **24**, 178–83. *163*.

WAWRIK, F., 1962. Zur Frage: Führt der Donaustrom autochthones Plankton? *Arch. Hydrobiol.* Suppl. **27**, 28–35. *94*, *100*.

WEATHERLEY, A. H., 1963. Zoogeography of *Perca fluviatilis* (Linnaeus) and *Perca flavescens* (Mitchill) with special reference to the effects of high temperature. *Proc. zool. Soc. Lond.* **141**, 557–76. *301*, *322*.

WEBSTER, D. A., and WEBSTER, P. C., 1943. Influence of water current on case weight of the caddisfly *Goera calcarata* Banks. *Can. Ent.* **75**, 105–8. *141*.

WEHRLE, 1942. Algen in Gebirgsbächen am sudostrande des Schwarzwaldes. *Beitr. naturk. Forsch. Oberrheingeb.* **7**, 128–286. *70*.

WEIBEL, S. R., *et al.*, 1964. Urban land runoff as a factor in stream pollution. *J. Wat. Pollut. Control Fed.* **36**, 914–24. *445*.

WEISE, J. G., 1957. The spring cave-fish, *Chologaster papilliferus*, in Illinois. *Ecology*, **38**, 195–204. *365*.

—— 1961. The ecology of *Urnatella gracilis* Leidy: phylum Endoprocta. *Limnol. Oceanogr.* **6**, 228–30. *207*.

WEISEL, G. F., and NEWMAN, H. W., 1951. Breeding habits, development and early life history of *Richardsonius balteatus*, a northwestern minnow. *Copeia*, **1951**, 187–94. *359*.

WELMAN, J. B., 1948. *Preliminary Survey of the Freshwater Fisheries of Nigeria.* Government Printer, Lagos. *86*, *302*, *351*, *354*.

WENE, G., 1940. The soil as an ecological factor in the abundance of aquatic chironomid larvae. *Ohio J. Sci.* **40**, 193–9. *250*.

—— and WICKLIFF, E. L., 1940. Modification of a stream bottom and its effect on the insect fauna. *Can Ent.* **72**, 131–5. *239*, *250*.

WENT, A. E. J., and FROST, W. E., 1942. River Liffey Survey V. Growth of brown trout (*Salmo trutta* L.) in alkaline and acid waters. *Proc. R. Ir. Acad.* **B48**, 67–84. *332*, *343*.

WESENBERG-LUND, C., 1911. Biologische Studien über netzspinnende, campodeoide Trichopterenlarven. *Int. Revue ges. Hydrobiol Hydrogr.* Suppl. **3,** 1–64. *187–8.*

—— 1943. *Biologie der Süsswasserinsekten.* Gyldendalske Boghandel, Copenhagen and Springer, Berlin. *122, 160.*

WESTLAKE, D. F., 1961. Aquatic macrophytes and the oxygen balance of running water. *Verh. int. Verein. theor. angew. Limnol.* **14,** 499–504. *417.*

—— 1966. A model for quantitative studies of photosynthesis by higher plants in streams. *Int. J. Air Wat. Pollut.* **10,** 883–96. *417.*

WETZEL, R. G., 1963. Primary productivity of periphyton. *Nature, Lond.* **197,** 1026–7. *414.*

—— 1964. Primary productivity of aquatic macrophytes. *Verh. int. Verein. theor. angew Limnol.* **15,** 426–36. *416.*

—— 1965. Techniques and problems of primary productivity measurement in higher aquatic plants and periphyton. *Memorie Ist. ital. Idrobiol.* Suppl. **18,** 249–67. *414, 416.*

WHITEHEAD, H., 1935. An ecological study of the invertebrate fauna of a chalk stream near Great Driffield, Yorkshire. *J. Anim. Ecol.* **4,** 58–78. *192, 292.*

WHITEHOUSE, J. W., and LEWIS, B. G., 1966. The separation of benthos from stream samples by flotation with carbon tetrachloride. *Limnol. Oceanogr.* **11,** 124–6. *243.*

WHITFORD, L. A., 1956. The communities of algae in the springs and spring streams of Florida. *Ecology,* **37,** 433–42. *399.*

—— 1960*a*. The current effect and growth of fresh-water algae. *Trans. Am. microsc. Soc.* **79,** 302–9. *63–4, 66.*

—— 1960*b*. Ecological distribution of fresh-water algae. *Spec. Publs Pymatuning Lab. Fld Biol.* **2,** 2–10. *53–4, 60–1, 66.*

—— and SCHUMACHER, G. J., 1961. Effect of current on mineral uptake and respiration by a fresh-water alga. *Limnol. Oceanogr.* **6,** 423–5. *64.*

—— —— 1964. Effect of a current on respiration and mineral uptake in *Spirogyra* and *Oedogonium. Ecology,* **45,** 168–70. *64.*

WHITLEY, L. S., 1962. New bottom sampler for use in shallow streams. *Limnol. Oceanogr.* **7,** 265–6. *238.*

WHITNEY, A. N., and BAILEY, J. E., 1959. Detrimental effects of highway construction on a Montana stream. *Trans. Am. Fish. Soc.* **88,** 72–3. *447.*

WHITNEY, R. J., 1939. The thermal resistance of may-fly nymphs from ponds and streams. *J. exp. Biol.* **16,** 374–85. *168.*

WICKLIFF, E. L., 1940. Natural productivity of fish and crayfish in riffles. *Trans. N. Am. Wildl. Conf.* **5,** 149–53. *255, 350.*

WIEBE, A. H., 1927. Biological survey of the Upper Mississippi River with special reference to pollution. *Bull. Bur. Fish., Wash.* **43, 2,** 137–67. *99.*

WIGGINS, G. B., 1966. The critical problem of systematics in stream ecology. *Spec. Publs Pymatuning Lab. Fld Biol.* **4,** 52–8. *114, 159.*

WILDING, J. L., 1940. A new square-foot aquatic sampler. *Spec. Publs limnol. Soc. Am.* **4,** 4 pp. *238.*

WILLIAMS, A. B., 1954. An explanation for the distribution of a North American crayfish. *Ecology,* **35,** 573–4. *445.*

WILLIAMS, L. G., 1964. Possible relationships between plankton-diatom species numbers and water quality estimates. Ibid., 45, 809–23. *99, 107.*

WILLIAMS, L. G., 1966. Dominant planktonic rotifers of major waterways of the United States. *Limnol. Oceanogr.* **11,** 83–91. *99, 104, 109, 110.*

WILLIAMS, R. W., 1960. A new and simple method for the isolation of fresh-water invertebrates from soil samples. *Ecology,* **41,** 573–4. *242.*

WILLIAMS, T. R., 1961. The diet of freshwater crabs associated with *Simulium neavei* in East Africa I. Crabs from west and east Uganda collected by the Cambridge East African Expedition, 1959. *Ann. trop. Med. Parasit.* **55,** 128–31. *193.*

—— 1962. Idem II. The diet of *Potamon berardi* from Mount Elgon, Uganda. Ibid. **56,** 362–7. *193.*

—— and OBENG, L., 1962. A comparison of two methods of estimating changes in *Simulium* larval populations, with a description of a new method. Ibid. **56,** 259–61. *239.*

—— *et al.*, 1961. Size of particles ingested by *Simulium* larvae. *Nature, Lond.* **189,** 78. *186.*

WILLIAMS, W. P., 1965. The population density of four species of freshwater fish, roach (*Rutilus rutilus* (L.)), bleak (*Albu*ı *nus alburnus* (L.)), dace (*Leuciscus leuciscus* (L.)) and perch (*Perca fluviatilis* L.) in the River Thames at Reading. *J. Anim. Ecol.* **34,** 173–85. *339, 428.*

WINBERG, G. G., 1956. Rate of metabolism and food requirements of fishes (Russian). *Nauk. Trudy Beloruss. Gosud. Univ. imeni,* **5, 1,** 253 pp. (Fish. Res. Bd Can. Tranl. Ser. 194, 1960). *319, 420.*

WINGFIELD, C. A., 1939. The function of the gills of mayfly nymphs from different habitats. *J. exp. Biol.* **16,** 363–73. *166.*

WINKLER, O., 1956. On the benthic macrofauna of several brooks in the environments of Horská Kvilda (Šumava-mountains) (Czech: English summary). *Zool. Listy,* **5 (19), 4,** 367–86. *219, 391.*

—— 1963. Beitrag zur Kenntnis der Bodenfauna der oberen Moldau vor der Errichtung der Talsperre in Lipno. *Acta Univ. Carol. Biol.* **1963,** 85–101. *208, 215, 391.*

WINN, H. E., 1958*a*. Observations on the reproductive habits of darters (Pisces–Percidae). *Am. Midl. Nat.* **59,** 190–212. *356, 359.*

—— 1958*b*. Comparative reproductive behavior and ecology of fourteen species of darter (Pisces–Percidae). *Ecol. Monogr.* **28,** 155–91. *306, 311, 349, 351, 356.*

—— 1960. Biology of the brook stickleback, *Eucalia inconstans* (Kirtland). *Am. Midl. Nat.* **63,** 424–38. *357.*

—— and PICCIOLO, A. R., 1960. Communal spawning of the glassy darter *Etheostoma vitreum* (Cope). *Copeia,* **1960,** 186–92. *305, 360.*

WINTERBOURN, M. J., 1966. Studies on New Zealand stoneflies 2. The ecology and life history of *Zelandoperla maculata* (Hare) and *Aucklandobius trivacuatus* (Tillyard)–(Gripopterygidae). *N.Z. Jl Sci. Technol.* **9,** 312–23. *272.*

WISELY, B., 1962. Studies on Ephemeroptera. II. *Coloburiscus humeralis* (Walker); ecology and distribution of the nymphs. *Trans. R. Soc. N.Z., Zool.* **2,** 209–20. *206.*

WISSMEYER, A., 1926. Nahrungsuntersuchungen bei Ephemeridenlarven. *Arch. Hydrobiol.* **16,** 668–98. *194.*

WITT, A., and MARZOLF, R. C., 1954. Spawning and behavior of the longear sunfish, *Lepomis megalotis megalotis. Copeia,* **1945,** 188–90. *357.*

WOLF, P., 1950. A trap for the capture of fish and other organisms moving downstream. *Trans. Am. Fish. Soc.* **80,** 41–5. *262.*

WOLFE, L. S., and PETERSON, D. G., 1958. A new method to estimate levels of infestations of black-fly larvae (Diptera: Simuliidae). *Can. J. Zool.* **36**, 863–7. *239*.

WOODS, C. S., 1963. *Native and Introduced Freshwater Fishes*. Reed, Wellington–Auckland. *329*.

WOODS, W., 1965*a*. Physical and chemical limnology of the Upper Ohio River. *Spec. Publs Pymatuning Lab. Fld Biol.* **3**, 4–44. *44*.

—— 1965*b*. Primary production measurements in the Upper Ohio River. Ibid. **3**, 66–78. *413, 415, 418*.

WRIGHT, M., and SHOUP, C. S., 1945. Dragonfly nymphs from the Obey River and adjacent streams in Tennessee. *J. Tenn. Acad. Sci.* **20**, 266–78. *219*.

WU, Y. F., 1931. A contribution to the biology of *Simulium*. *Pap. Mich. Acad. Sci.* **13**, 543–99. *148, 157, 163*.

WURTZ, C. B., 1960. Quantitative sampling. *Nautilus*, **73**, 131–5. *236*.

YOUNT, J. L., 1956. Factors that control species numbers in Silver Springs, Florida. *Limnol. Oceanogr.* **1**, 286–95. *73*.

YOUNG, F. N., 1954. *The Water Beetles of Florida*. Florida University Press, Gainsville, 238 pp. *145, 193, 210*.

YUKHIMENKO, S. S., 1963. The food of the Amur sturgeon *Acipenser schrencki* Brandt and the kaluga *Huso dauricus* (Georgi) in the lower Amur River (Russian). *Vop. Ikhtiol.* **3**, 311–18. *369*.

ZACHARIAS, O., 1898. Das Potamoplakton. *Zool. Anz.* **21**, 41–8. *94*.

ZAHAR, A. R., 1951. The ecology and distribution of black-flies (Simuliidae) in south-east Scotland. *J. Anim. Ecol.* **20**, 33–62. *197, 200, 202, 231, 273, 283, 285–7*.

ZAHNER, R., 1959. Über die Bindung der mitteleuropäischen *Calopteryx*-Arten (Odonata, Zygoptera) an den Lebensraum des strömenden Wassers I. Der Anteil der Larven an der Biotopbindung. *Int. Revue ges. Hydrobiol. Hydrogr.* **44**, 51–130. *163, 201, 233, 387*.

—— 1960. Idem II. Der Anteil der Imagines an der Biotopbindung. Ibid. **45**, 101–23. *230*.

ZARBOCK, W. M., 1951. Life history of the Utah sculpin, *Cottus bairdi semiscaber* (Cope). *Trans. Am. Fish. Soc.* **81**, 249–59. *356, 367*.

ZELINKA, M., 1950. K. Poznání Zvířeny Horských Potaků Slezských Beskyd (English summary). *Přírodov. Sb. ostrav. Kraje.* **11**, 3–28. *231, 391*.

ZIMMERMAN, P., 1961*a*. Experimentelle Untersuchungen über die ökologische Wirkung der Strömgeschwindigkeit auf die Lebensgemeinschaften des fliessenden Wassers. *Schweiz. Z. Hydrol.* **23**, 1–81. *63, 200*.

—— 1961*b*. Experimentelle Untersuchungen über den Einfluss der Strömungsgeschwindigkeit auf die Fliesswasserbiozönose. *Verh. int. Verein. theor. angew. Limnol.* **14**, 396–9. *63, 200*.

—— 1962. Der Einfluss der Strömung auf die Zusammensetzung der Lebensgemeinschaften im Experiment. *Schweiz. Z. Hydrol.* **24**, 408–11. *63*.

Index of organisms

This index has the two-fold object of enabling readers to find where organisms are referred to in the text and enabling them to identify the groups to which unfamiliar plants and animals belong. After each generic name the common name, if there is one, is given in brackets, and the name of the higher taxon to which the genus belongs. If the family to which the organism belongs occurs in the index its name is given, and reference to that name will indicate the higher group (order or class) which includes the family. Thus one or more references to the index will lead to the name of a higher taxon which should be familiar to all biologists. For example, *Diura* is shown as Perlodidae, Perlodidae as a family of the order Plecoptera, and the Plecoptera as stoneflies of the class Insecta; on the other hand *Brachyptera* is shown directly as Plecoptera because the family Taeniopterygidae, to which it belongs, does not feature in the index.

Most anglicized and common names are indexed in their scientific form, with appropriate cross-references, for example, 'stoneflies, *see* Plecoptera', but the word used on pages listed against Plecoptera may be stonefly, plecopteran or Plecoptera.

The taxonomic systems used are somewhat arbitrary as the aim has been clarification rather than classification. It has, for instance, proved convenient to retain the old divisions of the fishes and so to distinguish the teleosts, as Teleostei, from the lower bony fishes, and no attempt has been made to enter into the controversy as to whether the Bacillariophyceae should really be elevated to Bacillariophyta. Similarly no taxonomical judgements have been made, as for example, on the validity of the genus *Acentrella* which many entomologists regard as an artificial taxon. Indeed, where it has proved convenient, genera which are now regarded as subgenera have been retained; for example in *Ictalurus* and *Hybobsis*, but where this has been done the accepted generic name is shown in the index, e.g. *Ameiurus* (*Ictalurus*, bullheads, Ictaluridae). In the same way, no judgement has been made on the correct spellings of names for which alternatives exist, e.g. *Limnaea*, *Lymnaea* and *Bithynia*, *Bythinia*, in both of which the first alternative has been used. This is in no sense a taxonomical treatise and an important objective of this index is to permit the location of organisms in familiar groups—to inform zoologists that

Neidium is a diatom and botanists that *Neureclipsis* is a polycentropid caddis-worm.

Numbers shown in italics are those of pages on which the organisms are illustrated.

Subject index